# Inflammatory Bowel Disease
# 2nd Edition

# RIVER PUBLISHERS SERIES IN RESEARCH AND BUSINESS CHRONICLES: BIOTECHNOLOGY AND MEDICINE

Volume 6

Combining a deep and focused exploration of areas of basic and applied science with their fundamental business issues, the series highlights societal benefits, technical and business hurdles, and economic potentials of emerging and new technologies. In combination, the volumes relevant to a particular focus topic cluster analyses of key aspects of each of the elements of the corresponding value chain.

Aiming primarily at providing detailed snapshots of critical issues in biotechnology and medicine that are reaching a tipping point in financial investment or industrial deployment, the scope of the series encompasses various specialty areas including pharmaceutical sciences and healthcare, industrial biotechnology, and biomaterials. Areas of primary interest comprise immunology, virology, microbiology, molecular biology, stem cells, hematopoiesis, oncology, regenerative medicine, biologics, polymer science, formulation and drug delivery, renewable chemicals, manufacturing, and biorefineries.

Each volume presents comprehensive review and opinion articles covering all fundamental aspect of the focus topic. The editors/authors of each volume are experts in their respective fields and publications are peer-reviewed.

For a list of other books in this series, visit www.riverpublishers.com

http://www.riverpublishers.com/series.php?msg=Research_and_Business_Chronicles:_Biotechnology_and_Medicine

# Inflammatory Bowel Disease
# 2nd Edition

Editors

**Wilton Schmidt Cardozo**
**Carlos Walter Sobrado**

Routledge
Taylor & Francis Group

LONDON AND NEW YORK

**Published 2016 by River Publishers**
River Publishers
Alsbjergvej 10, 9260 Gistrup, Denmark
www.riverpublishers.com

**Distributed exclusively by Routledge**
4 Park Square, Milton Park, Abingdon, Oxon OX14 4RN
605 Third Avenue, New York, NY 10158

First published in paperback 2024

*Inflammatory Bowel Disease, 2nd Edition* / by Wilton Schmidt Cardozo, Carlos Walter Sobrado.

*Routledge is an imprint of the Taylor & Francis Group, an informa business*

Publisher's Note
The publisher has gone to great lengths to ensure the quality of this reprint but points out that some imperfections in the original copies may be apparent.

While every effort is made to provide dependable information, the publisher, authors, and editors cannot be held responsible for any errors or omissions.

ISBN: 978-87-93379-19-0 (hbk)
ISBN: 978-87-7004-468-4 (pbk)
ISBN: 978-1-003-33855-0 (ebk)

DOI: 10.1201/9781003338550

## WILTON SCHMIDT CARDOZO

Coordinator of the Multidisciplinary Assistance Group on Ostomy and Inflammatory Bowel Disease at the Padre Bento Hospital Complex in Guarulhos. Member of the Brazilian Society of Coloproctology (SBCP), of the Brazilian Federation of Gastroenterology (FBG) and of the Brazilian College of Digestive Surgery (CBCD).

## CARLOS WALTER SOBRADO

PhD and MSc in Surgery from FMUSP. President, Department of Coloproctology at the São Paulo Medical Association (APM). Member of the Brazilian Society of Coloproctology (SBCP), of the Brazilian College of Surgeons (CBC) and of the Brazilian College of Digestive Surgery (CBCD). Fellowship in Coloproctology awarded from the University of Texas. Former President of the São Paulo State Association of Coloproctology (ACESP) and of the Brazilian Society of Coloproctology (SBCP).

## COLLABORATORS

### ADÉRSON OMAR MOURÃO CINTRA DAMIÃO

PhD, Assistant Professor, Department of Gastroenterology, University of São Paulo Medical School (FMUSP). Member of the Group of Intestinal Diseases and of the Gastroenterology Research Laboratory (LIM-07), Service of Clinical Gastroenterology, HC-FMUSP.

### ANDRÉ ZONETTI DE ARRUDA LEITE

Specialist degree in Gastroenterology from HC-FMUSP. Medical PhD from FMUSP. Postdoctor from Case Western Reserve University. Physician-Researcher at the LIM-07, Gastroenterology Department, HC-FMUSP.

### ANDREA VIEIRA

Specialist, PhD and MSc in Gastroenterology from the Faculty of Medical Sciences at São Paulo's Holy House of Mercy (FCMSCSP). Lecturer, Division of Gastroenterology, Department of Medicine, FCMSCSP.

### ANGELA HISSAE MOTOYAMA CAIADO

Specialist in Abdominal Radiology. Assistant Physician at the Institute of Radiology, HC-FMUSP, and at Fleury Medicina e Saúde. Full Member of the Brazilian College of Radiology and of the Radiological Society of North America (RSNA).

### ARTUR A. PARADA

Coordinator of the Gastrointestinal Endoscopy Service at Hospital 9 de Julho, São Paulo, and of Centro de Diagnóstico e Terapêutica Endoscópica, in São Paulo. Full Member of the Brazilian Society of Digestive Endoscopy (SOBED). Former president of the SOBED (2006-2008).

### AUDREY KRÜSE ZEINAD VALIM

Doctorate student in Medical Sciences, FMUSP. Assistant Physician at the Hematology Service, HC-FMUSP – Group of Hemostasis.

### BRUNO CÉSAR DA SILVA

MSc in Medicine and Health from the Federal University of Bahia (UFBA). Residency in Gastroenterology from the Clinics Hospital, Federal University of Pernambuco (HC-UFPE). Member of the Brazilian Study Group of Inflammatory Bowel Diseases (GEDIIB).

### CLÁUDIA GOLDENSTEIN SCHAINBERG

Collaborating Professor for the Division of Rheumatology, Department of Internal Medicine, FMUSP. Member of the American College of Rheumatology, of the Group of Research in Psoriasis and Psoriatic Arthritis (GRAPPA), member of the American Academy of Pediatrics and of the Committee of Spondyloarthritis/Brazilian Society of Rheumatology (SBR).

### CLÁUDIA PINTO MARQUES SOUZA DE OLIVEIRA

Gastroenterology PhD, Postdoctor and Post-doctoral Professor from the Department of Gastroenterology at FMUSP. Associate Professor at the Department of Gastroenterology, FMUSP.

## CLÁUDIA UTSCH BRAGA

Physician with a degree from the Federal University of Minas Gerais (UFMG). Specialist degree in Gastroenterology and Endoscopy from the Federal University of São Paulo (Unifesp). Graduate student in Gastroenterology at Unifesp.

## CLAUDIO FIOCCHI

Specialist in Gastroenterology. Professor of Molecular Medicine at the Lerner Research Institute, USA. Member of the Gastroenterology Division at the Digestive Disease Institute, Cleveland Clinic, Cleveland, USA.

## CLEIDE RODRIGUES DE CASTRO

Specialist in Hospital Psychology. Member of the Multidisciplinary Assistance Group on Ostomy and Inflammatory Bowel Disease at the Padre Bento Hospital Complex in Guarulhos (GAMEDII).

## CRISTINA FLORES

Physician with a degree from the Federal University of Rio Grande do Sul (UFRGS). Specialist degree in Digestive Endoscopy from the SOBED. PhD and MSc in Gastroenterology from UFRGS. Full Member of the GEDIIB.

## CYNTHIA MARIA BASSOTTO CURY MELLO

Lawyer with a specialist degree in Civil Procedure Law from the University Center of Faculdades Metropolitanas Unidas (FMU). Member of the Legal Department at the Brazilian Association of Ulcerative Colitis and Crohn's Disease (ABCD), at the National Association for Assistance of Diabetics (Anad), and at the National Federation of Diabetes Associations and Entities (Fenad).

## CYRLA ZALTMAN

Specialist in Gastroenterology. MSc in Gastroenterology from the Federal University of Rio de Janeiro (UFRJ). PhD in Gastroenterology from the Paulista School of Medicine at Unifesp (EPM-Unifesp). Postdoctoral degree from the Hospital Clinic, Barcelona, Spain. Associate Professor, Division of Gastroenterology, Department of Internal Medicine, UFRJ. Regional Director of the ABCD. Full Member of the SOBED, FBG and the GEDIIB (Scientific Director).

### DARIO ARIEL TIFERES

PhD in Radiology from EPM-Unifesp. Radiologist, Department of Diagnostic Imaging at EPM-Unifesp, and at Fleury Medicina e Saúde.

### EDUARDO DE CASTRO HUMES

Specialist degree in Psychiatry from FMUSP and the Brazilian Association of Psychiatry (ABP). Member of the ABP, of the American Psychiatric Association (APA) and of the São Paulo Psychodrama Society (SOPSP).

### ELBIO ANTONIO D'AMICO

Assistant Hematologist at the Hematology and Hemotherapy Service, HC-FMUSP. Post-doctoral Professor, FMUSP. Collaborating Professor at FMUSP.

### ELOÁ MARUSSI MORSOLETTO

Head of the Department of Gastroenterology and Digestive Endoscopy at the São Vicente Hospital, in Curitiba. General Secretary, GEDIIB. Full Member of the FBG and the SOBED.

### ELVINO BARROS

Physician with a degree from the Federal University of Health Sciences of Porto Alegre (UFCSPA). Specialist degree in Nephrology from the Brazilian Society of Nephrology (SBN). MSc in Nephrology from UFRGS. PhD in Nephrology from Unifesp. Associate Professor, Division of Nephrology, Department of Internal Medicine, UFGRS.

### FÁBIO VIEIRA TEIXEIRA

Specialist in General Surgery from the Brazilian College of Surgeons (CBC), in Digestive Surgery from the Brazilian College of Digestive Surgery (CBCD), and in Coloproctology from the Brazilian Society of Coloproctology and the GEDIIB. MSc in Surgery from the São Paulo State University, School of Medicine, Botucatu Campus (FM/Unesp). PhD in Surgery from Mayo Clinic Scottsdale, Arizona, USA, and Unesp. Visiting Professor for the Division of Surgical Gastroenterology, Department of Surgery, FM/Unesp. Full Member of the CBC, the CBCD, of the Brazilian Society of Coloproctology (SBCP) and of the GEDIIB.

### FLÁVIO ANTONIO QUILICI

PhD and MSc in Surgery from the State University of Campinas (Unicamp). Full Professor of Digestive Surgery at the Pontifical Catholic University of Campinas (PUC-Campinas). Full Member of the FBG, SOBED, SBCP and of the CBCD. President of the Gastroenterology Society of São Paulo (SGSP). Former president of the SBCP and the SOBED.

### FLÁVIO FEITOSA

Graduate student at the Department of Gastroenterology, FMUSP.

### GENOILE OLIVEIRA SANTANA

PhD and MSc in Medicine and Health from UFBA. Coordinator of the IBD Outpatient Clinic at the Professor Edgard Santos Teaching Hospital (HUPES)/ UFBA. Member of the Board, GEDIIB.

### GUILHERME CUTAIT DE CASTRO COTTI

Assistant Physician at the Octavio Frias de Oliveira Cancer Institute/FMUSP.

### GUILHERME MARQUES ANDRADE

Specialist degree in Internal Medicine from the Clinics Hospital, Faculty of Medicine, University of São Paulo – HC-FMUSP. Gastroenterologist at the Clinics Hospital, Faculty of Medicine, University of São Paulo – HC-FMUSP.

### HUMBERTO SETSUO KISHI

Pathologist at the Division of Anatomic Pathology, HC-FMUSP, and at Laboratório Diagnóstika de São Paulo.

### IDBLAN CARVALHO DE ALBUQUERQUE

Assistant at the Coloproctology Service, Heliópolis Hospital, São Paulo. Responsible for the IBD Outpatient Clinic at the Coloproctology Service, Heliópolis Hospital, São Paulo. Full Member of the SBCP.

### JAIME ROIZENBLATT

Specialist degree in Ophthalmology from FMUSP. Postdoctor from the University of California, USA. Assistant Physician at HC-FMUSP.

### JARBAS FARACO MALDONADO LOUREIRO

Colonoscopist at Sírio-Libanês Hospital and at the Oswaldo Cruz German Hospital in São Paulo. Member of the Franco-Brazilian Center of Echo-endoscopy (CFBEUS), of the FCMSCSP and the SOBED.

### JULIANO COELHO LUDVIG

Specialist degree in Clinic Gastroenterology from the FBG, and specialist degree in Digestive Endoscopy from the SOBED. Regional Director of the ABCD in Santa Catarina.

### LISANDRA CAROLINA MARQUES QUILICI

Specialist degree in Coloproctology from the SBCP, and specialist degree in Digestive Endoscopy from the SOBED. Digestive Surgeon at the PUC-Campinas Hospital and Maternity.

### LUCIANA DOS SANTOS

Pharmacist with a degree from the Lutheran University of Brazil (ULBRA). Specialist degree in Hospital Pharmacy from the Institute of Hospital Administration and Health Sciences (IACHS). MSc in Pharmaceutical Sciences from UFRGS. Clinical Pharmacist at the Clinics Hospital of Porto Alegre (HCPA).

### LUCIANE REIS MILANI

MSc in Sciences from USP. Specialist degree in Gastroenterology from the FBG, and specialist degree in Digestive Endoscopy from the SOBED.

### MAGALY GEMIO TEIXEIRA

Post-doctoral Professor, FMUSP. Responsible for the Inflammatory Disease Outpatient Clinic at HC-FMUSP.

### MARCELLO MENTA SIMONSEN NICO

Dermatologist. Associate Professor at the Dermatology Department, FMUSP. Supervising Physician at the Dermatology Division, HC-FMUSP. Responsible for the Outpatient Clinic of Diseases of the Oral Mucosa, Dermatology Division, HC-FMUSP.

### MARCO ANTÔNIO ZERÔNCIO

Physician with a degree from the Federal University of Rio Grande do Norte (UFRN). Biochemist with a degree from the University of Maine, USA. Residency in Internal Medicine and Gastroenterology from the Hospital Foundation of the Federal District (FHDF). Specialist degree in Gastroenterology from the FBG, and specialist degree in Digestive Endoscopy from the SOBED. Coordinator of the IBD Outpatient Clinic at the Medical School, Potiguar University (UnP). Member of the GEDIIB.

### MARCOS CARDOSO RESENDE

Radiologist, Collaborating Physician at the Department of Diagnostic Imaging, EPM-Unifesp.

### MARIA DE LOURDES TEIXEIRA DA SILVA

Specialist degree in Enteral and Parenteral Nutrition from the Brazilian Society of Enteral and Parenteral Nutrition (SBNPE). MSc in Gastroenterology from the Brazilian Institute for Studies and Research in Gastroenterology (IBEPEGE), São Paulo. Director of the Enteral and Parenteral Nutrition Support Group (Ganep).

### MARIA IZABEL LAMOUNIER DE VASCONCELOS

Specialist degree in Clinical Nutrition from the São Camilo Faculty, and specialist degree in Enteral and Parenteral Nutrition from the SBNPE. MSc in Experimental Nutrition from USP.

### MARINA ROIZENBLATT

Medical residency in Ophthalmology, EPM-Unifesp.

### MARJORIE ARGOLLO

Physician with a degree from UFBA. Specialist in Clinical Gastroenterology and Digestive Endoscopy from Unifesp. Graduate student at Unifesp.

### MAYDE SEADI TORRIANI

Pharmacist and specialist in infection control in hospital pharmacy from UFRGS. Specialist degree in Hospital Administration from the Institute of Hospital Administration and Health Sciences, PUC-RS, and specialist degree in Hospital Pharmacy from the Brazilian Society of Hospital Pharmacy (SBRAFH). MSc in Internal Medicine from the Medical Sciences Graduate Program, UFRGS Medical School. Head of the Sector of Medicine Management and Logistics at the HCPA. Financial and administrative director of the Brazilian Society of Oncology Pharmacists (SOBRAFO) (2012-2013 and 2014-2015).

### MILTON RUIZ ALVES

Associate Professor at FMUSP. Professor in the Ophthalmology Graduate Program at FMUSP. Head of the Department of Cornea and External Eye Disease at the Ophthalmology Outpatient Clinic, HC-FMUSP.

### ORLANDO AMBROGINI JUNIOR

Specialist and MSc in Gastroenterology, and PhD in Internal Medicine from EPM-Unifesp. Gastroenterology Professor and Co-responsible for the Department of Bowel Diseases at EPM-Unifesp. Full Member of the GEDIIB and of the FBG. International Member of the AGA.

### PATRÍCIA LIMA JUNQUEIRA

Hematologist, HC-FMUSP. Preceptor at the Hematology Service, HUPES/UFBA. Physician at the Hematology Service, Hematology and Oncology Center of Bahia. Former Assistant at the Hematology Service, São Paulo Cancer Institute (ICESP), HC-FMUSP.

### PAULA B. POLETTI

Coordinator of the Gastrointestinal Endoscopy Service at the São Paulo State Public Servant's Hospital. Assistant at the Gastrointestinal Endoscopy Service at the 9 de Julho Hospital, São Paulo.

### PAULO ALBERTO FALCO PIRES CORRÊA

Specialist degree in General Surgery from HC-FMUSP. Surgeon and Colonoscopist at Sírio-Libanês Hospital. Charter Member of the Brazilian Society of Videosurgery (Sobracil). Full Member of the SBCP and the SOBED. Associate Member of the CBCD.

### PAULO GUSTAVO KOTZE

MSc in Surgical Clinic from PUC-PR. Head of the Coloproctology Service at the Cajuru Teaching Hospital (SeCoHUC)/PUC-PR. Full Member of the SBCP.

### PAULO LISBOA BITTENCOURT

Specialist in Gastroenterology with a focus on Hepatology and Digestive Endoscopy. PhD in Gastroenterology from USP. Full Member of the FBG.

### RAQUEL FRANCO LEAL

Specialist degrees in Coloproctology from Unicamp and the SBCP. PhD and MSc in Surgery from Unicamp. Postdoctor from the University of Chicago and the University of Barcelona (Institut d'investigacions Biomèdiques August Pi i Sunyer). Lecturer at the Coloproctology Service, School of Medical Sciences, Unicamp.

### RAQUEL GUERRA DA SILVA

Pharmacist with a degree from UFRGS. MSc degree student on the Postgraduate Program in Pharmaceutical Assistance at UFRGS.

### RENATA EMY OGAWA

Physician with a degree from FMUSP. Specialist in Radiology and Diagnostic Imaging from the Institute of Radiology (INRAD), HC-FMUSP.

### RENÉRIO FRÁGUAS JUNIOR

Specialist in Psychiatry and PhD in Psychiatric Medicine from FMUSP. Associate Professor at FMUSP. Executive Director at Center for Studies of the Psychiatry Institute (CEIP), HC-FMUSP.

### RICARDO ROMITI

PhD in Dermatology from Ludwig-Maximilians-Universität Müchen, Germany. Lecturer at the Dermatology Department, HC-FMUSP, responsible for the Psoriasis Outpatient Clinic.

### ROBERTO EL IBRAHIM

Pathologist with a degree from FMUSP. Founding Director, Laboratório Diagnóstika de São Paulo.

### RODOLFO DELFINI CANÇADO

Specialist degree in Hematology and Hemotherapy from the Brazilian Association of Hematology, Hemotherapy and Cell Therapy (ABHH). PhD and MSc in Health Sciences from FCMSCSP. Adjunct Professor at FCMSCSP.

Physician; Hematologist at the Hematology and Hemotherapy Service, São Paulo's Holy House of Mercy, and at São Paulo's Samaritan Hospital.

### ROGÉRIO SAAD HOSSNE

PhD-Professor at the Surgery and Orthopedics Department, Unesp Medical School, Botucatu campus. Surgery Coordinator at the IBD Outpatient Clinic, Unesp Medical School, Botucatu campus. Full Member of the SBCP, CBCD and of the CBC. Member of the GEDIIB.

### RUSSELL D. COHEN

Professor of Medicine at Pritzker School of Medicine, USA. Director of the Inflammatory Bowel Disease Center, USA. Co-Director of the Advanced IBD Fellowship, The University of Chicago Medicine, USA.

### SABRINA SISTO ALESSI CÉSAR

Assistant Physician, Dermatology Department, HC-FMUSP.

### SENDER JANKIEL MISZPUTEN

PhD in Gastroenterology from EPM-Unifesp. Associate Professor, Division of Gastroenterology, Department of Internal Medicine, EPM-Unifesp. President of the GEDIIB. Vice-President of the SGSP. National Honorary, National Academy of Medicine.

### SHEYLA BATISTA BOLOGNA

Odontologist with a degree from the School of Dentistry, USP (FOUSP). PhD in Sciences from FMUSP.

### SILVIA VANESSA LOURENÇO

Dentist Surgeon and Oral Pathologist. Associate Professor, Division of General Pathology, FOUSP.

### THIAGO FESTA SECCHI

Member of the Gastrointestinal Endoscopy Service at the 9 de Julho Hospital, São Paulo. Former president of the SOBED in the State of São Paulo.

### VERA LUCIA SDEPANIAN

MSc in Pediatrics from EPM-Unifesp, and MSc in Pediatric Gastroenterology and Nutrition from the International University of Andaluzia, Spain. Medical PhD from EPM-Unifesp. Postdoctor in Pediatric Gastroenterology from the University of Maryland, USA. Adjunct Professor and Head of the Division of Pediatric Gastroenterology at EPM-Unifesp. Supervisor of Medical Residency in Pediatric Gastroenterology, EPM-Unifesp. Vice-President of São Paulo's Pediatric Gastroenterology, Hepatology and Nutrition Association (APPGHN), and of the Department of Gastroenterology of São Paulo's Pediatrics Society (SPSP).

### WILSON ROBERTO CATAPANI

MSc in Gastroenterology and PhD in Internal Medicine from EPM-Unifesp. Post-PhD in Gastroenterology from the University of Edinburgh, UK. Full Professor of Gastroenterology, Department of Internal Medicine at the ABC Medical School. Fellowship awarded from the American College of Gastroenterology.

# CONTENTS

**EXHIBITS**

# Series Note

Inflammatory bowel diseases refers to a group of inflammatory diseases of the colon and small intestine, notably comprising Crohn's disease and ulcerative colitis. As chronic diseases, Crohn's disease and ulcerative colitis have long vexed pharmaceutical product developers in their quest for curative treatments. In this second edition of their monograph, the authors perform a deep dive in the pathobiology, diagnosis, and treatment of inflammatory bowel diseases. Particularly, they explore in a methodical manner every critical aspects not only of the best practice of today to best manage these conditions in adults, pregnant women, and in children, but also the various complications of these diseases in other organs such as the eye, skin, or liver, as well as in other disease areas such as in oncology, haematology, and rheumatology. What is more, the psychological or psychiatric impacts that these conditions may exert on the quality of daily life also are revisited. At a time when numerous pharmaceutical companies have exited the field of gastro-intestinal therapeutic product research and development, it is even more important for researchers and practitioners in the field to exhaustively grasp the fundamental and applied knowledge developed to this date regarding these ailments. Indeed, the lessons already learned using conventional treatments including small molecules, biologics, or surgical interventions might very well be complemented, alone or in combination, with emerging disease-modifying concepts such as microbiome and fecal microbiota transplantation approaches, or mesenchymal stem cell therapies. That path is now open.

Alain Vertès, Basel, Switzerland
Pranela Rameshwar Rutgers, USA
Paolo di Nardo, Roma, Italy

# Foreword

JULIÁN PANÉS

Inflammatory bowel disease (IBD) imposes significant challenges to patients' life from the time of diagnosis, and in a number of patients a complicated course of the diseases results in successive challenges to their personal, familial, social, and professional development.

IBD becomes also a challenge for the treating health care professionals, physicians, surgeons and nurses included, and for basic scientists. The disease results from an abnormal immune response to components of the intestinal lumen, mainly flora, in genetically susceptible individuals, which are triggered and modulated by environmental factors. But each one of these components: genetics, microbiome, immune response and environmental factors remain to be fully elucidated.

At the clinical level, diagnosing IBD possess significant challenges. In world regions with a high prevalence of infectious colitis, the differential diagnosis between these and IBD is not an easy task. In patients with purely colonic IBD distinction between Crohn's disease and ulcerative colitis needs frequently expert assessment of clinical, endoscopic, histologic and radiologic data. Challenges in disease management are not only related to the treatment of complicated disease, but to the development of a treatment plan for every patient, tailored to the patient's needs and expectations, by making an optimal use of available medical and surgical options, and adopting measures to prevent disease, and treatment-related complications. The wealth of knowledge in the field is rapidly increasing, and there is an increasing need of master works that not only summarize up-to-date information, but also provide a balanced perspective of the clinical relevance of each new discovery.

I felt highly honored when I was invited to preface the second edition of *Doença Inflamatória Intestinal*, edited by Dr. Wilton Schmidt Cardozo and Dr. Carlos Walter Sobrado. The handbook gathers the best knowledge in medical care, presented under the perspective of top clinical experience, by a group of authors of high stature in the field. The chapters of the handbook cover basic aspects of disease pathophysiology, present an extensive review of aspects of diagnosis that includes endoscopy, histology and cross-sectional imaging, best use of different drugs classes and therapeutic strategies, the role of surgery, and other complementary aspects of IBD care.

I can only congratulate the editors and authors for this outstanding work, and encourage the readers to implement the recommendations provided in the handbook as the basis for the best care for our patients.

JULIÁN PANÉS. M.D., PH.D.
Chief of Department of Gastroenterology
Hospital Clínic de Barcelona,
Barcelona, Spain

# Foreword

## ANGELITA HABR-GAMA

It is my pleasant responsibility and a high distinction to preface the first and second edition of the book Inflammatory Bowel Disease compiled by Wilton Schmidt Cardozo and Carlos Walter Sobrado, longtime friends of mine. This is proof of the success resulting from the first edition, valued by experts in the fields of gastroenterology, digestive surgery and endoscopy, coloproctology, nutrology and psychology.

The "inflammatory bowel disease" classification applies primarily to two diseases: ulcerative colitis and Crohn's disease, chronic conditions similar in many aspects and different in so many others, particularly with regard to clinical manifestations, such as disease progression and response to medical or surgical treatment.

Almost all aspects of inflammatory bowel disease are still partially unknown and, therefore, very controversial. The increasing incidence in many countries, especially of Crohn's disease, draws attention to similar presentations despite ethnic, environmental, nutritional, genetic and social differences. Its predominance in the younger population, the highest prevalence in certain families, the questionable causal possibilities (bacterial, viral and immunological), the genetic influence of immunoregulation in the host's defenses, new treatment modalities, and more, are issues of great importance, worthy of deep analysis.

The organizers made the correct choice of authors for each chapter, since all of them not only excel in their areas of expertise, but also enjoy high prestige in the national scientific community, with extensive experience in the subject. The original chapters have been updated, and eleven new ones were added.

I read with interest the text resulting from this outstanding second edition and found that the knowledge contained in this book reflects the state of the art regarding inflammatory bowel disease, as well as demonstrates its application in all fields of medicine, particularly in Gastroenterology and Proctology.

The organizers succeeded in designing this book, which until now was missing in the national literature. As a result, it provides information and concepts that will greatly contribute to the dissemination and standardization of diagnosis and treatment of inflammatory bowel disease among Brazilian physicians.

I am thankful for the honorable invitation to write this preface to the second edition and, thus, to display my opinion on its relevance and scientific content as well as on the social and professional impact that it represents.

Last, it is up to me to congratulate both creators and accomplishers for their dedication, perseverance and work in order to provide readers with knowledge acquired over time.

ANGELITA HABR-GAMA

Professor Emeritus, Faculty of Medicine, University of São Paulo. President of the Brazilian Association of Intestinal Cancer Prevention (Abrapeci). Honorary Member of the American Surgical Association (ASA), of the American College of Surgeons (ACS), of the European Surgical Association (ESA), the American Society of Colon and Rectal Surgeons (ASCRS), the Royal College of Surgeons (RCS) – England, and of the American Society for Radiation Oncology (Astro).

# THE HISTORY OF INFLAMMATORY BOWEL DISEASE

FLÁVIO ANTONIO QUILICI
LISANDRA CAROLINA MARQUES QUILICI

## THE HISTORY OF ULCERATIVE COLITIS

In the 4th century BC, Greek physician Hippocrates of Kos, in his book *Corpus Hippocraticum peri Syriggon*, and physicians Areteus of Cappadocia and Soranus of Ephesus in the 2nd century AD, among others, described a type of chronic diarrhea associated with bloody stools and ulceration in the colon, with characteristics that were different from other types of diarrhea known at the time, and constituting a clinical illness that resembled ulcerative colitis (UC).[1]

However, UC was first described as a disease entity by Samuel Wilks, from Guy's Hospital in London in 1859, with the report of the autopsy of the first case of the illness, sent by letter to the Medical Times and Gazette. The relative delay in its discovery was due to the fact that the diarrhea was caused by infection.[2] Wilks, aided by Moxon in 1875 also defined UC as a specific pathological entity, which they called inflammation of the large intestine or idiopathic colitis. After that, UC was characterized as a disease distinct from other – dysenteric – inflammatory bowel diseases (IBD) already well known in the nineteenth century, and started to be taught in classes and books.[3]

Habershon, in 1862, described in his book on *Diseases of the Abdomen* the classic pseudopolyps of UC:

> These intestinal ulcerations gradually tend to coalesce until there is destruction of the entire mucosal surface, except for a few isolated areas that become strongly congested and resemble polypoid growth.[1]

Sir William Allchin, in 1885, made the distinction between UC and specific forms of colitis, and said that:

> It is important that the word *dysentery* is not limited to tropical inflammatory diseases or applied as an adjective to any form of diarrhea or colon ulceration.[1]

The possibility of surgical treatment of this still obscure disease only emerged in the late nineteenth century, with a colostomy being performed in a female patient with UC for irrigation of the inflamed colon by Mayo-Robson, from Leeds, in 1893.[1]

Only in 1931, Sir Arthur Hurst made a full description of the pathological and endoscopic features through sigmoidoscopy. Special attention to UC as an important IBD was given only in the last half of the twentieth century.

UC and Crohn's disease (CD) were confirmed as different pathogenic entities in 1961 by Lockhart-Mummery and Morson, who presented a detailed description of the clinical and pathological findings of these two diseases, including the characterization of segmental nature, and granulomatous inflammation.[4]

The first publications about UC in Brazil occurred in the late 1950s, with cases reported by Passarelli in 1959.[1] Later, Faustino Porto published studies on UC and DC, showing their various aspects. His vast experience culminated in the publication of the first Brazilian book on IBD in 1990 with the collaboration of gastroenterologists Sylvia da Silveira Mello Vargas and Eduardo Lopes Pontes.[3,5,6] Based on the experience gained by these authors, especially in the idiopathic IBD outpatient clinic at Hospital Universitário Clementino Fraga, Universidade Federal do Rio de Janeiro (UFRJ), they showed the epidemiological, pathophysiological and clinical aspects of IBDs in Rio de Janeiro.

## THE HISTORY OF CROHN'S DISEASE

The disease now known as Crohn's disease (CD) is not a new condition. Reports on it are found since Ancient Greece and Alexandria, when it was probably mistaken for enterocolitis caused by parasitic diseases. Soranus of Ephesus (170 AD) described in his book a condition very similar to CD. It did take long, however, for this to become a recognized clinical disease.[1,7]

In 1813, Charles Combe and William Saunders published in the *Medical Transactions* of the Royal College of Physicians of London, the illustrated case of a patient with ileal stricture and intense inflammatory process (Figure 1.1).

In 1859, Samuel Wilks proposed that idiopathic colitis should be considered a different disease from specific epidemic dysentery and, in 1882, N. Moore was one of the first to describe and publish in the *Transactions of the Pathological Society of London,* the microscopic and macroscopic findings of CD in a patient with intestinal obstruction and presence of intense chronic cell infiltrating inflammatory process. For final recognition of the ileal form as a clinical disease, three publications were relevant:

- the first was that by T. K. Dalziel, a surgeon from Edinburgh, in 1913;
- the second was the work by Eli Moschcowitz and A. O. Wilensky, from Israelite Hospitals Mount Sinai (in New York) and Beth (in Boston), in 1923;
- and finally, the study by Crohn, Ginzburg and Oppenheimer, from the Mount Sinai Hospital, in 1932.

For the colonic form, the most important publication was that by Lockhart-Mummery and Morson, from St. Mark's Hospital in London, in 1960.

In a study published by Kennedy Dalziel (Figure 1.2), in 1913, in the *British Medical Journal*, which had little impact at the time, a retrospective of 9 patients operated on at the Edinburgh Hospital was presented, with 2 deaths. The disease affected jejunum and ileum, as well as transverse and sigmoid colon. Dalziel differentiated it from intestinal tuberculosis and reported the unfavorable prognosis, except in case of early detection and surgery. His work was interrupted by World War I, when he retired to his farm in the Scottish countryside.

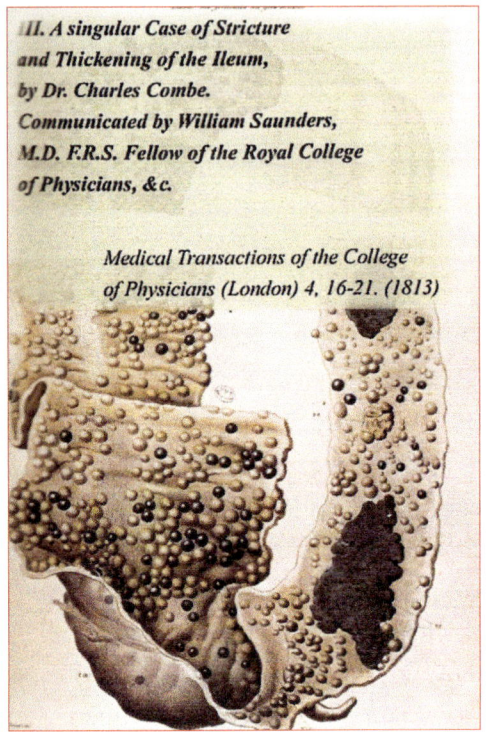

**Figure 1.1** Illustration of the 1813 publication, by Combe and Saunders, in the *Medical Transactions* of the Royal College of Physicians of London, showing a patient with stricture of the terminal ileum and intense inflammatory process.

The publication by Moschcowitz and Wilensky in the *American Journal of Medical Sciences* did not have a great repercussion, either. The study reported the presence of nonspecific granulomatous inflammation with giant intestinal cells in an ileocecal surgical specimen, differentiating it from hyperplastic appendicitis and noting the absence of bacteria and caseation.[1]

What marked the history of this condition was the work entitled *Regional ileitis: a pathologic and clinical entity*, published in 1932, with a positive reception worldwide.[8] The study was a collaboration of some eminent doctors (Figure 1.3):

**Figure 1.2** Publication by K. Dalziel in 1913, in the *British Medical Journal*, with a retrospective of nine cases operated by him at the Edinburgh Hospital on account of disease involving jejunum and ileum, as well as transverse and sigmoid colon.

**Figure 1.3** Picture taken at a congress in 1969, held by the American Gastroenterology Association (AGA), with Gordon D. Oppenheimer, Burril B. Crohn and Leon Ginzburg.

- Burrill B. Crohn: clinical gastroenterologist responsible for the diagnosis, surgical indication and monitoring of the 14 cases evaluated in this study, all with terminal ileitis. The patients operated at Mount Sinai Hospital in New York were referred from his private clinic;
- Alexander A. Berg: Head of Surgery at the Mount Sinai Hospital and responsible for 14 successful surgical resections;
- Leon Ginzburg: Berg's assistant and responsible for the pathological assessment of the 14 ileocecal surgical specimens;
- Gordon D. Oppenheimer: pathologist and, like Ginzburg, interested in IBD, particularly in granulomatous disease. Responsible for histopathological analyses of the study.

Berg was the surgeon who insisted on the unity and collaboration of the three authors; he refused the authorship credit, claiming it is a clinical study (a fact confirmed by Crohn himself), and also suggested that the order of authors should be alphabetical. Thus, on December 11th, 1931, Crohn wrote to the American Gastroenterology Association (AGA) informing that he had an important scientific contribution to present as open subject (*abstract*) at the meeting in May 1932, as he supposed to have discovered a new bowel disease referred to as terminal ileitis by the authors. Such presentation took place on May 2nd, 1932 in Atlanta and was followed by its publication, also in 1932, in AGA's journal (*Transactions of the AGA – Annals*). The study was also published by *JAMA* (Figure 1.4).[8,9]

The condition became known as Crohn's disease because of numerous publications, mainly European, which insisted on citing it that way. The most important were editorials in the *Lancet*, written by his London friend, surgeon Brian Brooke.

In Brazil, it was Berardinelli who in 1943 published the first case of regional ileitis, 11 years after the article by Crohn, Oppenheimer and Ginzburg.[8,9] In 1951, H. Rappaport noted that more than half of 100 patients with regional ileitis also had similar lesions in the colon, which, however, were less severe. In 1952, C. Wells showed that segmental colitis or gastrointestinal granulomatous disease, terms used by Crohn and Hurst, was, in fact, the

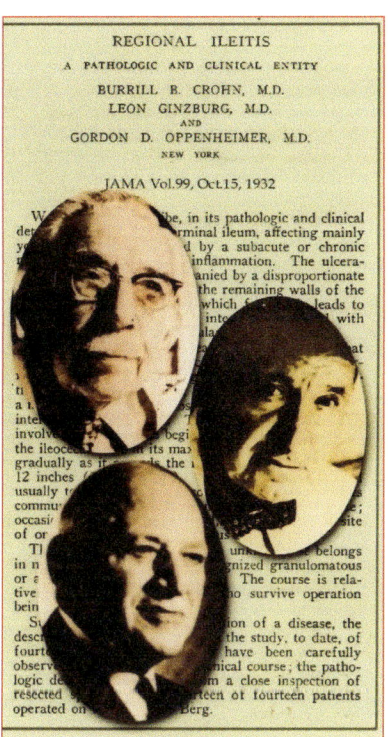

**Figure 1.4** Illustration of the study published in *JAMA*, volume 88, on October 15th, 1932, and photos of the authors.

colon being affected by CD. His work was published in the *Annals of the Royal College of Surgeons of England.*

The concept of colonic disease for CD and its distinction as a separate entity from UC occurred after a crucial publication in 1960 by *sir* H. E. Lockhart-Mummery and Basil Morson, from St. Mark's Hospital in London, who submitted a detailed description of the clinical and histological nature of the inflammatory process of both, differentiating their findings, including segmental disease and granulomatous inflammation.

## REFERENCES

1.  Quilici FA. Retocolite ulcerativa. São Paulo: Lemos, 2002.

2.  Senagore AJ. Immunologic aspects of inflammatory bowel disease. In: Mazier WP, Levien DH, Luchtefeld MA, Senagore AJ. Surgery of the colon, rectum and anus. Filadélfia: Saunders, 1995; p.837-40.

3.  Porto JAF, Pontes EL, Vargas SSM. Doenças inflamatórias intestinais idiopáticas. Rio de Janeiro: Guanabara Koogan, 1990.

4.  Morson BC, Dawson IMP. Inflammatory disorders. In: Gastro-intestinal pathology. Oxford: Blackwell Scientific Publ, 1990; p.477-781.

5.  Pontes JF. Retocolite ulcerativa idiopática. In: Dani R, Castro LP (eds.). Gastroenterologia clínica. 2.ed. Rio de Janeiro: Guanabara Koogan, 1988.

6.  Porto JAF. Clínica das doenças intestinais. Rio de Janeiro: Atheneu, 1976.

7.  Quilici FA. Coloproctologia: estórias da história. Rev Br Coloproct 1994; 14(1):43-5.

8.  Crohn BB, Ginzburg L, Oppenheimer GD. Regional ileitis: a pathologic and clinical entity. JAMA 1932; 99:1223.

9.  Crohn BB, Yarnis H. Regional ileitis. 2.ed. Nova York: Grune & Stratton, 1958.

## BIBLIOGRAPHY

1.  Alexander-Williams J. Historical review. In: Allan R, Rhodes J, Hanauer SB. Inflammatory bowel diseases. 3.ed. Nova York: Churchill Livingstone, 1997.

2.  Allan RN, Rhodes JM, Hanauer SB, Keighley MRB, Alexander-Williams J, Fazio VW. Inflammatory bowel diseases. 3.ed. Nova York: Churchill Livingstone, 1997.

3.  Calkins BM, Mendeloff AI. Epidemiology of idiophatic inflammatory bowel disease. In: Kirner JB, Shorter RG. Inflammatory bowel disease. 4.ed. Baltimore: Williams & Wilkins, 1995; p.31-8.

4.  Corman ML. Ulcerative colitis. In: Corman ML. Colon and rectal surgery. Filadélfia: Lippincott, 1998; p.1079-192.

5.  Elson CO, McCabe RP. The immunology of inflammatory bowel disease. In: Inflammatory bowel disease. In: Kirsner JB, MaCAbe RP (eds.). Inflammatory bowel disease. 4.ed. Baltimore: Williams & Wilkins, 1995; p.203-51.

6.  Fenoglio-Preiser CM, Lantz PE. Idiophatic inflamatory bowel disease. In: gastrointestinal pathology. An atlas and text. Nova York: Raven Press, 1989; p.427-33.

7.  Forbes A. Clinician's guide to inflammatory bowel disease. Londres: Chapman & Hall, 1998.

8.  Freitas JA, Tacla M. Retocolite ulcerativa. In: Dani R (ed.). Gastroenterologia essencial. Rio de Janeiro: Guanabara-Koogan, 1998; p.326-36.

9.  Habr-Gama A. Doença inflamatória intestinal. São Paulo: Atheneu, 1997.

10. Keighley MRB. Ulcerative colites. In: Keighley MRB, Williams N. Surgery of the anus, rectum and colon. Londres: Saunders, 1993.

11. Kirsner JB. Inflammatory bowel disease. 5.ed. Filadélfia: Saunders, 2000.

12. Lashner BA, Kirsner JB. The epidemiology of inflammatory bowel disease: are we learning anything new? Gastroenterology 1992; 103:696-8.

13. Lockart-Mummery HE. Crohn's disease: anal lesions. Dis Colon Rectum 1975; 18:200.

14. Maratka Z. Pathogenesis and aethiology of inflammatory bowel disease. In: de Dombal FT, Nyren J, Boucher IDA, Watkinson G. Inflammatory bowel disease. Oxford: Some, 1986; p.29-65.

15. McConnel RB. Genetics factors. In: Inflammatory disease of the bowel. Bath: Brooke & Wilkinson, 1980; p.8-14.

16. Quilici FA, Tolentino M. Viagem ao processo inflamatório intestinal. Unimagem – Videotec 1998.

17. Quilici FA, Reis Neto JA. Atlas de proctologia. São Paulo: Lemos, 2000.

18. Quilici FA. Doença inflamatória intestinal – Guia prático. Rio de Janeiro: Elsevier, 2007.

19. Rawet V. Retocolite ulcerativa. In: Habr-Gama A (ed.). Doença inflamatória intestinal. São Paulo: Atheneu, 1997; p.9-20.

20. Reis Neto JA, Reis JA Jr. Retocolite ulcerativa. In: Cruz GMG (eds.). Coloproctologia – Terapêutica. São Paulo: Revinter, 2000; p.2044-55.

21. Wilks S. Morbid appearances in the intestine of Miss Banks. Londres: Medical Times and Gazette Churchill, 1959.

# EPIDEMIOLOGY OF INFLAMMATORY BOWEL DISEASE

ANDRÉ ZONETTI DE ARRUDA LEITE
GUILHERME MARQUES ANDRADE

Inflammatory bowel disease (IBD) is a broad term used to designate Crohn's disease (CD) and ulcerative colitis (UC), both characterized by chronic inflammation of the intestine.[1,2] These diseases differ in terms of location and involvement of the layers of the intestine, but also in pathogenesis, which has not yet been fully elucidated. There is a consensus that genetic factors make the individual susceptible to developing the disease, and environmental factors are responsible for the initiation and modulation, including diet, hygienic conditions, health and composition of the gut flora.[3]

An important impact factor is that the disease affects young people. CD has a peak incidence at age 20-30 years, while UC affects people between 30 and 40 years with a questionable second peak later in life.[4,5] As to gender, UC appears to be slightly more common in men (60%), while CD is 20-30% more common in women, especially in high incidence areas.[6]

IBD is present worldwide, but its distribution both regarding incidence and prevalence, is not homogeneous.[7] In Western countries, the prevalence is between 8 and 214 cases/100,000 inhabitants for UC and 21 to 294/100,000 for CD.[6,8] While in some countries the number of cases is small, in others the disease has become a public health problem, especially because of the high morbidity and high treatment cost. In the first year that follows diagnosis, there is an increased mortality of 10 and 50% in UC and CD, respectively, which is particularly caused by infections, cancer and respiratory diseases. However, registry studies suggest that the mortality rate is similar to the general population for UC and slightly higher in CD.[8,9]

Traditionally, IBD is more common in Caucasians who live in urban and industrialized areas, such as in the countries of North America and northern Western Europe, compared to developing countries in continents such as Asia, Africa and South America. However, just as seen with other autoimmune diseases, the difference between the regions has declined in recent years, mainly because of the increase of the disease in areas with low prevalence. This is in parallel with the increasing urbanization and development of these areas.[10]

The incidence of a disease around the world is always compared based on distinct socioeconomic, ethnic and cultural conditions, including food and social habits, different conditions of access to medical services, and the availability of access to information on the population's health (Figure 2.1). These disparities do not explain the increase in IBD due to chance, migration or genetic changes, which would be better explained by changes in the environment to which these people belong.

Changes that may be directly or indirectly related to the increased number of cases of IBD[11] are: access to heated water, use of toothpaste, smaller families and dwellings with lower population density, reduced parasitic diseases, consumption of refrigerated foods, exposure to pathogens later in life, and vaccination. All of these factors alter the composition and amount of microorganisms with which people come into contact throughout life, particularly during childhood. The assessment of such impact is even more complicated considering that cases of IBD can occur indirectly, caused by changes in gut flora, many years before the onset of the disease.

Corroborating this information, recent studies have shown that the use of antibiotics for a short period causes changes in gut flora that remain for 2 years after the end of treatment.[12] Furthermore, when antibiotics are used in the first year of life, stage in which the child's gut flora is being formed, the

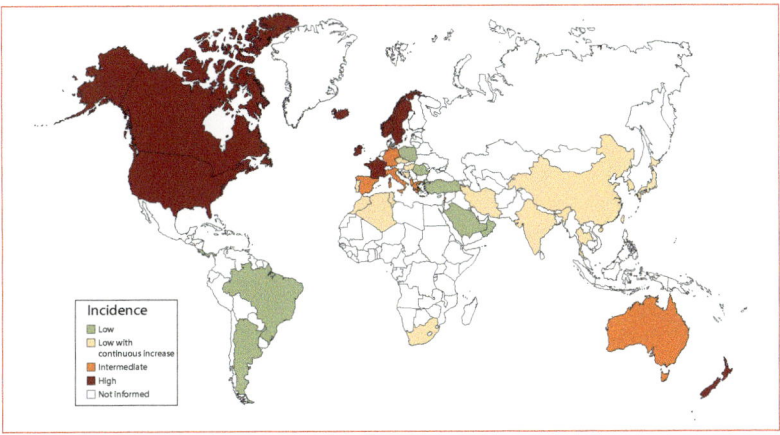

**Figure 2.1** Global map of IBD: red represents an annual incidence greater than 10/10[5]; orange is an incidence of 5 to 10/10[5]; green is an incidence below 4/10[5]; yellow is low incidence with continuous increase.

Source: adapted from Cosnes et al., 2011.[6]

risk of developing IBD in adolescence increases up to 5 times compared to those not treated with antibiotics at this age.[13]

Significant increase in the onset of the disease is also observed only in the descendants of immigrants from regions with low incidence of IBD moving to high-incidence areas, noting that in addition to the changes, the time when the individual comes into contact with the new environment is an additional factor to the development of IBD.[14,15] In addition to the increased incidence of the disease, there was a change of CD ratio in relation to the UC cases associated with the place of origin for similar proportions to the immigration area, suggesting that the type of IBD is also influenced by the conditions of the local environment.

Currently in Brazil, 84% of the population lives in cities, reaching 94% in the South and Southeast regions. The country has a high rate of miscegenation, with 44% of the population self-declared as mixed (interracial), according to 2008 data from the Brazilian Census Bureau (IBGE), and has been undergoing noticeable changes in eating habits, especially evidenced by the large increase in obesity in the general population. Recent changes during the last decades, with improvements in hygiene and sanitary conditions, greater urbanization, and the change of traditional diets for processed foods, led to an increase in the number of cases of IBD.

Epidemiological studies in Brazil are few and generally restricted to certain regions of the country. Federal government data show that the northern region is the one with the lowest number of hospitalizations related to IBD (1.16/100,000 inhabitants/year), followed by the Northeast (2.17/100,000 inhabitants/year), Southeast (2.42/100,000 inhabitants/year), South (3.07/100,000 inhabitants/year) and Mid-West (3.32/100,000 inhabitants/year). If these data are regarded as incidence rate sources, Brazil would be close to Asian countries like Singapore (1.06/100,000) and China (3.44/100,000).[16] However, it is likely that the South and Southeast regions are underestimated, due to the higher proportion of patients treated through the private health system.

Brazilian data[17,18] show that the disease follows the demographic pattern of the region studied regarding skin color (Table 2.1). Two studies[19,20] conducted in the State of São Paulo have shown that the increase in incidence occurred in recent decades and the change in ratio between CD and UC cases was similar to that seen in other developing countries. Given these changes, we highlight the importance and the need for a public policy aimed at better patient education about the disease and better training of health staff for identification and management of new cases of IBD.

## FINAL CONSIDERATIONS

Improved socioeconomic conditions changed the profile of the disease in Brazil. Thus, an epidemiological study nationwide to assess the incidence and prevalence of IBD in the country and allow the development of suitable measures for treatment of these patients is necessary.

**Table 2.1** Comparison of the demographics of the general population to patients with IBD by region in Brazil

| Skin color | Mato Grosso* | Mato Grosso[17] | São Paulo* | São Paulo[18] |
|---|---|---|---|---|
| White | 38% | 43% | 64% | 68% |
| Mixed | 50% | 48% | 28% | 22% |
| Black | 10% | 8% | 6% | 8% |
| Other | 2% | 1% | 2% | 2% |

*Data from the Brazilian Census Bureau, IBGE.

## REFERENCES

1. Baumgart DC, Sandborn WJ. Inflammatory bowel disease: clinical aspects and established and evolving therapies. Lancet 2007; 369:1641-57.

2. Baumgart DC, Carding SR. Gastroenterology 1 Inflammatory bowel disease: cause and immunobiology. Lancet 2007; 369:1627-40.

3. Schirbel A, Fiocchi C. Inflammatory bowel disease: established and evolving considerations on its etiopathogenesis and therapy. J Dig Dis 2010; 11(5):266-76.

4. Burisch J, Munkholm P. Inflammatory bowel disease epidemiology. Curr Opin Gastroenterol 2013; 29(4):357-62.

5. Gower-Rousseau C, Vasseur F, Fumery M, Savoye G, Salleron J, Dauchet L et al. Epidemiology of inflammatory bowel diseases: new insights from a French population-based registry (EPIMAD). Dig Liver Dis 2013; 45:89-94.

6. Cosnes J, Gower-Rousseau C, Seksik P, Cortot A. Epidemiology and natural history of inflammatory bowel diseases. Gastroenterology 2011; 140(6):1785-94.

7. Baumgart DC, Bernstein CN, Abbas Z, Colombel JF, Day AS, D'Haens G et al. IBD around the world: comparing the epidemiology, diagnosis, and treatment: proceedings of the World Digestive Health Day 2010 – Inflammatory Bowel Disease Task Force Meeting. Inflamm Bowel Dis 2011; 17(2):639-44.

8. Hovde Ø, Moum BA. Epidemiology and clinical course of Crohn's disease: results from observational studies. World J Gastroenterol 2012; 18(15):1723-31.

9. Molodecky NA, Soon IS, Rabi DM, Ghali WA, Ferris M, Chernoff G et al. Increasing incidence and prevalence of the inflammatory bowel diseases with time, based on systematic review. Gastroenterology 2012; 142(1):46-54.e42; quiz e30.

10. Logan I, Bowlus CL. The geoepidemiology of autoimmune intestinal diseases. Autoimmun Rev 2010; 9(5):A372-8.

11. Krishnan A, Korzenik JR. Inflammatory bowel disease and environmental influences. Gastroenterol Clin North Am 2002; 31(1):21-39.

12. Jernberg C, Löfmark S, Edlund C, Jansson JK. Long-term impacts of antibiotic exposure on the human intestinal microbiota. Microbiology 2010; 156(Pt 11):3216-23.

13. Shaw SY, Blanchard JF, Bernstein CN. Association between the use of antibiotics in the first year of life and pediatric inflammatory bowel disease. Am J Gastroenterol 2010; 105(12):2687-92.

14. Pinsk V, Lemberg DA, Grewal K, Barker CC, Schreiber RA, Jacobson K. Inflammatory bowel disease in the South Asian pediatric population of British Columbia. Am J Gastroenterol 2007; 102(5):1077-83.

15. Probert CS, Jayanthi V, Hughes AO, Thompson JR, Wicks AC, Mayberry JF. Prevalence and family risk of ulcerative colitis and Crohn's disease: an epidemiological study among Europeans and south Asians in Leicestershire. Gut 1993; 34(11):1547-51.

16. Ng SC, Tang W, Ching JY, Wong M, Chow CM, Hui AJ et al. Incidence and phenotype of inflammatory bowel disease based on results from the Asia-pacific Crohn's and colitis epidemiology study. Gastroenterology 2013; 145(1):158-65.e2.

17. Souza MM de, Belasco AGS, Aguilar-Nascimento JE de. Perfil epidemiológico dos pacientes portadores de doença inflamatória intestinal do estado de Mato Grosso. Rev Bras Coloproctol 2008; 28(3).

18. Poli DD. Impacto da raça e ancestralidade na apresentação e evolução da doença de Crohn no Brasil. São Paulo: Universidade de São Paulo, 2007.

19. Souza MH, Troncon LEA, Rodrigues CM, Viana CFG, Onofre PHC, Monteiro RA et al. Evolução da ocorrência (1980-1999) da doença de Crohn e da retocolite ulcerativa idiopática e análise das suas características clínicas em um hospital universitário. Arq Gastroenterol 2002; 39(2):98-105.

20. Victoria C, Sassak L, Nunes H. Incidence and prevalence rates of inflammatory bowel diseases, in midwestern of São Paulo State, Brazil. Arq Gastroenterol 2009; (1):20-5.

# ETIOPATHOGENESIS OF INFLAMMATORY BOWEL DISEASE

CLAUDIO FIOCCHI

## INTRODUCTION

The gastrointestinal tract is affected by several chronic inflammatory conditions, but few have a serious impact as inflammatory bowel disease (IBD), demonstrated by an ever expanding number of publications on IBD.[1-4] Both forms of IBD, Crohn's disease (CD) and ulcerative colitis (UC), are characteristically chronic, significantly impair quality of life, require prolonged medical attention, and represent a major burden to society overall. What makes IBD particularly challenging is its still unknown cause, its unpredictable presentations and symptoms, less than optimal treatments, and a continuous rise in incidence and prevalence in many areas of the world. These uncertainties have long been recognized, and the notion that IBD may actually represent a constellation of diseases or syndromes rather than the single entities of CD and UC has been taken into consideration at least since the 1970s.[5] Many theories have been proposed to explain IBD, ranging from infectious to psychosomatic, social, metabolic, vascular, genetic, allergic, autoimmune, and immune-mediated.[3,4,6-8] Some of these theories have been abandoned while others have gained support, showing an evolution in knowledge and belief that developed over the last three decades in parallel with the acquisition of new data from studies on epidemiology, genetics, the intestinal flora and the immune response. Each of these areas is in constant evolution as reflected by continuous progress in how epidemiology, genetics, the intestinal microbiota and the immune response contribute to the emergence of CD and UC. Another important concept is the realization that none of these components alone can explain IBD, but it is their integration that actually determines whether IBD will appear and with which features. An example of how this integration may lead to IBD has been recently reported in the literature. Infection of mice carrying the CD susceptibility gene *ATG16L1* with a murine virus induces intestinal pathology resembling that of CD, these changes being dependent on tumor necrosis factor (TNF)-α and interferon (IFN)-γ and preventable by treatment with broad spectrum antibiotics.[9] This demonstrates that the interplay of genes, infectious agents, environmental factors and commensal microbiota not only induces an IBD-like condition but also determines its clinical phenotype. Each component of IBD pathogenesis is a potential therapeutic target, and their perceived importance has a direct impact on the evolution of treatment for IBD patients. So, at the moment we are at a stage where the research and clinical IBD community is focused on what are considered the four pillars of IBD pathogenesis, but knowledge about the mutual interaction among each of the pillars is far from complete. This chapter discusses the basic components of CD and UC pathogenesis and touches upon how new knowledge of epidemiology, genetics, the intestinal microbiota and the immune response help improve and expand therapeutic options for IBD patients.[10]

## ENVIRONMENTAL FACTORS

There is abundant evidence showing that IBD is increasingly recognized worldwide.[11] Such increase began after World War II, a time when IBD was primarily known in North America and Northern Europe. Since then, IBD started to be diagnosed more frequently in Central and Western Europe, Japan, and Australia, then in South America and Eastern Europe, and in the last decade IBD is emerging in the Asia-Pacific area, with a steady increase in the incidence of IBD in various Asian countries.[12-15] The clinical characteristics of IBD in Asia show both similarities and differences with IBD in Western populations: in Asia UC is more common than CD, and the clinical course tends to be milder, with fewer complications and less need for surgical procedures.[15-17] This picture resembles IBD at the time of its emergence in Europe and North America over half a century ago, when the disease was also milder and with fewer complications. Strong support for the importance of environmental factors in the conditioning of IBD comes from studies that have analyzed the risk of IBD in populations migrating from low to high risk areas.[18] Subjects that move from low risk areas, such as India or Indochina, to high risk areas, such as North America or Northern Europe, have a risk that is directly proportional to the age they had at the time of migration: if they move as adults their risk remain essentially the same as that of their native countries; on the other hand, if they move as infants or young children the risk of developing IBD is the same, or even higher, than that of natives indigenous to the high risk areas. Thus, individuals with a fix genetic make-up increase their odds of developing IBD only when their genes are challenges by the factors present in more developed, high prevalence regions of the world. In other words, same genes, different environments, increased odds.

The unresolved key question in regard to IBD epidemiology is obviously what causes IBD to emerge in new areas of the world, and a large number of risk factors have been proposed as possible culprits: cigarette smoking, diet, oral contraceptives, appendectomy, antibiotics, infections and vaccinations, and perinatal and childhood factors have all been suspected. Recent studies are increasingly stressing the importance of the diet in shaping the gut microbiome.[19] As discussed below, the gut microbiome is not only crucially involved in IBD pathogenesis, but also in the education of the systemic and mucosal immune systems.[20] Consequently, qualitative or quantitative alterations of the gut microbiota composition in the first years of life – what is now considered a "sensitive period" – may determine whether robust and effective immunity will ensue or, instead, a weak and defective immune response that makes the individual prone to autoimmunity or chronic inflammatory conditions – including IBD – later in life. Indirect but compelling evidence in support of this scenario is found in epidemiological studies of populations where the time and number of administration of antibiotics in early life was correlated to the risk of developing IBD later on.[21] The earlier the use and the higher the number of antibiotics, the higher the risk of IBD in adulthood; a troubling correlation with important clinical implications.

With the exception of cigarette smoking, none of the other factors is supported by evidence strong enough to be considered as true risk factors requiring avoidance or life style modifications for therapeutic purposes. The link between IBD and smoking is well established, reproducible, and dictates clinical management, even though a pathophysiological connection between tobacco and intestinal inflammation is still missing. Modulation of immune responses, changes in cytokine levels, alterations in mucus composition, vascular and pro-thrombotic effects, changes in gut permeability as well as other effects have all been suggested as possible reasons for the effect of smoking in IBD, but none has been confirmed.[22] Still intriguing and unexplained is the fact that cigarette smoking has an opposite effect on each form of IBD: while smoking worsens the clinical course of CD with more symptoms, more relapses, more hospitalizations and more surgeries, it has a protective effect on the clinical course of UC.[22] This clearly points out fundamental differences in the pathogenesis of CD and UC.

The evolution of IBD and the diversity of factors linked to its epidemiology have suggested the "hygiene hypothesis" as a reason to explain the increasing incidence around the world. This hypothesis was originally proposed by Strachan in 1989 and proposes that the lack of proper exposure to common infections early in life negatively affects the development of the immune system, which becomes less "educated" and less prepared to deal with multiple new challenges later in life.[23] This hypothesis is indirectly supported by evidence of improved health in parts of the world where IBD is emerging, with less infectious and parasitic diseases, improved sanitary conditions, safer water and foods, vaccinations, etc., and the acquisition of dietary and other habits from Western societies where IBD is more prevalent.[24,25] This makes IBD a disorder of modern lifestyle, a position shared with other chronic allergic and inflammatory conditions such as asthma, psoriasis, multiple sclerosis and rheumatoid arthritis.[26] The true importance of the hygiene hypothesis to explain IBD is still uncertain, and this theory can be criticized because the majority of studies in this area used indirect and retrospective data.[27] On the other hand, circumstantial evidence favoring the hygiene hypothesis is considerable, although complicated by an almost endless list of environmental factors that have been associated with increasing frequency of IBD around the globe.[28] The potential mechanisms underlying the hygiene hypothesis are still not understood, but defective innate immunity and imbalances of effector and regulatory T cell functions mediating adaptive immunity are likely involved.

## GENETIC FACTORS

Among the various components of IBD pathogenesis, none has progressed as dramatically as genetics. The well-known familial occurrence of IBD has for a long time suggested that CD and UC could have a genetic basis,[29] but only recently objective evidence supporting this notion has been secured. After the discovery of the association of variants of the NOD2 gene with ileal CD in 2001,[30,31] and variants of the interleukin (IL)-23 receptor gene with both CD and UC in 2006,[32] the number of IBD genetic associations has undergone a true explosion. This is due primarily to the implementation of genome-wide association studies (GWAS) by national and international consortia of investigators that utilize massive combined databases encompassing thousands of well-phenotyped CD and UC patients.[33] This wealth of information has uncovered increasingly large numbers of genetic associations with CD, UC or both.[34] Early studies focused primarily on CD gave the initial impression that this condition had a stronger genetic connotation than UC simply because of the high number of susceptibility loci discovered in CD,[35] but recent reports show that large numbers of genetic loci are also associated with UC.[36] In addition, some loci may be associated with early-onset IBD in children,[37] demonstrating the potential of genetic analyses to differentiate the various stages of disease progression.

It should be noted that most reported genetic associations are not pinpointing single gene defects that underlie specific pro-inflammatory pathways, but primarily loci containing multiple candidate genes that remain to be singled out by more refined DNA screening approaches. A recent meta-analysis of CD and UC GWAS provides a comprehensive view of genetic associations in IBD that includes 163 genetic loci: of these, 110 are associated with both forms of IBD, 30 appear to be CD-specific and 23 UC-specific, and while most of them confer risk, a few confer protection.[38] The associations detected by this IBD GWAS meta-analysis are enriched for genes involved in primary immunodeficiencies, T cell function, modulation of cytokine production, and mycobacterial diseases. Many of the IBD *loci* are also found in other immune-mediated conditions such as psoriasis, multiple sclerosis, rheumatoid arthritis, lupus, type I diabetes, and celiac disease, suggesting that many immune-mediated disease have shared disease mechanisms.

In regard to genetic heterogeneity, it is important to note the existence of genetic compartmentalization of IBD-associated variants in different populations around the globe. While several variants are found in Western populations of Caucasian and Jewish patients, many of them are not present in oriental populations, like the Japanese and Chinese, which lack an association with *NOD2* and autophagy genes. In fact, there appears to be reproducible differences in genetic background between Westerners and Asians, with the exception of the polymorphisms of the *TNFSF15* gene, which encodes for TL1A, a TNF superfamily member. Such polymorphisms are present in Asian, European and North American IBD patients,[39-41] making *TNFSF15* an IBD susceptibility gene shared among racially and ethnically diverse populations. Considering that almost a decade has elapsed since the discovery of the *NOD2* gene variants in CD and we still do not have a clear understanding of how such variants cause intestinal inflammation, the massive number of other gene variants poses a formidable challenge to basic IBD investigators' intent to unveil the mechanistic bases of the various CD- and UC-associated gene variations.[42,43]

In spite of this complexity, some pathophysiologically relevant defects have emerged that can pave the way to a better understanding of the pro-inflammatory pathways in CD and UC and potentially the development of specific interventions. For instance, CD patients with *NOD2* gene defects have an impaired ability to recognize and process bacterial products, and this may lead to an inappropriately long or ineffective immune response. Some CD patients have variants of the *ATG16L1* and *IRGM* autophagy genes, implying a defective capacity to process cell degradation products as well as bacteria, which may lead to insufficient elimination of pro-inflammatory stimuli.[44] In addition, some CD patients with *ATG16L1* gene variants display defects in the Paneth cell granule exocytosis pathway,[45] which may impair their ability to secrete endogenous antibacterial peptides and control the quantity or quality of intestinal bacteria. Polymorphisms of the *TLR4* gene exist in both CD and UC,[46] further reinforcing the notion of potentially defective innate immunity pathways with which these patients recognize and respond to bacteria. Variants in the *XBP1* gene, involved in the response to endoplasmic reticulum stress, are found in some IBD patients,[47] suggesting an impaired ability of cells to respond to a variety of cellular stress signals.

In addition to identifying pro-inflammatory pathways, the study of genetic abnormalities in IBD can also potentially provide information that is directly relevant to clinical diagnosis and management. In fact, recent reports indicate that combined gene analyses may help create a molecular classification of IBD,[48] discriminate between CD and UC,[49] predict disease risk and severity in CD patients,[50] and forecast response to infliximab.[51] However, despite obvious progress in the field of IBD genetics, it is unlikely that identification of genetic abnormalities alone will help explain all aspects of IBD because gene expression is modulated by many other factors.[52] Prime among these factors are other genes and the environment, and active investigation of gene-gene interactions and gene-environment interactions in IBD is currently under way.[53,54]

## MICROBIAL FACTORS

The possibility that IBD represents chronic inflammation directed against microbial agents has been considered since the initial reports describing UC and CD. Over many years, repeated attempts have been made to identify common and uncommon organisms of bacterial, viral or fungal origin, but most have been dismissed due to lack of valid or reproducible evidence. One agent that raised a great deal of controversy is *Mycobacterium avium* subspecies *paratuberculosis* (MAP), which was originally identified in the mid 1980s.[55] This mycobacterium has been the center of much debate, with some reports supporting and other reports denying its possible etiological role in CD. The last study that attempted to provide a definitive answer to this possibility has been a large clinical trial in which patients with CD were given a combination of three anti-mycobacterial antibiotics or placebo, and followed-up for evaluation of remission and clinical activity.[56] At the end of two years, no evidence of sustained benefit was detected, strongly suggesting that the elimination of MAP does not significantly affect the course of CD, a response that should be interpreted as denying an etiological role of MAP in this form of IBD. Another microorganism that is still under active investigation is adherent-invasive *Escherichia coli* (AIEC). In the late 1990s, a French group described the isolation of *E. coli* strains from ileal mucosa of CD patients with a capacity to adhere to and invade intestinal epithelial cells.[57] The same group later showed that AIECs are specifically associated with the ileal mucosa in CD while being uncommon in control or UC tissues.[58] It is not entirely clear whether these AIECs are pathogens or commensals, but an argument against an etiological role in CD is that treatment with antibiotics effective against coliforms fails to cure CD patients.[59]

The consistent failure to identify true pathogens coupled with progress in the investigation of antimicrobial immune responses in humans with IBD as well as several animal models of experimental IBD has drawn the attention of investigators to the normal enteric microbiota as a possible inducer of chronic intestinal inflammation. Based on a large number of reports, it is now established that essentially all animals raised under germ-free conditions, i.e., in the complete absence of a commensal flora, do not develop experimental intestinal inflammation regardless of strain and genetic background or method used to induce inflammation.[60] In humans with IBD the existence of an anti-intestinal microbe immune reactivity has been known for a long time, as shown by the presence of several serum antibodies against a variety of microorganisms, including anti-*Saccharomyces cerevisiae* (ASCA), anti-outer membrane protein C (anti-OmpC), anti-CBir1 flagellin (anti-CBir1), and anti-I2. These antibodies are currently used as CD biomarkers[61], and the higher the number of detected antibodies, the more likely are CD patients to have a complicated disease course.[62,63] These findings establish a close connection between the magnitude of the immune response against enteric microbial antigens and IBD pathogenesis, although this conclusion is applicable to CD but not necessarily to UC.

It is widely agreed that an immune response to antigens of the gut microbiota, the so called pathogen-associated molecular patterns (PAMPs), is central to IBD pathogenesis. If so, then the question arises of whether such immune response is directed at the gut microbiota as a whole, specific subgroups or strains, or only selected microbes. The development of immune reactivity to enteric bacteria is an entirely physiological phenomenon that starts immediately after birth, when the immature intestinal immune system begins to be exposed to a variety of microbial antigens and creates a life-long state of tolerance against them.[64] A variety of reports show that unique bacterial strains, such as segmented filamentous bacteria, are essential to shape the T helper (Th) cell repertoire of the gut,[65] providing additional proof of the intimate relationship between the microbiota and intestinal immunity. A generally accepted principle is that there is a loss of the state of tolerance in IBD, and the local immune system mounts an anti-microbiota inflammatory response that is translated as IBD at the clinical level.[66,67] One of the major obstacles to understand why and how bacterial tolerance is lost is the lack of a better knowledge of the composition of the normal human intestinal microbiota. Despite a large number of reports, it is still unclear what the exact make-up of the normal human gut flora is and how it changes under pathological conditions.[68,69] In addition, it is becoming increasingly clear that, even though bacterial communities grossly maintain a relatively stable composition in humans, they vary across space and time.[70] Moreover, there is accumulating evidence that bacteria do not always function as inciters of an immune response but some actually promote an anti-inflammatory status.[71,72] The number of bacteria associated with the mucosa layer is dramatically increased in IBD patients,[73,74] but its qualitative composition is still poorly defined. There is some consensus that there is a reduced diversity of fecal microbiota in IBD when compared to controls, that enterobacteria are significantly more frequent in CD, and that specific differences can be detected when comparing the fecal flora of CD, UC and healthy control subjects.[75-78]

Another aspect directly relevant to how the normal or IBD host modulates its response to bacterial challenges is the role of spontaneously produced anti-bacterial peptides, such as defensins. Enteric α-defensins are part of the anti-microbial arsenal produced in the mammalian small intestine by Paneth cells, specialized secretory cells localized at the bottom of the crypts. These antimicrobial peptides not only defend against pathogenic bacteria, such as *Salmonella*, but also control the balance among the various bacterial populations and contribute to local homeostasis.[79] Therefore, any defect in defensin production or function could impact on flora composition and potentially contribute to IBD. There is some evidence that α-defensin production is reduced in ileal CD,[80] and that in colonic CD there is reduced mucosal antimicrobial activity, a finding consistent with low antibacterial peptide expression.[81] Thus, several questions remain to be answered before we reach a better understanding of host-commensal microbiota interactions in health and IBD.[82]

A number of studies employing molecular analysis of the gut microbiota composition have recently uncovered two aspects of particular relevance to IBD pathogenesis: the first is the greater than anticipated complexity of the human gut flora and the second is unexpected number of factors that can modify microbiota composition and function. In regard to the first, metagenomic sequencing of human gut microbial genes shows that they are 150 times larger than the human genome, representing around 1.000 prevalent bacterial species.[83] This enormous number of microbial genes makes it a domineering modulator of overall health and disease. Investigation of how changes in microbial composition, such as the ones cited above, contribute to IBD is a formidable challenge under intense investigation. In regard to the second, more recent studies are revealing how the diet and other environmental factors can dramatically alter the gut microbiota. Children from African communities ingesting primitive diets and children from European urban center eating western foods harbor totally inverse proportions of bacteroidetes and firmicutes in the stools.[84] Normal volunteers fed with a high protein and animal fat diet preferentially develop *Bacteroides* in their bowel, while *Prevotella* is dominant if fed with a carbohydrate diet.[85] Maltodextrin, a ubiquitous food additive used as a sugar substitute, promotes adherence of bacteria to intestinal epithelial cells while inhibiting autophagy, resulting in excessive bacterial growth,[86] a typical feature of IBD. These are just a few examples of a tremendously important area of research that will help us to better understand the actual role of the gut microbiota in IBD pathogenesis.

## IMMUNE FACTORS

Until the emergence of genetics and the renewed interest in microbiology, immunology had practically dominated the investigation of IBD pathogenesis. As stated in the introduction, it is now clear that no single pathogenic component can trigger or maintain the disease. On the other hand, immune response is the effector arm that mediates inflammation, and understanding its function in the gastrointestinal tract and its derangement in CD and UC is fundamental to unravel the mechanisms of chronic gut inflammation and understand how to control them to induce disease remission. To provide an overview of how immunology contributes to IBD pathogenesis this section is divided and discussed according to the main cell types that respectively mediate innate and adaptive immunity. In addition, more recently identified systems that contribute to modulate the overall immune and inflammatory response will also be discussed, including "sterile inflammation", inflammasome, epigenome and microRNAs (miRNAs).

## Innate immunity

Innate immunity represents the first line of defense against invading microbes and other noxious agents; it develops in a few minutes to a few hours, is largely non-specific, and has no immunological memory.[87] Innate immune response to the gut microbiota is presently considered a central event in IBD pathogenesis.[88] Two main types of immune cells mediate innate immunity: macrophages and dendritic cells.

In the normal intestine, macrophages are conditioned by the mucosal microenvironment to express a non-inflammatory phenotype translated by a downregulated expression of innate immunity receptors and limited production of pro-inflammatory cytokines.[89] In contrast, in IBD-affected tissues, mucosal macrophages display an activated phenotype and are phenotypically heterogeneous.[90] Macrophages derived from newly recruited peripheral blood monocytes still express the monocytic CD14 marker but are primed for the production of various pro-inflammatory cytokines such as IL-1α IL1-β and TNF-α.[91,92] In CD, these CD14+ pro-inflammatory macrophages are increased in number and produce more IL-23 and TNF-α than those in normal and UC mucosa, and contribute to the production of IFN-γ by local T cells.[93] Recent studies performed in support of the theory that CD is an immunodeficiency[94] claim that secretion of pro-inflammatory cytokines by CD macrophages is severely impaired because of excessive internal lysosomal degradation.[95] This would result in diminished recruitment of neutrophils, impaired clearance of bacteria, and granuloma formation.[96] This proposition is intriguing, but it is hard to reconcile with the well-known activated phenotype of mucosal macrophages in CD, the high tissue levels of pro-inflammatory cytokines, including TNF-α, and the abundance of neutrophils in actively inflamed mucosa.

Intestinal dendritic cells (DCs) are antigen-presenting cells crucially involved in the initiation and regulation of local innate immune phenomena but also have a role in adaptive immunity.[97] Like macrophages, their function is modulated by the mucosal microenvironment and they can function to provide protection and defense, induce tolerance or mediate inflammation.[98] Their number is relatively small, but they are extremely diverse in phenotype and function, and these characteristics make their study in humans rather difficult, explaining the very limited amount of information on human mucosal DCs. In IBD, DCs are activated, their expression of microbial receptors is enhanced, and they produce increased levels of pro-inflammatory cytokines like IL-12 and IL-6.[99]

### Adaptive immunity

Adaptive immunity represents the second line of defense against invading microbes and a vast array of other antigens; it develops in a few hours to several days, is antigen-specific, and has immunological memory.[100] Two types of immune cells mediate innate immunity: B cells (humoral immunity) and T cells (cell-mediated immunity).

B cell-mediated antibody production in active IBD is increased both in the circulation and at the mucosal levels. There are changes in IgM, IgG and IgA synthesis and secretion by both peripheral blood as well as mucosal plasma cells in both UC and CD,[101] and in the affected mucosa there is an increased production of monomeric IgA, which is normally predominant in the circulation.[102] The patterns of antibody class production differ in UC and CD, particularly in regard to IgG production: in UC there is a disproportional increase in IgG1 secretion, while in CD IgG1, IgG2 and IgG3 are increased compared to control cells but in a proportional fashion.[103] At the moment limited attention is being given to B cell immunity in IBD, but a renew in interest may occur if some of the new biologicals that specifically induce B cell depletion, like rituximab,[104] turn out to be effective in the management of CD or UC. So far, this has not been the case.[105]

After the original identification of the CD4+ T helper 1 (Th1) and 2 (Th2) subsets in mice,[106] and the subsequent demonstration in humans,[107] the field of Th cell differentiation, type and function has undergone a substantial evolution. In addition to IFN-γ-producing Th1 cells and IL-4-, IL-5- and IL-13-producing Th2 cells, new Th subsets have joined the field. A few years ago IL-17-producing Th17 cells were identified as a distinct new Th subset in both mice and humans;[108] then dual IFN-γ- and IL-17-producing Th1/Th17 cells were characterized, including in CD mucosa;[109] recently two new subsets of CD4+ effector Th cells were described based on the production of their respective cytokines, i.e., Th9 and Th22 cells, whose function is still poorly understood;[110] and more recently Th cells producing both IL-17 and IL-4, a dual Th17 and Th2 pattern, have been identified.[111] In addition to the realization that Th cells are very heterogeneous, evidence is now emerging that Th cells may be quite plastic, questioning whether the various CD4+ Th cells are actually terminally differentiated cells or are cells that retain the capacity to continuously differentiate and shift from one subset to another depending on the host's need to mount the most appropriate and efficient immune response.[112] In addition to Th cells, another major subset is made up of T regulatory (Treg) cells whose function is to monitor the immune response and prevent an excessive and potentially harmful immune activation.[113,114] Recent data provide support for the novel and unanticipated notion that Th17 and Treg cells share common pathways, suggesting developmental and functional links between Treg and Th17 cells.[115,116] All these new findings are both exciting and intriguing, but add enormous complexity to the understanding how the various CD4+ T cells subsets operate and relate to each other. When viewed under a specific disease perspective, such complexity is likely to become even greater, as in the case for IBD.

T cells, and mucosal CD4+ Th cells in particular, have been the focus of IBD pathogenesis for many years, and their study has provided key information on cell-mediated immunity in both CD and UC, and helped define distinctive patterns of immunoregulatory and effector function as well as cytokine secretion profiles in each type of IBD.[117] There is good evidence that CD has a dominant Th1 component as shown by an elevated production of IFN-γ and IL-12 by lamina propria mononuclear cells,[118,119] in addition, in CD there is also a considerable production of IL-17 by Th17 cells and dual IFN-γ- and IL-17-producing mucosal Th cells,[109,120] but also IL-21 that regulates IL-17 production.[121] Taken together, these findings support the generally accepted view of CD as a Th1-like condition. In contrast, UC is considered an atypical Th2 response based on the observation of increased IL-5 and IL-13 production by Th cells and of IL-13 by NK T cells in the inflamed mucosa.[122] Of note, IL-13 induces cytotoxicity, apoptosis and impairs epithelial barrier function,[123] events that may explain some key features of UC pathogenesis. However, Th17 cells are also present in UC mucosa, although in lower numbers that in CD mucosa.[120] The number of cytokines produced by Th cells in both forms of IBD is not limited to ones cited above, and many new soluble immunoregulatory and pro-inflammatory mediators are likely to play important roles in the immunopathogenesis of CD and UC.[124] Less information is available on Tregs in IBD, but the function or number of CD4+CD25+FoxP3+ and CD8+ Tregs may be impaired in CD and UC,[125-127] perhaps contributing to maintenance of inflammation.

### Novel contributors to immunity and inflammation in IBD

In addition to PAMPs, which induce classical microbial inflammation, a new class of molecules called damage-associated molecular patterns (DAMPs) has been indentified that mediate what is called "sterile inflammation",[128] i.e. an inflammatory response triggered in the absence of microbial elements.[129] DAMPs are intracellular elements released upon necrotic cells death, such as nucleic acids (DNA and RNA), nuclear proteins, heat shock proteins, IL-1α, etc. Being intracellular, these components are normally not seen by the immune system but, once liberated in the tissue microenvironment, they incite an innate immune response largely mediated by the same receptors that recognize PAMPs. This certainly occurs when the gut mucosa is damaged or ulcerated as commonly seen in both CD and UC. The investigation of DAMPs in IBD is still at an early stage, but it is worth noting that stool calprotectin, one of the most frequently used markers of IBD activity, is a complex of S100A8/S100A9, two prototypical DAMPs.[130]

Inflammasomes are a group of cytosolic protein complexes that recognize exogenous, microbial, stress and endogenous danger signals and respond by activating the enzyme caspase-1, which in turn leads to the production of the pro-inflammatory cytokines IL-1β and IL-18.[131] What makes the inflammasome particularly attractive and biologically relevant in IBD is its increasingly obvious importance to regulate the crosstalk between mucosal immune system and microbiota.[132] Presently, the role of the inflammasome in IBD is still unclear because studies in animal models disagree on the proinflammatory *vs.* protective role of inflammasomes,[133] and studies in patients with IBD are yet to be reported.

Epigenetics can be variably defined, but a simple definition is the sum of changes in gene expression – the epigenome – not due to alterations in the DNA sequence that are potentially reversible or heritable.[134] Epigenetic modifications, mediated primarily by enzymatic activities, exert transcriptional regulation on the inflammatory response,[135] a finding relevant to chronic disorders including IBD. The investigation of epigenetics in IBD is just beginning, but preliminary evidence indicates the crucial role of epigenetic modifications in predisposing, initiating, maintaining or even promoting inheritance of IBD.[136,137]

Newly discovered modulators of immunity and inflammation are the microRNAs (miRNAs), short single stranded non-coding RNAs that function primarily by silencing gene expression.[138] Recent reports describe the differential and downregulated expression of several miRNAs in both human and experimental IBD. These studies suggest that loss of the regulatory function of miRNAs in IBD may be an important contributor to the inflammatory response of both UC and CD, and re-establishing their expression to proper levels may yield novel ways to control gut inflammation.[139]

## CONCLUSIONS AND THERAPEUTIC IMPLICATIONS

Before discussing how a better knowledge of IBD etiopathogenesis may improve our treatment options, an additional concept must be considered. Traditionally IBD has been viewed as a condition diagnosed at some point in time in the life of CD or UC patients. What we see at that time is, however, only a glimpse of a process that started long before the clinical diagnosis and that will last for the rest of the patient's life. During its long progression, IBD is not necessarily the same. It is hard to conceive that the very same triggers and mechanisms mediating gut inflammation in a young child are exactly the same that still mediate the disease several decades later in the same subject. Even routine clinical evidence supports this possibility, as symptoms and response to therapy – or lack of response – also evolve and change with time. Thus, it seems more logical to see IBD as a dynamic condition with different phases progressing from an early to a late stage, each stage being dependent on different mechanisms and requiring different therapeutic strategies for optimal results. During this progression, primary and secondary factors are likely to play concomitant or sequential roles in the development of gut inflammation and create the clinical picture that we clinically define as IBD (Figure 3.1).

From a practical standpoint and for the benefit of the patients, the central question is which of the components of IBD pathogenesis should be targeted for optimal clinical results. As of today, environmental and genetic factors are not yet amenable to manipulation, but the other two components of IBD pathogenesis, i.e. the intestinal microbiota and the immune system, are suitable to therapeutic intervention. In fact, the past and recent advances achieved in the treatment of CD and UC are based on the use of antibiotics and probiotics to manipulate the gut flora, and anti-inflammatory medications, biologicals or other drugs to block or neutralize cytokines, receptors or signaling molecules that mediate the action of inflammatory cells.[140-142] This pathophysiology-based approach has allowed the generation of numerous monoclonal antibodies, recombinant cytokines, small molecules, genetically modified organisms, devices and cell-based therapies that offer an unprecedented wide range of treatment options for physicians and patients alike. Table 3.1 provides an overview of biological and other miscellaneous agents for therapy of IBD. It should be noted, however, that some of them have no proven therapeutic efficacy or are still awaiting clinical trials to establish whether or not they will become useful agents in IBD.

Although altering the flora or controlling the immune response is certainly helpful to IBD patients, it is currently impossible to predict whether a beneficial response will occur, the type of response observed, which patients will improve with what form of treatment, or which patients will fail to respond to the same or alternate medications. One of the reasons why we still fail to cure IBD is because current therapeutic approaches separately target single components of IBD pathogenesis. If CD and UC are complex diseases where environmental, genetic, microbial and immune factors must come together to trigger gut inflammation, modifying or blocking any one of them individually will not correct the abnormalities in the other factors. For at least a couple of decades IBD therapy has been heavily focused on controlling inflammation by suppressing excessive immune reactivity with immunosuppressant drugs or biologics. There is no question that this approach has greatly benefitted the patients, but it has not altered the natural course of IBD, with recurrent flare-ups and continuous need for maintenance therapy. Even when new agents are developed based on logical and pathophysiology-based pathways that make full sense in face of present knowledge of IBD pathogenesis, failures can occur. This is well portrayed by clinical trials with IL-10 and anti-IL-17 antibodies. Administration of IL-10, a potent immunosuppressive cytokine, to patients with active CD did not result in clinical improvement.[143] More recently, administration of secukinumab, a human anti-IL-17 monoclonal antibody, in mild to moderate CD patients not only failed to induce remission, but actually worsened the patients' clinical condition.[144] When we look at all the new biological therapies developed in little over a decade, only 9 are or appear to be effective out of a total of more than twenty.[145] These failures underscore the practical fact that, despite major advances in the understanding of IBD, we still do not have a grasp on the disease as a whole and reinforce the notion that targeting the immune system alone will not yield a complete cure of CD or UC.

Thus, it is clear that more studies and new approaches are needed to precisely categorize individual subgroups of IBD patients that are not only defined at the clinical phenotypic level but also at genotypic, microbial and immune level.[48] This will create specific "biological signatures" unique to each CD or UC patient so that each subject can be given a rational, highly customized therapeutic approach that will target the specific defects or aberrations underlying his or her intestinal inflammatory pathways. This scenario is not as farfetched as one may imagine because integration of knowledge and information is already occurring[146] with rapid progress in "system biology" that will eventually replace the study of single molecules and pathways in isolation from all the others, an artificial and ineffective way to look at disease pathogenesis.[147,148]

**Table 3.1** Pathophysiology-based mechanisms of action of biological and other miscellaneous agents for therapy of IBD

| Agent | Mechanism | Molecular class |
|---|---|---|
| Infliximab, Adalimumab, Certolizumab, Golimumab | Neutralization of TNF-α activity | Monoclonal antibody |
| Ustekinumab, ABT-874 | Blockade of the IL-12/IL23 pathway | Monoclonal antibody |
| AIN457 | Neutralization of IL-17 activity | Monoclonal antibody |
| Fontolizumab | IFN-γ activity | Monoclonal antibody |
| Natalizumab | Blockade of α4β1 and α4β7 integrins; inhibition of leukocyte adhesion | Monoclonal antibody |
| Vedolizumab | Blockade of α4β7 integrin; inhibition of leukocyte adhesion | Monoclonal antibody |
| PF-00547659 | Blockade of human mucosal addressin cell adhesion molecule-1 (MAdCAM); inhibition of leukocyte homing | Monoclonal antibody |
| Rituximab | Depletion of B cells | Monoclonal antibody |
| Tocilizumab | Blockade of the IL-6 pathway | Monoclonal antibody |
| MDX-1100 | Inhibition of IP-10 (CXCL10 chemokine) activity | Monoclonal antibody |
| Visilizumab | Binding to the T cell CD3 receptor and induction of apoptosis | Monoclonal antibody |
| Daclizumab, Basiliximab | Blockade of the IL-2 receptor (CD25) | Monoclonal antibody |
| Abatacept | Inhibition of T cell activation | CTLA4 ECD-Fc, mutated IgG1Fc |
| Interleukin-10 | Immunosuppression | Recombinant cytokine |
| Interferon β-1a | Anti-inflammatory | Recombinant cytokine |
| Sargramostin (GM-CSF) | Immunostimulation | Recombinant cytokine |
| Alicaforsen | Inhibition of leukocyte adhesion | Antisense oligonucleotide |

*(continues)*

**Table 3.1** Pathophysiology-based mechanisms of action of biological and other miscellaneous agents for therapy of IBD

| Agent | Mechanism | Molecular class |
|-------|-----------|-----------------|
| Tacrolimus (FK-506) | Inhibition of T cell activation and IL-2 production | Small molecule |
| Rosiglitazone | Binding to PPARγ and inhibition of NF-κB activity | Small molecule |
| SCI2267 | Blockade of the pyrimidine synthesis pathway | Small molecule |
| OREI001 | Inhibition of angiotensin converting enzyme 2 | Small molecule |
| Tetomilast | Inhibition of phosphodiesterase activity and leukocyte proinflammatory activity | Small molecule |
| CP-690 550 | Inhibition of Janus kinase 3 (JAK3), leukocyte activation and cytokine production | Small molecule |
| AEB071 | Inhibition of protein kinase C and T cell activation | Small molecule |
| HE3286 | Inhibition of NF-κB activity | Small molecule |
| AG011 | Secretion of the immunosuppressive cytokine IL-10 by *Lactococcus lactis* | Genetically modified probiotic |
| *Trichuris suis* | T helper cell modulation | Helminth eggs |
| Leukapheresis | Removal of granulocytes, monocytes and macrophages | Adsorptive extracorporeal device |
| Stem cell transplantation | Resetting of immune homeostasis | Cell-based therapy |

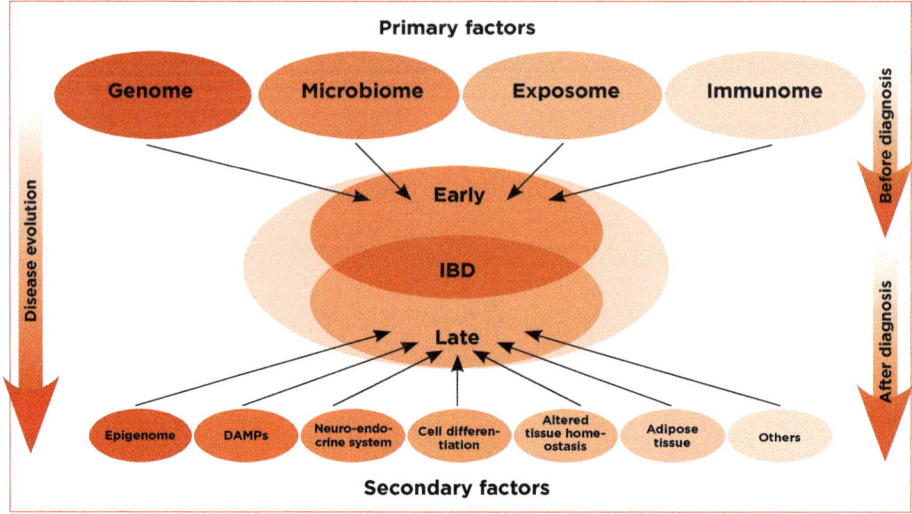

**Figura 3.1** Model of the time-dependent progression of IBD pathogenesis from early to late disease.

DAMPs: damage-associated molecular patterns.

## REFERENCES

1.  Fiocchi C. Inflammatory bowel disease pathogenesis: therapeutic implications. Chinese J Dig Dis 2005; 6:6-9.

2.  Scaldaferri F, Fiocchi C. Inflammatory bowel disease: progress and current concepts of etiopathogenesis. J Dig Dis 2007; 8:171-8.

3.  Kaser A, Zeissing S, Blumberg RS. Inflammatory bowel disease. Annu Rev Immunol 2010;28:573-621.

4.  Xavier RJ, Podolsky D. Unraveling the pathogenesis of inflammatory bowel disease. Nature 2007;448:427-34.

5.  Arend P, Martini GA. Ulcerative colitis. Etiologic unity or polyetiologic syndrome? Am J Proctol 1970;21:331-6.

6.  Wen Z, Fiocchi C. Inflammatory bowel disease: autoimmune or immune-mediated pathogenesis? Clin Develop Immunol 2004;11:195-204.

7.  Korzenik JR. Past and current theories of etiology of IBD. Toothpaste, worms, and refrigerators. J Clin Gastroenterol 2005;39(Supp.2):S59-S65.

8. Fiocchi C. Susceptibility genes and overall pathogenesis of inflammatory bowel disease: where do we stand? Dig Dis 2009; 27:226-35.

9. Cadwell K, Patel KK, Maloney NS et al. Virus-plus-susceptibility gene interaction determines Crohn's disease Atg16L1 phenotypes in intestine. Cell 2010;141:1135-45.

10. Baumgart D, Sandborn WJ. Inflammatory bowel disease: clinical aspects and established and evolving therapies. Lancet 2007;369:1641-57.

11. Loftus EV. Clinical epidemiology of inflammatory bowel disease: incidence, prevalence, and environmental influences. Gastroenterology 2004;126:1504-17.

12. Ouyang Q, Tandon R, Goh K-L et al. The emergence of inflammatory bowel disease in the Asian Pacific region. Curr Opin Gastroenterol 2005;21:408-13.

13. Thia KT, Loftus Jr. EV, Sandborn WJ, Yang SK. An update on the epidemiology of inflammatory bowel disease in Asia. Am J Gastroenterol 2008;103(12):3167-82.

14. Goh KL, Xiao S-D. Inflammatory bowel disease: a survey of the epidemiology in Asia. J Dig Dis 2009;10:1-6.

15. Wang YF, Ouyang Q, Hu RW. Progression of inflammatory bowel disease in China. J Dig Dis 2010;11:76-82.

16. Hilmi I, Singh R, Ganesananthan S et al. Demography and clinical RCUrse of ulcerative colitis in a multiracial Asian population: a nationwide study from Malaysia. J Dig Dis 2009;10:15-20.

17. Ahuja V, Tandon RK. Inflammatory bowel disease in the Asia-Pacific area: a comparison with developed RCUntries and regional differences. J Dig Dis 2010;11:134-47.

18. Shanahan F. The gut microbiota-a clinical perspective on lessons learned. Nat Rev Gastroenterol Hepatol 2012;9:609-14.

19. Muegge BD, Kuczynski J, Knights D et al. Diet drives convergence in gut microbiome functions across mammalian phylogeny and within humans. Science 2011;332:970-4.

20. Dominguez-Bello MG, Blaser MJ, Ley RE et al. Development of the human gastrointestinal microbiota and insights from high-throughput sequencing. Gastroenterology 2011;140:1713-9.

21. Hviid A, Svanstrom H, Frisch M. Antibiotic use and inflammatory bowel diseases in childhood. Gut 2011;60:49-54.

22. Birrenbach T, Bocker U. Inflammatory bowel disease and smoking: a review of epidemiology, pathophysiology, and therapeutic implications. Inflamm Bowel Dis 2004;10:848-59.

**23.** Strachan D. Hay fever, hygiene, and household size. Br Med J 1989;299:1259-60.

**24.** Yazdanbakhsh M, Kremsner PG, vanRee R. Allergy, parasites, and the hygiene hypothesis. Science 2002;296:490-4.

**25.** Feillet H, Bach J-F. Increased incidence of inflammatory bowel disease: the price of the decline of infectious burden? Curr Opin Gastroenterol 2004;20:560-4.

**26.** Bernstein CN, Shanahan F. Disorders of a modern lifestyle: reconciling the epidemiology of inflammatory bowel diseases. Gut 2008;57:1185-91.

**27.** Koloski NA, Bret L, Radford-Smith G. Hygiene hypothesis in inflammatory bowel disease: a critical review of the literature. World J Gastroenterol 2008;14:165-73.

**28.** Lakatos PL. Environmental factors affecting inflammatory bowel disease: have we made progress? Dig Dis 2009;27:215-25.

**29.** Orholm M, Munkholm P, Langholz E, et al. Familial occurence of inflammatory bowel disease. N Engl J Med 1991;324:84-8.

**30.** Hugot J-P, Chamaiilard M, Zouali H et al. Association of NOD2 leucine-rich repeat variants with susceptibility to Crohn's disease. Nature 2001;411:599-603.

**31.** Ogura Y, Bonen DK, Inohara N et al. A frameshift mutation in Nod2 associated with susceptibility to Crohn's disease. Nature 2001;411:603-6.

**32.** Duerr RH, Taylor KD, Brant SR et al. A genome-wide association study identifies IL23R as an inflammatory bowel disease gene. Science 2006;314:1461-3.

**33.** Wellcome Trust Case Control Consortium. Genome-wide association study of 14,000 cases of seven common diseases and 3,000 shared controls. Nature 2007;447:661-673.

**34.** Cho JH, Brant SR. Recent insights into the genetics of inflammatory bowel disease. Gastroenterology 2011;140:1704-12 e2.

**35.** Barrett JC, Hansoul S, Nicolae DL et al. Genome-wide association defines more than 30 distinct susceptibility loci for Crohn's disease. Nat Genet 2008;40:955-62.

**36.** McGovern DPB, Gardet A, Torkvist L et al. Genome-wide association identifies multiple ulcerative colitis susceptibility loci. Nat Genet 2010;42:332-337.

**37.** Imielinski M, Baldassano RN, Griffiths A et al. Common variants at five new loci associated with early-onset inflammatory bowel disease. Nat genet 2009;41:1335-40.

**38.** Jostins L, Ripke S, Weersma RK et al. Host-microbe interactions have shaped the genetic architecture of inflammatory bowel disease. Nature 2012;491:119-24.

**39.** Yamazaki K, McGovern D, Ragoussis J et al. Single nucleotide polymorphisms in TNFSF15 confer susceptibility to Crohn's disease. Hum Mol Genet 2007;14:3499-3506.

40. Thiebaut R, Kotti S, Jung C et al. TNFSF15 polymorphisms are associated with susceptibility to inflammatory bowel disease in a new European cohort. Am J Gastroenterol 2009;104:384-91.

41. Michelsen KS, Thomas LS, Taylor KD et al. IBD-associated TL1A gene (TNFSF15) haplotypes determine increased expression of TL1A protein. PLoS One 2009;4:e4719.

42. Massey D, Parkes M. Common pathways in Crohn's disease and other inflammatory diseases revealed by genomics. Gut 2007;56:1489-92.

43. Cho JH. The genetics and immunopathogenesis of inflammatory bowel disease. Nat Rev Immunol 2008;8:458-66.

44. Deretic V. Links between autophagy, innate immunity, inflammation and Crohn's disease. Dig Dis 2009;27:246-51.

45. Cadwell K, Stappenbeck TS, Virgin HW. Role of autophagy and autophagy genes in inflammatory bowel disease. Curr Top Microbiol Immunol 2009;335:141-67.

46. Franchimont D, Vermeire S, ElHousni H et al. Deficient host-bacteria interactions in inflammatory bowel disease? The toll-like receptor (TLR)-4 Asp299gly polymorphism is associated with Crohn's disease and ulcerative colitis. Gut 2004;53:987-92.

47. Kaser A, Lee A-H, Franke A et al. XBP1 links ER stress to intestinal inflammation and confers genetci risk for human inflammatory bowel disease. Cell 2008;134:743-56.

48. Cleynen I, Mahachie John JM, Henckaerts L et al. Molecular reclassification of Crohn's disease by cluster analysis of genetic variants. PLoS One 2010;5:e12952.

49. von Stein P, Lofberg R, Kuznetsov NV et al. Multigene analysis can discriminate between ulcerative colitis, Crohn's disease, and irritable bowel syndrome. Gastroenterology 2008;134:1869-81.

50. Weersma RK, Stokkers PC, van Bodegraven AA et al. Molecular prediction of disease risk in a large Dutch Crohn's disease cohort. Gut 2009;58:388-95.

51. Arijs I, Li K, Toedter G et al. Mucosal gene signatures to predict response to infliximab in patients with ulcerative colitis. Gut 2009;58:1612-9.

52. Manolio TA, Collins FS, Cox NJ et al. Finding the missing heritability of complex diseases. Nature 2009;461:747-53.

53. Wang MH, Fiocchi C, Ripke S et al. A novel approach to detect cumulative genetic effects and genetic interactions in Crohn's disease. Inflamm Bowel Dis 2013;19:1799--808.

54. Wang MH, Fiocchi C, Zhu X, Ripke S, Kamboh MI, Rebert N et al. Gene-gene and gene-environment interactions in ulcerative colitis. Hum Genet 2014; 133(5):547-58.

**55.** Chiodini RJ, Kruiningen HJV, Thayer WR et al. Possible role of mycobacteria in inflammatory bowel disease. I. An unclassified Mycobacterium species isolated from patients with Crohn's disease. Dig Dis Sci 1984;29:1073-9.

**56.** Selby W, Pavli P, Crotty B et al. Two-year combination antibiotic therapy with clarithromycin, rifabutin, and clofazimine for Chron's disease. Gastroenterology 2007;132:2313-9.

**57.** Darfeuille-Michaud A, Neut C, Barnich N et al. Presence of adherent Escherichia coli strains in ileal mucosa of patients with Crohn's disease. Gastroenterology 1998;115:1405-13.

**58.** Darfeuille-Michaud A, Boudeau J, Bulois P et al. High prevalence of adherent-invasive Escherichia coli associated with ileal mucosa in Crohn's disease. Gastroenterology 2004;127:412-21.

**59.** Rolhion N, Darfeuille-Michaud A. Adherent-invasive Escherichia coli in inflammatory bowel disease. Inflamm Bowel Dis 2007;13:1277-83.

**60.** Elson CO, McCracken VJ, Dimmit RA et al. Experimental models of inflammatory bowel disease reveal innate, adaptive and regulatory mechanisms of host dialogue with the microbiota. Immunol Rev 2005;206:260-76.

**61.** Beaven SW, Abreu MT. Biomarkers in inflammatory bowel disease. Curr Opin Gastroenterol 2004;20:318-27.

**62.** Mow WS, Vasiliauskas EA, Lin Y-C et al. Association of antibody responses to microbial antigens and complications of small bowel Crohn's disease. Gastroenterology 2004;126:414-24.

**63.** Dubinsky MC, Kugathasan S, Mei L et al. Increased immune reactivity predicts aggressive complicating Crohn's disease in children. Clin Gastroenterol Hepatol 2008;6:1105-11.

**64.** Duerkop BA, Vaishnava S, Hooper LV. Immune responses to the microbiota at the intestinal mucosal surface. Immunity 2009;31:368-76.

**65.** Ivanov II, Littman DR. Segmented filamentous bacteria take the stage. Mucosal Immunol 2010;3:209-12.

**66.** Sartor RB. Microbial influences in inflammatory bowel disease. Gastroenterology 2008;134:577-94.

**67.** Round JL, Mazmanian SK. The gut microbiota shapes intestinal immune responses during health and disease. Nat Rev Immunol 2009;9:313-23.

**68.** Guarner F, Malagelada J-R. Gut flora in health and disease. Lancet 2003;361:512-9.

**69.** Guarner F. The intestinal flora in inflammatory bowel disease: normal or abnormal? Curr Opin Gastroenterol 2005;21:414-8.

**70.** Costello EK, Lauber CL, Hamady M et al. Bacterial community variation in human body habitats across space and time. Science 2009;326:1694-7.

**71.** Mazmanian SK, Round JL, Kasper DL. A microbial symbiosis factor prevents intestinal inflammatory disease. Nature 2008;453:620-5.

**72.** Sokol H, Pigneur B, Watterlot L et al. Faecalibacterium prausnitzii is an anti-inflammatory commensal bacterium identified by gut microbiota analysis of Crohn disease patients. Proc Natl Acad Sci U S A 2008;105:16731-6.

**73.** Schultz C, VandenBerg FM, Kate FWT et al. The intestinal mucus layer from patients with inflammatory bowel disease harbors high numbers of bacteria compared with controls. Gastroenterology 1999;117:1089-97.

**74.** Swidsinski A, Ladhoff A, Pernthaler A et al. Mucosal flora in inflammatory bowel disease. Gastroenterology 2002;122:44-54.

**75.** Seksik P, Rigottier-Gois L, Gramet G et al. Alterations of the dominant faecal bacterial groups in patients with Crohn's disease of the colon. Gut 2003;52:237-42.

**76.** Ott SJ, Musfeldt M, Wenderoth DF et al. Reduction in diversity of the colonic mucosa associated bacterial microflora in patients with active inflammatory bowel disease. Gut 2004;53:685-93.

**77.** Manichanh C, Rigottier-Gois L, Bonnaud E et al. Reduced diversity of faecal microbiota in Crohn's disease revealed by a metagenomic analysis. Gut 2006;55:205-11.

**78.** Sokol H, Seksik P, Rigottier-Gois L et al. Specificities of the fecal microbiota in inflammatory bowel disease. Inflamm Bowel Dis 2006;12:106-11.

**79.** Menendez A, Ferreira RBR, Finlay BB. Defensins keep the peace too. Nat Immunol 2010;11:49-50.

**80.** Wehkamp J, Salzman NH, Porter E et al. Reduced Paneth cell $\alpha$-defensins in ileal Crohn's disease. Proc Natl Acad Sci USA 2005;102:18129-34.

**81.** Nuding S, Fellermann K, Wehkamp J et al. Reduced mucosal antimicrobial activity in Crohn's disease of the colon. Gut 2007;56:1240-7.

**82.** Sartor RB. Key questions to guide a better understanding of host-commensal microbiota interactions in intestinal inflammation. Mucosal Immunol 2011;4:127-32.

**83.** Qin J, Li R, Raes J et al. A human gut microbial gene catalogue established by metagenomic sequencing. Nature 2010;464:59-65.

**84.** De Filippo C, Cavalieri D, Di Paola M et al. Impact of diet in shaping gut microbiota revealed by a comparative study in children from Europe and rural Africa. Proc Natl Acad Sci U S A 2010;107:14691-6.

**85.** Wu GD, Chen J, Hoffmann C et al. Linking long-term dietary patterns with gut microbial enterotypes. Science 2011;334:105-8.

**86.** Nickerson KP, McDonald C. Crohn's disease-associated adherent-invasive Escherichia coli adhesion is enhanced by exposure to the ubiquitous dietary polysaccharide maltodextrin. PLoS One 2012;7:e52132.

**87.** Medzhitov R, Janeway C. Innate immunity. N Engl J Med 2000;343:338-44.

**88.** Abraham C, Medzhitov R. Interactions between the host innate immune system and microbes in inflammatory bowel disease. Gastroenterology 2011;140:1729-37.

**89.** Smith PD, Ochsenbauer-Jambor C, Smythies LE. Intestinal macrophage: unique effector cells of the innate immune system. Immunol Rev 2005;206:149-59.

**90.** Selby WS, Poulter LW, Hobbs S et al. Heterogeneity of HLA-DR positive histiocytes of human intestinal lamina propria: a combined histochemical and immunological analysis. J Clin Pathol 1983;36:379-84.

**91.** Rugtveit J, Brandtzaeg P, Halstensen TS et al. Increased macrophage subsets in inflammatory bowel disease: apparent recruitment from peripheral blood monocytes. Gut 1994;35:669-74.

**92.** Rugtveit J, Nilsen EM, Bakka A et al. Cytokine profiles differ in newly recruited and resident subsets of mucosal macrophages from inflammatory bowel disease. Gastroenterology 1997;112:1493-505.

**93.** Kamada N, Hisamatsu T, Okamoto S et al. Unique CD14 intestinal macrophages contribute to the pathogenesis of Crohn disease via IL-23/IFN-gamma axis. J Clin Invest 2008;118:2269-80.

**94.** Korzenik JR, Dieckegraefe BK. Is Crohn's disease an immunodeficiency? A hypothesis suggesting possible early events in the pathogenesis of Crohn's disease. Dig Dis Sci 2000;45:1121-9.

**95.** Smith AM, Rahman FZ, Hayee B et al. Disordered macrophage cytokine secretion underlies impaired acute inflammation and bacterial clearance in Crohn's disease. J Exp Med 2009;206:1883-97.

**96.** Casanova J-L, Abel L. Revisiting Crohn's disease as a primary immunodeficiency of macrophages. J Exp Med 2009;206:1839-43.

**97.** Rescigno M, diSabatino A. Dendritic cells in intestinal homeostasis and disease. J Clin Invest 2009;119:2441-50.

**98.** Bilsborough J, Viney JL. Gastrointestinal dendritic cells play a role in immunity, tolerance, and disease. Gastroenterology 2004;127:300-9.

**99.** Hart AL, Al-Hassi HO, Rigby RJ et al. Characteristics of intestinal dendritic cells in inflammatory bowel disease. Gastroenterology 2005;129:50-65.

**100.** Hoebe K, Janssen E, Beutler B. The interface between innate and adaptive immunity. Nat Immunol 2004;5:971-4.

**101.** MacDermott RP, Nash GS, Bertovich MJ et al. Alterations of IgM, IgG, and IgA synthesis and secretion by peripheral blood and intestinal mononuclear cells from patients with ulcerative colitis and Crohn's disease. Gastroenterology 1981;81:844-52.

**102.** MacDermott RP, Nash GS, Bertovich MJ et al. Altered patterns of secretion of monomeric IgA and IgA subclass 1 by intestinal mononuclear cells in inflammatory bowel disease. Gastroenterology 1986;91.379-85.

**103.** Scott MG, Nahm MH, Macke K et al. Spontaneous secretion of IgG subclasses by intestinal mononuclear cells: differences between ulcerative colitis, Crohn's disease, and controls. Clin Exp Immunol 1986;66:209-15.

**104.** Perosa F, Prete M, Racanelli V et al. CD20-depleting therapy in autoimmune diseases: from basic research to the clinic. J Intern Med 2010;267:269-77.

**105.** Goetz M, Atreya R, Ghalibafian M et al. Exacerbation of ulcerative colitis after rituximab salvage therapy. Inflamm Bowel Dis 2007;13:1365-8.

**106.** Mosmann TR, Cherwinski H, Bond MW et al. Two types of murine helper T cell clone. I. Definition according to profiles of lymphokine activities and secreted proteins. J Immunol 1986;136:2348-57.

**107.** Romagnani S. Human Th1 and Th2 subsets: doubt no more. Immunol Today 1991;12:256-7.

**108.** Annunziato F, Romagnani S. Do studies in human better depict Th17 cells? Blood 2009;114:2213-9.

**109.** Annunziato F, Cosmi L, Santarlasci V et al. Phenotypic and functional features of human Th17 cells. J Exp Med 2007;204:1849-61.

**110.** Annunziato F, Romagnani S. Heterogeneity of human effector CD4+ T cells. Arthritis Res Ther 2009;11:257.

**111.** Cosmi L, Maggi L, Santarlasci V et al. Identification of a novel subset of human circulating memory CD4(+) T cells that produce both IL-17A and IL-4. L Allergy Clin Immunol 2010;125:222-30.

**112.** O'Shea JJ, Paul WE. Mechanisms underlying lineage commitment and palsticity of helper CD4[+] T cells. Science 2010;327:1098-102.

**113.** Jiang H, Chess L. An integrated view of suppressor T cell subsets in immunoregulation. J Clin Invest 2004;114:1198-208.

**114.** Feuerer M, Hill JA, Mathis D et al. Foxp3+ regulatory T cells: differentiation, specification, and subphenotypes. Nat Immunol 2009;10:689-95.

**115.** Weaver CT, Harrington LE, Mangan PR et al. Th17: an effector CD4 T cell lineage with regulatory T cell ties. Immunity 2006;24:677-88.

**116.** Weaver CT, Hatton RD. Interplay between the $T_H 17$ and Treg cell lineages: a (co-) evolutionary perspective. Nat Rev Immunol 2009;9:883-9.

**117.** Strober W, Fuss IJ. Proinflammatory cytokines in the pathogenesis of inflammatory bowel diseases. Gastroenterology 2011;140:1756-67 e1.

**118.** Parronchi P, Romagnani P, Annunziato F et al. Type 1 T-helper cell predominance and interleukin-12 expression in the gut of patients with Crohn's disease. Am J Pathol 1997;150:823-32.

**119.** Monteleone G, Biancone L, Marasco R et al. Interleukin 12 is expressed and actively released by Crohn's disease intestinal lamina propria mononuclear cells. Gastroenterology 1997;112:1169-78.

**120.** Fujino S, Andoh A, Bamba S et al. Increased expression of interleukin 17 in inflammatory bowel disease. Gut 2003;52:65-70.

**121.** Monteleone G, Monteleone I, Fina D et al. Interleukin-21 enhances T-helper cell type I signaling and interferon-gamma production in Crohn's disease. Gastroenterology 2005;128:687-94.

**122.** Fuss IJ, Heller F, Boirivant M et al. Nonclassical CD1d-restricted NK T cells that produce IL-13 characterize an atypical Th2 response in ulcerative colitis. J Clin Invest 2004;113:1490-7.

**123.** Heller F, Florian P, Bojarski C et al. Interleukin-13 is the key effector Th2 cytokine in ulcerative colitis that affects epithelial tight junctions, apoptosis, and cell restitution. Gastroenterlogy 2005;129:550-64.

**124.** Fantini MC, Monteleone G, MacDonald TT. New players in the cytokine orchestra of inflammatory bowel disease. Inflamm Bowel Dis 2007;13:1419-23.

**125.** Makita S, Kanai T, Oshima S et al. CD4$^+$CD25$^{bright}$ T cells in human intestinal lamina propria as regulatory cells. J Immunol 2004;173:3119-3130.

**126.** Maul J, Loddenkemper C, Mundt P et al. Peripheral and intestinal regulatory CD4+CD25+$^{high}$ T cells in inflammatory bowel disease. Gastroenterology 2005;128: 1868-78.

**127.** Brimnes J, Allez M, Dotan I et al. Defects in CD8+ regulatory T cells in the lamina propria of patients with inflammatory bowel disease. J Immunol 2005;174:5814-22.

**128.** Rubartelli A, Lotze MT. Inside, outside, upside down: damage-associated molecular-pattern molecules (DAMPs) and redox. Trends Immunol 2007;28:429-36.

**129.** Rock KL, Latz E, Ontiveros F et al. The sterile inflammatory response. Annu Rev Immunol 2010;28:321-42.

**130.** Foell D, Wittkowski H, Roth J. Monitoring disease activity by stool analyses: from occult blood to molecular markers of intestinal inflammation and damage. Gut 2009;58.859-68.

**131.** Strowig T, Henao-Mejia J, Elinav E et al. Inflammasomes in health and disease. Nature 2012;481:278-86.

**132.** Elinav E, Henao-Mejia J, Flavell RA. Integrative inflammasome activity in the regulation of intestinal mucosal immune responses. Mucosal Immunol 2013;6:4-13.

**133.** Hao LY, Liu X, Franchi L. Inflammasomes in inflammatory bowel disease pathogenesis. Curr Opin Gastroenterol 2013;29:363-9.

**134.** Esteller M. Epigenetics in cancer. N Engl J Med 2008;358:1148-59.

**135.** Medzhitov R, Horng T. Transcriptional control of the inflammatory response. Nat Rev Immunol 2009;9:692-703.

**136.** Scarpa M, Stylianou E. Epigenetics: Concepts and relevance to IBD pathogenesis. Inflamm Bowel Dis 2012;18:1982-96.

**137.** Pekow JR, Kwon JH. MicroRNAs in inflammatory bowel disease. Inflamm Bowel Dis 2012;18:187-93.

**138.** Rebane A, Akdis CA. MicroRNAs: essential players in the regulation of inflammation. J Allergy Clin Immunol 2013;132:15-26.

**139.** Coskun M, Bjerrum JT, Seidelin JB et al. MicroRNAs in inflammatory bowel disease-pathogenesis, diagnostics and therapeutics. World J Gastroenterol 2012;18:4629-34.

**140.** Korzenik JR, Podolsky DK. Evolving knowledge and therapy of inflammatory bowel disease. Nat Rev Drug Discovery 2006;5:197-209.

**141.** Neurath MF, Finotto S. Translating inflammatory bowel disease research into clinical medicine. Immunity 2009;31:357-61.

**142.** Rutgeerts P, Vermeire S, Van Assche G. Biological therapies for inflammatory bowel diseases. Gastroenterology 2009;136:1182-97.

**143.** Schreiber S, Fedorak RN, Nielsen OH et al. Safety and efficacy of recombinant human interleukin 10 in chronic active Crohn's disease. Gastroenterology 2000;119:1461-72.

**144.** Hueber W, Sands BE, Lewitzky S, Vandemeulebroecke M, Reinisch W, Higgins PD et al. Secukinumab, a human anti-IL-17A monoclonal antibody, for moderate to severe Crohn's disease: unexpected results of a randomised, double-blind placebo-controlled trial. Gut 2012; 61(12):1693-700.

**145.** Danese S. New therapies for inflammatory bowel disease: from the bench to the bedside. Gut 2012;61:918-32.

**146.** Embrace the complexity. Nat Immunol 2009;10(4):325.

**147.** Fraser IDC, Germain RN. Navigating the network: signaling cross-talk in hematopoietic cells. Nat Immunol 2009;10:327-31.

**148.** van der Greef J, McBurney RN. Innovation: Rescuing drug discovery: in vivo systems pathology and systems pharmacology. Nat Rev Drug Discov 2005;4:961-7.

# INITIAL CLINICAL EVALUATION AND FOLLOW-UP

ANDREA VIEIRA

The initial evaluation of a patient with inflammatory bowel disease (IBD) is perhaps the most important. That is when the doctor has the first impression of both patient and disease, and the patient establishes a trusting relationship with the doctor. This is a *sine qua non* for acceptance and adherence to treatment of this chronic and incurable disease.

The starting point of this first visit is to verify whether the diagnosis of IBD is correct (evaluate all differential diagnoses) and if there is any possibility of uncertainty between the two entities - Crohn's disease (CD) and ulcerative colitis -, very common in clinical practice. When in doubt, it's worth investigating the patient once again, carrying out additional tests that may be necessary.[1]

The phenotypic expression of the disease must be understood and the level of inflammatory activity should be evaluated; that is, if the patient has CD, it is important whether the disease is located in the small intestine and/or colon, and which the predominant form is: inflammatory, stricturing or penetrating/fistulizing. The same reasoning should be applied in ulcerative colitis: disease limited to the rectum, the left colon or pancolitis.[2]

When questioning about the various systems in the body, it is important to ask about oral, eye, rheumatologic, dermatologic, hepatobiliary, lung, urinary, genital and perianal abnormalities. Even in patients who deny any symptoms, the physician must ensure that a visual assessment has been performed.[1]

Some indexes can be used in order to grade inflammatory activity, as shown in the Annexes to the end of this book. Factors related to greater severity should be questioned, such as dependence on corticosteroids, need for previous hospitalizations, number of surgeries, use of narcotics and anti-diarrheal agents, and impact of the disease on labor activities.[3]

The second step is to review all medications the patient already used or still uses. The assessment should include whether these drugs were helpful, the presence of side effects, adherence, time of use, dosage, the way these patients purchased the medication, how the medicine was stored, etc. Patients should also be asked about the use of alternative therapies.[1]

Knowledge about the patient, including their occupation, family relationships, their fears, their difficulties and anxieties, traumas, childhood, other conditions, family history (whether more members of the family have IBD), habits and addictions leads to a secure and complete relationship between doctor and patient.[2]

The patient's physical examination should be as detailed as possible, detailing: general status, temperature, calculation of body mass index (BMI), assessment of the oral cavity, eyes, ears, neck, respiratory and cardiovascular systems, and a detailed assessment of the abdominal and perianal regions, with digital rectal examination being performed. All joints should be palpated to rule synovitis, the skin should be observed carefully in search of conditions such as erythema nodosum, pyoderma gangrenosum, allergies, etc. Figure 4.1 shows a patient with pyoderma gangrenosum.

A review of recent laboratory tests should be included in the initial assessment, especially complete blood count, electrolytes, renal function, liver function and inflammatory activity, such as C-reactive protein (CRP), iron kinetics and dosage of folic acid and vitamin B. Endoscopic and imaging tests should undergo careful evaluation.

It is important to assess the need to include the patient in colonoscopic screening for suspicious lesions, which depends upon the age of the patient, the extent of disease, the presence of sclerosing cholangitis, and disease progression time. It is also important to take steps to diagnosis, prevention and treatment of osteoporosis. All patients, particularly those who will use immunosuppressive drugs and/or biological agents, must be questioned about the immunization schedule, gynecological status (women) and screening for tuberculosis (previous history, contact, Mantoux reaction and chest X-ray).[1,4]

In the initial evaluation, aspects that relate to patient education should also be highlighted, including understanding of the disease, the need for chronic use of medications, need to conduct periodic tests, the negative impact of smoking on CD, and the harmful effect non-steroidal anti-inflammatory drugs (NSAIDs). The main initial evaluation points are summarized in Box 4.1.

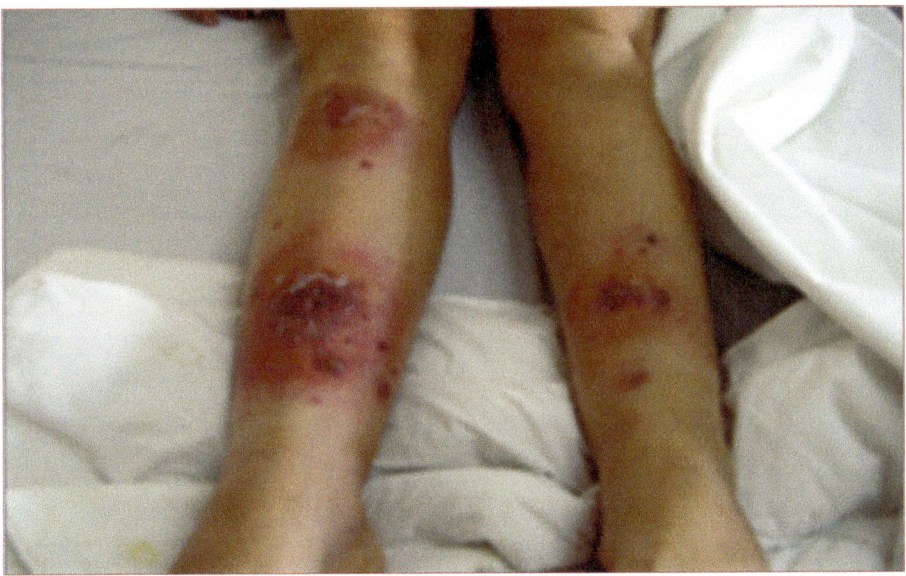

**Figure 4.1** Pyoderma gangrenosum.

| **Box 4.1** initial evaluation checklist |
|---|
| Check if the diagnosis is correct |
| Phenotype and extent of the disease |
| Extra-intestinal manifestations |
| Previous drugs |
| Alternative therapies |
| Personal and family history |
| Complete physical examination |
| Laboratory tests |
| Disease activity |
| Screening for colon cancer |
| Screening for osteoporosis |
| Vaccination schedule |
| Gynecological evaluation (screening for HPV) |
| Patient education (smoking, NSAIDs, disease information) |

HPV: human papillomavirus; NSAIDs: nonsteroidal anti-inflammatory drugs.

In the follow-up of patients with IBD, once clinical and symptomatic control of the disease (remission) is achieved, maintenance therapy is initiated, in order to keep the patient free of symptoms and with an acceptable quality of life. Unfortunately, even under maintenance therapy, both diseases are characterized by periods of relapse due to exacerbation of the inflammatory process.[5] Thus, determining the degree of inflammatory activity is of great importance to monitor clinical progress and adjust the therapy. The patients' symptoms may be indicative of inflammation and disease activity, but are subjective and often influenced by other non-inflammatory factors, such as stricture by scarring, fibrosis and malabsorption of bile salts or micro- or macronutrients.[5,6]

Some studies consider the laboratory markers the gold standard for assessing inflammation. Classically, ESR, CRP, WBC and platelet count, hemoglobin, serum iron and albumin have been described as parameters to evaluate inflammation. In general, ESR indirectly measures the concentration of plasma proteins in the acute phase, but this value is influenced by the size, weight and number of red blood cells and other plasma constituents, such as immunoglobulins.[5]

CRP is also an acute phase protein produced by hepatocytes and regulated by interleukin-6, interleukin-1 or tumor necrosis factor (TNF). However, 20% of the population does not produce this protein due to genetic polymorphism. Important CRP association with the response to anti-TNF therapy has been demonstrated (the higher the CRP values, the better the response to anti-TNF). Patients with nonspecific ulcerative colitis (UC) that early in the diagnosis have high levels of CRP are at high risk for colectomy. There is good correlation between CRP and endoscopic activity and radiological signs of mesenteritis. Moreover, CRP has demonstrated association with disease relapse – in cases of monitored patients who begin to show increased CRP, infectious processes, complications or recurrent disease should be discarded.[5]

Tests such as white blood cell count, platelet count, serum iron dosage, albumin and hemoglobin are not efficient as markers of inflammatory activity, as they are influenced by numerous factors.[5]

Some authors consider colonoscopy with biopsy the best method to assess inflammation and its location, extent and severity; however, besides being an invasive method, there is a risk of complications.[6]

Several studies have described fecal markers as powerful biomarkers for inflammation of the intestinal mucosa in patients with IBD. Neutrophil granule proteins have been marked and studied as indicators of inflammation, including lactoferrin, calprotectin, and more.[6-9]

Lactoferrin is a glycoprotein bound to iron, resistant to proteolysis, secreted by most mucosal membranes. It is the largest component in the secondary granules of polymorphonuclear leukocytes, which are the first representatives of the acute inflammatory response. Other hematopoietic cells such as monocytes and lymphocytes do not contain lactoferrin. During intestinal inflammation, leukocytes infiltrate the mucosa, resulting in increased fecal excretion of lactoferrin.[9]

Calprotectin is a calcium binding protein and represents 5% of the total protein and 60% of cytosolic protein in neutrophils. It has bacteriostatic and fungistatic properties and is found in the feces in a concentration 6 times higher than in plasma.[9]

Some studies have compared mainly fecal lactoferrin and calprotectin with activity indices and/or colonoscopic/histological evaluation to verify intestinal inflammation in patients with IBD. The results of these studies are promising, showing that these markers are useful to detect inflammation and differentiate it from other diseases, in addition to predicting relapse in up to 1 year.[7-9]

In a study in which fecal calprotectin was analyzed both during active disease as the disease into remission, normal values of this marker were correlated with mucosal healing.[10] A smaller number of studies has evaluated the role of fecal biomarkers to monitor the response to therapy. In general, but especially with biological therapy, the levels of fecal markers fall among patients who obtain endoscopic response, which indicates that the normalization of these levels has great correlation with mucosal healing.[11,12]

Another role of fecal markers is the ability to predict future recurrence.[13-16] In CD, calprotectin and lactoferrin demonstrated sensitivity and specificity to predict relapse rates at 1 year, from 69 to 90% and from 43 to 83%, respectively.[13]

Table 4.1 shows the results of a study comparing clinical and endoscopic activity indices, CRP, lactoferrin, calprotectin and histological activity in patients with CD and ulcerative colitis.

| Table 4.1 Correlation of clinical and endoscopic indices, CRP and fecal markers with histology in the follow-up of patients with IBD[7] | | | | | |
|---|---|---|---|---|---|
| | Sensitivity | Specificity | PPV | NPV | Accuracy |
| CDAI | 25% | 100% | 100% | 44% | 53% |
| CDEIS | 92% | 93% | 96% | 87% | 92% |
| MMDAI | 75% | 100% | 100% | 63% | 83% |
| CRP | 58% | 73% | 81% | 46% | 63% |
| Lactoferrin | 90% | 92% | 96% | 83% | 91% |
| Calprotectin | 77% | 100% | 100% | 68% | 85% |

PPV: positive predictive value; NPV: negative predictive value; CDAI: Crohn's Disease Activity Index; CDEIS: Crohn's Disease Endoscopic Index of Severity; MMDAI: Modified Mayo Disease Activity Index.

Source: Vieira et al., 2009.[7]

## REFERENCES

1. Sands BE. From symptom to diagnosis: clinical distinctions among various forms of intestinal inflammation. Gastroenterology 2004; 126(6):1518-32.

2. Beaugerie L, Seksik P, Nion-Larmurier I, Gendre JP, Cosnes J. Predictors of Crohn's disease. Gastroenterology 2006; 130(3):650-6.

3. Canavan C, Abrams KR, Hawthorne B, Drossman D, Mayberry J. Long-term prognosis in Crohn's disease: factors that affect quality of life. Aliment Pharmacol Ther 2006; 23(3):377-85.

4. Itzkowitz SH, Present DH. Consensus conference: colorectal cancer screening and surveillance in inflammatory bowel disease. Inflamm Bowel Dis 2005; 11(3):314-21.

5. Biancone L, De Nigris F, Del Vecchio Blanco G, Montelione I, Vavassori P, Geremia A et al. Review article: monitoring the activity of Crohn's disease. Aliment Pharmacol Ther 2002; 16(Suppl.4):29-33.

6. Konikoff M, Denson LA. Role of fecal calprotectin as a biomarker of intestinal inflammation in inflammatory bowel disease. Inflamm Bowel Dis 2006; 12:524-34.

7. Vieira A, Fang CB, Rolim EG, Klug WA, Steinwurz F, Rossini LGB et al. Inflammatory bowel disease activity assessed by fecal calprotectin and lactoferrin: correlation with laboratory parameters, clinical, endoscopic and histological indexes. BMC Research Notes 2009; 2:221.

8. Schoepfer AM, Beglinger C, Straumann A, Trummler M, Vavricka SR, Bruegger LE et al. Fecal calprotectin correlates more closely with the simple endoscopic score for Crohn's disease (SES-CD) than CRP, Blood Leukocytes and the CDAI. AM J Gastroenterol 2010; 105:162-9.

9. Desai D, Faubion WA, Sandborn WJ. Review article: biological activity markers in inflammatory bowel disease. Aliment Pharmacol Ther 2007; 25:247-55.

10. Roseth AG, Aadland E, Grzyb K. Normalization of faecal calprotectin: a predictor of mucosal healing in patients with inflammatory bowel disease. Scand J Gastroenterol 2004; 39:1017-20.

11. Sipponen T, Savilahti E, Karkkainen P, Kolho KL, Nuutinen H, Turunen U et al. Fecal calprotectin, lactoferrin, and endoscopic disease activity in monitoring anti-TNFalpha therapy for Crohn's disease. Inflamm Bowel Dis 2008; 14:1392-8.

12. Sipponen T, Bjorkesten CG, Farkkila M, Nuutinen H, Savilahti E, Kolho KL. Faecal calprotectin and lactoferrin are reliable surrogate markers of endoscopic response during Crohn's disease treatment. Scand J Gastroenterol 2010; 45:325-31.

13. Gisbert JP, Bermejo F, Perez-Calle JL, Taxonera C, Vera I, McNicholl AG et al. Fecal calprotectin and lactoferrin for the prediction of inflammatory bowel disease relapse. Inflamm Bowel Dis 2009; 15:1190-8.

14. Costa F, Mumolo MG, Ceccarelli L, Bellini M, Romano MR, Sterpi C et al. Calprotectin is a stronger predictive marker of relapse in ulcerative colitis than in Crohn's disease. Gut 2005; 54:364-8.

15. Van Rheenen PF, Van de Vijver E, Fidler V. Faecal calprotectin for screening of patients with suspected inflammatory bowel disease: diagnostic meta-analysis. BMJ 2010; 341:1-11.

16. Mendoza JL, Abreu MT. Biological markers in inflammatory bowel disease: practical consideration for clinicians. Gastroentérologie Clinique et Biologique 2009; 33(3):s158-73.

# CLASSIFICATION AND ACTIVITY INDICES

JULIANO COELHO LUDVIG

## INTRODUCTION

After a short time since the first edition of the book *Doença Inflamatória Intestinal*, by editors Wilton Cardozo and Carlos Sobrado, the scientific journey into the intriguing world of inflammatory bowel disease (IBD) remains a major challenge for doctors and scientists at the beginning of the 21st century. Sedimentation of concepts related to diagnosis, propedeutics, surveillance and therapy highlighted the need for more reliable indices and scores.

For this new edition, two new indices were added, the Lemann score for Crohn's disease (CD) and the ulcerative colitis endoscopic index of severity (UCEIS).

Scores and classification systems are the main tools that allow the formation of a reliable basis for data pairing and analysis. In the literature, there is a variety of them, which ultimately hinder a proper evaluation among different studies. Thus, standardizing systems is crucial. Ideally, they should be easy to use and adaptable to different situations, as well as validated.

In 1975, Farmer et al. published the first CD classification, location being the main criterion.[1] Later, Greenstein proposed the perforation factor as the main criterion.[2] By the early 1990s, the so-called Rome Classification used as parameters location, presentation form, extent and history of surgery, resulting in the possibility of more than 750 combinations of subgroups.[3]

In 1998, during the World Congress of Gastroenterology, a new system was introduced, the Vienna Classification.[4] Based on the age at diagnosis (<40 or ≥ 40 years), location (terminal ileum, colon, ileocolonic and upper gastrointestinal [GI] tract) and behavior (non-stricturing, non-penetrating; stricturing, and penetrating), this combination allows only 24 subgroups. Gender, ethnicity and family history are also considered as secondary data (Table 5.1).

| Table 5.1 Vienna and Montreal Classifications for CD | | |
|---|---|---|
| | **Vienna** | **Montreal** |
| Age at diagnosis | A1 – below 40 years | A1 – below 16 years |
| | A2 – above 40 years | A2 – between 17 and 40 years |
| | | A3 – above 40 years |
| Location | L1 – ileal | L1 – ileal |
| | L2 – colonic | L2 – colonic |
| | L3 – ileocolonic | L3 – ileocolonic |
| | L4 – upper GI tract | L4 – isolated upper GI tract |
| Behavior | B1 – non-stricturing, non-penetrating | B1 – non-stricturing, non-penetrating |
| | B2 – stricturing | B2 – stricturing |
| | B3 – penetrating | B3 – penetrating |
| | | P – perianal disease |

GI: gastrointestinal.

However, over the years, new factors have been integrated into the dynamics of the disease, many serving as parameters for diagnosis and treatment. So, in 2005, an updated version was presented at the World Congress. The Montreal Classification did not replace the three main parameters, but changes were made in each of them (Table 5.1). In addition, a sub-classification for cases of ulcerative colitis (UC) was incorporated, involving extension and degree of clinical activity[5] (Tables 5.2 and 5.3).

| Table 5.2 Montreal Classification for UC location | |
|---|---|
| **Extension** | **Anatomy** |
| E1 – Ulcerative proctitis | Involvement limited to the rectum (proximal extent of inflammation is distal to the rectosigmoid junction) |
| E2 – Left side UC | Involvement extends up to the splenic flexure |
| E3 – Extensive UC | Involvement extends beyond the splenic flexure |

| Table 5.3 Montreal Classification of severity of UC | |
|---|---|
| **Severity** | **Definition** |
| S0 – Clinical remission | Asymptomatic |
| S1 – Mild UC | Passage of four or fewer stools/day (with or without blood), absence of any systemic illness, and normal inflammatory markers |
| S2 – Moderate UC | Passage of more than four stools per day but with minimal signs of systemic toxicity |
| S3 – Severe UC | Passage of at least six bloody stools daily, pulse rate of at least 90 bpm, temperature of at least 37.5°C (99.5°F), hemoglobin below 10.5 g/100 mL, and ESR $\geq$ 30 mm/h |

ESR: erythrocyte sedimentation rate.

As important as the classification, proper stratification of the presentation is fundamental for the management of disease. Therefore, various systems for analysis of CD and UC have also been proposed.

## CROHN'S DISEASE (CD)

Evaluation systems available for CD measure parameters of clinical and endoscopic activity, and perianal disease. For clinical evaluation, the most used are the Crohn's Disease Activity Index (CDAI) and the Harvey-Bradshaw Index. For evaluation and classification of perianal lesions, the Perianal Disease Activity Index (PDAI) is used. However, these indexes record the activity of the disease in a given period, and do not provide long-term damage information. That is the proposal of the Lemann score: showing the impact of the disease in the long term.

### Crohn's Disease Activity Index (CDAI)

Developed in the 1970s, this disease severity assessment system uses 8 objective and subjective variables. Based on a scale from 0 to 600 points, the disease can be classified as clinical response or remission (decrease from 70 to 100 points suggests clinical response; while total ≤ 150 points suggests remission), and moderate activity with CDAI score > 220 or severe with a value of CDAI > 450[6-8] (Table 5.4).

It is the most widely used worldwide to evaluate the clinical activity. Due to its prospective validity, it is considered the gold standard of clinical evaluation, particularly in clinical trials. Nevertheless, it has limitations. The fact of including subjective variables can give rise to different interpretations; in addition, the CDAI requires time (at least 7 days) for complete evaluation and it does not consider fistulizing disease as a sign of disease activity.[7,9]

### Harvey-Bradshaw Index

This is a simplified version of the CDAI, validated and easy to use, which excludes the need for 1-week follow-up and weight measurement. Categories assessed are: wellness sensation, abdominal pain, number of daily liquid bowel movements, presence of abdominal mass, and CD complications (arthralgia, uveitis/iritis,

| **Table 5.4** Crohn's Disease Activity Index (CDAI)[6] | |
|---|---|
| | **Multiplied by** |
| Number of liquid stools in the previous week | 2 |
| Abdominal pain (none = 0; mild = 1; moderate = 2; severe = 3). Consider the total sum of individual data in the previous week | 5 |
| General condition (excellent = 0; well = 1; poor = 2; very poor = 3; terrible = 4). Consider the total sum of individual data in the previous week | 7 |
| No. of associated symptoms/signs – list by categories: a) Arthritis/arthralgia; b) Iritis/uveitis; c) Erythema nodosum/pyoderma gangrenosum/ aphthous stomatitis; d) Anal fissure, fistula or perirectal abscess; e) Other fistula; f) Fever | 20 (maximum score = 120) |
| Use of antidiarrheal drugs (no = 0; yes = 1) | 30 |
| Abdominal mass (none = 0; dubious = 2; definite = 5) | 10 |
| Hematocrit deficit: For Male = 47 Ht; For Female = 42 Ht (subtract instead of adding if the patient's Ht is > the standard) | 6 |
| (Weight/Usual weight) × 100 Weight*: percentage below the expected (subtract instead of adding if the patient's weight is higher than expected) | 1 |
| Total (Crohn's Disease Activity Index) = < 150 = Remission 150 to 250 = Mild 250 to 350 = Moderate > 350 = Severe | |

erythema nodosum, aphthous stomatitis, pyoderma gangrenosum, anal fissure, fistula, and abscess). Similar to CDAI, it has limitations due to the subjectivity of the analysis with respect to intensity of abdominal pain and the sense of well-being (Table 5.5).[10]

## Perianal Disease Activity Index (PDAI)

The perianal lesions received their own evaluation system because of the frequency of their occurrence and the important impact on the quality of life of patients with CD; also because they have their own characteristics and because they are measured very superficially in the CDAI and the Harvey-Bradshaw Index. The PDAI evaluates five variables based on a grading scale

| Table 5.5 Harvey-Bradshaw index of inflammatory activity[10] | |
|---|---|
| | **Score** |
| General well-being (very well = 0; slightly below par = 1; poor = 2; very poor = 3; terrible = 4) | 0 to 4 |
| Abdominal pain (none = 0; mild = 1; moderate = 2; severe = 3) | 0 to 3 |
| Number of liquid stools per day | n$^{o}$/day |
| Abdominal mass (none = 0; dubious = 1; definite = 2; definite and tender = 3) | 0 to 3 |
| Complications: arthralgia/arthritis, uveitis/iritis, erythema nodosum, aphthous stomatitis, pyoderma gangrenosum, anal fissure, fistula, abscess etc. | Score 1 per item |

< 8 = inactive/mild; 8 to 10 = mild/moderate; > 10 = moderate/severe.

and scores: discharge, pain/restriction of activities, restriction of sexual activity, type of perianal disease and degree of induration. Each category is rated according to a 5-point scale, ranging from no symptoms (score 0) to severe symptoms (score 4); higher scores indicate more severe disease.[11] It is an already validated index and currently considered the gold standard for assessing the severity of perianal disease (Table 5.6).

## Endoscopic assessment of disease activity

The endoscopic pattern of CD has been characterized and based on a series of different mucosal lesions: erythema, cobblestoning, aphthous stomatitis and ulcers of different sizes and depth, fistulae and stricture.[12]

The three main systems for endoscopic evaluation of CD activity are the Crohn's disease endoscopic index of severity (CDEIS), the Simple Endoscopic Score for Crohn's Disease (SES-CD) and the Rutgeerts Score.

### Crohn's Disease Endoscopic Index of Severity (CDEIS)

To measure endoscopic activity in CD, the French group GETAID developed in 1989 the CDEIS. Validation occurred in a large multicenter study and, soon after, became the standard for evaluation of endoscopic activity.[13] The calculation is based on aspects of mucosal injury, segments involved and selected independent variables. It correlates well with overall clinical

| Table 5.6 Perianal Disease Activity Index (PDAI) | | |
|---|---|---|
| **Perianal disease activity** | | **Score** |
| Discharge | No discharge | 0 |
| | Minimal mucus discharge | 1 |
| | Moderate mucus or purulent discharge | 2 |
| | Substantial discharge | 3 |
| | Gross fecal soiling | 4 |
| Pain/restriction of activities | No activity restriction | 0 |
| | Mild discomfort, no restriction | 1 |
| | Moderate discomfort, some limitation of activities | 2 |
| | Marked discomfort, marked limitation | 3 |
| | Severe pain, severe limitation | 4 |
| Restriction of sexual activity | No restriction of sexual activity | 0 |
| | Slight restriction of sexual activity | 1 |
| | Moderate limitation of sexual activity | 2 |
| | Marked limitation of sexual activity | 3 |
| | Unable to engage in sexual activity | 4 |
| Type of perianal disease | No perianal disease/skin tags | 0 |
| | Anal fissure or mucosal tear | 1 |
| | < 3 perianal *fistulae* | 2 |
| | > 3 perianal *fistulae* | 3 |
| | Anal sphincter ulceration or *fistulae* with significant undermining of the skin | 4 |
| Degree of induration | No induration | 0 |
| | Minimal induration | 1 |
| | Moderate induration | 2 |
| | Substantial induration | 3 |
| | Gross fluctuance/abscess | 4 |
| Total | | |

PDAI = Perianal Disease Activity Index.

aspect. However, being considered time consuming and complicated, it is not routinely used in daily clinical practice. Moreover, in cases of clinical remission induced by corticosteroids, it is known that there is a persistence of mucosal lesions.[14]

## Simple Endoscopic Score for Crohn's Disease (SES-CD)

In order to facilitate the use of an endoscopic evaluation system for activity in CD, the Simple Endoscopic Score for Crohn's Disease (SES-CD) was proposed recently (Table 5.7). An already validated instrument, it is based on four endoscopic variables, including size of the ulcer, extent, area affected and stricturing.[15] Concordance ranges from good to excellent. The SES-CD is the answer to two of the objectives of an ideal endoscopic score: it is easier and quicker to mark and calculate compared to CDEIS, and its results are reproducible and reliable.

## Rutgeerts Score

Designed to monitor the progress of post-surgical endoscopic recurrence of CD in the ileum, the scale measures the appearance of the mucosa and the number of aphthous ulcers.[16] Currently, it is considered the best post-operative evaluation system in CD (Table 5.8).

| Table 5.7 Simple Endoscopic Score for Crohn's Disease (SES-CD) | | | | |
|---|---|---|---|---|
| Variable | 0 | 1 | 2 | 3 |
| | | Aphthous ulcers | Large ulcers | Very large ulcers |
| Size of ulcers | None | 0.1 to 0.5 cm | 0.5 to 2 cm | > 2 cm |
| Ulcerated surface | None | < 10% | 10 to 30% | > 30% |
| Affected surface | None | < 50% | 50 to 75% | > 75% |
| Presence of structuring | None | Single, can be passed | Multiple, can be passed | Cannot be passed |

SES-CD = Simple Endoscopic Score for Crohn's Disease.

| Table 5.8 Rutgeerts Score | |
|---|---|
| Grade | Endoscopic findings |
| i0 | No ileal lesion |
| i1 | < 5 aphthous lesions, less than 5 mm in length |
| i2 | > 5 aphthous lesions with normal mucosa between the lesions OR skip areas of larger lesions OR lesions confined to the ileocolic anastomosis, less than 1 cm in length |
| i3 | Diffuse aphthous ileitis with diffusely inflamed mucosa |
| i4 | Diffuse ileitis, already with larger ulcers, nodules and/or narrowing |

## Crohn's Disease Digestive Damage Score (Lemann score)

Conducted by the International Program to Develop New Indexes for Crohn's Disease (IPINIC) group, the Lemann score was designed to assess the damage of long-term inflammation. This index considers criteria such as location, severity, extent, progression and reversibility, measured by imaging tests (computed tomography [CT] and magnetic resonance imaging [MRI]), and history of surgical resection. In addition to designing the possible progression of the disease, it can help evaluate responses to therapy.[17]

## ULCERATIVE COLITIS

Evaluation systems of UC include clinical, laboratory and endoscopic aspects.

### Truelove-Witts Severity Index

In 1995, Truelove and Witts published their first classification system for disease activity (Table 5.9).[18] Evaluating clinical and laboratory variables such as number of bowel movements, bleeding, fever, heart rate, hemoglobin value and erythrocyte sedimentation rate (ESR) test, the authors stratified the disease as mild, moderate and severe disease. Limitations such as lack of scientific validation and quantitative system make this score most often used to classify patients in general.[19]

| Table 5.9 Inflammatory activity index - severity of acute disease[18] | | | |
|---|---|---|---|
| | **Mild** | **Moderate** | **Severe** |
| 1. Number of bowel movements/day | ≤ 4 | 4 to 6 | > 6 |
| 2. Blood in the stool | ± | + | ++ |
| 3. Body temperature | Normal | Intermediate values | Average body temperature at night >37.5°C (99.5°F) or >37.8°C (100.04°F) in 2 out of 4 days |
| 4. Pulse | Normal | Intermediate | > 90 bpm |
| 5. Hemoglobin (g/dL) | > 10.5 | Intermediate | < 10.5 |
| 6. ESR (mm/first hour) | < 30 | Intermediate | > 30 mm, first hour |

ESR: erythrocyte sedimentation rate.

### Mayo Score

This is a system for assessing disease activity which includes endoscopic and clinical features (Table 5.10). Bowel habits, rectal bleeding, mucosal appearance on sigmoidoscopy, and personal clinical aspect are the monitored parameters.[20] Unlike the system used by Truelove and Witts, this score can track and measure changes in the clinical picture. Although not totally validated, it is the most commonly used system in clinical trials. Currently, clinical remission is defined as total score ≤ 2. Clinical response is defined as a decrease from baseline ≥ 3 points and ≥ 30%; decrease in the rectal bleeding subscore ≥ 1 point or total absence; and lastly, mucosal healing subscore defined as score of 0 or 1.[21]

### Ulcerative colitis endoscopic index of severity (UCEIS)

Despite the technological advances of recent years and the standardization of treatment and surveillance system, there is still no endoscopic standardization of patients with UC. Recently, Travis et al. proposed a new evaluation model based on the findings of vascular pattern, bleeding and erosion/ulcers seen on endoscopy. Despite not being validated yet, either, the UCEIS provides good correlation with the severity of the disease[22] (Table 5.11).

### System for absessing quality of life

Because they are chronic diseases and have high morbidity, the indices measuring quality of life in patients with IBD have been used with increasing frequency as a variable for evaluation in clinical trials. There are generic and specific systems to assess the well-being of people with CD and UC.

### Inflammatory Bowel Disease Questionnaire (IBDQ)

Used since 1989 as a specific system for evaluating the quality of life in patients with IBD, this questionnaire assesses, using 32 questions, four aspects of life of the patient: symptoms related to intestinal disorder, systemic symptoms, emotional and social status. The IBDQ uses a scoring system ranging from 32 to 224, so that higher values mean better well-being (Table 5.12).[23] There is good correlation with the CDAI, and it is validated and widely used in clinical trials of patients with CD; an increase of 16 points in the IBDQ indicates improvement in quality

| Score | Stool frequency | Rectal bleeding | Endoscopic findings | Global assessment |
|---|---|---|---|---|
| 0 | Normal amount | No blood seen | Normal or inactive disease (scarring) | Normal |
| 1 | Normal amount + 1-2 BM/day | Streaks of blood with stool – less than half of BM | Mild disease (erythema, ↓ of vascular pattern, mild friability) | Mild disease |
| 2 | Normal amount + 3-4 BM/day | Obvious blood with stools | Moderate disease (evident erythema, loss of vascular pattern, erosion) | Moderate disease |
| 3 | Normal amount + > 5 BM/day | Blood alone without stools | Severe disease (spontaneous bleeding, ulceration) | Severe disease |

Table 5.10 Disease Activity Index – Mayo Clinic Score[20]

| Score (points) | Disease severity |
|---|---|
| 0 to 2 | Normal – remission |
| 3 to 5 | Mild activity |
| 6 to 10 | Moderate activity |
| 11 to 12 | Severe activity |

| Score (points) | Disease severity |
|---|---|
| 0 to 2 | Normal – remission |
| 3 to 5 | Mild activity |
| 6 to 10 | Moderate activity |
| 11 to 12 | Severe activity |

Table 5.11 Ulcerative colitis endoscopic index of severity (UCEIS)

of life, and the value of 170 or more points is correlated with clinical remission.[24] There are also reports of its use in patients with UC, with good correlation with disease activity indices.[25]

### IBDQ Scoring

The questions composing each domain are not asked in an orderly fashion in the questionnaire, so that bias is avoided in the responses.

Each question within each domain has 7 possible answers. Each response option is worth its own number in points, with 1 being the worst quality of life, and 7 the best, adding to the total points achieved in each domain. The simple sum of all domains will result in the total score obtained by the patient.

The domains related to their respective questions are shown below:

1. Questions regarding the component "intestinal symptoms": 1, 5, 9, 13, 17, 20, 22, 24, 26, 29 (scores range from 10 to 70 points).
2. Questions regarding the component "systemic symptoms": 2, 6, 10, 14, 18 (scores range from 5 to 35 points).
3. Questions regarding the component "social aspects": 4, 8, 12, 16, 28 (scores range from 5 to 35 points).
4. Questions regarding the component "emotional aspects": 3, 7, 11, 15, 19, 21, 23, 25, 27, 30, 31 and 32 (scores range from 12 to 84 points).

**Table 5.12** IBDQ

**1. How frequent have your bowel movements been during the last 2 weeks?**

| |
|---|
| 1. As or more frequent than they have ever been |
| 2. Extremely frequent |
| 3. Very frequent |
| 4. Moderate increase in frequency |
| 5. Some increase |
| 6. Slight increase |
| 7. Normal, no increase in frequency of bowel movements |

**2. How often has the feeling of fatigue or being tired and worn out been a problem for you during the last 2 weeks?**

| |
|---|
| 1. All of the time |
| 2. Most of the time |
| 3. A good bit of the time |
| 4. Some of the time |
| 5. A little of the time |
| 6. Hardly any of the time |
| 7. None of the time |

**3. How often during the last 2 weeks have you felt frustrated, impatient, or restless?**

| |
|---|
| 1. All of the time |
| 2. Most of the time |
| 3. A good bit of the time |
| 4. Some of the time |
| 5. A little of the time |
| 6. Hardly any of the time |
| 7. None of the time |

*(continues)*

**Table 5.12** IBDQ

**4. How often during the last 2 weeks have you been unable to attend school or work because of your bowel problem?**

| |
|---|
| 1. All of the time |
| 2. Most of the time |
| 3. A good bit of the time |
| 4. Some of the time |
| 5. A little of the time |
| 6. Hardly any of the time |
| 7. None of the time |

**5. How much time during the last 2 weeks have your bowel movements been loose?**

| |
|---|
| 1. All of the time |
| 2. Most of the time |
| 3. A good bit of the time |
| 4. Some of the time |
| 5. A little of the time |
| 6. Hardly any of the time |
| 7. None of the time |

**6. How much energy have you had during the last 2 weeks?**

| |
|---|
| 1. No energy at all |
| 2. Very little energy |
| 3. A little energy |
| 4. Some energy |
| 5. A moderate amount of energy |
| 6. A lot of energy |
| 7. Full of energy |

*(continues)*

**Table 5.12** IBDQ

**7. How often during the last 2 weeks did you feel worried about the possibility of needing surgery because of your bowel problem?**

| |
|---|
| 1. All of the time |
| 2. Most of the time |
| 3. A good bit of the time |
| 4. Some of the time |
| 5. A little of the time |
| 6. Hardly any of the time |
| 7. None of the time |

**8. How often during the last 2 weeks have you had to delay or cancel a social engagement because of your bowel problems?**

| |
|---|
| 1. All of the time |
| 2. Most of the time |
| 3. A good bit of the time |
| 4. Some of the time |
| 5. A little of the time |
| 6. Hardly any of the time |
| 7. None of the time |

**9. How often in the past 2 weeks have you been troubled by cramps in your abdomen?**

| |
|---|
| 1. All of the time |
| 2. Most of the time |
| 3. A good bit of the time |
| 4. Some of the time |
| 5. A little of the time |
| 6. Hardly any of the time |
| 7. None of the time |

*(continues)*

**Table 5.12** IBDQ

**10. How often in the past 2 weeks have you felt generally unwell?**

| 1. All of the time |
| 2. Most of the time |
| 3. A good bit of the time |
| 4. Some of the time |
| 5. A little of the time |
| 6. Hardly any of the time |
| 7. None of the time |

**11. How often during the last 2 weeks have you been troubled because of fear of not finding a bathroom?**

| 1. All of the time |
| 2. Most of the time |
| 3. A good bit of the time |
| 4. Some of the time |
| 5. A little of the time |
| 6. Hardly any of the time |
| 7. None of the time |

**12. How much difficulty have you had, as a result of your bowel problems, doing leisure or sports activities you would liked to have done during the last 2 weeks?**

| 1. A great deal of difficulty; activities made impossible |
| 2. A lot of difficulty |
| 3. A fair bit of difficulty |
| 4. Some difficulty |
| 5. A little difficulty |
| 6. Hardly any difficulty |
| 7. No difficulty; no limit sports or leisure activities |

*(continues)*

**Table 5.12** IBDQ

**13. How often during the last 2 weeks have you been troubled by pain in the abdomen?**

1. All of the time

2. Most of the time

3. A good bit of the time

4. Some of the time

5. A little of the time

6. Hardly any of the time

7. None of the time

**14. How often during the past 2 weeks have you had problems getting a good night's sleep, or been troubled by waking up during the night?**

1. All of the time

2. Most of the time

3. A good bit of the time

4. Some of the time

5. A little of the time

6. Hardly any of the time

7. None of the time

**15. How often during the past 2 weeks have you felt depressed or discouraged?**

1. All of the time

2. Most of the time

3. A good bit of the time

4. Some of the time

5. A little of the time

6. Hardly any of the time

7. None of the time

(*continues*)

**Table 5.12** IBDQ

**16. How often during the past 2 weeks have you had to avoid attending events where there was no bathroom at hand?**

| |
|---|
| 1. All of the time |
| 2. Most of the time |
| 3. A good bit of the time |
| 4. Some of the time |
| 5. A little of the time |
| 6. Hardly any of the time |
| 7. None of the time |

**17. Overall, in the past 2 weeks, how much problem have you had with passing large amounts of gas?**

| |
|---|
| 1. A major problem |
| 2. A big problem |
| 3. A significant problem |
| 4. Some trouble |
| 5. A little trouble |
| 6. Hardly any trouble |
| 7. No trouble |

**18. Overall, in the last 2 weeks, how much of a problem have you had maintaining or getting to the weight you would like to be at?**

| |
|---|
| 1. A major problem |
| 2. A big problem |
| 3. A significant problem |
| 4. Some trouble |
| 5. A little trouble |
| 6. Hardly any trouble |
| 7. No trouble |

*(continues)*

**Table 5.12** IBDQ

**19. Many patients with bowel problems often have worries and anxieties related to their illness. These include worries about getting cancer, worries about never feeling better, and worries about having a relapse. In general, how often during the last 2 weeks have you felt worried or anxious?**

| |
|---|
| 1. All of the time |
| 2. Most of the time |
| 3. A good bit of the time |
| 4. Some of the time |
| 5. A little of the time |
| 6. Hardly any of the time |
| 7. None of the time |

**20. How much of the time during the last 2 weeks have you been troubled by a feeling of abdominal bloating?**

| |
|---|
| 1. All of the time |
| 2. Most of the time |
| 3. A good bit of the time |
| 4. Some of the time |
| 5. A little of the time |
| 6. Hardly any of the time |
| 7. None of the time |

**21. How often during the last 2 weeks have you felt relaxed and free of tension?**

| |
|---|
| 1. None of the time |
| 2. Hardly any of the time |
| 3. A little of the time |
| 4. Some of the time |
| 5. A good bit of the time |
| 6. Most of the time |

(continues)

**Table 5.12** IBDQ

| |
|---|
| 7. All of the time |

**22. How much time during the last 2 weeks have you had a problem with rectal bleeding with your bowel movements?**

| |
|---|
| 1. All of the time |
| 2. Most of the time |
| 3. A good bit of the time |
| 4. Some of the time |
| 5. A little of the time |
| 6. Hardly any of the time |
| 7. None of the time |

**23. How much time during the last 2 weeks have you felt embarrassed as the result of soiling, or because of an unpleasant odor caused by your bowel movement?**

| |
|---|
| 1. All of the time |
| 2. Most of the time |
| 3. A good bit of the time |
| 4. Some of the time |
| 5. A little of the time |
| 6. Hardly any of the time |
| 7. None of the time |

**24. How much of the time during the past 2 weeks have you been troubled by a feeling of having to go to the bathroom even though your bowels are empty?**

| |
|---|
| 1. All of the time |
| 2. Most of the time |
| 3. A good bit of the time |
| 4. Some of the time |
| 5. A little of the time |

*(continues)*

| Table 5.12 IBDQ |
| --- |
| 6. Hardly any of the time |
| 7. None of the time |
| **25. How much of the time during the last 2 weeks have you felt tearful of upset?** |
| 1. All of the time |
| 2. Most of the time |
| 3. A good bit of the time |
| 4. Some of the time |
| 5. A little of the time |
| 6. Hardly any of the time |
| 7. None of the time |
| **26. How much of the time during the last 2 weeks have you been troubled by accidental soiling of your underpants?** |
| 1. All of the time |
| 2. Most of the time |
| 3. A good bit of the time |
| 4. Some of the time |
| 5. A little of the time |
| 6. Hardly any of the time |
| 7. None of the time |
| **27. How much of the time in the past 2 weeks have you felt angry as a result of your bowel problems?** |
| 1. All of the time |
| 2. Most of the time |
| 3. A good bit of the time |
| 4. Some of the time |
| 5. A little of the time |

*(continues)*

| Table 5.12 IBDQ |
| --- |
| 6. Hardly any of the time |
| 7. None of the time |
| **28. To what extent has your bowel problem limited sexual activity during the last 2 weeks?** |
| 1. No sex as a result of Crohn's disease |
| 2. Major limitation as a result of Crohn's disease |
| 3. Moderate limitation as a result of Crohn's disease |
| 4. Some limitation as a result of Crohn's disease |
| 5. A little limitation as a result of Crohn's disease |
| 6. Hardly any limitation as a result of Crohn's disease |
| 7. No limitation as a result of Crohn's disease |
| **29. How much of the time during the last 2 weeks have you been troubled by feeling sick to your stomach?** |
| 1. All of the time |
| 2. Most of the time |
| 3. A good bit of the time |
| 4. Some of the time |
| 5. A little of the time |
| 6. Hardly any of the time |
| 7. None of the time |
| **30. How much of the time during the past 2 weeks have you felt irritable?** |
| 1. All of the time |
| 2. Most of the time |
| 3. A good bit of the time |
| 4. Some of the time |
| 5. A little of the time |
| 6. Hardly any of the time |
| 7. None of the time |

*(continues)*

| Table 5.12 IBDQ |
| --- |
| **31. How often during the last 2 weeks have you felt a lack of understanding from others?** |
| 1. All of the time |
| 2. Most of the time |
| 3. A good bit of the time |
| 4. Some of the time |
| 5. A little of the time |
| 6. Hardly any of the time |
| 7. None of the time |
| **32. How satisfied, happy, or pleased have you been with your personal life during the past 2 weeks?** |
| 1. Very dissatisfied, unhappy most of the time |
| 2. Generally dissatisfied, unhappy |
| 3. Somewhat dissatisfied, unhappy |
| 4. Generally satisfied, pleased |
| 5. Satisfied most of the time, happy |
| 6. Very satisfied most of the time, happy |
| 7. Extremely satisfied, could not have been more happy or pleased |

## REFERENCES

1. Farmer RG, Hawk WA, Turnbull RB. Clinical patterns in Crohn's disease. A statistical study of 615 cases. Gastroenterology 1990; 98:811-8.

2. Greenstein AJ, Lachman P, Sachar DB, Springhorn J, Heimann T, Janowitz HD et al. Perforating and non-perforating indications for repeated operations in Crohn's disease: evidence of two clinical formas. Gut 1988; 29:588-92.

3. Sachar DB, Andrews HÁ, Farmer RG, Pallone F, Pena AS, Prantera C et al. Proposed classification of patient subgroups in Crohn's disease. Gastroenterology Intl 1992; 5:141-54.

4. Gasche C, Scholmerich J, Brynskov J, D'Haens G, Hanauer SB, Irvine EJ et al. A simple classification of Crohn's disease: report of the working party for the world congresses of gastroenterology, vienna 1998. Inflammatory Bowel Diseases 1998; 1:8-15.

5.  Satsangi J, Silverberg MS, Vermeire S, Colombel JF. The Montreal classification of inflammatory bowel disease: controversies, consensus and implications. Gut 2006; 55:749-53.

6.  Best WR, Beckel JM, Singleton JW, Kern Jr. F. Development of a Crohn's disease activity index. National Cooperative Crohn's Disease Study. Gastroenterology 1976; 70(3):439-44.

7.  Sandborn WJ, Feagan BG, Hanauer SB, Lochs H, Löfberg R, Present D et al. A review of activity indices and efficacy endpoints for clinical trials of medical therapy in adults with Crohn's disease. Gastroenterology 2002; 122(2):512-30.

8.  Targan SR, Hanauer SB, van Deventer SJ, Mayer L, Present DH, Braakman T et al. A short-term study of chimeric monoclonal antibody cA2 to tumor necrosis fator alpha for Crohn's disease. Crohn's disease cA2 study group. N Engl J Med 1997; 337(15):1029-35.

9.  De Dombal FT, Softley A. IOIBD report no 1: observer variation in calculating indices of severity and activity in Crohn's disease. International Organization for the Study of Inflammatory Bowel Disease. Gut 1987; 28(4):474-81.

10. Harvey RF, Bradshaw JM. A simple index of Crohn's disease activity. Lancet 1980; 1(8167):514.

11. Irvine EJ. Usual therapy improves perianal Crohn's disease as measured by a new disease activity index. McMaster IBD Study Group. J Clin Gastroenterol 1995; 20:27-32.

12. Waye JD. The role of colonoscopy in the differential diagnosis of inflammatory bowel disease. Gastrointest Endosc 1977; 23(3):150-4.

13. Mary JY, Modigliani R. Development and validation of an endoscopic index of the severity for Crohn's disease: a prospective multicentre study. Groupe d'Etudes Therapeutiques des Affections Inflammatoires du Tube Digestif (GETAID). Gut 1989; 30(7):983-9.

14. Modigliani R, Mary JY, Simon JF, Cortot A, Soule JC, Gendre JP et al. Clinical, biological, and endoscopic picture of attacks of Crohn's disease. Evolution on prednisolone. Groupe d'Etude Therapeutique des Affections Inflammatoires Digestives. Gastroenterology 1990; 98(4):811-8.

15. Daperno M, Van Assche G, Bulois P. Development of Crohn's disease endoscopic score (CDES): a simple index to assess endoscopic severity of Crohn's disease. Gastroenterology 2002; 122:A216(Abstract).

**16.** Rutgeerts P, Geboes K, Vantrappen G, Beyls J, Koremans R, Hiele M. Predictability of the postoperative course of Crohn's disease. Gastroenterology 1990; 99(4):956-63.

**17.** Pariente B, Cosnes J, Danese S, Sandborn W, Lewin M, Fletcher JG et al. Development of the Crohn's Disease Digestive Damage Score, the Lémann Score. Inflamm Bowel Dis 2011; 17(6):1415-22.

**18.** Truelove SC, Witts LJ. Cortisone in ulcerative colitis; a final report on a therapeutic trial. Br Med J 1955; 2(4947):1041-8.

**19.** D'Haens G, Sandborn WJ, Feagan BG, Geboes K, Hanauer SB, Irvine EJ et al. A review of activity indices and efficacy end points for clinical trials of medical therapy in adults with ulcerative colitis. Gastroenterology 2007; 132(2):763-86.

**20.** Schroeder KW, Tremaine WJ, Ilstrup DM. Coated oral 5-aminosalicylic acid therapy for mildly to moderately active ulcerative colitis: a randomized study. N Engl J Med 1987; 317(26):1625-9.

**21.** Rutgeerts P, Sandborn WJ, Feagan BJ, Reinisch W, Olson A, Johanns J et al. Infliximab for induction and maintenance therapy for ulcerative colitis. N Engl J Med 2005; 353(23):2462-76.

**22.** Travis SPL, Schnell D, Krzeski P, Abreu MT, Altman DG, Colombel JF et al. Developing an instrument to assess the endoscopic severity of ulcerative colitis: the Ulcerative Colitis Endoscopic Index of Severity (UCEIS). Gut 2012; 61(4):535-42.

**23.** Guyatt G, Mitchell A, Irvine EJ, Singer J, Wiiliams N, Goodacre R et al. A new measure of health status for clinical trials in inflammatory bowel disease. Gastroenterology 1989; 96(3):804-10.

**24.** Irvine EJ, Feagan B, Rochon J, Archambault A, Fedorak RN, Groll A et al. Quality of life: a valid and reliable measure of therapeutic efficacy in the treatment of inflammatory bowel disease. Canadian Crohn's Relapse Prevention Trial Study Group. Gastroenterology 1994; 106:287-96.

**25.** Han SW, McColl E, Barton JR. Predictors of quality of life in ulcerative colitis: the importance of symptoms and illness representations. Inflamm Bowel Dis 2005; 11(1):24-34.

# CLINICAL MANIFESTATIONS IN INFLAMMATORY BOWEL DISEASE

WILTON SCHMIDT CARDOZO

CARLOS WALTER SOBRADO

## INTRODUCTION

Inflammatory bowel disease (IBD) comprises mainly two diseases: ulcerative colitis (UC) and Crohn's disease (CD).[1]

Ulcerative colitis is characterized by diffuse mucosal inflammation limited to the rectum and colon. It is classified according to the maximum extent of the inflammation observed by colonoscopy and can be divided, broadly, into distal or more extensive disease.[2,3] The distal disease refers to inflammation of rectal mucosa up to 15 cm from the dentate line (proctitis). When inflammation compromises the mucosa until the splenic flexure and eventually up to the distal transverse colon, it is called left-sided UC. When inflammation of the mucosa extends to the proximal transverse colon and beyond, it is called extensive UC (pancolitis) (Table 6.1).[4,5]

| Table 6.1 UC classification according to disease extension (Montreal, 2005) | |
| --- | --- |
| **Classification** | **Extension** |
| E1 – ulcerative proctitis | Involvement limited to the rectum |
| E2 – left-sided UC (distal colitis) | Involvement extends up to the splenic flexure |
| E3 – extensive UC (pancolitis) | Involvement proximal to the splenic flexure |

The rectum is where about 95% of UC cases occur and may extend proximally to involve the entire colon.[6]

The CD is characterized by transmural inflammation foci that can affect any part of the gastrointestinal tract, from the oral cavity to the anus and perianal region but the most affected locations are the ileum and proximal colon.[2,3,7,8] Segmental inflammation, interspersed with preserved areas of the mucosa, is a peculiar feature of the disease. The lesion may extend into all layers of the intestinal wall from mucosa to serosa, causing thickening of the wall and narrowing of the intestinal *lumen*.[4] It can be classified according to location (terminal ileum, ileocolonic, colonic, anorectal, upper gastrointestinal [GI] disease) or its behavior pattern (inflammatory, stricturing or fistulizing) (Table 6.2). It is important to emphasize that during the course of the disease there may be change from one pattern to another, especially from inflammatory to fistulizing or stricturing, which occurs in 50 to 60% of cases after 10 years of disease.

Approximately 5-10% of patients with nonspecific pathological changes in the colon are considered to have indeterminate colitis, for presenting clinical characteristics similar to IBD, however with no definitive diagnosis.[3,4]

Etiology of UC and CD is still unknown, although considerable progress has been made in recent years. It was suggested that the disease occurs in genetically predisposed individuals, due to a dysregulated immune response to unknown antigens (probably infectious or environmental, including the endogenous microflora), resulting in continuing inflammation mediated by the immune system.[2,4,9]

UC, as well as CD, has uniform distribution between genders and typically affects young people, with a peak of incidence between 15 and 30 years, although it can occur at any age. Approximately 10% of cases occur in individuals under the age of 18 years.[1] It presents a bimodal distribution of age, with a second small peak occurring in individuals at 50 to 70 years of age.[1]

UC and CD exhibit different clinical manifestations and progression, which are determined by several factors, such as location and extent of disease, degree

**Table 6.2** Classification of CD according to disease extension (Montreal, 2005)

| Age at diagnosis (A) | | | |
|---|---|---|---|
| **A1** | 16 years or younger | | |
| **A2** | 17 to 40 years | | |
| **A3** | Over 40 years | | |
| **Location (L)** | | **Upper GI tract (L4)** | |
| **L1** | Terminal ileum | **L1 + L4** | Terminal ileum + upper GI tract |
| **L2** | Colon | **L2 + L4** | Colon + upper GI tract |
| **L3** | Ileocolon | **L3 + L4** | Ileocolon + upper GI tract |
| **L4** | Upper GI tract | | |
| **Behavior (B)** | | **Perianal disease (p)** | |
| **B1** | Non-stricturing Non-penetrating | **B1p** | Non-stricturing + perianal |
| **B2** | Stricturing | **B2p** | Stricturing + perianal |
| **B3** | Penetrating | **B3p** | Penetrating + perianal |

GI: gastrointestinal.

of activity and severity of the inflammatory process, associated systemic and extraintestinal manifestations, complications of the disease, and the presence of comorbidities.[5,10]

Symptoms range from mild to severe during relapses and may disappear or decrease during remission. In general, the clinical picture depends on the segment involved in the intestinal tract. The most frequent extraintestinal manifestations include rheumatologic, dermatologic, ophthalmic, and hepatobiliary involvement, among others, which may or may not be related to disease activity, and may also be present before symptoms in the gastrointestinal tract appear.

Disease diagnosis is confirmed by clinical evaluation and a combination of tests based on biochemistry, endoscopy, radiology, histology and nuclear medicine (Table 6.3).[4,11]

| Table 6.3 Clinical, endoscopic, radiological and histological features of CD and UC | | |
|---|---|---|
| **Symptoms** | **CD** | **UC** |
| Abdominal pain | Prominent, frequent complaint in the lower right quadrant | Colic, especially in the lower left quadrant |
| Diarrhea | Common in adults, and may be absent in children | Common in adults; may alternate with constipation |
| Hematochezia | In 20 to 30% of patients, especially with distal disease | Usually related to disease activity |
| Abdominal mass | Right lower quadrant — inflamed ileum | Left lower quadrant, if sigmoid is inflamed in thin individuals |
| Insomnia | Occasional | Rare |
| Malnutrition | Common | Occasional |
| Abdominal bloating | Present | Only in severe disease |
| Obstructive symptoms | Common | Rare |
| Perianal disease/ fistula | In over 30% of patients | Rare |
| **Laboratory abnormalities** | | |
| Acute-phase proteins (CRP) | Common | Present in severe or extensive disease |
| Anemia | Common | Present in severe disease |
| Macrocytosis | Present (chronic ileal disease) | Rare |
| Hypoalbuminemia | Common | Present in severe disease |
| pANCA | + (colitis) | ++ (+++ in UC with PSC) |
| ASCA | ++ | + |
| **Complications/extraintestinal manifestations** | | |
| Abscess | Common in perianal disease, occasional in the abdomen | Rare |
| Toxic megacolon | Uncommon | Uncommon |
| Acute ileitis | Occasional | Absent |
| **Complications/extraintestinal manifestations** | | |
| Primary sclerosing cholangitis | Rare | Occasional (5 to 15%) |
| Hepatitis | Rare | Occasional |
| Erythema nodosum | Occasional | Rare |

*(continues)*

**Table 6.3** Clinical, endoscopic, radiological and histological features of CD and UC

| Symptoms | CD | UC |
|---|---|---|
| Pyoderma gangrenosum | Rare | Rare |
| Arthralgia/arthritis | Very common | Common |
| **Epidemiology** | | |
| Active Smoker | Common | Rare |
| Former smoker | Rare | Common |
| Passive smoker (children) | Common | Rare |
| Previous appendectomy | Occasional | Less than expected |
| **Endoscopic features** | | |
| Distribution | Any segment of the GI tract | Full extent of the rectum |
| Involvement of the small intestine | Common | Backwash ileitis |
| Rectal Involvement | 30 to 50% | Most of the time |
| Uniform continuous disease | Uncommon (Crohn's colitis) | All of the time |
| Longitudinal ulcers | Common | No |
| Cobblestone | Common | No |
| Normal mucosa within the inflamed areas | Common (skip lesions) | No |
| Strictures | Occasional | Rare (always suspect carcinoma) |
| Mucosal edema | Occasional | Common |
| Ulceration | Deep | Superficial and extensive |
| Circumferential inflammation | Rare | Common |
| **Imaging** | | |
| Thickening of the intestinal wall | Extensive | Moderate |
| Mesenteric lymph nodes | Common | Uncommon |
| Involvement of mesenteric fat | Common | No |

ASCA: anti-Saccharomyces cerevisiae antibodies; CD: Crohn's disease; PCR: C-reactive protein; GI: gastrointestinal; pANCA: perinuclear anti-neutrophil cytoplasmic antibodies; PSC: primary sclerosing cholangitis; UC: ulcerative colitis; +: association seen variably; ++: associated; +++: commonly associated; ++++: strongly associated.

This chapter presents the main intestinal clinical manifestations of IBD with emphasis on natural history, clinical presentation and physical examination of patients with UC and CD. Extraintestinal manifestations will be addressed in subsequent chapters.

## ULCERATIVE COLITIS (UC)

### Natural history

The onset can be insidious or abrupt, and evolution is usually chronic, with relapsing interspersed with periods of clinical remission.[12] The clinical picture is variable, both in relation to severity and clinical manifestations and prognosis.

The UC activity can be classified as mild (up to 4 bowel movements a day, with or without blood, no systemic involvement and normal ESR), moderate (4-6 bowel movements a day, with blood and with minimal systemic involvement) or severe (more than 6 bowel movements per day, with blood and evidence of systemic involvement, including fever, tachycardia, anemia and ESR above 30). Presence of toxic megacolon suggests poor prognosis (Table 6.4).[13]

Fulminant UC is defined when the patient has more than 10 bowel movements with blood, fever, tachycardia, need for blood transfusion, inflammatory activity tests very altered (e.g., ESR > 30 mm in the first hour), with or without toxic megacolon (dilation of the transverse colon > 6 cm) or perforation.[13]

| Table 6.4 UC Classification according to severity of acute disease (Truelove and Witts) | | | |
|---|---|---|---|
| | **Mild** | **Moderate** | **Severe** |
| 1. Number of bowel movements/day | ≤ 4 | 4 to 6 | > 6 |
| 2. Blood in the stool | ± | + | ++ |
| 3. Body temperature | Normal | Intermediate values | Average body temperature at night >37.5°C (99.5°F) or >37.8°C (100.04°F) in 2 out of 4 days |
| 4. Pulse | Normal | Intermediate | > 90 bpm |
| 5. Hemoglobin (g/dL) | > 10.5 | Intermediate | < 10.5 |
| 6. ESR (mm/first hour) | < 30 | Intermediate | > 30 mm, first hour |

ESR: erythrocyte sedimentation rate.

Truelove SC, Witts LJ. Br Med J 1995; 2:1041-8.

In clinical practice, during the interview, data related to intestinal and extraintestinal clinical manifestations, previous or current history of drug use, smoking or triggering factors of the disease, and family history of IBD should be obtained.[14]

UC is 2-6 times more common in non-smokers. This suggests that smoking protects against the development of disease. There is an increased risk of developing cancer in UC, compared to the general population, which is related to two factors: duration and extent of inflammation. The risk of colorectal cancer increases in patients over 10 years of disease progression and diagnosed with pancolitis.[15]

## Clinical Picture

The clinical manifestations of UC vary and depend on the anatomical extent of lesions, as well as disease intensity and severity. In general, the most common symptoms are diarrhea and bleeding in stool. Aspects of the mucosa in the rectum and/or colon by colonoscopy reflect the clinical progression of the disease.

The predominant symptom in UC, during the active non-complicated phase, is the passing of liquid stool containing mucus, blood and/or pus, associated or not with fecal matter. The number of stools is variable from 2 to 3 up to countless in 24 hours. In the acute phases, blood loss in bowel movements or as pure intestinal bleeding is common. This is a relatively common symptom during periods of exacerbation of the inflammatory process, reflecting the activity of the disease. The general clinical manifestations that commonly accompany UC are fever, loss of appetite, weakness, weight loss and anemia.[16]

In active phases, asthenia and weight loss reflect major losses such as water, electrolytes, proteins and blood from the gastrointestinal tract.

UC can be classified according to the severity of acute disease (mild, moderate and severe) clinical progress (fulminant acute, chronic continuous and chronic intermittent) and the extension of the inflammatory process (distal UC, UC in the left hemicolon and pancolitis).[5,12]

Currently, the Montreal classification is most often used to describe the extent of disease (defined by colonoscopy) (see Table 6.1); the Truelove and

Witts classification is used to characterize the disease in its mild, moderate and severe forms, according to the severity of the initial flare (see Table 6.4); and the Mayo score is used to assess the severity of disease (Table 6.5).

| Table 6.5 Disease Activity Index – Mayo Clinic score | | | | |
|---|---|---|---|---|
| Score | Number of bowel movements | Rectal bleeding | Endoscopic findings | Global assessment |
| 0 | Normal amount | No blood seen | Normal or inactive disease (scarring) | Normal |
| 1 | Normal amount + 1-2 BM/day | Streaks of blood with stool – less than half of BM | Mild disease (erythema, ↓ in vascular pattern, mild friability) | Mild disease |
| 2 | Normal amount + 3-4 BM/day | Obvious blood with stools | Moderate disease (evident erythema, loss of vascular pattern, erosion) | Moderate disease |
| 3 | Normal amount + > 5 BM/day | Blood alone without stools | Severe disease (spontaneous bleeding, ulceration) | Severe disease |

| Score (points) | Disease severity |
|---|---|
| 0 to 2 | Normal – remission |
| 3 to 5 | Mild activity |
| 6 to 10 | Moderate activity |
| 11 to 12 | Severe activity |

Source: Adapted from Schroeder et al., 1987.[17]

## Distal UC

Clinical symptoms of proctitis or proctosigmoiditis are usually considered mild or moderate in relation to gravity. Generally, there are rectal bleeding, change in frequency of bowel movements, defecation urgency, presence of feces with mucus and/or pus, and tenesmus.

The bleeding is bright red and can be quite alarming to the patient, as it occurs in most evacuations; however, the amount of blood loss is small, and the patient hardly ever has anemia. Clots and dark red rectal bleeding are rare.

The frequency of bowel movements is generally increased and must be differentiated from diarrhea, since stool consistency may be normal or just slightly loose. Nevertheless, 80% of patients present typical diarrhea, invariably with blood and pus. Fecal incontinence may be present, but it is unusual. Defecation urgency is so disturbing that quality of life is compromised.[14]

In about 1/3 of the cases, there is history of constipation. The complaint of tenesmus is very common, and its intensity correlates with the degree of rectal involvement.[18] It is common to feel rectal weight with permanent desire to evacuate. Although rare, patients may experience mild to moderate abdominal cramping preceding evacuations, located in the left iliac fossa and/or hypogastrium.

Due to the high number of bowel movements, the patient may experience liquid deposition in the perianal skin, causing painful irritation in the region. The vast majority of patients with active UC have mucus or mucus and pus associated with blood, around the stool.[18]

The past history may reveal previous episodes of rectal bleeding, tenesmus, urgency and pus in the stool, indicating a natural history of spontaneous remissions and relapses.[5,13]

### Left-sided UC and extensive UC

Symptoms observed in the disease that affects the left colon or the entire colon represent, in general, moderate or severe forms of the disease. Clinical manifestations considered most common are: episodes of diarrhea, rectal bleeding, mucus elimination, abdominal pain and weight loss. The clinical picture usually has an insidious onset, with an increase in the number of daily bowel movements and pasty or watery stools; in a few weeks, this condition may progress to evident diarrhea, with fresh blood, mucus and pus mixed or not in the stool.[14,19] Rectal bleeding is usually dark red and there can be clots. There may be fever, asthenia and anorexia.

In severe UC, blood is often mixed with pus, epithelial debris, mucus and fecal matter and may be associated with bloody diarrhea.

Diarrhea is one of the best indicators of disease severity. Large volumes indicate that the mucosa of the colon is impaired and sodium and water reabsorption is seriously impaired. The patient may have nocturnal diarrhea, which is also a good indicator of disease severity. Postprandial diarrhea is a common symptom. Patients with left-sided colitis commonly have solid stool in the right colon and eliminate small amounts of diarrhea from the affected intestine. Tenesmus and defecation urgency are more frequent in patients with rectal disease.

Abdominal pain is more intense than in the case of distal UC, but is uncommon during the stage of remission. Abdominal pain in severe colitis may appear along with active disease or may suggest complications of the disease, such as intestinal obstruction or subocclusion due to benign strictures, tumors and polyps.[14,19]

Patients with severe UC may be anemic and dyspneic. In acute colitis, caloric intake is reduced both by anorexia and the concern that food will trigger pain and diarrhea. It is common for patients with rapid weight loss to complain of swelling of ankles, malaise and lethargy.

Fulminant colitis is considered the severe form of UC, leading to severe deterioration of the patient's general condition, with more than 10 bowel movements with blood and pus daily, fever, abdominal pain, and anemia with need for blood transfusion. In this form, there may be serious complications such as strictures, massive bleeding, toxic megacolon, bowel perforation, and systemic complications.[12,13]

### Physical Examination

The general physical examination of the patient diagnosed with mild to moderate UC can present no significant clinical signs and, in most cases, the health status is good or reasonable. However, in severe UC, anemia and malnutrition signs can be found. Patients with systemic toxicity may develop fever, dehydration, mucocutaneous pallor, tachycardia and postural hypotension.

Abdominal examination is usually normal, but in uncomplicated acute illness, the patient usually refers abdominal pain on palpation, but with no defensive movement, preserved bowel sounds, which can be tympanitic.

Severe forms of the disease may present with abdominal pain on palpation, especially on the left side of the abdomen, more precisely in the lower left quadrant, where the sigmoid colon can be palpated.[20]

The perineal region may be scarred, but perianal disease is uncommon. Fissures and abscesses can occur occasionally. During the digital rectal examination, it is common to observe hypersensitivity of the anal canal, coarse graininess, ulcers, pseudopolyps and blood and/or pus in the examiner's glove.

## CROHN'S DISEASE (CD)

### Natural history

CD activity can greatly affect the natural history and it can have a deleterious effect on employment prospects and social and family life of patients.[21] The disease may progress to intermittent crises, alternating with phases of remission of variable duration, or as a progressive and continuous chronic form. The clinical presentation is largely dependent on the location and extent of the lesions and the presence of occasional complications.[13,19]

In an initial phase, the extent of lesions can be so small that the patient may not experience symptoms; but when the injuries affect a greater extent of the small intestine or colon, symptoms can be intense. The typical presentation includes the involvement of various segments of the gastrointestinal tract, and the clinical course of CD is often complicated by the formation of fistulas, perianal disease and strictures. Invariably, these complications also have an impact on the quality of life of patients.

The disease is limited to the terminal ileum in 47% of cases, colon in 28%, ileocolon in 21%, and upper gastrointestinal tract in 3% of cases.[13] It can be classified as inflammatory (70% of patients), stricturing (17% of patients) or penetrating/fistulizing - fistulas, abscesses, or both (13% of diagnosed cases).[13]

The genetic component is stronger in CD than in UC. Cigarette smoking increases the risk of CD, but decreases the risk of UC through mechanisms that are still unclear.[4] Rheumatologic, dermatologic, ophthalmic and hepatobiliary diseases, and other extraintestinal manifestations may be present. The extraintestinal manifestations are more common when the disease affects the colon, and may or may not be related to disease activity.[22] Long-term disease (> 8 to 10 years) is associated with an increased risk of colon carcinoma. At least 50% of patients can require surgical treatment in the first 10 years of disease, and approximately 70 to 80% can require surgery during their lifetime.[21]

Anamnesis is of utmost importance and must include data regarding the onset of symptoms, progress of symptoms, occurrence of an epidemiological factor for infectious pathology, occupation, family history, addictions, food history (including intolerance), drug history (antibiotics, non-steroidal anti-inflammatory drugs - NSAIDs) and comorbidities.

Clinical data based on history and physical examination also allow CD classification and serve to guide the choices for workup (laboratory, radiological, endoscopic and histopathological investigations) and treatment.[23] CD can be divided clinically in:

- **Mild to moderate CD:** outpatients able to tolerate oral feeding, without the occurrence of dehydration, toxicity, abdominal discomfort, painful mass, obstruction, or weight loss greater than 10%;
- **Moderate to severe CD:** patients who failed to respond to treatment or those with more prominent symptoms of fever, weight loss, abdominal pain, nausea or intermittent vomiting (without intestinal obstruction findings), or significant anemia;
- **Severe to fulminant CD:** patients with persistent symptoms despite the introduction of corticosteroids and/or biological therapy or individuals who present with fever, persistent vomiting, evidence of intestinal obstruction, peritoneal irritation, cachexia or abscess.

### Clinical Picture

CD can manifest as gastrointestinal symptoms, extraintestinal symptoms or a combination of both.[11]

The symptoms of CD are heterogeneous, but typically include: abdominal pain (in 70 to 85% of patients), diarrhea (70 to 75% of patients) and weight loss (in 60% of patients). Systemic symptoms of malaise, anorexia and fever are common.[24,25] The disease may progress to intestinal obstruction caused by strictures, fistulas (often perianal) or abscesses.

In general, patients with CD may have various clinical forms of the disease, according to localization of the lesions: exclusively in the small intestine; ileocolon; colon; anorectal area and, more rarely, damage to the esophagus, stomach or duodenum. Currently, the Montreal Classification (2005) is the most widely used for CD staging; however, it does not assess the clinical or endoscopic activity (see Table 6.2). To evaluate CD activity, the most used tool in clinical studies is the CDAI - Crohn's disease activity index (CDAI) (Table 6.6).

Approximately 1/3 of the cases present disease restricted to the terminal ileum, and more than half there is involvement of the terminal ileum and proximal colon. The isolated involvement of the large intestine is less frequent than the terminal ileum.[4] Anorectal/perineal involvement ranges from 8 to 80% in different medical publications and is often associated with colonic involvement; 5 to 10% of cases can present as an isolated manifestation of CD.[26]

The main clinical symptoms, according to the location of the lesions and the predominant activity, are described below.

### Ileocolonic CD

It is the most frequent anatomical form of the disease, and patients often have abdominal pain, usually in the right upper quadrant, with or without diarrhea and weight loss. Diarrhea, if present, generally has no bleeding, ranging between 5 and 6 bowel movements per day. In patients with inflammation or abscess in the ileocolic region, pain tends to be constant. These patients often present with fever, weight loss and malnutrition. The disease may progress to bowel subocclusion, caused by inflammation or stenosis in the region, leading to persistence and worsening of pain, which appears more widespread, intermittent, colicky and associated with borborygmi, abdominal distension, vomiting and occasional constipation.

The involvement of the right ureter in the inflammatory process may appear as urinary complaints.

When there is significant impairment of the terminal ileum, malabsorption of biliary salts and vitamin B12 can occur, which leads to choleretic diarrhea and megaloblastic anemia, respectively. Strictures can cause stasis in the

| **Table 6.6** Crohn's Disease Activity Index (IACD)[27] | |
|---|---|
| | **Multiplied by** |
| Number of liquid stools in the previous week | 2 |
| Abdominal pain (none = 0; mild = 1; moderate = 2; severe = 3). Consider the total sum of individual data in the previous week | 5 |
| General condition (excellent = 0; well = 1; poor = 2; very poor = 3; terrible = 4). Consider the total sum of individual data in the previous week | 7 |
| N. of associated symptoms/signs – list by categories: a) Arthritis/ arthralgia; b) Iritis/uveitis; c) Erythema nodosum/pyoderma gangrenosum/aphthous stomatitis; d) Anal fissure, fistula or perirectal abscess; e) Other fistula; f) Fever | 20 (maximum score = 120) |
| Use of antidiarrheal drugs (no = 0; yes = 1) | 30 |
| Abdominal mass (none = 0; dubious = 2; definite = 5) | 10 |
| Hematocrit deficit: For Male = 47 Ht; For Female = 42 Ht (subtract instead of adding if the patient's Ht is > the standard) | 6 |
| (Weight/Usual weight) × 100 Weight*: percentage below the expected (subtract instead of adding if the patient's weight is higher than expected) | 1 |
| Total (Crohn's Disease Activity Index) =< 150 = Remission 150 to 250 = Mild 250 to 350 = Moderate > 350 = Severe | |

Best WR et al., 1976.

small intestine, with consequent bacterial overgrowth and progression to steatorrhea. Enterocutaneous fistulas can occur, which are easy to diagnose clinically, as well as enterovesical and rectovaginal fistulas. Acute abdomen with peritonitis due to bowel perforation is unusual.

## CD of small intestine

In general, patients present with symptoms of epigastric colicky and intermittent pain, increased borborygmi, and frequent bouts of diarrhea, although of low intensity. Patients with extensive disease may have malabsorption, steatorrhea, anemia, weight loss and malnutrition. Diarrhea is a feature of active disease. Mild fever reflects the inflammatory process, while high persistent fever may mean suppurative complications.[19] Strictures may present with intestinal subocclusion, characterized by periumbilical colicky pain, moderate to strong, usually related to food intake. Enterocutaneous or enterovesical fistulas may be forms of presentation of disease in the small intestine. Hypocalcemia, hypoalbuminemia, hypomagnesemia and nephrolithiasis are possible complications.

## Colonic CD

Patients present with acute bouts of diarrhea, colicky abdominal pain, often in the hypogastric abdomen, preceding evacuations, episodes of fever and bloody stools. Symptoms can mimic active UC, although rectal bleeding is less common in CD.

## Anorectal CD

Anal and perianal manifestations of CD are very common, and the involvement of this region can be quite severe and debilitating. The clinical picture is characterized by complaints of anorectal pain, burning sensation, pain with bowel movements, perianal purulent discharge that stains clothes and in the case of rectovaginal fistula, exacerbation of symptoms is observed.

The perianal involvement in CD presents often with abscesses and fistulas, which are usually multiple, complex and recurrent. It can be also present with fissures, anal ulcer, fecal incontinence and anal stricture, rarely associated with cancer[28,29]

### CD of the esophagus, stomach and duodenum

It is extremely rare and usually causes clinical manifestations related to these locations. The most common symptoms are abdominal pain in the epigastric region, dyspepsia, often associated with anorexia, nausea, vomiting, and weight loss. Isolated esophageal involvement in CD may manifest with dysphagia caused by stricture, sore throat, heartburn, or chest pain. In more advanced cases, there may be esophago-bronchial or esophago-gastric fistulas.[28]

Patients with gastric involvement may present asymptomatic or mildly symptomatic, mimicking gastritis and mild endoscopic changes, such as aphthous ulcers. Advanced disease presents with vomiting and weight loss, and can progress with fistula into the colon with complaints of diarrhea, fecaloid vomiting and weight loss.[30]

When the duodenum is affected, the most common finding is also appearance of aphthous ulcers and thickening of mucosal folds. Fistulas are usually duodenocolic.[30]

### Physical Examination

The general physical examination of patients with CD ranges from normal to multiple abnormalities. The clinical signs seen are more revealing in CD than in UC, reflecting localized and transmural behavior of the inflammatory process.

Patients may present with anemia, malnutrition, weight loss and fever. When the disease occurs in childhood, there may be delayed growth and also delays in development. The examination of the oral cavity can show aphthous ulcers, gingivitis and glossitis. Examination of the abdomen may be painful to deep palpation, especially in the lower right quadrant (LRQ), with palpable masses, plastron or fistulas. Enterocutaneous fistulas are more common in younger individuals or after surgical procedures with intestinal involvement.

The inflammatory process can extend to structures adjacent to the intestine. In such case, an inflammatory mass may be palpated, usually consisting of adherent inflammatory bowel, with thickened mesentery and enlarged abdominal lymph nodes.

Abdominal distension with or without tumor in the right lower quadrant can be observed.

Perianal involvement in CD often presents with chronic abscess, which can be single, multiple, simple or complicated, perianal cellulitis and fistulas - usually multiple, complex and recurrent. CD can also present with inflammatory plicomas, which are frequent and usually do not cause significant symptoms, fissures that are often deep and painless, anal ulcer, fecal incontinence, and anal stricture, which is a common complication of chronic anal disease. These conditions are rarely associated with cancer.

The diagnosis of perianal fistulas in CD is quite obvious when it appears during the course of intestinal disease. However, in patients without history of CD, lesions should be suspicious when there are multiple and complex fistulas and abscesses, non-traumatic rectovaginal fistulas, fissures that are difficult to treat, and injuries in the perineal region.

Rectal examination may reveal thickened walls in the anal canal, variable stricture of the anal canal, and often, when the disease is extensive locally, the perception of irregularity of the surface of the anal canal and distal rectum. Rectovaginal fistula may occur in 3 to 10% of female patients with CD, more often in the middle of the rectovaginal septum.[4,28]

Extraintestinal manifestations with involvement of the joints, skin and eyes occur in about 30% of cases, both in UC and in CD.[23,31]

### REFERENCES

1. Hanauer SB. Inflammatory bowel disease: epidemiology, pathogenesis, and therapeutic opportunities. Inflamm Bowel Dis 2006; 12(Suppl.1):S3-9.

2. Kozuch PL, Hanauer SB. Treatment of inflammatory bowel disease: a review of medical therapy. World J Gastroenterol 2008; 14:354-77.

3. Mowat C, Cole A, Windsor A, Ahmad T, Arnott I, Driscoll R et al. Guidelines for the management of inflammatory bowel disease in adults. Gut 2011; 60:571-607.

4. Carter MJ, Lobo AJ, Travis SP. IBD Section, British Society of Gastroenterology. Guidelines for the management of inflammatory bowel disease in adults. Gut 2004; 53(Suppl.5):V1-16.

5. Sipahi AM, Santos FM. Doença inflamatória intestinal. In: Martins MA, Carrilho FJ, Alves VAF, Castilho EA, Cerri GG, Wen CL (eds.). Clínica médica. Barueri: Manole, 2009.

6. Kornbluth A, Sachar DB. Ulcerative colitis practice guidelines in adults: American College of Gastroenterology, Practice Parameters Committee. Am J Gastroenterol 2010; 105:501-24.

7. Van Assche G, Dignass A, Panes J, Beaugerie L, Karagiannis J, Allez M et al. The second European evidence-based Consensus on the diagnosis and management of Crohn's disease: Definitions and diagnosis. J Crohns Colitis 2010; 4:7-27.

8. Brasil. Ministério da Saúde. Secretaria de Atenção à Saúde. Portaria SAS/MS nº 711, de 17 de dezembro de 2010, Protocolo Clínico e Diretrizes Terapêuticas da Doença de Cronh. Brasília, 2010. Available at: portal.saude.gov.br/portal/arquivos/pdf/pcdt_doenca_de_crohn.pdf.

9. Cassinotti A, Ardizzone S, Porro GB. Adalimumab for the treatment of Crohn's disease. Biologics 2008; 2:763-77.

10. Sipahi AM, Damião AOM. Doença inflamatória intestinal: retocolite ulcerativa inespecífica e doença de Crohn. In: Federação Brasileira de Gastroenterologia. Condutas em gastroenterologia. São Paulo: Revinter, 2004.

11. Sipahi AM. Quadro clínico e diagnóstico da doença inflamatória intestinal. In: Quilici FA, Damião AOMC, Sipahi AM, Zaltman C, Flavio S, Magaly GT (eds.). Guia prático – Doença inflamatória intestinal. Rio de Janeiro: Elsevier, 2007.

12. Freitas JA, Tacla M. Retocolite ulcerativa. In: Dani R (ed.). Gastroenterologia essencial. 2.ed. Rio de Janeiro: Guanabara Koogan, 2001.

13. Baumgart DC, Sandborn WJ. Inflammatory bowel disease: clinical aspects and established and evolving therapies. Lancet 2007; 369:1641-57.

14. Rampton DS, Shanahan F. Fast facts: inflammatory bowel disease. 3.ed. Oxford: Health Press, 2008. Available at: www.fastfacts.com/_files/samplefiles/FF_IBD3e_sample.pdf.

15. Kiss DR, Teixeira MG, Gama AH. Retocolite ulcerativa. In: Rodrigues JJ, Machado MCC, Rasslan S (eds.). Clínica cirúrgica. Barueri: Manole, 2008.

16. Houli J, Medeiros Netto G. Retocolite ulcerative inespecífica. Rev Bras Colo-proctol 1984; 4:191-205.

17. Schroeder KW, Tremaine WJ, Ilstrup DM. Coated oral 5-aminosalicylic acid therapy for mildly to moderately active ulcerative colitis. A randomized study. N Engl J Med 1987; 317(26):1625-9.

18. Santos Jr. JCM. Retocolite ulcerativa – aspectos clínicos, diagnóstico e tratamento clínico. Parte 1. Rev Bras Colo-proctol 1999; 19:29-34.

**19.** Pereira AS, Filho RAP. Doença de Crohn. In: Mincis M (ed.). Gastroenterologia e hepatologia. 3.ed. São Paulo: Lemos Editorial, 2002.

**20.** Dassopoulos T, Hanauer S. Presentation and diagnoses of inflammatory bowel disease. In: Cohen RD (ed.). Inflammatory bowel disease: diagnosis and therapeutics. Nova Jersey: Humana Press, 2003.

**21.** Keighley MRB, Willians NS. Cirurgia do ânus, reto e colo. Barueri: Manole, 1998.

**22.** Damião AOMC, Habr-Gama A. Retocolite ulcerativa idiopática. In: Dani R, Paula-Castro L (eds.). Gastroenterologia clínica. Rio de Janeiro: Guanabara Koogan, 1993.

**23.** Brazilian Study Group of Inflammatory Bowel Diseases. Consensus guidelines for the management of inflammatory bowel disease. Arq Gastroenterol 2010; 7:313-25.

**24.** Marzinotto MAN, Leite AZA. Doença inflamatória intestinal: quadro clínico e diagnóstico. In: Zaterka S, Eisig JN (eds.). Tratado de gastroenterologia: da graduação à pós-graduação. São Paulo: Atheneu, 2011.

**25.** Dignass A, Van Assche G, Lindsay JO, Lémann M, Söderholm J, Colombel JF et al. The second European evidence-based Consensus on the diagnosis and management of Crohn's disease: Current management. J Crohns Colitis 2010; 4:28-62.

**26.** Podolsky DK. Inflammatory bowel disease. N Engl J Med 2002; 347:417-29.

**27.** Best WR, Becktel JM, Singleton JW, Kern F. Development of a Crohn's disease activity index. National Cooperative Crohn's Disease Study. Gastroenterology 1976; 70:439-44.

**28.** Teixeira MG, Gama AH. Doença de Crohn ileocolorretal. In: Cruz GMG (ed.). Coloproctologia: propedêutica nosológica. Rio de Janeiro: Revinter, 2000.

**29.** Magalhães AFN. Doença de Crohn. In: Dani R, Paula-Castro L (eds.). Gastroenterologia clínica. Rio de Janeiro: Guanabara Koogan, 1993.

**30.** Teixeira MG, Gama AH, Pinotti HV. Doença de Crohn. In: Pinotti HW. Tratado de clínica cirúrgica do aparelho digestivo. Rio de Janeiro: Guanabara Koogan, 1993.

**31.** Bernstein CN, Fried M, Krabshuis JH, Cohen H, Eliakim R, Fedail S et al. World Gastroenterology Organization Practice Guidelines for the diagnosis and management of IBD in 2010. Inflamm Bowel Dis 2010; 16(1):112-24.

## BIBLIOGRAPHY

**1.** Gama AH, Teixeira MG, Neto CB. Retocolite ulcerativa In: Pinotti HW. Tratado de clínica cirúrgica do aparelho digestivo. São Paulo: Atheneu, 1994.

**2.** Koltum WA. Inflammatory bowel disease: diagnosis and evaluation In: Wolff BG, Fleshman JW, Beck DE, Pemberton JH, Wexner SD (eds.). The ASCRS texbook of colon and rectal surgery. Nova York: Springer, 2007.

# ORAL MANIFESTATIONS OF INFLAMMATORY BOWEL DISEASE

MARCELLO MENTA SIMONSEN NICO
SHEYLA BATISTA BOLOGNA
SILVIA VANESSA LOURENÇO

## INTRODUCTION

The involvement of the oral cavity can occur at any time during the course of inflammatory bowel disease (IBD), both in ulcerative colitis (UC) and Crohn's disease (CD), and may even precede the intestinal symptoms for many years.[1] These oral lesions are often classified as specific (granulomatous on histopathology, reflecting the presence of the disease directly in the oral mucosa) or nonspecific (nongranulomatous, due to systemic inflammatory changes caused by disease activity).[2]

In a recent study of 10 patients with oral manifestations and IBD treated at the Dermatology Division of Hospital das Clínicas, Faculdade de Medicina da Universidade de São Paulo, the presence and proper diagnosis of the oral lesions led to clinical suspicion and subsequent confirmation of the disease in 5 cases,[3,4] demonstrating the importance of proper recognition of these lesions. In the other 5 patients, oral lesions appeared during exacerbations of previously diagnosed IBD.

## GRANULOMATOUS LESIONS

Granulomatous lesions are seen in CD (Figure 7.1) and can affect different portions of the gastrointestinal tract. Among these manifestations, linear ulcers and hyperplastic polypoid lesions are the most characteristic.[5-7]

Ulcers with linear configuration and hypertrophic edges are seen on the buccal mucosa. These lesions correspond clinically with the skin lesions in metastatic CD, which are also oval ulcers affecting large skin folds. They have histological appearance of noncaseating granulomas in the lamina *propria*.

Multiple papular or polypoid lesions involving the oral mucosa and vestibules are also characteristic. When grouped together, they give the mucosa a "cobblestone" aspect.[7] This is a rare manifestation, however highly suggestive of CD (Figure 7.2).

Granulomatous cheilitis is characterized clinically by firm edema, painless and persistent, on the lips, oral mucosa or even on the face. It is part of the so called orofacial granulomatosis, a term that encompasses mucocutaneous manifestations that can be observed in IBD, sarcoidosis, and Melkersson-Rosenthal syndrome.[7]

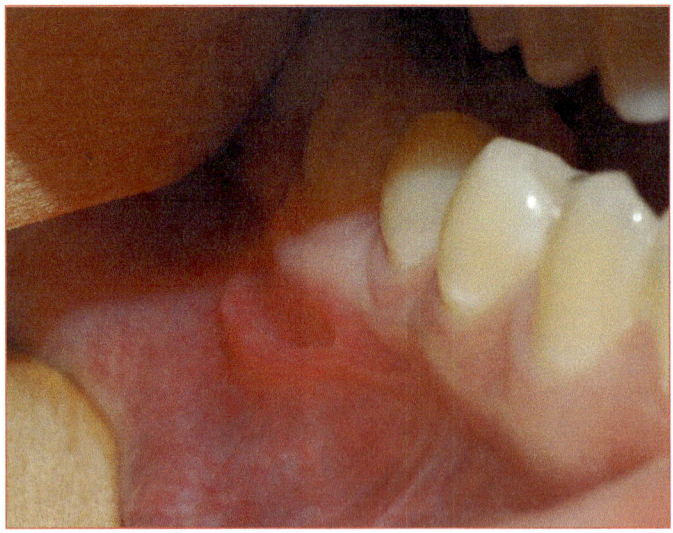

**Figure 7.1** Mucosal adherence with cicatricial aspect. On histopathology, granulomas are observed.

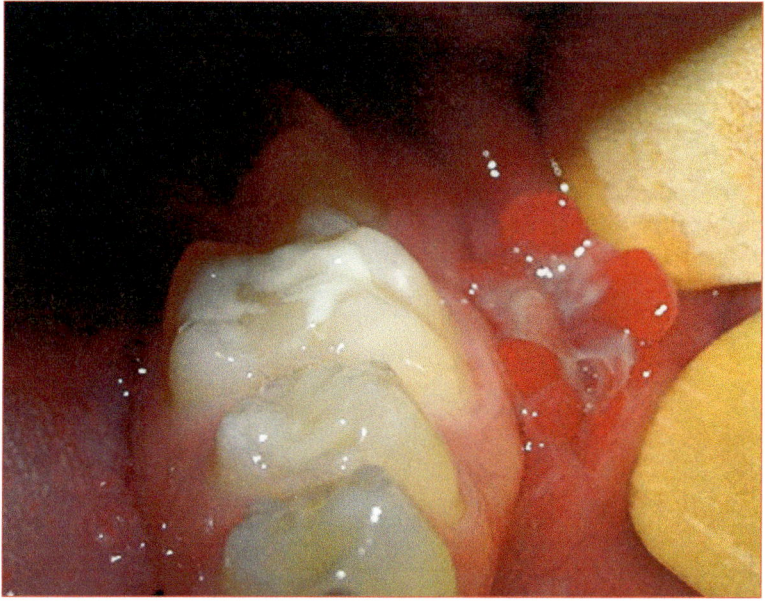

Figure 7.2 "Cobblestone" aspect of the mucosa and aphthous stomatitis in a case of CD.

## NONGRANULOMATOUS LESIONS

They can also occur in the presence of other diseases besides IBD, and therefore are not diagnostic. Its presence generally indicates a mucosal reactivity disorder that occurs in parallel with the systemic activity of the underlying disease.

Five to 10% of patients present ulcers during the active phase of intestinal disease.[8] Aphthous stomatitis lesions associated with IBD are clinically identical to common aphthous stomatitis, characterized by round ulcers with erythematous halo and a central fibrin pseudomembrane (Figure 7.3). They are always very painful, and may present clinically as minor aphthous ulcers, major aphthous ulcers or herpetiform aphthous ulceration.

Minor aphthous ulcers are characterized by the occurrence in small number and size. Major aphthous ulcers are deep, disabling and take longer to heal. Herpetiform ulcers, in turn, are small, multiple and clustered, with a

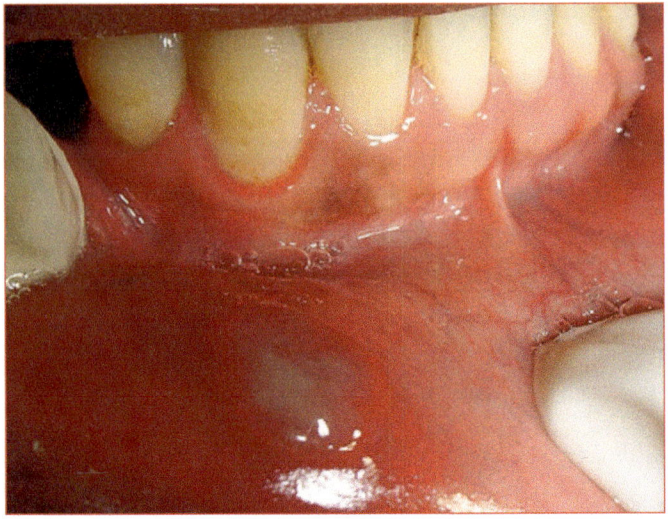

**Figure 7.3** Aphthous stomatitis in patient with CD.

recurrent and subintrant character. Not only these can occur in association with CD and UC, but they can also be seen alone or in diseases such as AIDS, Behcet's disease and other immunodeficiencies.[1] They often have good therapeutic response with the appropriate treatment of intestinal disease, but lesions should be properly diagnosed and differentiated from other ulcerative diseases of the mucosa.

Pyostomatitis vegetans is a rare condition, characterized by vegetating and pustular lesions of the mucosa. A strong association with IBD is well documented, especially with UC. The lesions consist of multiple sterile miliary pustules that are clustered on erythematous and edematous mucosa, resulting in a pattern of blisters and superficial ulcerations with elongated appearance, which is described in literature as "snail track"-like (Figure 7.4). These lesions may affect any area of the oral mucosa.[9] On histopathologic analysis, there is epithelial hyperplasia with formation of neutrophilic and eosinophilic abscesses; acantholysis may also occur. This latter phenomenon, together with the occasional detection of antiepithelial autoantibodies using direct and indirect immunofluorescence, gives some clinically typical cases of pyostomatitis vegetans, including those associated with IBD, laboratory features that cannot be differentiated from *pemphigus vulgaris*.[10]

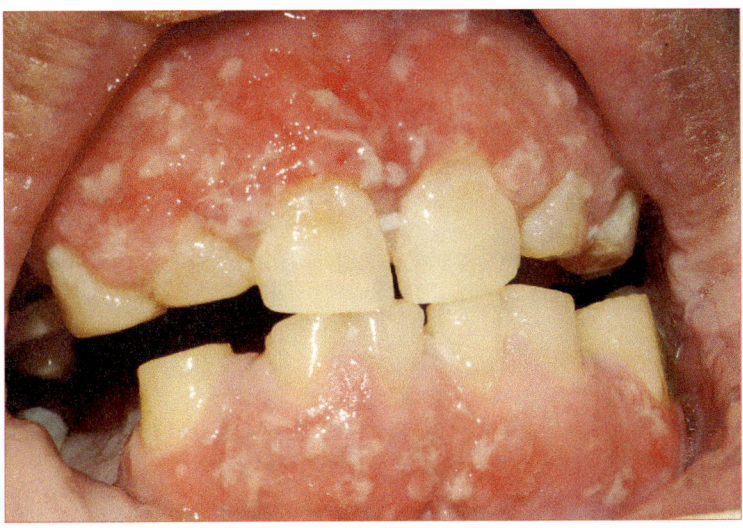

**Figure 7.4** Pyostomatitis vegetans: multiple confluent pustules in a case of UC.

## TREATMENT OF ORAL MANIFESTATIONS OF IBD

In most cases, conduct is guided by the general treatment of disease. Treatment of the associated intestinal disease is usually sufficient to control oral lesions.[9,11] In the absence of active IBD, corticosteroids and topical immunomodulators can be used, but most authors consider treatment of choice the use of systemic corticosteroids at medium to high dosages, with rapid remission.[1,12]

Whenever there is diagnostic or therapeutic difficulty in relation to oral manifestations, or if lesions persist despite general treatment, the patient should be referred to a specialist in oral mucosal diseases - most often, a dermatologist - who will assess the need for additional therapeutic measures.

## REFERENCES

1.  Trost LB, McDonnell JK. Important cutaneous manifestations of inflammatory bowel disease. Postgrad Med J 2005; 81:5.

2.  Harty S, Fleming P, Rowland M, Crushell E, McDermott M, Drumm B et al. A prospective study of the oral manifestations of Crohn's disease. Clin Gastroenterol Hepatol 2005; 3:886-91.

3. Lourenço SV, Hussein TP, Bologna SB, Sipahi AM, Nico MMS. Oral manifestations of inflammatory bowel disease: a review based on the observation of six cases. J Eur Acad Dermatol Venereol 2010; 27(2):204-7.

4. Pincelli TPH. Manifestações orais da doença inflamatória intestinal: estudo clínico--patológico retrospectivo. [Masters Dissertation]. São Paulo: Faculdade de Medicina da Universidade de São Paulo, 2010.

5. Dupuy A, Cosnes J, Revuz J, Delchier JC, Gendre JP, Cosnes A. Oral Crohn disease: clinical characteristics and long-term follow-up of 9 cases. Arch Dermatol 1999; 135:439-42.

6. Pittock S, Drumm B, Fleming P, McDermott M, Imrie C, Flint S et al. The oral cavity in Crohn's disease. J Pediatr 2001; 138:767-71.

7. Alawi F. Granulomatous diseases of the oral tissues: differential diagnosis and update. Dent Clin N Am 2005; 49:203-21.

8. Ruocco E, Cuomo A, Salerno R, Ruocco V, Romano M, Baroni A. Crohn's disease and its mucocutaneous involvement. Skinmed 2007; 6:179-85.

9. Ruiz-Roca JA, Berini-Aytés L, Gay-Escoda C. Pyostomatitis vegetans. Report of two cases and review of the literature. Oral Surg Oral Med Oral Pathol Oral Radiol Endod 2005; 99:447-54.

10. Nico MMS, Hussein TP, Aoki V, Lourenço SV. Pyostomatitis vegetans and its relation to inflammatory bowel disease, pyoderma gangrenosum, pyodermatitis vegetans, and pemphigus. J Oral Pathol Med 2012; 41:584-8.

11. Mehravaran M, Kémeny L, Husz S, Korom I, Kiss M, Dobozy A. Pyodermatitis-pyostomatitis vegetans. Br J Dermatol 1997; 137:266-9.

12. Danese S, Semeraro S, Papa A, Roberto I, Scaldaferri F, Fedeli G et al. Extraintestinal manifestation in inflammatory bowel disease. World J Gastroenterol 2005; 11:7227-36.

# DERMATOLOGIC MANIFESTATIONS OF INFLAMMATORY BOWEL DISEASE

RICARDO ROMITI
SABRINA SISTO ALESSI CÉSAR

## INTRODUCTION

Several extraintestinal manifestations related to inflammatory bowel disease (IBD) are described. Among them, the cutaneous manifestations stand out, since they occur frequently and may precede intestinal symptoms, serving as a warning to the doctor. Although known to be common, the incidence of cutaneous manifestations related to IBD can vary from 2 to 34%.[1-5] At diagnosis of IBD, the average incidence of cutaneous manifestations is around 10%;[1] in the course of the disease, however, a wide variety of dermatological alterations may occur.

It is estimated that about 1/3 of patients with IBD develop skin lesions.[3] Skin manifestations can be classified as granulomatous lesions specific of Crohn's disease (CD), associated diseases, and manifestations secondary to nutritional deficiency.[1]

## CD-SPECIFIC GRANULOMATOUS MANIFESTATIONS
### Ulcers and perianal/peristomal fistulas
Perianal disease is often the first manifestation of CD. In a study of patients with colonic and ileocolonic disease, it was observed in 7% of cases at

diagnosis and 45% during the course of disease.[1] The spectrum of lesions is variable, with the most common early lesions being perianal erythema, followed by aphthous ulcers in the anal canal and perianal fissures (Figure 8.1). Linear ulcers and vegetating lesions are also common. Aggressive forms include ulcers that destroy the anal sphincter, abscesses that progress to perianal or rectovaginal fistulas and scarring leading to deformity.

Fistulas may appear in the abdominal wall after surgery or, more rarely, in the navel. In patients with ostomy, complications can occur such as allergic contact dermatitis and ulcers.

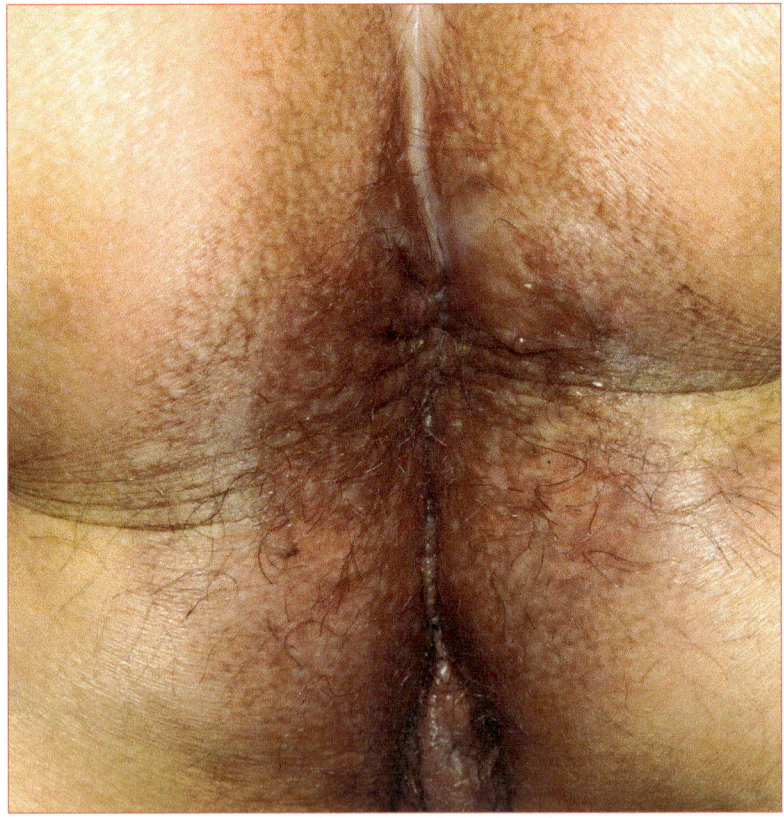

**Figure 8.1** Presence of erythema and fistulas in the perianal region.

### Metastatic CD

This term is applied when granulomatous lesions occur in sites distant from the gastrointestinal tract. They can occur in many regions, but are found mainly in the abdomen, lower extremities and skin folds.[1] They also occur in the penis and vulva, characterized by a painful edema.[6] Metastatic CD is more common in patients with colonic disease and usually is not related to disease activity. Clinically, the lesions manifest as subcutaneous nodules or erythematous plaques that can ulcerate.

Oral granulomatous lesions are discussed in Chapter 7 – Oral Manifestations of Inflammatory Bowel Disease.

### ASSOCIATED DISEASES

### Erythema nodosum

Erythema nodosum is the skin lesion that appears most often, in conjunction with symptoms of active bowel disease. It is characterized by painful nodules on the extensor surface of the legs, more palpable than visible, progressing with residual contusiform aspect (Figure 8.2). Its prevalence in patients with IBD is around 3 to 8%,[1-5] and it is most common in females and patients with large intestine involvement and arthritis.[1-5]

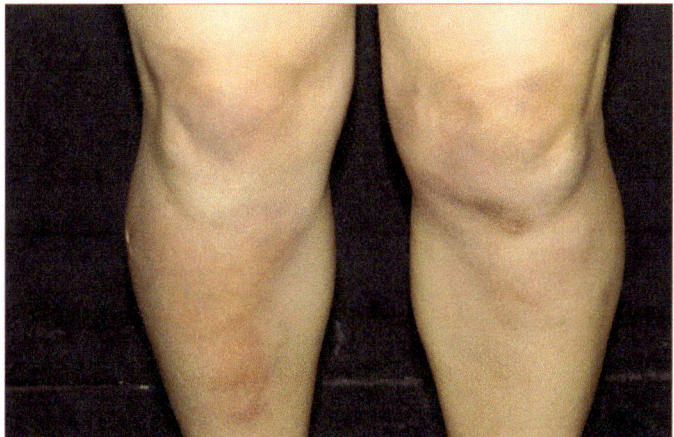

**Figure 8.2** Erythematous nodules with contusiform aspect on the extensor surface of the legs.

This manifestation tends to occur for the first time during the first 2 years of the clinical course of disease and may recur in approximately half of cases.[1] Histological examination is required to confirm diagnosis, and lesions usually respond to treatment with systemic corticosteroids.

## Pyoderma gangrenosum

Pyoderma gangrenosum is a chronic and disabling skin disorder, which can be sometimes more incapacitating than IBD itself. This severe dermatosis occurs in 1 to 2% of patients and its correlation with the activity of the intestinal disease is still controversial, despite the fact that it sometimes coincides with IBD exacerbations, particularly in the case of colonic disease.[1-5]

Gangrenous pyoderma seems to be more common in colitis than in CD.[2] Lesions are usually multiple and occur on the extensor surfaces of the lower extremities, but may also be found elsewhere. They begin as erythematous pustules and nodules that develop into irregular painful burrowing ulcers, with violaceous edges, and with a granular base (Figure 8.3). They can appear in skin around stoma and at sites of trauma (pathergy).[7,8] Pathology reveals neutrophilic dermatosis and should be performed for disease confirmation. In most cases, the use of systemic corticosteroids and other immunomodulators such as cyclosporine and azathioprine is effective.[8]

## Sweet's syndrome

The neutrophilic dermatosis is characterized by erythematous plaques and nodules infiltrated in the face, neck, trunk and extremities, accompanied by fever, leukocytosis and neutrophilia (Figure 8.4). It is a rare manifestation of IBD, with about 40 cases described in the literature. All reported cases were from patients with colonic involvement and the majority related to females. It can occur before, during or after the diagnosis of IBD, but its appearance is more common with active disease.[9-11] It usually responds to corticosteroids.

## Hidradenitis suppurativa

Hidradenitis suppurativa is a chronic inflammatory skin disease characterized by the appearance of nodules, abscesses, and painful and recurrent fistulas in areas containing a high concentration of apocrine glands, especially the

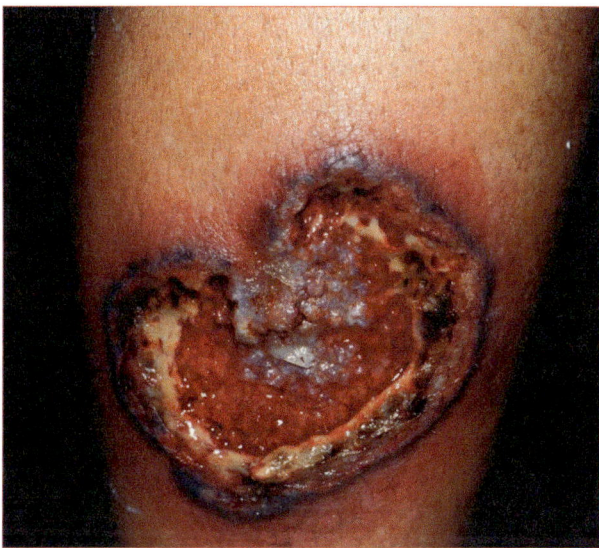

**Figure 8.3** Burrowing ulcers with erythematous and violaceous edges, with a granular base in the leg.

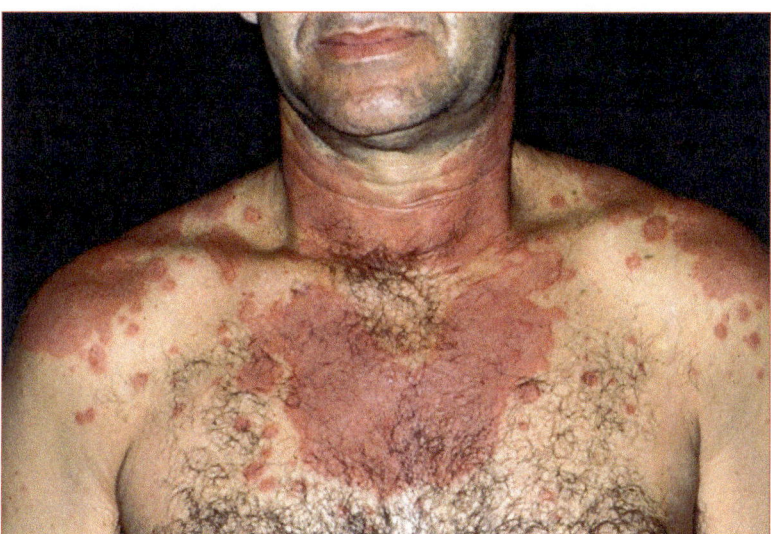

**Figure 8.4** Erythematous infiltrated papules and plaques on the trunk.

groin and axilla, leading to scarring and severe impairment of quality of life of affected individuals (Figure 8.5). This dermatosis affects especially young adults and women, and its pathogenesis remains unknown.[12-14] The occurrence of hidradenitis suppurativa and CD has been reported in over 37 cases in the literature.[12] The largest series published cases showed the presence of CD in 38% of patients with hidradenitis suppurativa.[14] More recently, the relationship with ulcerative colitis was also demonstrated, although to a lesser extent. Current treatment options include topical and oral antibiotics, immunosuppressive agents such as azathioprine, corticosteroids and more recently anti-TNFa in severe cases.[12-14] Studies indicate that hidradenitis associated with inflammatory diseases responds less to immunosuppressive treatment.[13] Surgical excision of recurrent nodules, or even large areas of involvement, is also possible.[12]

Some other associations have been observed, such as that of IBD and psoriasis. The latter occurs in 1 to 2% of the world population, and in 7 to 11% of patients with IBD, which suggests a genetic link between the two. However, there is no relationship between the course of psoriasis and the activity of intestinal disease.[1-4]

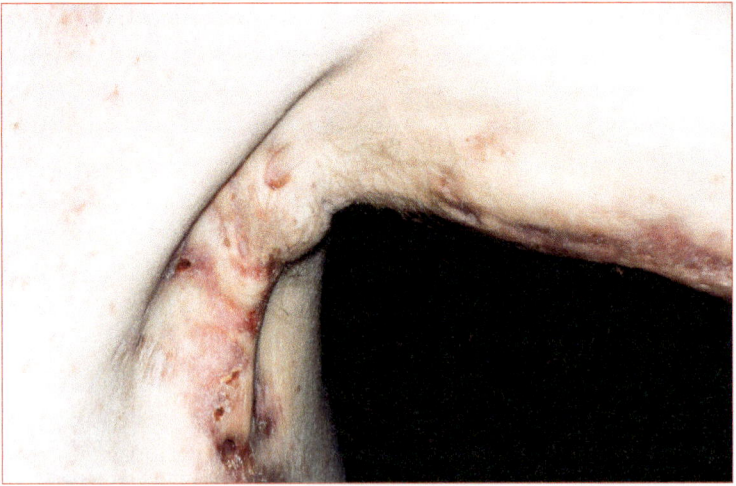

**Figure 8.5** Nodules, fistulas and scars in the axilla.

Vitiligo is also most common in patients with IBD compared with the general population, and an autoimmune hypothesis has been proposed to explain this relationship.[1]

Connective tissue diseases, such as polymyositis, lupus erythematosus and scleroderma, have also been reported in association with IBD.[1]

Established therapy for the treatment of IBD can per se lead to different and sometimes severe cases of drug eruptions. Sulfasalazine can cause a potentially fatal drug eruption variant, known by the acronym DRESS (drug rash with eosinophlia and systemic symptoms), which is characterized by fever, rash, face edema, lymph node enlargement and systemic involvement (Figure 8.6). Anti-TNFs, in turn, may be responsible for manifestations such as psoriasis, alopecia *areata*, lichen *planus* and possibly symptoms of toxic epidermal necrolysis (TEN).

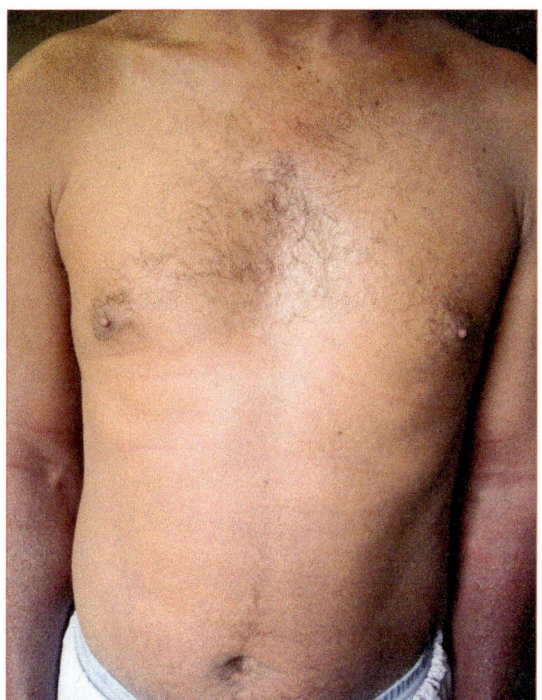

**Figure 8.6** Diffuse rash and peeling in the trunk and abdomen.

## MANIFESTATIONS SECONDARY TO NUTRITIONAL DEFICIENCY
### Acrodermatitis enteropathica

A rare disease associated with a zinc-deficient diet, parenteral nutrition or malabsorption. Clinically, it manifests as psoriasiform erythema with vesicles, pustules and crusts around the orifices (perioral, perigenital, perianal) and/or on extensor surfaces.[1]

## REFERENCES

1.  Veloso RV. Review article: skin complications associated with inflammatory bowel disease. Aliment Pharmacol Ther 2004; 20(Suppl.4):50-3.

2.  Tromm A, May D, Almus E, Voigt E, Greving I, Schwegler U et al. Cutaneous manifestations in inflammatory bowel disease. Gastrointerol 2001; 39(2):137-44.

3.  Yüksel I, Basar O, Ataseven H, Ertuğrul I, Arhan SS et al. N. Mucocutaneous manifestations in inflammatory bowel disease. Inflamm Bowel Dis 2009; 15(4):546-50.

4.  Repiso A, Alcántara M, Muñoz-Rosas C, Rodríguez-Merlo R, Pérez-Grueso MJ, Carrobles JM et al. Extraintestinal manifestations of Crohn's disease: prevalence and related factors. Rev Esp Enferm Dig 2006; 98(7):510-7.

5.  Farhi D, Cosnes J, Zizi N, Chosidow O, Seksik P, Beaugerie L et al. Significance of erythema nodosum and pyoderma gangrenosum in inflammatory bowel diseases: a cohort study of 2402 patients. 281 Medicine 2008; 87(5):280-93.

6.  Ulrike L, Horn LC, Witzigmann H, Pohl K, Mossner J, Keim V. Crohn's disease of the vulva. Med Clin 1998; 93:492-6.

7.  Mendoza JL, Garcia-Paredes J, Peña AS, Cruz-Santamaria DM, Iglesias C, Diaz-Rubio M. El espectro de las dermatitis neutrofilicas en la enfermedad de Crohn. Rev Esp Enferm Dig 2003; 95(3):229-32.

8.  Menachem Y, Gotsman I. Clinical manifestation of pyoderma gangrenosum associated with inflammatory bowel disease. Imag 2004; (6):88-90.

9.  Ahmadi S, Powell F. Pyoderma gangrenosum: uncommon presentations. Clin Dermatol 2005; 23:612-20.

10. Massud A, Duerksen D. Ulcerative colitis and Sweet's syndrome: a case report and review of the literature. Can J Gastroenterol 2008; 22(3):296-8.

11. Lear JT, Atherton MT, Byrne JPH. Neutrophilic dermatoses: pyoderma gangrenosum and Sweet's syndrome. Post Grad Med 1997; 73:65-8.

12. van der Zee HH, van der Woude CJ, Florencia EF, Prens EP. Hidradenitis suppurativa and inflammatory bowel disease: are they associated? Results of a pilot study. Br J Dermatol 2010; 162:195-7.

13. Machet L, Samimi M, Delage M, Paintaud G, Maruani A. Systematic review of the efficacy and adverse events associated with infliximab treatment of hidradenitis suppurativa in patients with coexistent inflammatory diseases. J Am Acad Dermatol 2013; 69(4):649-50.

14. Scheinfeld N. Diseases associated with hidranitis suppurativa: part 2 of a series on hidradenitis. Dermatol Online J 2013; 19(6):2.

# OPHTHALMOLOGIC MANIFESTATIONS OF INFLAMMATORY BOWEL DISEASE

MILTON RUIZ ALVES

JAIME ROIZENBLATT

MARINA ROIZENBLATT

## INTRODUCTION

Inflammatory bowel disease (IBD) is a group of chronic, recurrent and inflammatory diseases, which is a combination of genetic, immunological and environmental factors associated with the lifestyle of a patient. According to the latest theories, IBDs result of excessive pro-inflammatory response of bacterial populations in the gastrointestinal (GI) tract.[1] The extra-intestinal manifestations of IBD occur in 25-36% of patients[2] and may involve more commonly joints, skin, bulb of the eye and attachments, biliary tract, neural tissue (among other) and bone marrow.[2,3] Current studies focusing on ophthalmology reported that 43 to 60% of IBD patients showed ocular manifestations.[4,5] A study conducted by Felekis showed that silent ophthalmological manifestations occur fairly often and require early treatment in order to prevent more serious problems in the future.[4]

Ocular manifestations in IBD occur in approximately 10% of patients with Crohn's disease (CD) and a lesser percentage in cases of ulcerative colitis (UC).[6] Ocular involvement, as a rule, occurs in parallel to intestinal inflammatory manifestations;[7,8] rarely ocular manifestations precede the symptoms of IBD.[9,10]

Ocular involvement in IBD happens more frequently when at least one other extraintestinal manifestation is present.[11] In patients with CD and arthritis, ocular involvement incidence increases to 33%.[11] In this group, there is also increased prevalence of HLA-B27 histocompatibility antigen.[12]

Another risk factor for onset of ocular inflammation in patients with CD includes the manifestation of colitis or ileocolitis.[13]

## OCULAR MANIFESTATIONS

Ocular manifestations of IBD are divided into three groups: primary, secondary and coincident. The primary manifestations are associated with increased activity of IBD and respond to treatment directed to the intestinal condition (e.g., systemic corticosteroids or surgical excision of the affected bowel tissue). They can affect both segments of the bulb of the eye, anterior and posterior, as well as the orbit content, but the impairment of the anterior segment is most frequently described.[4] The involvement of the anterior segment is manifested by episcleritis (Figure 9.1), scleritis, uveitis (acute iritis, chronic iridocyclitis or panuveitis) and keratopathy.[2,4,6,14,15] Acute episcleritis relates to active CD and can be used as a marker of IBD activity, which does not occur with scleritis and uveitis.[11,14]

Other manifestations include eyelid edema, blepharitis and cataract. Episcleritis occurs in up to 29% of cases of IBD and scleritis, in up to 18% of patients. Repeated episodes of scleritis can lead to perforation of the sclera (*scleromalacia perforans*). Uveitis may be present in up to 18% of cases of IBD.[9,16]

Keratopathy appears in two forms: by the presence of epithelial or subepithelial infiltrates, seen as small elevated grayish-white dots, and subepithelial infiltrates conferring an aspect of clouding or scarring of the cornea (Figure 9.2). It is most prominent inferiorly, and tends to be bilateral and symmetric. It does not stain with fluorescein, and is first visible at the inner periphery of the cornea, 2 to 3 mm from the corneal limbus. There are case reports of unilateral keratopathy characterized by white peripheral subepithelial infiltrates (two with UC and another with CD).[17]

Keratopathy does not affect the vision, because the lesions spare the central area of the cornea. It is speculated that some kind of autoimmune inflammation or hypersensitivity at the periphery of the cornea is responsible for the biomicroscopic aspect revealed in the examination of the infiltrates. However,

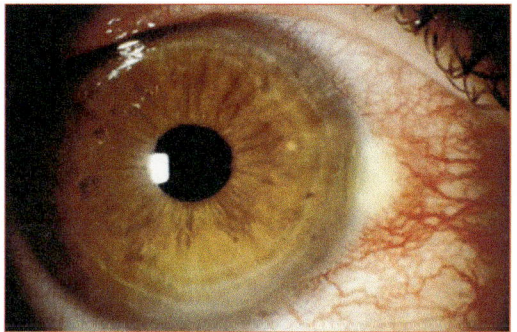

**Figure 9.1** Episcleritis in patient with CD.

**Figure 9.2** Subepithelial keratopathy in patient with CD.

histopathology and electron microscopy of material obtained from biopsy of the affected cornea did not clarify its pathogenesis.[18] Primary complications associated with the posterior segment and orbit include macular edema, central serous retinopathy, proptosis of the eye caused by orbital myositis,[14,19] ischemic optic neuropathy, optic neuritis, sheathing of retinal vessels (Figure 9.3), vascular occlusions and retinal pigment epithelial changes, resulting in less than 1% in most studies.[9,20] Nevertheless, recent studies in which more detailed eye examinations were made, including systematic examination with fluorescein angiography, show high incidence (30.8%) manifestations of the posterior segment.[4]

Secondary ocular manifestations of IBD derive from primary gastrointestinal complications or medication, or therapy used in the treatment of these intestinal complications such as inadequate diet or resection of affected bowel tissue, leading to malabsorption syndrome or malnutrition.[21] In this case, vitamin A deficiency may be caused by reduced intake or insufficient absorption of vitamin A, which can lead to decreased

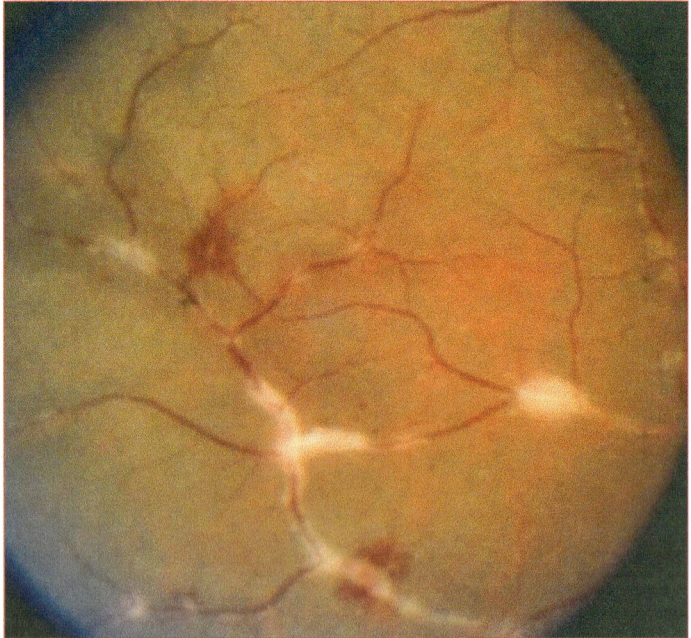

**Figure 9.3** Retinal vasculitis.

production of the tear film (xerophthalmia), and night blindness.[21] Other complications include cataracts, presumably resulting from uveitis and the use of corticosteroids; escleromalacia as a result of repeated episodes of scleritis; optic disk edema caused by peripapillary scleritis; and dry keratitis, found in 43% of patients using mesalazine.[21,22]

Ocular complications do not depend on the extent of bowel involvement and tend to occur in the first years of IBD.[11,23]

Coincident ocular manifestations are not considered complications or actual consequences, but ocular abnormalities commonly found in the general population. Thus, it is difficult to establish a causal relationship between them and IBD. The list of overlapping manifestations includes conjunctivitis, dacryoadenitis, recurrent corneal erosion, corneal ulcer, glaucoma and subconjunctival hemorrhage. Interestingly, in the same way in which the incidence and severity of IBD vary according to the region of the planet, incidence, range and severity of ocular manifestations are also variable.[24,25]

One publication described, in a 13-year old boy with CD, bilateral conjunctival injection and conjunctival nodules with gelatinous appearance, arranged circumferentially. Biopsy of nodules revealed granulomatous inflammation with caseation and conjunctival hyperemia, and nodules responded to systemic treatment of IBD.[26] This fact may suggest that some cases of conjunctivitis may be associated with CD, and therefore be classified as primary complications, and not coincident.

Changes resulting from the treatment of IBD may also occur. Persistent corneal endothelial deposits were described, associated with the use of rifabutin, a broad spectrum antibiotic used to treat mycobacterial infection in patients positive for human immunodeficiency virus (HIV) and, more recently, to treat CD.[27] Corneal deposits were described in the treatment of HIV positive patients but have not yet been reported in patients treated for CD.[27]

## TREATMENT

Treatment planning for ocular manifestations in IBD starts with proper evaluation by a gastroenterologist and continues, in the case of ocular manifestations, with their classification as primary, secondary or coincidental changes.

Local treatment of anterior uveitis and scleritis is done by using topical steroid, and, in special circumstances, with subtenon injection of corticosteroids. The treatment of necrotizing scleritis sequels can be done with grafts including scleral, periosteum, pericardium, and amniotic membrane, and with a combination of amniotic membrane and pericardium graft.[28,29] Systemic treatment with corticosteroids can be made if approved by a gastroenterologist.

The use of systemic corticosteroids is indicated in the presence of severe inflammation nonresponsive to local therapy, optic neuropathy or orbital disease. When adequate control of inflammation is not obtained, medications such as azathioprine and methotrexate or other immunosuppressive agents should be considered, especially in HLA-B27 positive patients. New drugs, such as mycophenolate mofetil - an immunosuppressant which acts in the synthesis of purines, preferably on lymphocytes T and B, inhibiting their antigenic stimulation and recruitment at inflammation sites - and immunomodulatory agents - which target the inflammatory cytokines, particularly infliximab, a anti-tumor necrosis factor alpha (TNF-alpha) monoclonal antibody, and adalimumab - have been used successfully in patients unresponsive to conventional therapy.[30,31]

The possibility of ocular complications arising from the use of the so called anti-TNF-alpha agents in the treatment of IBD should also be mentioned. Several cases of optic neuropathy caused by infliximab and adalimumab were described in patients with rheumatoid arthritis.[32,33] It is unclear, however, whether the anti-TNF-alpha agents are inducers of demyelinating optic neuropathy and accompanying multiple sclerosis, or if the demyelinating presentations are isolated events.[34]

## REFERENCES

1. Whelan RA, Hartmann S, Rausch S. Nematode modulation of inflammatory bowel disease. Protoplasma 2012; 249(4):871-86.

2. Danzi JT. Extraintestinal manifestations of idiopathic inflammatory bowel disease. Arch Intern Med 1988; 148(2):297-302.

3. Present DH. Extraintestinal manifestations. Mt Sinai J Med 1983; 50(2):126-32.

4. Felekis T, Katsanos K, Kitsanou M, Trakos N, Theopistos V, Christodoulou D et al. Spectrum and frequency of ophthalmologic manifestations in patients with inflammatory bowel disease: a prospective single-center study. Inflamm Bowel Dis 2009; 15(1):29-34.

5. Yilmaz S, Aydemir E, Maden A, Unsal B. The prevalence of ocular involvement in patients with inflammatory bowel disease. Int J Colorectal Dis 2007; 22(9):1027-30.

6. Petrelli EA, McKinley M, Troncale FJ. Ocular manifestations of inflammatory bowel disease. Ann Ophthalmol 1982; 14(4):356-60.

7. Palli D, Trallori G, Saieva C, Tarantino O, Edili E, D'Albasio G et al. General and cancer specific mortality of a population based cohort of patients with inflammatory bowel disease: the Florence Study. Gut 1998; 42(2):175-9.

8. Stenson WF. Inflammatory bowel disease. In: Yamada T, Alpers DH, Laine L et al. (eds.). Textbook of gastroenterology. 3.ed. Philadelphia: Lippincott Williams & Wilkins, 1999. p.1775-839.

9. Ghanchi FD, Rembacken BJ. Inflammatory bowel disease and the eye. Surv Ophthalmol 2003; 48(6):663-76.

10. Lyons JL, Rosenbaum JT. Uveitis associated with inflammatory bowel disease compared with uveitis associated with spondyloarthropathy. Arch Ophthalmol 1997; 115(1):61-4.

11. Hopkins DJ, Horan E, Burton IL, Clamp SE, de Dombal FT, Goligher JC. Ocular disorders in a series of 332 patients with Crohn's disease. Br J Ophthalmol 1974; 58(8):732-7.

12. Mallas EG, Mackintosh P, Asquith P, Cooke WT. Histocompatibility antigens in inflammatory bowel disease. Their clinical significance and their association with arthropathy with special reference to HLA-B27 (W27). Gut 1976; 17(11):906-10.

13. Salmon JF, Wright JP, Murray AD. Ocular inflammation in Crohn's disease. Ophthalmology 1991; 98(4):480-4.

**14.** Knox DL, Schachat AP, Mustonen E. Primary, secondary and coincidental ocular complications of Crohn's disease. Ophthalmology 1984; 91(2):163-73.

**15.** Razumova I, Ambartsumian AR. Acute uveitis in nonspecific ulcerative colitis. Vestn Oftalmol 2009; 125(6):29-31.

**16.** Banares A, Jover JA, Fernandez-Gutierrez B et al. Patterns of uveitis as a guide in making rheumatologic and immunologic diagnoses. Arthritis Rheum 1997; 40(2):358-70.

**17.** Schulman MF, Sugar A. Peripheral corneal infiltrates in inflammatory bowel disease. Ann Ophthalmol 1981; 13(1):109-11.

**18.** van Vliet AA, van Balen AT. Corneal pathology in Crohn's disease: electron microscopic study of a case. Ophthalmologica 1985; 190(2):72-6.

**19.** Caramoy A, Lappas A, Fauser S, Kirchhof B. Central scotoma and blurred vision in a patient with Crohn's disease. Ophthalmologe 2009; 106(9):836-8.

**20.** Ernst BB, Lowder CY, Meisler DM, Gutman FA. Posterior segment manifestations of inflammatory bowel disease. Ophthalmology 1991; 98(8):1272-80.

**21.** Stoller GL, Kaiura TL, Florakis GJ. Inflammatory bowel disease and other systemic inflammatory diseases. In: Krachmer JH, Mannis MJ, Holland EJ (eds.). Cornea: fundamentals, diagnosis and management. 2.ed. Philadelphia: Elsevier Mosby, 2005.

**22.** Cury DB, Moss AC. Ocular manifestations in a community-based cohort of patients with inflammatory bowel disease. Inflamm Bowel Dis 2010; 16(8):1393-6.

**23.** Wright R, Lumsden K, Luntz MH, Sevel D, Truelove SC. Abnormalities of the sacro-iliac joints and uveitis in ulcerative colitis. Q J Med 1965; 34:229-36.

**24.** Katsanos KH, Christodoulou DK, Michael M et al. Inflammatory bowel disease-related dysplasia and cancer: a referral center study in northwestern Greece. Eur J Intern Med 2005; 16(3):170-5.

**25.** Shivananda S, Lennard-Jones J, Logan R, Fear N, Price A, Carpenter L et al. Incidence of inflammatory bowel disease across Europe: is there a difference between north and south? Results of the European Collaborative Study on Inflammatory Bowel Disease (EC-IBD). Gut 1996; 39(5):690-7.

**26.** Blase WP, Knox DL, Green WR. Granulomatous conjunctivitis in a patient with Crohn's disease. Br J Ophthalmol 1984; 68(12):901-3.

**27.** Williams K, Ilari L. Persistent corneal endothelial deposits associated with rifabutin therapy for Crohn's disease. Cornea 2010; 29(6):706-7.

**28.** Alves MR, Matayoshi S. Transplante escleral. In: Hofling-Lima AL, Nishiwaki-Dantas MC, Alves MR (eds.). Doenças externas oculares e córnea. Rio de Janeiro: Guanabara Koogan, 2008.

**29.** Lazzaro DR. Repair of necrotizing scleritis in ulcerative colitis with processed pericardium and a Prokera amniotic membrane graft. Eye Contact Lens 2010; 36(1):60-1.

**30.** D'Haens G, Daperno M. Advances in medical therapy for Crohn's disease. Curr Gastroenterol Rep 2002; 4(6):506-12.

**31.** Sfikakis PP. The first decade of biologic TNF antagonists in clinical practice: lessons learned, unresolved issues and future directions. Curr Dir Autoimmun 2010; 11:180-210.

**32.** Faillace C, de Almeida JR, de Carvalho JF. Optic neuritis after infliximab therapy. Rheumatol Int 2013; 33(4):1101-3.

**33.** Foroozan R, Buono LM, Sergott RC, Savino PJ. Retrobulbar optic neuritis associated with infliximab. Arch Ophthalmol 2002; 120(7):985-7.

**34.** Bensouda-Grimaldi L, Mulleman D, Valat JP, Autret-Leca E. Adalimumab-associated multiple sclerosis. J Rheumatol 2007; 34(1):239-240; discussion 240.

# RHEUMATOLOGICAL MANIFESTATIONS OF INFLAMMATORY BOWEL DISEASE

CLÁUDIA GOLDENSTEIN SCHAINBERG

## INTRODUCTION

Arthritis was initially associated with inflammatory bowel disease (IBD) by Bargen in 1929, being distinguished from rheumatoid arthritis in the 1950s. Currently, IBD is considered within the spectrum of spondyloarthritis (SpA) for presenting synovitis, enthesitis and genetic association with HLA-B27. The preferred inflammatory involvement occurs in ligament and tendon insertions, and entheses, associated with axial joint involvement or peripheral asymmetrical of the spine. Up to 70% of patients with SpA have inflammatory intestinal lesions, consistent with Crohn's disease (CD) in about 26% of those undergoing ileocolonoscopy. In fact, 6 to 10% of patients with SpA develop IBD, and there is a strong correlation between the presence of intestinal and articular inflammation.[1,2]

Some authors believe that joint involvement is the initial manifestation of IBD and intestinal disorders with multiple symptoms may emerge years later. Approximately 36 to 46% of patients with IBD have extra-intestinal manifestations, the most frequent being rheumatic (22 to 33%),[3,4] and significantly more common in patients with disease confined to the colon. In ulcerative colitis (UC), the prevalence of joint involvement is up to 62%.[5]

Less commonly, arthritis may occur in Whipple's disease, celiac disease and after intestinal bypass. The prevalence of SpA in patients with IBD is variable: 18 to 45% meet the criteria for SpA, 3 to 9.9% for ankylosing spondylitis (AS); about 14% develop one or more clinical manifestations of SpA without meeting diagnostic criteria, and some of these (up to 24%) have asymptomatic sacroiliitis.[6,7]

Genetic susceptibility, abnormal antigen presentation and recognition of self, the presence of autoantibodies against specific antigens common to the colon and extracolonic tissues and increased intestinal permeability in addition to response against microorganisms possibly through molecular mimicry, are pathogenic autoimmune mechanisms common to IBD and SpA.

## CLINICAL PICTURE

In general, intestinal symptoms precede or coincide with rheumatological manifestations, although joint stiffness worsening is more associated with bowel inflammatory activity in UC than in CD. Involvement can be peripheral and/or axial including sacroiliitis with or without spondylitis, similar to that of AS. Other periarticular manifestations such as enthesopathy, tendinitis, periostitis, dactylitis, clubbing and granulomatous lesions in the joints and bones, are less common, as well as secondary osteoporosis and osteomalacia.[8]

## PERIPHERAL ARTHRITIS

Peripheral arthritis occurs in 17 to 20% of cases and is more common in CD (10 to 20%)[9] than in UC (2 to 7%). In 1998, a retrospective study of 1,459 patients showed peripheral arthritis in 10% of patients with CD and 6% of patients with UC.[10] Then, another study revealed arthritis in 10%, enthesitis in 7%, and history of arthritis in 29% of patients with IBD.[11]

The Oxford group[10] has classified the peripheral arthritis of IBD (also called peripheral enteropathic arthropathy) without axial involvement as pauciarticular or polyarticular arthropathy. In the oligo- or pauciarticular form, symptoms are usually acute and self-limited, involvement is asymmetric and migratory, in both large and small joints, and the lower limbs are more affected. Episodes last from 6 to 10 weeks, but relapses are frequent and tend to coexist with IBD relapses; occurrence of erythema nodosum and uveitis is frequent. In 31% of cases, arthropathy may appear up to 3 years before the diagnosis of IBD is made. The polyarticular form tends to have a more chronic, non-erosive course, although it can be destructive; its course is independent of IBD exacerbations and the coexistence of other extraintestinal manifestations is rare, except for uveitis.[8] Enthesopathy, particularly affecting the Achilles tendon or plantar fascia insertion are common. Usually, intestinal symptoms precede by many years or coexist with joint manifestations. In a prospective study that included 123 patients with SpA, 6% of them developed CD from 2 to 9 years after the first joint symptom. In UC, there seems to be a temporal relationship between arthritis and gut inflammation relapses.[8]

## Axial involvement

Unlike the peripheral arthritis, axial involvement (spondylitis and/or sacroiliitis) tends to precede intestinal manifestations and have chronic course; HLA-B27 is found in 50% of patients. Symptoms are almost the same as in AS without extra-articular manifestations, with low back pain at night, inflammatory in nature, improvement with walking, and progressive limitation of movement in all axes of the column.[12] Clinical course is independent of the intestinal inflammatory activity, and intestinal surgery does not alter the course of SpA.

The prevalence of axial disease is 10 to 20% in the case of sacroiliitis and 7 to 12% for AS,[9,13,14] and it is more common in CD (5 to 22%) than in UC (2 to 6%). Studies from reference centers showed that 30% of patients with IBD had inflammatory low back pain, 33% had abnormal Schober test results, and 30% had unilateral or bilateral sacroiliitis grade I or II. Also, 35% met the criteria established by the European Spondyloarthropathy Study Group for SpA[11] and 10% met the criteria for AS, revealing higher prevalence than for population studies, and probably reflecting a referral bias.[7,11,14,15] Unlike what happens with idiopathic AS, which commonly begins before 40 years of age and affects more men than women (2.5H:1M), IBD-associated spondylitis can occur at any age and in equal proportion between men and women.[16]

### Enthesitis

Enthesopathy is an important manifestation of spondyloarthritis in general. In CD and UC, it can occur at a frequency of 5.4 to 50% of cases, most commonly affecting the lower limbs, the Achilles tendon or plantar fascia insertion, as well as patellar tendon of the knee.

### Other extraintestinal manifestations

Anterior uveitis and episcleritis may be present in 3.5 to 11.8% of patients with UC or CD, and may manifest before, during or after the diagnosis of IBD. However, acute or chronic posterior uveitis is also reported.

Mucocutaneous lesions, such as oral aphthae, erythema nodosum and pyoderma gangrenosum, may be present. Erythema nodosum is reported in 3 to 15% of patients and, in general, is associated with peripheral arthritis and uveitis. Pyoderma gangrenosum, in turn, is rare, occurring in 0.5 to 4% of cases, especially in the lower limbs of women, and often leave scars after resolution of ulceration.

### Diagnostic evaluation

Hypochromic microcytic anemia (due to chronic inflammation and intestinal bleeding), leukocytosis, thrombocytosis and increased inflammatory activity tests (C-reactive protein and erythrocyte sedimentation rate – ESR) are common. There is no specific serological marker; antinuclear antibodies and rheumatoid factor are usually absent, while perinuclear ANCA antibodies are positive in up to 60% of patients with UC and in some patients with CD, and they seem to be directed against lactoferrin autoantigens.[17] Stool cultures allow acute cases of rheumatoid arthritis secondary to pathogens such as *Salmonella* or *Yersinia* to be ruled out. Synovial fluid is inflammatory, but sterile. Antibodies markers of IgA anti-*Saccharomyces cerevisiae* (ASCA), but not IgG, were demonstrated in a study of 87 patients with SpA positive for HLA-B27 and no gastrointestinal symptoms, especially in those presenting AS, although its association with the development of IBD in patients with SpA remains undefined. Sacroiliitis and spondylitis in IBD are associated with the presence of HLA-B27, although in lower frequencies than in AS (33 *versus* 71%). In addition, patients with AS and HLA-B27-negative have a higher risk of developing IBD than those with HLA-B27.[11]

The Oxford study[10] has shown an association of pauci-articular arthritis with HLA-B27 (27% *versus* 7% in controls), HLA-B35 (32 *versus* 15%) and HLA-DRB1 0103 (33 *versus* 3%), while the polyarticular form was associated with HLA-B44 (62 *versus* 30%). The presence of the shared epitope associated with synovitis in patients with IBD without sacroiliitis has also been demonstrated.[8] Radiological evaluation reveals edema in soft tissues in phases of active inflammation and uncommon erosions in patients with self-limited arthritis, while in persistent arthritis there may be joint erosion and loss of joint space if the hips are affected. Changes typical of AS may be observed, including vertebral body squaring, osteopenia and formation of marginal and bilateral syndesmophytes, as well as ossification of apophyseal joints, although the frequency of asymmetric sacroiliitis may be higher. Computed tomography and magnetic resonance imaging can show early lesions in the spine and sacroiliac joints, especially of inflammatory nature, and enthesitis can be documented by ultrasound.

## Other rheumatological manifestations

The combination of IBD and Sjögren's syndrome, rheumatoid arthritis, inflammatory myopathy, Takayasu arteritis and fibromyalgia is less common. Secondary osteoporosis may occur in patients receiving high doses of glucocorticosteroids and for longer periods, so that preventive measures such as supplementation of calcium and vitamin D must be prescribed.

## Treatment

Treatment should be customized according to the severity of the clinical picture. When necessary, surgical resolution appears to control arthritis in UC better than in CD. Relative rest, physiotherapy and intra-articular glucocorticoid injections may be enough to control mild oligoarthritis. Non steroidal anti-inflammatory drugs (NSAIDs) can be used cautiously and limited to the minimum time and the lowest possible effective dose, since they allow symptomatic and inflammatory control; however, they do not prevent joint destruction and can even flare-up the IBD, causing ulcers in the small intestine and colon, which can make the occasional use of corticosteroids necessary.

Sulfasalazine and 5-aminosalicylic acid are useful in the control of IBD and arthritis, particularly in UC, whereas, in the CD arthritis, the results are controversial. These drugs have no effect on progression of aggressive arthritis, and its use in axial involvement is marginal. Furthermore, they do not prevent the development of IBD in patients with SpA. Methotrexate, azathioprine, 6-mercaptopurine, cyclosporine and leflunomide can have a beneficial effect on peripheral arthritis and other extra-intestinal manifestations, although there are no controlled studies demonstrating their effectiveness.[18] Immunobiological agents, such as TNF-alpha blockers, improve the intestinal manifestations of patients with SpA and IBD, particularly infliximab and adalimumab in CD.[19] Currently, infliximab is the drug of choice for treating patients with active AS associated with IBD, being approved for use in children and adults with IBD. Other TNF-alpha blockers, such as etanercept, which is directed against the TNF receptor, do not work well in IBD, while golimumab and certolizumab have shown good results.[20] Given the pathogenicity relationship between intestinal and joint inflammation in SpA, other biological agents seem to be promising, such as blockers of interleukins (IL-10, IL-11, IL-6), adhesion molecules, kinases and integrins (alpha-4 and alpha-4 beta-7), as well as TLR modulation.[21] The use of probiotics to modulate the gut flora using bacteria and their products has been proposed for patients with persistent arthralgia in the early stages of the disease before chronic damage is established via IFN-mediated response induced by TLR9.[22]

## REFERENCES

1.  De Vos M, Mielants H, Cuvelier C, Elewaut A, Veys E. Long-term evolution of gut inflammation in patients with spondyloarthropathy. Gastroenterology 1996; 110:1696-703.

2.  Leirisalo-Repo M, Turunen U, Stenman S, Helenius P, Seppälä K. High frequency of silent inflammatory bowel disease in spondylarthropathy. Arthritis Rheum 1994; 37:23-31.

3.  Repiso A, Alcántara M, Muñoz-Rosas C, Rodríguez-Merlo R, Pérez-Grueso MJ, Carrobles JM et al. Extraintestinal manifestations of Crohn's disease: prevalence and related factors. Rev Esp Enferm Dig 2006; 98:510-7.

4.  Paredes JM, Barrachina MM, Román J, Moreno-Osset E. Joint disease in inflammatory bowel disease. Gastroenterol Hepatol 2005; 28:240-9.

5.  Scarpa R, del Puente A, DíArienzo A, di Girolamo C, della Valle G, Panarese A et al. The arthritis of ulcerative colitis: clinical and genetic aspects. J Rheumatol 1992; 19:373-7.

6.  Turkcapar N, Toruner M, Soykan I, Aydintug OT, Cetinkaya H, Duzgun N et al. The prevalence of extraintestinal manifestations and HLA association in patients with inflammatory bowel disease. Rheumatol Int 2006; 26:663-8.

7.  Salvarani C, Vlachonikolis IG, van der Heijde DM, Fornaciari G, Macchioni P, Beltrami M et al. Musculoskeletal manifestations in a population-based cohort of inflammatory bowel disease patients. Scand J Gastroenterol 2001; 36:1307-13.

8.  Holden W, Orchard T, Wordsworth P. Enteropathic arthritis. Rheum Dis Clin North Am 2003; 29:513-30.

9.  Gravallese EM, Kantrowitz FG. Arthritic manifestations of inflammatory bowel disease. Am J Gastroenterol 1988; 83:703-9.

10. Orchard TR, Wordsworth BP, Jewell DP. Peripheral arthropathies in inflammatory bowel disease: their articular distribution and natural history. Gut 1998; 42:387-91.

11. de Vlam K, Mielants H, Cuvelier C, De Keyser F, Veys EM, De Vos M. Spondyloarthropathy is underestimated in inflammatory bowel disease: prevalence and HLA association. J Rheumatol 2000; 27:2860-5.

12. Weiner SR, Clarke J, Taggart N, Utsinger PD. Rheumatic manifestations of inflammatory bowel disease. Semin Arthritis Rheum 1991; 20:353.

13. Dekker-Saeys BJ, Meuwissen SG, Van Den Berg-Loonen EM, De Haas WH, Agenant D, Tytgat GN. Ankylosing spondylitis and inflammatory bowel disease. II. Preva- lence of peripheral arthritis, sacroiliíte, and ankylosing spondylitis in patients suffering from inflammatory bowel disease. Ann Rheum Dis 1978; 37:33-5.

14. Palm O, Moum B, Ongre A, Gran JT. Prevalence of ankylosing spondylitis and other spondyloarthropathies among patients with inflammatory bowel disease: a population study (the IBSEN study). J Rheumatol 2002; 29:511-5.

15. Wordsworth P. Arthritis and inflammatory bowel disease. Rheumatol Rep 2000; 2:87-8.

16. De Keyser F, Elewaut D, De Vos M, De Vlam K, Cuvelier C, Mielants H et al. Bowel inflammation and the spondyloarthropathies. Rheum Dis Clin North Am 1998; 24:785-813, ix-x.

**17.** Török HP, Glas J, Gruber R, Brumberger V, Strasser C, Kellner H et al. Inflammatory bowel disease-specific autoantibodies in HLA-B27-associated spondyloarthropathies: increased prevalence of ASCA and pANCA. Digestion 2004; 70:49-54.

**18.** Padovan M, Castellino G, Govoni M, Trotta F. The treatment of the rheumatological manifestations of the inflammatory bowel diseases. Rheumatol Int 2006; 26:953-8.

**19.** Van den Bosch F, Kruithof E, De Vos M, De Keyser F, Mielants H. Crohn's disease as- sociated with spondyloarthropathy: effect of TNF-alpha blockade with infliximab on articular symptoms. Lancet 2000; 356:1821-2.

**20.** Behm BW, Bickston SJ. Tumor necrosis factor-alpha antibody for maintenance of remission in Crohn's disease. Cochrane Database Syst Rev 2008; CD006893.

**21.** Van Assche G, Vermeire S, Rutgeerts P. Focus on mechanisms of inflammation in inflammatory bowel disease sites of inhibition: current and future therapies. Gastroenterol Clin North Am 2006; 35:743-56.

**22.** Karimi O, Peña AS. Indications and challenges of probiotics, prebiotics, and synbi- otics in the management of arthralgias and spondyloarthropathies in inflammatory bowel disease. J Clin Gastroenterol 2008; 42(Suppl.3):S136-41.

# HEMATOLOGICAL MANIFESTATIONS IN INFLAMMATORY BOWEL DISEASE

RODOLFO DELFINI CANÇADO

## INFLAMMATORY BOWEL DISEASE (IBD) BACKGROUND

Inflammatory bowel disease (IBD) is a group of intestinal disorders that share a variety of symptoms such as chronic or recurrent diarrhea, abdominal pain, fever and anemia. Two major forms of IBD can be distinguished: Crohn's disease (CD) and ulcerative colitis (UC). CD can affect any part of the gastrointestinal (GI) tract although it generally starts in the terminal ileum, while UC is restricted to the colon and rectum.[1,2]

Although the GI tract is mainly affected in most cases of IBD, both UC and CD are systemic disorders that are often associated with extra-intestinal manifestations of the skin (pyoderma gangrenosum, erythema nodosum), joints (arthralgia, peripheral arthritis, or spondyloarthritis) or eyes (uveitis or conjunctivitis). Furthermore, IBD runs together with chronic inflammatory diseases of other tissues such as psoriasis, primary sclerosing cholangitis, and ankylosing spondylitis.[1,2]

## PREVALENCE OF ANEMIA IN IBD

Anemia is one of the most common systemic complications of IBD. Prevalence rates between 6-73% have been described for CD and 3-74% for UC. This wide range may be due to different criteria (hemoglobin cut-off levels) and different study populations (e.g. hospitalized patients *versus* outpatients). Although the incidence of anemia seems to be decreasing over the past few years, particularly due to the more efficient therapy with fewer adverse events, it is still the most common systemic complication of IBD.[1-6]

### THE MULTIFACTORIAL ORIGIN OF ANEMIA IN IBD

The most common cause of anemia in IBD is iron deficiency (ID), which is found in up to 90% of IBD patients. The underlying cause of ID is considered to be primarily due to chronic intestinal blood loss through ulcerations, which leads to a negative iron balance and thereby to the development of iron-deficiency anemia (IDA), usually associated with dietary restrictions and malabsorption (in part as a result of inflammation).[1-6]

Furthermore, mediators of intestinal inflammation such as tumor necrosis factor (TNF)-α or interferon (INF)-δ may also affect erythropoiesis and iron metabolism, leading to anemia of chronic disease (ACD). Prevalence rate of ACD in IBD patients vary between 11% and 42%. IDA and ACD are frequently combined in this type of patient.[3,7-9]

Many other causes of anemia may exist in IBD but are generally less frequent, such as vitamin $B_{12}$ deficiency, particularly associated with extensive ileal resection in patients with CD; folic acid deficiency, various pharmacologic drugs that are used for treatment of IBD (e.g. azathioprine, mesalazine) may interfere with erythropoiesis; and hemolytic anemia due to a possible shared autoimmune mechanism of anemia. All the conditions that may contribute to anemia frequently overlap (Table 11.1).[1,4,6]

### CLINICAL CONSEQUENCES AND THE IMPACT OF ANEMIA ON THE QUALITY OF LIFE OF IBD PATIENTS

The signs and symptoms induced by anemia are dependent upon the degree of anemia and the rate at which it has progressed, as well as the oxygen demands of the patient (Table 11.2). Symptoms are much less likely with anemia that progresses slowly, because there is time for multiple homeostatic forces to adjust to a reduced oxygen carrying capacity of the blood.[1-6]

**Table 11.1** Causes of anemia in patients with IBD

| Etiology of anemia | |
|---|---|
| Common | Iron deficiency anemia (IDA) |
| | Anemia of chronic disease (ACD) |
| Occasional | Vitamin B12 deficiency |
| | Folate deficiency |
| | Drug-induced (sulfasalazine, thiopurines) |
| | Renal insufficiency |
| | Alcoholism |
| | Hypothyroidism |
| Exceptional | Hemolytic anemia |
| | Hemoglobinopathies |
| | Bone marrow diseases (leukemias, lymphomas, myelodysplastic syndrome) |
| | Bone marrow infiltrations by cancer (prostate, breast, etc) |
| | Aplastic anemia (often drug-induced) |

**Table 11.2** Common symptoms and signs of anemia

| |
|---|
| Fatigue, lassitude, muscle cramps |
| Physical weakness |
| Paleness, brittle nails, pica |
| Shortness of breath, exertional dyspnea, dyspnea at rest |
| Palpitations, tachycardia, postural dizziness |
| Headache, vertigo, tinnitus |
| Hypomenorrhea/amenorrhea |
| Reduced libido |
| Sleeping disorder, depression, lethargy, confusion, syncope |
| Persistent hypotension and potentially life-threatening complications such as congestive failure, angina, arrhythmia, and/or myocardial infarction, and death |

Anemia also reduces the ability to perform normal daily activities. Chronic fatigue is a common symptom of IBD with anemia and associated with significant physical, emotional, psychological, and social consequences, with virtually every aspect of daily life being affected.

The repercussion of anemia on quality of life (QoL) in both general patients and specifically in patients with IBD is substantial. Moreover, anemia may impair quality of life, cognitive function and the ability to work even in the absence of specific symptoms.[4-6]

Several studies in anemic patients with malignancies and renal insufficiency provide evidence that a correction of anemia is accompanied with an improvement in energy and activity levels as well as overall QoL. Interestingly, it appears that the largest improvement in QoL occurs when Hb levels increase from 11 to 12 g/dL.[1-6]

Therefore, anemia in IBD is not just a laboratory marker but a pathological condition which needs a specific diagnostic and therapeutic approach.

## DIAGNOSIS OF ANEMIA IN PATIENTS WITH IBD

Anemia workup should be initiated if hemoglobin (Hb) levels are below normal. The lowest Hb levels below which anemia is defined as present are those proposed by the World Health Organization (non-pregnant women, <12.0 g/dL; men <13.0 g/dL).[10] The minimum workup includes serum ferritin, transferrin saturation and C-reactive protein (CRP). More extensive workup (including vitamin $B_{12}$, folic acid, haptoglobin, lactate dehydrogenase, creatinine, reticulocyte and a differential white blood cell count) should be performed if these investigations do not identify the cause of anemia, or if a certain therapeutic intervention is unsuccessful.

Most patients with IBD have mild to moderate anemia (Hb above 10.0 g/dL), but in the presence of bleeding episodes, Hb concentration may decrease further. In order to provide the appropriate treatment, it is essential to distinguish between predominantly IDA and IDA with ACD.[4-6]

While ACD is mostly normochromic and normocytic, IDA or IDA/ACD more frequently presents as microcytic and hypochromic anemia. Other aids to the differentiation between the two conditions include quantification of reticulocyte Hb and the percentage of hypochromic red cells, which indicate the availability of iron for erythroid progenitors, as well as determination of serum hepcidin. [4-6]

ACD without ID is uncommon and only seen after excessive intravenous iron replacement therapy. It is characterized by typical changes of body iron homeostasis. While serum ferritin (the iron storage protein) levels are low (<30 µg/L) in patients with IDA, they are normal or increased in patients with ACD. This is due to two factors. One is that increased ferritin levels reflect iron retention within monocytes and macrophages; the other is that ferritin expression is induced by inflammation and, therefore, ferritin levels do not exactly reflect the amount of stored iron in patients with IBD unlike what is seen in subjects without inflammation. This is the reason why the international guidelines recommend using 100 µg/L as the lower ferritin cut-off in active IBD (see Table 11.3 and Figure 11.1). [4-6]

Hemoglobin, serum ferritin and CRP should be used for laboratory screening. For patients in remission or mild disease, measurements should be performed every 6 to 12 months. In outpatients with active disease such measurements should be performed at least every 3 months. Patients at risk for vitamin $B_{12}$ or folic acid deficiency (e.g., small bowel disease or resection) need proper surveillance. Serum levels of vitamin $B_{12}$ and folic acid should be measured at least annually, or if macrocytosis is present.[4]

| Table 11.3 Diagnosis of anemia in IBD | |
|---|---|
| **Type of anemia** | **Definition** |
| Iron deficiency anemia | In patients with no evidence of inflammation: anemia* and serum ferritin < 30 µg/L or TSAT < 20% |
| | In patients with inflammation: anemia* and serum ferritin < 100 µg/L or TSAT < 20% |
| Anemia of chronic disease | In patients with inflammation: anemia* and serum ferritin ≥ 100 µg/L or TSAT < 20% |

#Absence of biochemical (CRP, leukocyte count) or clinical evidence (diarrhea, hematochezia, endoscopic findings) of inflammation.

*WHO definitions of anemia: men Hb <13.0 g/dL, non-pregnant women Hb <12.0 g/dL; TSAT, transferrin saturation.

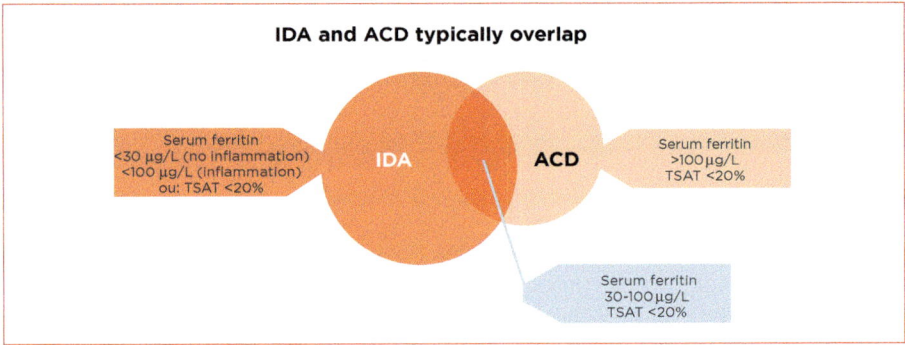

**Figure 11.1** Iron deficiency anemia (IDA) and anemia of chronic disease (ACD) typically overlap.

TSAT, transferrin saturation.

## THE ROLE OF TREATMENT OF IBD IN THE CORRECTION OF ANEMIA

A general correlation exists between disease activity and the depth of anemia.[9] Active disease can cause anemia because of multiple aforementioned factors. Therefore, the most important measure for IBD anemia treatment is the treatment of the underlying disease. Although apparently obvious, sometimes this step is missed in actual clinical practice. Moreover, the long term effect to alleviate anemia depends on whether bowel inflammation itself can be adequately treated. Every effort to accomplish this has to be undertaken in order to preclude recurrent anemia.[4-6,9]

## THE GOALS OF ANEMIA TREATMENT IN PATIENTS WITH IBD

The goals of anemia treatment in patients with IBD are: correct anemia and replenish body stores (increase Hb level and serum ferritin above lower threshold of normal, respectively), prevent a further fall in Hb, avoid use of blood transfusion, relieve symptoms related to anemia and improve quality of life.[1-6]

## WHEN TO START IRON SUPPLEMENTATION IN IBD ANEMIC PATIENTS?

It should not be assumed that some level of anemia is a normal finding in IBD patients and consequently need not be treated. Iron supplementation should be started for all patients with anemia (Hb <13 g/dL in males, and <12 g/dL in females), if detected.[1-6] The decision to initiate therapy depends on symptoms, etiology, and severity of anemia, rate of change, comorbidity, and potential adverse effects of therapy.[4]

## HOW FAR AND WHEN TO STOP IRON SUPPLEMENTATION IN IBD ANEMIC PATIENTS?

All patients should receive enough iron supplementation to correct anemia (Hb >13 g/dL in males, and >12 g/dL in females) and replenish body stores (serum ferritin >30 μg/L or >100 μg/L with or without inflammation, respectively; and TSAT > 20%).[1-6]

An appropriate response to iron therapy is defined as Hb concentration increase by at least 2 g/dL within 2-4 weeks or reaching normality within 4 weeks of treatment. If the therapeutic response is inappropriate, treatment should be checked (in terms of dose taken and compliance), intensified (e.g. oral iron switched to IV iron treatment), changed (addition of erythropoietic agents), or the cause of anemia should be reevaluated (possibly with the assistance of a hematologist).[4-6]

Therefore, treatment response should be evaluated on a regular basis with Hb measurement every 4 weeks in asymptomatic patients and sooner in symptomatic patients in order to adjust treatment accordingly. When monitoring oral iron supplementation, serum ferritin >100 μg/L indicates appropriate iron stores. On the other hand, serum ferritin is not useful for monitoring the effect of intravenous (IV) iron supplementation as the values will be "falsely" high. On these situations, we should use TSAT (>50% indicates iron overload), reminding that the risk of iron overload is considered very low in a population with ongoing blood loss.[4-6]

## HOW TO TREAT IBD ANEMIC PATIENTS: ORAL OR PARENTERAL IRON THERAPY?

Oral iron supplementation is usually the first choice for the treatment of IDA because of its effectiveness, relatively safety profile and low cost. But unfortunately in many ID conditions, oral iron is a less than ideal treatment mainly because of gastrointestinal adverse events (AEs) as well as a long course needed to resolve anemia and replenish body iron stores. Lack of compliance with a prescribed course of oral iron is common, and even in compliant patients, insufficient duodenal iron absorption fails to compensate for the iron need, particularly in the presence of ongoing blood losses or inflammatory conditions.

Intestinal absorption of oral iron in patients with IBD is compromised due to many causes such as disease activity (increased hepcidin levels may inhibit iron absorption from the GI tract), aggravation of the intestinal inflammation from the non-absorbed iron (which potentially increases IBD activity), slow response (compared with IV iron), particularly in patients with persistent blood loss exceeding the capacity of intestinal absorption of iron, and the high prevalence of gastrointestinal adverse events leading to poor tolerance and compliance to iron oral therapy.[1,4,5]

IBD patients with anemia rarely have an iron deficit less than 1,000 mg, demonstrating a clear limitation in oral iron because as only 10-20 mg of oral iron can be absorbed a day. [1,4,5,11-13]

Therefore, since absorption and efficacy of oral iron are compromised in patients with IBD, oral iron, if used, should be recommended at low doses (e.g. 50-100 mg of elemental iron daily), response and tolerance should be monitored, and treatment changed to IV if necessary.

The efficacy and safety of IV iron, particularly iron sucrose, for the treatment of IDA in the general population has been demonstrated in numerous studies.[14-18] Although the experience with IV in IBD is more limited, it is similarly encouraging.

In 2000, iron sucrose (IS) (Venofer) was approved in the United States, although long being used in Europe. By far, the largest experience in the published literature is with this formulation. IS can be safety administered as a 30-45 minute infusion in doses of 100 or 200 mg (diluted in 100 or 250 mL of 0.9% NaCl, respectively) and the maximum weekly dose should not exceed 600 mg. If higher-than-recommended doses are not infused, AEs are rarely observed. The cumulative dose of IV iron that is needed to replenish stores can be calculated using Ganzoni's formula: Iron deficit (mg) = weight (kg) *versus* [target Hb − actual Hb (g/dL)] *versus* 0.24 + deposit iron (500 mg).

It has been shown that IV iron sucrose is more effective (in terms of faster and prolonged response) than oral iron supplements, has a better safety profile which might positively influence the compliance of IBD patients, and is an important alternative option to blood transfusion in a variety of clinical as well as surgery settings leading to significant reduction of blood transfusion requirement. The incidence of serious life-threatening anaphylaxis with IS is 0.002% *versus* 0.6-2.3% and 0.04% with high-molecular-weight iron dextran and ferric gluconate, respectively. Moreover, fatal hypersensitivity reactions have not been reported with IS.[14]

The initial therapeutic strategy of IDA in IBD patients would be based on the level of Hb. Patients with hemoglobin >10 g/dL would initiate treatment with oral iron, while in those with levels ≤10 g/dL, the IV route would be of choice. IV iron should also be prescribed to patients with hemoglobin >10 g/dL when intolerance to the oral formulation is present. In summary, the established indications for the use of IV iron are: moderate to severe anemia (Hb <10 g/dL), need for quick recovery in mild anemia, intolerance or inappropriate response to oral iron, severe intestinal disease activity, concomitant therapy with an erythropoietic agent and patient preference.[14-18]

Although in IBD iron sucrose is the most used IV formulation, there are other new IV iron preparations which theoretically could be used, with an extremely low incidence of AEs, and in particular severe AEs, but data in the specific IBD population is still lacking.[14] The experience with low-molecular-weight iron dextran is rather more extensive and encouraging and also a new molecule, iron carboxymaltose, merits mention because its pharmacokinetic characteristics and preliminary clinical experience seems very promising, and data was obtained directly from an IBD population.[14-16]

## ROLE OF ERYTHROPOIETIN IN THE TREATMENT OF ANEMIA IN IBD

As previously mentioned, anemia in IBD patients results primarily from ID because of chronic intestinal blood loss. However, intestinal inflammation is mediated by overproduction of cytokines, which may contribute to the development of anemia in chronic disease, accompanied by inadequate erythropoietin (EPO) production. Thus, IBD-associated anemia is a unique example of a combination of IDA and ACD. Since it was first used in chronic renal failure, recombinant human EPO has been shown to be effective to treat anemia that accompanies several chronic diseases. During the last few years, several studies have evaluated the efficacy of EPO in IBD patients, reporting encouraging results. Nevertheless, as the cost of EPO is much higher than the cost of IV iron, the latter formulation should be considered first-line therapy in patients with severe anemia, and EPO therapy should be considered only for patients with low EPO levels or who are unresponsive to IV iron. One must not forget to exclude or correct other causes of anemia in IBD patients before administering EPO. Finally, EPO should be reserved for patients in whom aggressive management of IBD (including immunosuppressive therapy) has not suppressed inflammation, which underlines the idea that EPO is an adjunct – and not an alternative – to appropriate treatment of IBD.[19-21]

In order to optimize the effect of EPO, this drug should always be combined with IV iron supplementation, because functional ID, defined as an inappropriate availability of iron for erythropoiesis despite normal body iron stores, is likely to develop. In the particular case of Crohn's disease, folic acid and vitamin $B_{12}$ status should also be checked and deficiencies adequately corrected.[19-21]

### ADMINISTRATION OF VITAMIN SUPPLEMENTATION

Measurements of serum folate and vitamin $B_{12}$ levels have many limitations and are not always reliable. In the presence of macrocytosis or unexplained anemia, especially in patients with ileal resection, serum homocysteine and methylmalonic acid levels should be measured.[4]

### ROLE OF BLOOD TRANSFUSION IN THE TREATMENT OF ANEMIA IN IBD

Blood transfusions are widely used as an immediate intervention for rapid correction of severe or life-threatening anemia. The decision on whether to administer blood should not, therefore, be based only on Hb levels, but should also take into account clinical status (including the rate of bleeding, hemodynamic status, age, co-morbidities) and are best judged by the treating physician.[4,22]

Management should be directed at diagnosing and stopping intestinal bleeding. Blood transfusion is no substitute for the treatment of IDA with IV iron, possibly in combination with EPO.[4]

## NEW THERAPEUTIC STRATEGIES IN THE TREATMENT OF ANEMIA IN IBD

Bergamaschi et al.[23] presented data on the therapeutic effects of anti-TNF-α treatment with infliximab on the resolution of anemia in a subgroup of patients with Crohn's disease (CD). They found that patients who responded to treatment started to have an improvement in their anemia within 2 weeks after the first infusion of infliximab. This was paralleled by a significant improvement of the CD activity index. Patients responding to therapy were found to have an increase in endogenous EPO levels over time while ferritin levels decreased, which may indicate iron mobilization and also reflected by a slight increase of transferrin saturation. The data suggest that infliximab neutralizes the inhibitory effects of TNF-α on EPO production, and increases the availability of iron for erythropoiesis. Mechanistically, this may be explained by reduced cytokine-induced formation of ferritin and hepcidin and subsequent improvement of intestinal iron absorption and iron release from macrophages via ferroportin-mediated iron export.

We certainly need more detailed information on the underlying mechanisms by which infliximab improves anemia in many, but not all, IBD patients, but we should also consider whether other therapies may be effective. For example, since interleukin-6 is the major inflammation-driven inducer of hepcidin identified so far, therapy with anti-interleukin-6 may be an even more promising approach to resolve anemia in IBD patients. New therapeutic strategies may also emerge from our expanding knowledge on the pathophysiology of ACD and the underlying regulatory circuits of iron homeostasis. Such new therapies may include the neutralization of hepcidin to overcome the retention of iron within monocytes/macrophages, modifiers of erythropoietin and/or erythropoietin receptor sensitivity, and new hormones and cytokines/anti-cytokines, which can effectively stimulate erythropoiesis and/or counterbalance iron restriction under inflammatory conditions.[1-6]

## CONCLUSION

Anemia is quite common in inflammatory bowel disease (IBD) and some degree of anemia should not be considered a normal finding in IBD patients, which consequently does not need to be treated. On the contrary, anemia should be aggressively diagnosed, investigated, and treated.

Anemia in IBD is often complex and likely to be multifactorial in origin, frequently being the result of a combination of IDA and ACD.

Although in many cases anemia parallels the clinical activity of the disease, many patients in remission have anemia, and iron, vitamin $B_{12}$ and/or folic acid deficiency.

Anemia, as well as iron deficiency without anemia, has important consequences in the clinical status and quality of life of the patient.

Oral iron is poorly absorbed because of the inflammatory status associated with IBD and can lead to gastrointestinal intolerance and failure of treatment. The use of oral iron preparations is limited to patients with mild anemia who tolerate oral iron therapy.

Intravenous route of iron administration is the preferable way to treat IDA in patients with IBD and should be considered, since it does not carry the risk of potentiating IBD symptoms and provides fast correction of anemia and iron depletion.

## REFERENCES

1. Weiss G, Gasche C. Pathogenesis and treatment of anemia in inflammatory bowel disease. Haematologica 2010; 95(2):175-1.

2. Kulnigg S, Gasche C. Systematic review: managing anaemia in Crohn's disease. Aliment PharmacolTher 2006; 24:1507-23.

3. Gasche C, Lomer MC, Cavill I, Weiss G. Iron, anaemia, and inflammatory bowel diseases. Gut 2004;53(8):1190-7.

4. Gasche C, Berstad A, Befrits R, Beglinger C, Dignass A, Erichsen K, Gomollon F, Hjortswang H, Koutroubakis I, Kulnigg S et al. Guidelines on the diagnosis and management of iron deficiency and anemia in inflammatory bowel diseases. Inflamm Bowel Dis 2007;13:1545-53.

5. Gisbert JP, Gomollón F. Common misconceptions in the diagnosis and management of anemia in inflammatory bowel disease. Am J Gastroenterol 2008;103:1299-1307.

6. Gomollon F, Gisbert JP. Anemia and inflammatory bowel diseases. World J Gastroenterol 2009;15(37):4659-65.

7. Weiss G, Goodnough LT. Anemia of Chronic Disease. N Engl J Med 2005; 352:1011-23.

8.  Nemeth E, Tuttle MS, Powelson J, Vaughn MB, Donovan A, Ward DM et al. Hepcidin regulates cellular iron efflux by binding to ferroportin and inducing its internalization. Science. 2004;306(5704):2090-3.

9.  Semrin G, Fishman DS, Bousvaros A, Zholudev A, Saunders AC, Correia CE, Nemeth E, Grand RJ, Weinstein DA. Impaired intestinal iron absorption in Crohn's disease correlates with disease activity and markers of inflammation. Inflamm Bowel Dis 2006;12:1101-6.

10. WHO, UNICEF, UNU. Iron deficiency anaemia: assessment, prevention, and control. A guide for programme managers. Geneva: World Health Organization, 2001. WHO/NHD/01.3.

11. De Silva AD, Mylonaki M, Rampton DS. Oral iron therapy in inflammatory bowel disease: usage, tolerance, and efficacy. Inflammatory Bowel Diseases 2003; 9:316-20.

12. Erichsen K, Ulvik RJ, Nysaeter G, Johansen J, Ostborg J, Berstad A, Berge RK, Hausken T. Oral ferrous fumarate or intravenous iron sucrose for patients with inflammatory bowel disease. Scand J Gastroenterol 2005;40:1058-65.

13. Schröder O, Mickisch O, Seidler U, de Weerth A, Dignass AU, Herfarth H, Reinshagen M, Schreiber S, Junge U, Schrott M et al. Intravenous iron sucrose versus oral iron supplementation for the treatment of iron deficiency anemia in patients with inflammatory bowel disease – a randomized, controlled, open-label, multicenter study. Am J Gastroenterol 2005;100:2503-9.

14. Auerbach M, Ballard H, Glaspy J. Clinical update: intravenous iron for anaemia. Lancet 2007;369:1502-4.

15. Bodemar G, Kechagias S, Almer S, Danielson BG. Treatment of anaemia in inflammatory bowel disease with iron sucrose. Scand J Gastroenterol 2004;39:454-8.

16. Kulnigg S, Stoinov S, Simanenkov V, Dudar LV, Karnafel W, Garcia LC, Sambuelli AM, D'Haens G, Gasche C. A novel intravenous iron formulation for treatment of anemia in inflammatory bowel disease: the ferric carboxymaltose (FERINJECT) randomized controlled trial. Am J Gastroenterol 2008;103:1182-92.

17. Lindgren S, Wikman O, Befrits R, Blom H, Eriksson A, Grännö C et al. Intravenous iron sucrose is superior to oral iron sulphate for correcting anaemia and restoring iron stores in DII patients: a randomized, controlled, evaluator-blind, multicentre study. Scand J Gastroenterol 2009;44(7):838-45.

18. Kulnigg S, Teischinger L, Dejaco C, Waldhör T, Gasche C. Rapid recurrence of DII-associated anemia and iron deficiency after intravenous iron sucrose and erythropoietin treatment. Am J Gastroenterol 2009;104(6):1460–7.

19. Gasché C, Dejaco C, Waldhoer T, Tillinger W, Reinisch W, Fueger GF et al. Intravenous iron and erythropoietin for anemia associated with Crohn disease. A randomized, controlled trial. Ann Intern Med 1997;126(10):782–7.

20. Tsiolakidou G, Koutroubakis IE. Stimulating erythropoiesis in inflammatory bowel disease associated anemia. World J Gastroenterol 2007;13:4798–806.

21. Kulnigg S, Teischinger L, Dejaco C, Waldhör T, Gasche C. Rapid recurrence of DII-associated anemia and iron deficiency after intravenous iron sucrose and erythropoietin treatment. Am J Gastroenterol 2009;104:1460–7.

22. Marik PE, Corwin HL. Efficacy of red blood cell transfusion in the critically ill: a systematic review of the literature. Crit Care Med 2008;36(9):2667–74.

23. Bergamaschi G, Di Sabatino A, Albertini R, Ardizzone S, Biancheri P, Bonetti E et al. Prevalence and pathogenesis of anemia in inflammatory bowel disease. Influence of anti-tumor necrosis factor-alpha treatment. Haematologica 2010;95(1):199–205.

# THROMBOEMBOLIC MANIFESTATIONS OF INFLAMMATORY BOWEL DISEASE

ELBIO ANTONIO D'AMICO

AUDREY KRÜSE ZEINAD VALIM

PATRÍCIA LIMA JUNQUEIRA

## INTRODUCTION

Venous and arterial thrombotic events are important causes of morbidity and mortality in patients with ulcerative colitis (UC) and Crohn's disease (CD).[1] The association between inflammatory bowel disease (IBD) and venous and arterial thrombosis has been known since 1936, when Bargen and Barken, studying 1,500 patients with chronic ulcerative disease described the occurrence of episodes of arterial thrombosis and thrombophlebitis, which were sufficiently extensive and severe in 1.2% of cases.[2]

To date, the mechanisms that connect IBD to thrombotic events and the best thromboprophylactic measures for these patients are not precisely understood. It is speculated that thrombotic events could be implicated in the pathogenesis of IBD.[3]

## EPIDEMIOLOGY

Although the risk of thromboembolism is reported as 3 to 4 times higher in patients with UC or CD, in comparison with the general population,[1,3,4] the actual incidence of these events in patients with these diseases is not well established. The rates vary according to the type of study, and range from

1 to 10% in clinical studies and from 39 to 41% in necropsy studies.[4,5] The fact that many of these studies were conducted at a time when treatment was more invasive, with a higher frequency of surgical procedures, and without the use of mechanical and/or pharmacological thromboprophylactic measures, explains the wide variation in incidence of thromboembolism in these analyzes.[3]

The risk of recurrent venous thromboembolism (VTE) (deep vein thrombosis and pulmonary embolism) is described between 10 and 13%, regardless of the clinical treatment, with mortality rates of 8 to 25%.[1,6] The association between IBD and puerperal period implies higher risk of VTE, probably due to the hypercoagulable state of blood that develops during pregnancy and continues until the puerperium. However, while the predictive value calculated for mothers with CD was 6.1, predictive value in UC was 8.[4,1]

As for arterial thrombotic events, a study with 17,487 patients with IBD and 69,948 matched controls showed that these patients had increased risk for acute mesenteric ischemia (RR = 11.2), with no increased risk for acute myocardial infarction (AMI) and transient ischemic attack. However, women younger than 40 years showed higher risk for ischemic stroke (RR = 2.1), while those aged between 40 and 59 years had a higher risk for myocardial infarction (RR = 1.6). The risk of any arterial occlusive event was the same for CD and UC.[6]

Although the reported risk of thromboembolic complications is equal both in CD and in UC,[4] other authors report a higher frequency in UC. A population-based cohort study revealed incidence of VTE at 40:10,000 persons per year in CD, and 50:10,000 persons per year in UC.[3]

There is an inverse relationship between the age at diagnosis of IBD and the risk of VTE. Thus, in patients aged less than 40 years hospitalized with IBD, the predictive value for VTE was 4.3 in UC and 3.3 in CD compared to hospitalized patients without IBD. In patients older than 40 years, the presence of UC was associated with a slightly higher risk (predictive value - PV = 1.5), whereas this was not observed in CD.[1]

The phase and extent of IBD also seem to have influence on the occurrence of thromboembolic episodes. Thus, in the active phase of the disease, vaso-occlusive events are seen more often, although 1/3 of these complications may occur during the quiescent period.[1,3] A study that analyzed 98 patients with IBD and thromboembolism showed that most CD patients had colonic disease, while those with UC had extensive disease.[3]

## PATHOGENESIS

Etiology of thrombotic events in IBD is multifactorial. The vaso-occlusive events are considered as resulting from one or more defects occurring in the body's defense mechanisms against thrombosis. The coagulation abnormalities have been classified under the heading of Virchow's triad, described by Rudolf Virchow in 1866. This triad can be summarized as:

- defects in blood flow, resulting in stasis;
- abnormalities in the normal balance between procoagulant and anticoagulant proteins, leading to activation of clotting proteins;
- endothelial abnormalities that end up in change of endothelial anticoagulant features to a procoagulant standard.[7]

Functional changes in the intestinal vasculature during acute inflammatory and chronic phases of CD and UC have been described, which could contribute to the increased risk of thromboembolism. Studies in chronically inflamed UC and CD lesions show significant changes in microvascular physiology and function compared with healthy vessels and intestinal areas not affected by IBD, besides increased microvascular strictures and significant reduction of perfusion, suggesting chronic intestinal ischemia.[8] These microvascular and perfusion changes could contribute to abnormal repair of the intestinal mucosa in CD and UC.[9]

The presence of activating factors, inherited and acquired, resulting in a hypercoagulable and prothrombotic state in IBD, has been investigated. Several acquired factors are described, including inflammation, surgery, prolonged immobilization, central venous catheter, dehydration, use of corticosteroids, oral contraceptives, smoking, presence of antiphospholipid antibodies and hyperhomocysteinemia.[1] Probably the most important is the inflammatory process. Its interaction with coagulation is well established, with inflammation activating the coagulation and the latter modulating the inflammatory activity.[10]

For this purpose, a study was conducted to investigate the incidence of thrombotic events in patients with inflammatory diseases, including UC, CD, celiac disease and rheumatoid arthritis. Patients with IBD had a higher risk for developing thromboembolic events; however, this was not observed in patients with other chronic inflammatory diseases, which led to the hypothesis that thromboembolic events correspond to a specific extraintestinal manifestation of IBD.[11]

There are few studies evaluating the presence of antiphospholipid antibodies in patients with IBD. Chiarantini et al. showed that approximately one third of these patients had positive tests for the presence of lupus anticoagulant and/or antiphospholipid antibodies, but apparently without any correlation with thrombotic tendencies.[12] High concentrations of homocysteine are most often found in patients with IBD than the general population.[13] However, this may be due to low plasma levels of folic acid, vitamin B6 and vitamin B12, which is a common finding in these patients. It has been shown that vitamin supplementation normalizes high plasma concentrations of homocysteine.[13]

There are also few studies on laboratory markers of inherited thrombophilia (G20210A mutation of the prothrombin gene, factor V Leiden, antithrombin deficiency, protein C deficiency and protein S deficiency), which showed no increased prevalence of these inherited abnormalities in patients with IBD compared to the general population.[1,14,15] Therefore, the vast majority of data points to a more important contribution of acquired prothrombotic factors, compared with congenital, to the development of vaso-occlusive events, particularly in the active phase of IBD.[1]

Other mechanisms could be implicated as responsible for the hypercoagulable state of IBD. Some authors report the presence of platelet hyperaggregability in the active phase of the disease,[12] and others report increased microparticles also during the acute phase.[13]

## CLINICAL SYMPTOMS

The vast majority of thrombotic events occurring in patients with CD or UC correspond to the vaso-occlusive venous episodes. DVT of the lower limbs and pulmonary embolism represents 75% of thrombotic events in these patients, the incidence of deep vein thrombosis being approximately 3 times higher than the incidence of pulmonary embolism.[1] The diagnosis of VTE in patients with IBD is challenging; they can be asymptomatic or present few specific signs and symptoms. Thus, the suspicion threshold should be reduced, with agile diagnostic procedures and early treatment, due to the high mortality of these events in this patient population.

Other locations of venous thrombosis are described, such as cerebral venous sinuses, retinal vein, portal venous system, vena cava and hepatic veins.[1] The association between UC and migratory thrombophlebitis is also important because of the possibility of etiological differential diagnosis.[3] Portal venous thrombosis has been diagnosed more frequently after the broader use of ultrasound and computed tomography. These vaso-occlusive processes affect most commonly the portal vein and the superior mesenteric vein, and are more seen after total retocolectomia with ileal pouch, which emphasizes the importance of surgery as a thrombotic risk factor.[4] These events are also common in patients dependent on corticosteroids and those not treated with antithrombotic heparin prophylaxis. It is estimated that portal vein thrombosis occurs in 6% of patients after total rectal colectomy and ileal pouch for IBD, and 4.8% of patients undergoing total colectomy of the superior mesenteric vein.[4] After limited ileocolic resections, venous thrombotic events of the portal system are uncommon.[4]

Episodes of arterial thrombosis are less common, but can affect arteries in the brain, retina, carotid, coronary, visceral organs, kidneys, aorta and extremities.[1,6] The most common events affect the small intestine, colon and lower limbs.[1,6] Arterial events in visceral organs are often initially interpreted as a clinical exacerbation of CD or UC. However, as they are associated with high mortality rate, episodes of significant abdominal pain associated with absence of abdominal signs and hypovolemia should be investigated quickly with abdominal angiography/computed tomography, in order to confirm the diagnosis, assess its extent and propose treatment measures.[1]

## TREATMENT AND CLINICAL COURSE

The treatment of VTE in patients with IBD is not different from the treatment used in the general population. However, treatment decisions are challenging for reasons that include lack of standardization, the risk of intestinal bleeding and the possibility of thrombotic recurrence or death if treatment is inadequate.[1,4]

The main question is regarding the duration of anticoagulant treatment with anti-vitamin K medications (warfarin). Theoretically, in cases with potentially reversible risk factors, treatment should be maintained until the aggravating or triggering factor is removed.[1] So it is important to reduce the severity of the inflammatory process with adequate pharmacological treatment, reduce the use of corticosteroids, avoid immobilization and hospitalizations, minimize the use of central venous catheters, discontinue use of oral contraceptives and smoking, as well as introducing proper vitamin replacement (B6, B12 and folate).[1] In the case of hospitalization, prophylactic use of low molecular weight heparin, classical heparin, fondaparinux or mechanical measures (intermittent pneumatic compression of the lower limbs and elastic stockings) is recommended when pharmacological prophylaxis is contraindicated.[16] Patients with venous thromboembolic event and temporary or reversible risk factor should be treated for at least 3 months.[17] In thrombosis of the portal venous system, it is assumed that a 6-month treatment duration is appropriate.[1] In cases of arterial thrombosis, treatment with thrombolytic or fibrinolytic agents, or surgical procedures (thromboembolectomy or arterial bypass) can be considered.[1]

## REFERENCES

1. Fabio FD, Lykoudis P, Gordon PH. Thromboembolism in inflammatory bowel disease: an insidious association requiring a high degree of vigilance. Sem Thromb Hemost 2011; 37(3):220-5.

2. Bargen JA, Barker NW. Extensive arterial and venous thrombosis complicating chronic ulcerative colitis. Arch Int Med 1936; 58(1):17-31.

3. Irving PM, Pasi KJ, Rampton DS. Thrombosis and inflammatory bowel disease. Clin Gastroenterol Hepatol 2005; 3:617-28.

4. Fabio FD, Obrand D, Satin R, Gordon PH. Intra-abdominal venous and arterial thromboembolism in inflammatory bowel disease. Dis Colon Rectum 2009; 52(2):336-42.

5. Zitomersky NL, Verhave M, Trenor CC. Thrombosis and inflamatory bowel disease: a call for improved awareness and prevention. Inflamm Bowel Dis 2011; 17:458-70.

6.   Ha C, Magowan S, Accortt NA, Chen J, Stone CD. Risk of arterial thrombotic events in inflammatory bowel disease. Am J Gastroenterol 2009; 104:1445-51.

7.   Dvorak HF, Rickles FR. Malignancy and hemostasis. In: Colman RW, Marder VJ, Clowes AW, George JN, Goldhaber SZ (eds.). Hemostasis and thrombosis basic principles and clinical practice. 5.ed. Philadelphia: Lippincott Williams & Wilkins, 2006.

8.   Hatoum OA, Binion DG. The vasculature and inflammatory bowel disease. Contribution to pathogenesis and clinical pathology. Inflamm Bowel Dis 2005; 11(3):304-13.

9.   Hatoum OA, Miura H, Binion DG. The vascular contribution in the pathogenesis of inflammatory bowel disease. Am J Physiol Heart Circul Physiol 2003; 285(5):H1791-6.

10.  Petäjä J. Inflammation and coagulation. An overview. Thrombosis Research 2011; 127(Suppl.2):S34-7.

11.  Miehsler W, Reinisch W, Vahc E, Osterode W, Tilinger W, Feichtenschlager T et al. Is inflammatory bowel disease an independent and specific risk factor for thromboebolism? Gut 2004; 53(4):542-8.

12.  Chiarantini E, Valanzano R, Liotta AA, Cellai AP, Fedi S, Ilari I et al. Hemostatic abnormalities in inflammatory bowel disease. Thrombosis Research 1996; 82(2):137-46.

13.  Danese S, Papa A, Saibeni S, Repici A, Malesci A, Vecchi M. Inflammation and coagulation in inflammatory bowel disease: the clot thickens. Am J Gastroenterol 2007; 102:174-86.

14.  Spina L, Saibeni S, Battaglioli T, Peyvandi F, Franchis RD, Vecchi M. Thrombosis in inflammatory bowel diseases: role of inherited thrombophilia. Am J Gastroenterol 2005; 100:2036-41.

15.  Tsiolakidou G, Koutroubakis IE. Thrombosis and inflammatory bowel disease – The role of genetic risk factors. World J Gastroenterol 2008; 14(28):4440-4.

16.  Gurts WH, Bergqvist D, Pineo GF, Heit JA, Samama CM, Lassen MR et al. Prevention of venous thromboembolism. American College of Chest Physicians evidence-based clinical practice guidelines. 8.ed. Chest 2008; 133:381S-453S.

17.  Kearon C, Kahn SR, Agnelli G, Goldhaber S, Raskob GE, Comerota AJ. Antithrombotic therapy for venous thromboembolic disease: American College of Chest Physicians evidence-based clinical practice guidelines. 8.ed. Chest 2008; 133:454-545.

# HEPATIC MANIFESTATIONS OF INFLAMMATORY BOWEL DISEASE

CLÁUDIA PINTO MARQUES SOUZA DE OLIVEIRA

PAULO LISBOA BITTENCOURT

## INTRODUCTION

Several hepatobiliary manifestations are observed in patients with inflammatory bowel disease (IBD) (Table 13.1), particularly primary sclerosing cholangitis (PSC), drug-induced hepatitis, non-alcoholic fatty liver disease (NAFLD) and gallstones. Liver enzyme abnormalities were found in about one third of patients with IBD. However, chronic liver disease is diagnosed in only 6% of patients, most often PSC.[1,2]

| Table 13.1 Hepatobiliary manifestations in IBD |
| --- |
| Hepatic amyloidosis |
| Liver abscess |
| Primary biliary cirrhosis |
| PSC-associated cholangiocarcinoma |
| Sclerosing cholangitis |
|   PSC |
| Small-duct (pericholangitis) |

*(continues)*

| **Table 13.1** Hepatobiliary manifestations in IBD |
| --- |
| Cholelithiasis |
| Extra-hepatic cholestasis |
| Choledocolithiasis |
| Acute pancreatitis |
| NAFLD |
| Malnutrition |
| Use of corticosteroids |
| IBD inflammatory activity |
| AIH, AIH/PSC overlap syndrome |
| Immunosuppressive-induced hepatitis B (reactivation) |
| Granulomatous hepatitis |
| Drug-induced hepatotoxicity: sulfasalazine, azathioprine, 6-mercaptopurine, methotrexate |
| infliximabe |
| Portal vein thrombosis |

PSC: primary sclerosing cholangitis; NAFLD: non-alcoholic fatty liver disease; IBD: inflammatory bowel disease; AIH: autoimmune hepatitis.

Gallstones, granulomatous hepatitis, liver abscess, portal vein thrombosis and hepatic amyloidosis are more frequently observed in Crohn's disease (CD), while PSC, autoimmune hepatitis (AIH) and AIH/PSC overlap syndrome are more prevalent in ulcerative colitis (UC).[1]

## PRIMARY SCLEROSING CHOLANGITIS

PSC is a chronic cholestatic liver disease of autoimmune etiology, characterized by inflammation and fibrosis of intra- and extra-hepatic bile ducts, with variable clinical course and slow progression to cirrhosis. The estimated prevalence of PSC in Europe is 2 to 7 cases per 100,000 habitants, and it is a relatively rare disease in Brazil. Its frequency is increased in patients with IBD, occurring in 2 to 7.5% of patients with UC and in 3.4% of patients with CD. In patients with PSC, in turn, UC and CD are observed in 60 to 80% and 13% of cases, respectively.

In these patients, the phenotype of IBD is different from that observed in patients with IBD without PSC. The following features are more frequently observed in patients with UC and PSC:

- oligosymptomatic course of disease;
- pancolitis with inflammatory activity more pronounced in proximal colonic segments;
- backwash ileitis;
- rectum paradoxically spared;
- increased risk of colorectal neoplasia;
- increased risk of inflammation of the ileal reservoir and peristomal varices in patients undergoing surgery.[3]

The initial manifestation of the disease ranges from nonspecific symptoms of fatigue, weakness and weight loss, to more characteristic symptoms of cholestatic jaundice, choluria, acholic stools and or itching or recurrent cholangitis. Approximately two thirds of cases are currently diagnosed by elevated liver enzymes found in routine examinations. The disease mainly affects men in adulthood; however, the finding of the disease in children is not uncommon.[4]

Patients with PSC usually present predominant elevation of alkaline phosphatase (ALP) and gamma-glutamyl transpeptidase (GGT). Aminotransferase levels may be increased, often 2-3 times the normal value. Hypergammaglobulinemia is observed in half of the patients at the expense of increased IgG. Autoantibodies, particularly anti-nuclear and anti-smooth muscle antibodies, and atypical pANCA may be observed, but they are not considered serologic markers of disease.

The presence of PSC should be investigated in all patients with IBD and unexplained cholestasis in the absence of antimitochondrial antibody (AMA). The gold standard for disease diagnosis is the finding of mural abnormalities and/or multiple diffuse strictures of the intra- and/or extra-hepatic biliary tree, alternating with normal or dilated segments (Figure 13.1) in the absence of secondary causes of sclerosing cholangitis (Table 13.2).

| **Table 13.2** Secondary causes of sclerosing cholangitis |
| --- |
| Familial immunodeficiency |
| Immunodeficiency, undefined |
| HIV-associated cholangiopathy: CMV, *Cryptosporidium*, idiopathic |
| Cystic fibrosis |
| Langerhans cell histiocytosis |
| Hodgkin's disease |
| Biliary surgery or trauma |
| Choledocolithiasis |
| Intra-arterial injection of fluxoridine |
| Ischemic injury of biliary tract (post-transplant/vasculitis) |
| IgG4-associated cholangitis/pancreatitis |

HIV: human deficiency virus; CMV: cytomegalovirus.

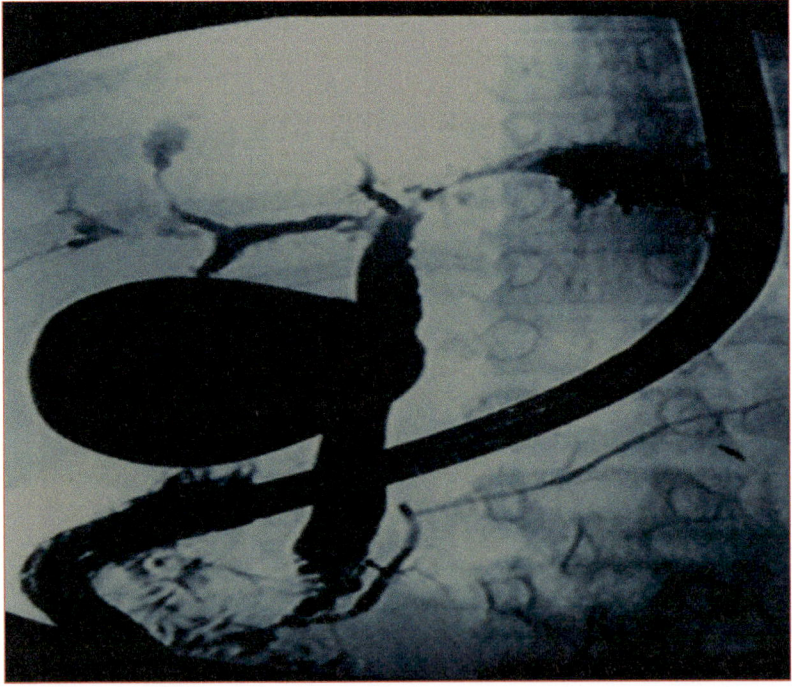

**Figure 13.1** Cholangiographic aspect of PSC on ERCP: mural abnormalities associated with intra- and/or extra-hepatic biliary tree strictures and sacculations.

PSC: primary sclerosing cholangitis; ERCP: endoscopic retrograde cholangiopancreatography.

The diagnosis of PSC can be made by endoscopic retrograde cholangiopancreatography (ERCP) or magnetic resonance cholangiopancreatography (MRCP), the latter being regarded as a method of choice for not being invasive. Liver biopsy can help in the diagnosis of patients with normal biliary ducts on bioimaging. In this context, the finding of pericholangitis characterizes the presence of small-duct PSC, an entity with a more benign clinical course and less risk of progression to cirrhosis and cholangiocarcinoma.[3] The diagnostic algorithm for PSC and other cholestatic diseases is summarized in Figure 13.2.[5]

The course of disease is variable and average survival after diagnosis is 12 years. Presentation can be further complicated by the occurrence of intermittent exacerbations of both cholangitis and cholangiocarcinoma in 10 to 30% and 10 to 15% of cases, respectively. The treatment of PSC aims at controlling symptoms and complications of cholestasis, such as pruritus, fatigue, osteoporosis and water-soluble vitamin deficiencies. Cholangiocarcinoma is highly suspected in patients with PSC and IBD, as well as colorectal cancer, and screening for dysplasia and/or cancer in shorter intervals is highly recommended, through colonoscopy with biopsy every 1 to 2 years.[3] There is no specific treatment which is satisfactory for PSC. Use of ursodeoxycholic acid (UDCA) was associated with biochemical improvement but had no impact on disease survival.

Endoscopic treatment can be considered for patients with recurrent cholangitis and dominant strictures. On the other hand, liver transplantation is regarded as the only effective treatment in the presence of refractory pruritus, recurrent acute cholangitis, progressive jaundice and liver failure.

## DRUG-INDUCED LIVER INJURY (DILI)

Medications used to treat IBD can often be associated with severe DILI, and these are common causes of asymptomatic elevation of liver enzymes.[1,2] Sulfasalazine and, less frequently, 5-ASA can cause hepatocellular or cholestatic acute hepatitis, and also granulomatous hepatitis. Azathioprine, in turn, can trigger acute hepatitis, usually cholestatic, veno-occlusive disease, nodular regenerative hyperplasia and hepatic peliosis.[6] Cholestatic or hepatocellular acute hepatitis has also been reported with 6-mercaptopurine.

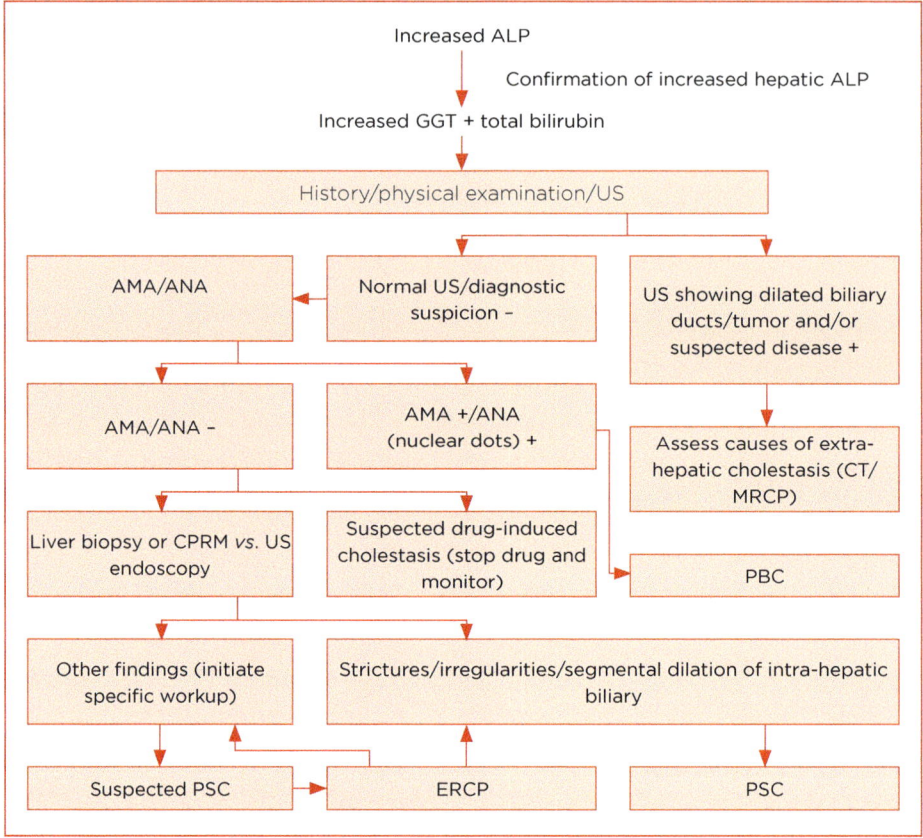

**Figure 13.2** Algorithm for diagnosis of PSC and other cholestatic diseases.

ALP: alkaline phosphatase; GGT: gamma-glutamyl transpeptidase; US: ultrasonography; ANA: antinuclear antibody; PBC: primary biliary cirrhosis; AMA: antimitochondrial antibody; PSC: primary sclerosing cholangitis; ERCP: endoscopic retrograde cholangiopancreatography; CT: computed tomography; MRCP: magnetic resonance cholangiopancreatography.

Source: adapted from EASL, 2009.[5]

Use of infliximab or adalimumab can also be associated with a mild elevation of aminotransferases. However, cases of acute hepatitis with severe progression have been reported.[1] The use of methotrexate can be associated with development of macrovesicular steatosis and hepatic fibrosis. The risk of progression to cirrhosis is dose-dependent, and is greater after 2 years of treatment using daily doses. Hepatotoxicity occurs with doses above 2 g. The risk is higher when the total cumulative dose is greater than 4 g in alcoholic, diabetic or obese patients.

Control with liver biopsies is recommended in case of prolonged treatment with methotrexate, using a total cumulative dose of 1,5 to 2 g, in addition to abstinence from alcoholic beverages.[1,2,6] Cyclosporine, in turn, has been associated with the occurrence of cholestatic acute hepatitis.[1]

## NON-ALCOHOLIC FATTY LIVER DISEASE

Steatosis or non-alcoholic steatohepatitis can be observed in up to 35% of biopsies from patients with IBD.[1] They are usually associated with disease activity in IBD, malnutrition, use of corticosteroids and proctocolectomy with ileal reservoir.

## GALLSTONES

Gallstones can be observed in 11% of CD patients and in 13 to 34% of those with ileitis or ileal resection. Complications such as acute biliary pancreatitis and choledocholithiasis can also occur.

Differential diagnosis of acute pancreatitis in patients with IBD must include drug causes, particularly azathioprine, 6-mercaptopurine and 5-ASA.

## MISCELLANEOUS

Other causes of hepatic disease in patients with IBD include AIH and primary biliary cirrhosis (PBC). Approximately 50% of patients with IBD and AIH present with AIH/PSC overlap syndrome, which requires a customized diagnostic and therapeutic approach. The diagnosis of PBC is easily established in the presence of AMA (see Figure 13.2).

Granulomatous hepatitis rarely derives from hepatic involvement of CD or the use of 5-ASA. Hepatic amyloidosis is also uncommon, and has been observed in 0.9 and 0.07% of cases of CD and UC, respectively.

Hepatobiliary manifestations, particularly PSC, should be investigated at regular intervals in all IBD patients. In suspected cases of the disease, the algorithm for investigation proposed in Figure 13.2 can be adopted. Due to the risk of hepatotoxicity, patients treated with azathioprine, 6-mercaptopurine and methotrexate should undergo liver enzyme investigation every 1 to 3 months. 3- to 6-month intervals are recommended in the case of biological agents. Prior to the use of these agents or corticosteroids, patients should be tested for hepatitis B virus (HBV). In the presence of active infection or in cases of asymptomatic HBV, consultation with a hepatologist for treatment or prevention of disease reactivation is advised, preferably with nucleoside analogues. In cases of suspected drug-induced hepatotoxicity, other underlying causes of liver injury must be ruled out, particularly by investigating a possible history of alcohol abuse, and tests that include abdominal ultrasonography and serum autoantibodies. The use of potentially hepatotoxic drugs must be stopped. In the absence of biochemical improvement, liver biopsy is suggested.

Liver biopsy should also be performed in case of chronic use of methotrexate with cumulative doses above 1.5 g, regardless of the biochemical pattern of aminotransferases.

## REFERENCES

1. Navaneethan U, Shen B. Hepatopancreatobiliary manifestations and complications associated with inflammatory bowel disease. Review Inflamm Bowel Dis 2010; 16:1598-619.

2. Mendes FD, Levy C, Enders FB, Loftus Jr. EV, Angulo P, Lindor KD. Abnormal hepatic biochemistries in patients with inflammatory bowel disease. Am J Gastroenterol 2007; 102:344.

3. Chapman R, Fevery J, Kalloo A, Nagorney DM, Boberg KM, Shneider B et al. American Association for the Study of Liver Diseases. Diagnosis and management of primary sclerosing cholangitis. Hepatology 2010; 51:660-78.

4.  Bittencourt PL, Palacios SA, Cançado EL, Carrilho FJ, Porta G, Kalil J et al. Susceptibility to primary sclerosing cholangitis in Brazil is associated with HLA-DRB1*13 but not with tumour necrosis factor alpha-308 promoter polymorphism. Gut 2002; 51:609-10.

5.  European Association for the Study EASL Clinical Practice guidelines: management of cholestatic liver diseases. J Hepatol 2009; 51:237-67.

6.  Bittencourt PL, Farias AQ, Silva LC. Fígado e drogas. In: Mattos AA, Dantas W (orgs.). Compêndio de hepatologia. 2.ed. São Paulo: Fundação Byk, 2001.

# COMPLICATIONS OF INFLAMMATORY BOWEL DISEASE

IDBLAN CARVALHO DE ALBUQUERQUE
PAULO GUSTAVO KOTZE

## ULCERATIVE COLITIS

Ulcerative colitis (UC) is a chronic disease characterized by diffuse inflammation of the colon mucosa, the rectum being affected in 95% of cases.[1] The therapeutic approach takes into account disease location and behavior, previous therapeutic response and extraintestinal manifestations, but especially inflammatory activity.[1-3] Disease activity is the main indicator for the use of drugs by mouth or intravenously, and also for surgical treatment.[4] Box 14.1 shows the main complications of UC.

| **Box 14.1** Distribution of the main complications of UC |
| --- |
| Severe colitis refractory to intravenous corticosteroids |
| Perforation |
| Toxic megacolon |
| Severe intestinal bleeding |
| Suspected diagnosis of cancer |

Severe acute UC should be medically treated quickly and effectively. If there is no response to therapy in up to 24 to 48 hours, immediate surgery should be indicated.[4,5] In cases with partial response to clinical treatment, surgical indication must take place within five days. When early surgical approach is adopted, the mortality rate is less than 3%.[5,6]

### Severe colitis refractory to intravenous corticosteroids

Severe intravenous corticosteroid-refractory colitis (SIVCRC) is the condition in which the patient is hospitalized and treated with intravenous corticosteroids for at least 1 day with little improvement (Figure 14.1). The clinical picture is characterized by abdominal pain and more than 10 bowel movements a day, associated with severe impairment of general condition.[7] At this point, it is prudent to evaluate the use of immunosuppressive drugs (cyclosporine and tacrolimus), anti-TNF-alpha (infliximab and adalimumab) or even perform a complete resection of the colon.[5,7] The decision to perform a colectomy in case of SIVCRC is difficult and should be discussed with the multidisciplinary team, the patient and the family.

### Perforation

The perforation of the large intestine as a complication of UC should be promptly diagnosed and treated, but even with an immediate treatment,

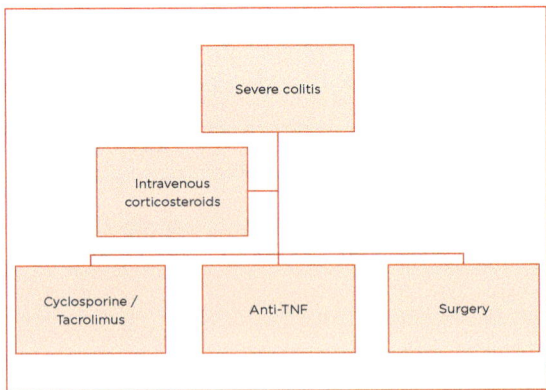

**Figure 14.1** Algorithm for treatment of severe UC.

the mortality rate is over 50%.[4] This clinical condition is associated with colonoscopy in the active phase of the disease or as a complication of toxic colitis.[4,6] Treatment consists of resection of the affected area without primary anastomosis due to the severity of the patient.

## Toxic megacolon

Toxic megacolon is a medical emergency with a high mortality rate, which arises as a complication of ulcerative colitis, Crohn's colitis, salmonellosis-associated colitis, ischemic colitis, or *C. difficile* colitis. It can occur as a worsening of disease, but in over 60% of patients this can be the first manifestation of UC.[1]

By definition, toxic megacolon is an acute colitis characterized by pain and increased abdominal volume, fever, tachycardia, and leukocytosis. Plain abdominal radiographs show dilation of the total length or segments of the colon.[5] Triggering factors include the use of anti-diarrheal medication and opiates, as well as the patient undergoing barium enema.

Medical treatment includes correction of electrolyte disturbances, use of broad-spectrum antibiotics and intravenous corticosteroids, and preparation for imminent surgery should be considered.[5,7] After 24 hours, in the absence of therapeutic response, exploratory laparotomy (EL) is indicated. This aggressive approach reduces the incidence of perforation (associated with 20% mortality compared to 4% in its absence). The performance of total proctocolectomy (TPC) or total colectomy with closure of the distal rectum is controversial.[8]

TPC is intended to remove all the mucosa of the large intestine and in the hands of a skilled professional, can be performed with low morbidity.[2] On the other hand, total colectomy is a smaller procedure, but leaves the rectal stump sick. In addition, after approximately two years, more than 30% progress with clinical intractability.[1,8]

## Severe rectal bleeding

Severe rectal bleeding in UC is rare; however, it accounts for about 10% of all emergency colectomies performed in individuals diagnosed with UC.[4]

### Suspected diagnosis of cancer

There is a positive relationship between chronic inflammation of the mucosa of the large intestine in UC and colorectal adenocarcinoma. The main risk factors are the extent of disease at diagnosis and disease duration (over 8 years).[1]

Colorectal cancer in UC does not follow the adenoma-adenocarcinoma sequence, because it derives from flat intestinal mucosa. These tumors, therefore, either display infiltrative behavior or are located in the submucosa, which makes diagnosis by colonoscopy and biopsy difficult. This is the reason why 20% of patients have advanced disease at diagnosis.[4,8]

## CROHN'S DISEASE

Crohn's disease (CD) is a chronic inflammatory disease of transmural behavior,[9] which affects the wall of the different segments of the digestive tract, resulting from immune dysfunction determined by genetic and environmental factors.[10]

The natural history of CD shows that, at first, the disease has an inflammatory behavior and over time it progresses to the fibrostricturing and penetrating forms, which often require surgery.[1,10,11]

Treatment is traditionally done with salicylates, corticosteroids, antibiotics, and immunosuppressants.[1,10] However, this therapeutic approach has undergone significant innovations due to the development of new therapies (immunobiological drugs), a better understanding of disease behavior and the definition of predictors of severity.

Despite advances in therapy, the number of patients operated for intractable or complicated disease (Figure 14.2) remained almost unchanged and, throughout life, up to 80% of these patients will be operated on.[12] Although laparotomy is the preferred approach, laparoscopic surgery gives good results in selected cases.[13,14] The abdominal cavity being accessed, diseased areas should be resected, and the decision to perform primary anastomosis or derivative stoma is evaluated based on local conditions, nutritional status and the use of corticosteroids, among other factors.[10,12,13]

## Peritoneal abscesses

Intracavitary abscesses are complications that occur in 10-30% of CD patients. A few decades ago, the presence of this complication in CD patients was synonymous with EL for drainage of collections, and resection of affected segment(s).[13] Using ultrasound or computed tomography (CT), it is possible to establish the location, volume and even treat this complication (Box 14.2).[10]

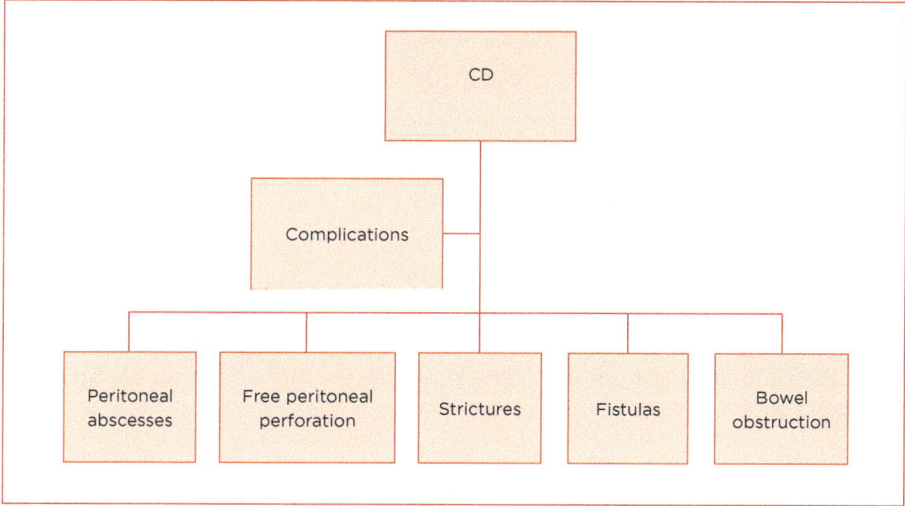

**Figure 14.2** Main complications of CD.

| Box 14.2 How to treat abdominal abscesses in CD |
|---|
| **Percutaneous drainage** |
| One or a few abscesses |
| Location favorable for puncture |
| Stable patients |
| Pediatric patients |

In addition to the intraperitoneal abscesses in CD, abscesses of the psoas muscle and liver are common. The treatment of these abscesses in atypical location is based on the combination of antibiotics and imaging-guided percutaneous drainage.[15] EL should be performed in cases of multiple or prolonged abscesses in the absence of clinical improvement following percutaneous drainage, and in cases where there is surgical indication for intestinal complications of CD such as perforations and intestinal occlusion.[12,13]

### Free peritoneal perforation

Free perforation with purulent or fecal peritonitis development is a rare complication in CD patients and usually occurs near the stricture area. In these situations, immediate surgical intervention with resection of the affected segment is recommended.[13]

The decision whether to perform an anastomosis depends on local conditions (degree of peritoneal cavity contamination, appearance of the ends of the bowel to be anastomosed, and presence or not of distal obstruction), as well as the patient's general condition. Intestinal resection followed by proximal end stoma or a loop stoma is a safe conduct and presents low level of mortality.[16]

### Strictures

Patients with stricture caused by CD experience pain, increased abdominal volume and postprandial vomiting associated with weight loss. Stricture is the scar repair of chronic and transmural inflammation and can occur in segments not previously operated (surgery-naive) or previous intestinal anastomoses (Table 14.1).[1,10] Stricture with proximal dilatation is easily diagnosed by intestinal transit examinations or CT scan of the abdomen and pelvis.[17]

The treatment includes resection procedures with or without anastomosis based on local conditions and the patient's clinical condition, enteroplasty,[18] and endoscopic dilation with hydrostatic balloon.[17,19] The therapeutic approach to stricture of the small intestine caused by CD should take into account the experience of the surgeon and the availability of endoscopic instruments. Colon stricture should always be resected, due to the increased risk of colorectal cancer (Figure 14.3).

| Table 14.1 Classification of strictures caused by CD | |
|---|---|
| **Strictures** | **Characteristics** |
| According to number | Single |
| | Multiple |
| According to extent | Short |
| | Long (> 10 cm) |

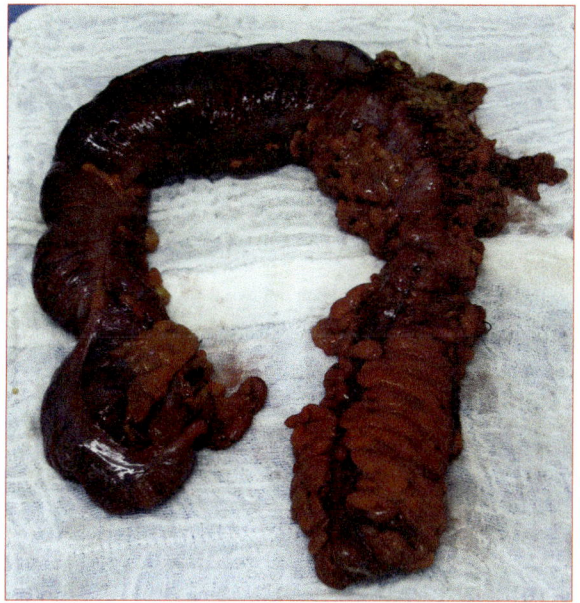

**Figure 14.3** Subtotal colectomy due to left colon stricture caused by CD.

## FISTULAS

Penetrating CD can affect any organ adjacent to the site of inflammation. The sensitivity and specificity of CT of the abdomen (Figure 14.4) and pelvis and magnetic resonance imaging (MRI) favor the choice of these methods to evaluate cases of penetrating disease.[1,2,10]

Treatment consists of resection of the affected segment and suturing of the secondarily inflamed organ. In cases of fistula of the small intestine to the colon, resection of a short segment is advised, as even the simple suture of the

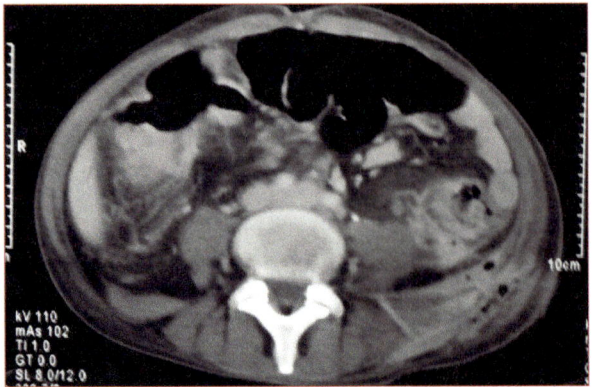

**Figure 14.4** Abdominal CT with a diagnosis of enterocutaneous fistula caused by penetrating CD.

large intestine is associated with a high rate of wound dehiscence.[10,12] Surgical treatment of external fistulas comprises resection of the affected segment and curettage of the fistula in the abdominal wall.

Data presented in the literature report the clinical treatment of enterocutaneous fistulas with biologic therapy (TNF-alpha inhibitors); however, asymptomatic internal fistulas should not be treated.

### Bowel obstruction

Obstruction of the gastrointestinal (GI) tract in CD occurs mainly in the small intestine, and may be caused by acute inflammation of the intestinal wall with subsequent narrowing of the *lumen*, chronic fibrosis of preexisting inflammatory strictures, or by extrinsic compression caused by adjacent inflammation or abdominal abscess.[1,15] Diagnosis of intestinal obstruction caused by CD is performed based on clinical history associated with contrast radiological evaluation.

Because of the inflammatory component of strictures in CD, the initial treatment is clinical, with oral fasting and intravenous hydration associated with intravenous corticosteroids. The use of antibiotics is warranted by evidence of infection or for the prevention of bacterial translocation. Limited clinical response after 48 hours indicates the need for surgery (Figure 14.5).[13] Special attention should be given to previously operated patients, due to the possibility of adhesions and associated malignancy.

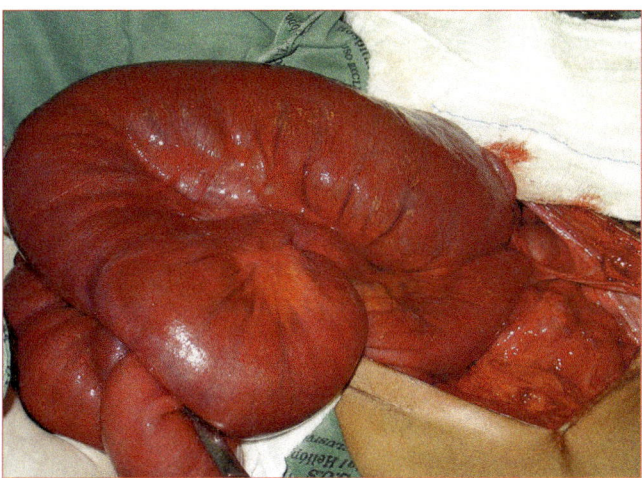

**Figure 14.5** Distal ileal obstruction caused by CD.

## REFERENCES

1. Baumgart DC, Sandborn WJ. Inflammatory bowel disease: clinical aspects and established and evolving therapies. Lancet 2007; 369:1641-57.

2. Koltum WA. The future of surgical management of inflammatory bowel disease. Dis Colon Rectum 2008; 51(6):813-7.

3. Solberg IC, Lygren I, Jahnsen J, Aadland E, Høie O et al. Clinical course during the first 10 years of ulcerative colitis: results from a population-based inception cohort (IBSEN Study). Scandinavian J Gastroenterology 2009; 44:431-40.

4. Hanauer SB, Lim WC, Sparrow M. Medical management of inflammatory bowel disease. In: Wolff BG, Fleshman JA, Beck DE, Pemberton JH, Wexner SD (eds.). The ASCRS textbook of colon and rectal surgery. New York: Springer Science, 2007. p. 555-66.

5. Travis SPL, Stange EF, Lémann M, Oresland T, Bemelman WA, Chowers Y et al. European evidence-based consensus on the management of ulcerative colitis: current management. J Crohns Colitis 2008; 24-62.

6. Fleshner PR, Schoetz Jr. DJ. Surgical management of ulcerative colitis. In: Wolff BG, Fleshman JA, Beck DE, Pemberton JH, Wexner SD (eds.). The ASCRS Textbook of Colon and Rectal Surgery. New York: Springer Science, 2007. p. 567-83.

7. Kornbluth A, Sachar DB. Ulcerative colitis practice guidelines in adults: American College of Gastroenterology, Practice Parameters Committee. Am J Gastroenterol 2010; 501-23.

8. Nivatvongs S. Ulcerative colitis. In: Gordon PH, Nivatvongs S. Principles and practice of surgery of the colon, rectum and anus. 3.ed. New York: Taylor & Francis, 2007. p. 755-818.

9. Crohn BB, Ginzburg L, Oppenheimer MD. Regional ileitis: a pathologic and clinical entity. JAMA 1932; 99(6):1323-9.

10. Dignass A, Van Assche G, Lindsay JO, Lémann M, Söderholm J, Colombel JF. The second European evidence-based consensus on the diagnosis and management of Crohnís disease: Current management. J Crohns and Colitis 2010; 28-62.

11. Schoepfer AM, Dehlavi MA, Fournier N, Safroneeva E, Straumann A, Pittet V et al. Diagnostic delay in Crohn's disease is associate with a complicated disease course and increased operation rate. Am J Gastroenterol 2013; 108:1744–53.

12. Strong SA. Surgery for Crohnís disease. In: Wolff BG, Fleshman JA, Beck DE, Pemberton JH, Wexner SD (eds.). The ASCRS textbook of colon and rectal surgery. New York: Springer Science, 2007. p. 584-600.

13. Nivatvongs S, Gordon PH. Crohn's disease. In: Gordon PH, Nivatvongs S. Principles and practice of surgery of the colon, rectum and anus. 3.ed. New York: Taylor & Francis, 2007. p. 819-908.

14. Stewart DB, Messaris E. Early experience with single-site laparoscopic surgery for complicated ileocolic Crohn's disease at a tertiary-referral center. Surg Endosc 2012; 26:777–82.

15. Beaugerie L, Seksik P, Nion-Larmurier I, Gendre JP, Consnes J. Predictors of Crohn's disease. Gastroenterology 2006; 130:650-6.

16. Gordon PH, MacDonald J, Cataldo PA. Intestinal stomas. In: Gordon PH, Nivatvongs S. Principles and practice of surgery of the colon, rectum and anus. 3.ed. New York: Taylor & Francis, 2007. p. 1031-79.

17. Ferlitsch A, Reinisch W, Püspök A, Dejaco C, Schillinger M, Schöfl R et al. Safety and efficacy of endoscopic balloon dilatation for treatment of Crohn's disease strictures. Endoscopy 2006; 38:483-7.

18. Hotokezaka M, Ikeda T, Uchiyama S, Hayakawa S, Tsuchiya, Chijiwa K. Side-to-side-to-end stricturoplasty for Crohn's disease. Dis Colon Rectum 2009; 52:1882-6.

**19.** Atreja A, Aggarwal A, Dwivedi S, Rieder F, Lopez R et al. Safety and efficacy of endoscopic dilation for primary and anastomotic Crohn's disease strictures. JCC 2014; 8:392–400.

## BIBLIOGRAPHY

**1.** Lichtenstein GR, Olson A, Travers S, Diamond RH, Chen DM, Pritchard ML et al. Factors associated with the development of intestinal strictures or obstructions in patients with Crohn's disease. Am J Gastroenterol 2006; 101:1030-8.

# LABORATORY DIAGNOSIS IN INFLAMMATORY BOWEL DISEASE

RAQUEL FRANCO LEAL

## INTRODUCTION

Inflammatory bowel diseases (IBD) are considered multifactorial and polygenic diseases, and Crohn's disease (CD) and nonspecific ulcerative colitis (UC) are its most important phenotypes. The differential diagnosis of these diseases before the manifestation of initial symptoms can be a real challenge. In some cases, even after long-term disease progression, it is not possible to distinguish between CD and UC due to the presence of features common to both diseases, especially when there is colorectal involvement alone.

Clinical presentation and colonoscopic examination with evaluation of the distal ileum and histopathological characteristics are sufficient, in most cases, to establish the differential diagnosis of IBD. The twenty percent of patients, including those with indeterminate colitis and colorectal disease, who present with endoscopic findings of CD and UC, are the ones in which serological markers and specific laboratory tests can be employed.

Until recently, serological markers were employed only in cases of indeterminate colitis and research protocols. However, these markers are now recognized as important tools for determining prognosis and response to therapy.[1] The best known serological markers are antineutrophil cytoplasmic antibodies (ANCA) and the antibody against *Saccharomyces cerevisiae* (ASCA), but several others have been studied and used.[2]

Once diagnosis is established, markers such as erythrocyte sedimentation rate (ESR), C-reactive protein (CRP), fecal calprotectin and lactoferrin can be used to assess disease activity. This chapter will discuss serological markers and laboratory tests used to assist in the differential diagnosis and follow up of patients with CD and UC.

## SEROLOGICAL MARKERS FOR THE DIFFERENTIAL DIAGNOSIS OF IBD

The advent of serological markers has provided a greater understanding of the complex pathophysiology of the disease and the connection between the innate and adaptive immune systems in IBD. In practical terms, IBD is characterized by a decreased intestinal barrier effect. This occurs in various forms, such as via the decreased production of defensins, which are important peptides that act against several micro-organisms and are part of the innate immune system in the intestinal epithelium. Thus, microbial antigens, most often bacterial, enter the lamina propria more easily, leading to the amplified activation of the innate and adaptive immune system cells. Bacterial antigens are responsible for activating the so called toll-like receptors (TLR), which are more highly expressed in IBD. Thus, immune cells are more sensitive to the action of these antigens, activating pro-inflammatory pathways.

Concomitantly, antibodies are produced and act as serological markers.[3] The best known, as previously mentioned, are ANCA and ASCA. ANCA autoantibody production is triggered by bacterial antigens. It is present in 65 to 70% of patients with UC, and constitutes one of the few markers of this disease. The other antibodies, in turn, are more efficient markers of CD. As for the ASCA antibody, it is directed against *Saccharomyces cerevisiae* and 55 to 70% of CD patients express this protein.

Other antibodies include OmpC, I2, CBir1-flagellin, A4-Fla2 flagellin and Fla-X. OmpC originates from the antigen formed by membrane surface proteins of *E. coli*; the I2 marker reacts against *P. aeruginosa* and CBir1-flagellin antibody is directed against flagella of commensal bacteria and is detected by studying the expression of TLR5, a receptor that interacts with this antigen.[4,5] A4-Fla2 and Fla-X flagellins are newly discovered and some CD patients are seropositive. In a prospective study including 252 patients with CD, the authors found that 59% were positive for A4-Fla2, and 57% for Fla-X, so that 76% of the overall sample had localized disease in the small intestine.[6] Another study[7] revealed that pouch-anal anastomosis for UC positive for ASCA IgG and CBir-1 was related to the development of fistulas and CD in the ileal reservoir. The identification of this group of patients at high risk of complications allows early and more aggressive manners to prevent ileal pouch failure.[7]

More recently, studies have been evaluated with anti-glycan or anti-saccharide antibodies, i.e. antibodies against saccharide components of the cell membrane of micro-organisms (bacteria, fungi and viruses), which are present in a variable percentage of patients with CD (10-28%, except for g-ASCA, whose sensitivity is 46-60%). The most well known antibodies are gASCA, ALCA, ACCA, AMCA, anti L and anti-C. In addition to assisting the diagnosis of CD, these markers may predict disease progression (e.g., gASCA and AMCA are signs of short duration disease; gASCA and ALCA are markers of disease at a young age; ACCA, long-term illness; anti-L and anti-C, colonic involvement). Although the sensitivity is not high for all these markers, specificity is slightly higher (about 40%).[1] The finding of antibodies against these saccharides suggests a connection between the innate and adaptive immune systems, reflecting loss of tolerance to the commensal flora, which is considered a hallmark of the immunopathogenic process in IBD.

Separately, serological markers have low sensitivity in distinguishing between CD and UC; however, it has been shown that the use of panels of markers increases this sensitivity.[8] The addition of anti-L and anti-C to gASCA and pANCA increases sensitivity for differentiating between UC and CD (p <0.001)[9] and, in addition, 33-56% of patients seronegative for gASCA can be identified by anti-glycan antibodies (ALCA, ACCA and AMCA).[10]

A panel composed by auto-antibodies together with antibodies reacting against bacterial antigens and genetic variant and molecules related to inflammation and angiogenesis has been studied.[11] More specifically, this panel evaluates auto-antibody positivity for ASCA, ANCA, pANCA markers; bacterial antigens, such as OmpC, CBir1, A4-Fla2, FlaX; genetic markers, including ATG16L1, NKX2-3, ECM1, STAT3; and inflammatory markers such as VEGF, CRP, SAA, ICAM-1 and VCAM-1. Great accuracy in distinguishing IBD from other diseases was observed. Furthermore, CD could be differentiated from UC, presenting 88.9% sensitivity and 81% specificity for CD and 97.7% sensitivity and 83.5% specificity for UC.[11]

The application of these antibodies in predicting a therapeutic response is a goal pursued for future clinical practice. Genome-wide association studies (GWAS) have presented a predictive model of non-responsiveness to anti-TNF-alpha, especially in children, and signalized a 15 times greater risk of non-response to anti-TNF-alpha when the children had a clinical diagnosis of UC. There was also positivity for pANCA and another BRWD1 marker.[12]

## LABORATORY TESTS AND MARKERS OF INFLAMMATORY ACTIVITY IN THE FOLLOW-UP OF PATIENTS WITH IBD

The importance of monitoring inflammatory activity in IBD is becoming increasingly clear. Endoscopic aspects have been established to investigate mucosal healing.[13] However, the development and employment of serological and fecal markers, which can predict disease activity and are correlated with endoscopic findings, are very useful and less invasive.

ESR, serum leukocyte and CRP are known to predict disease activity, as well as being accessible to laboratory examination and less expensive for monitoring disease activity in IBD patients; however, these are nonspecific.

Fecal markers comprise a heterogeneous group of proteins produced by the intestinal mucosa affected by the inflammatory process. Lactoferrin, calprotectin and polymorphonuclear neutrophil elastase are neutrophil-derived proteins capable of differentiating disease as inactive or quiescent. None of these markers is better than endoscopic findings of disease activity, but they have better accuracy compared to CRP. Alpha-1-antitrypsin is a fecal protease-inhibiting protein produced in the liver, macrophages and in the intestinal epithelium. The 72-hour-clearance test is a useful method for quantification of protein loss and determining the activity of the disease. Another anti-protease is alpha-2-macroglobulin, which is associated with CD activity, but its use is not proven for UC.[14,15]

The association of serological and fecal markers, in addition to the clinical disease activity index, enables increased accuracy in determining and predicting exacerbation phases of the disease and monitoring response to treatment. The implementation of fecal markers in clinical practice has been increasingly common in many countries, providing important tools for monitoring these patients.

## CLOSING REMARKS

Laboratory tests and fecal and serum markers should not be used alone in the diagnosis and monitoring of patients with IBD, but are important adjunct tools to endoscopic, histological and radiological findings. Furthermore, additional studies are needed to elucidate the role of serological and fecal markers in the context of IBD, and thus provide further support for their routine use.

## REFERENCES

1. Dotan I. New serologic markers for inflammatory bowel disease diagnosis. Dig Dis 2010; 418-23.

2. Vermeire S, Van Assche G, Rutgeerts P. Laboratory markers in IBD: useful, magic or unnecessary toys? Gut 2006; 426-31.

3. Tsianos EV, Katsanos K. Do we really understand what the immunological disturbances in inflammatory bowel disease mean? World J Gastroenterol 2009; 521-25.

4. Papp M, Norman GL, Altorjay I, Lakatos PL. Utility of serological markers in inflammatory bowel diseases: gadget or magic? World J Gastroenterol 2007; 2028-36.

5. Turkay C, Kasapoglu B. Noninvasive methods in evaluation of inflammatory bowel disease: where do we stand now? An update. Clinics 2010; 221-31.

6. Iskandar HN, Ciorba MA. Biomarkers in inflammatory bowel disease: current practices and recent advances. Transl Res 2012; 313-25.

7. Coukos JA, Howard LA, Weinberg JM, Becker JM, Stucchi AF, Farraye FA. ASCA IgG and CBir antibodies are associated with the development of Crohn's disease and fistulae following ileal pouch-anal anastomosis. Dig Dis Sci 2012; 57(6):1544-53.

8.  Peyrin-Biroulet L, Standaert-Vitse A, Branche J, Chamaillard M. IBD serological panels: facts and perspectives. Inflamm Bowel Dis 2007; 1561-66.

9.  Seow CH, Stempak JM, Xu W, Lan H, Griffiths AM, Greenberg GR et al. Novel anti-glycan antibodies related to inflammatory bowel disease diagnosis and phenotype. Am J Gastroenterol 2009; 1426-34.

10. Malickova K, Lakatos PL, Bortlik M, Komarek V, Janatkova I, Lukas M. Anticarbohydrate antibodies as markers of inflammatory bowel disease in a Central European cohort. Eur J Gastroenterol Hepatol 2010; 144-50.

11. Plevy SE, Stockfisch TP, Lockton S, Lisa J. Combined serologic, genetic, and inflammatory markers can accurately differentiate non-IBD, Crohn's disease, and ulcerative colitis patients. Gastroenterology 2012; 142(5):S41.

12. Dubinsky M. Can serologic markers help determine prognosis and guide therapy? Dig Dis 2010; 424-28.

13. Panaccione R, Colombel JF, Louis E, Peyrin-Biroulet L, Sandborn WJ. Evolving definitions of remission in Crohn's disease. Inflamm Bowel Dis 2013; 19:1645-53.

14. Langhorst J, Elsenbruch S, Koelzer J, Rueffer A, Michalsen A, Dobos GJ. Noninvasive markers in the assessment of intestinal inflammation in inflammatory bowel diseases: performance of fecal lactoferrin, calprotectin, and PMN-elastase, CRP, and clinical indices. Am J Gastroenterol 2008; 162-9.

15. Gisbert JP, Bermejo F, Pérez-Calle JL, Taxonera C, Vera I, McNicholl AG et al. Fecal calprotectin and lactoferrin for the prediction of inflammatory bowel disease relapse. Inflamm Bowel Dis 2009; 1190-98.

# ENDOSCOPIC DIAGNOSIS OF INFLAMMATORY BOWEL DISEASE

PAULO ALBERTO FALCO PIRES CORRÊA
JARBAS FARACO MALDONADO LOUREIRO

## INTRODUCTION

Regarding the nonspecific inflammatory bowel diseases (NS-IBD), it is very important to emphasize that it is not always possible to establish a definitive diagnosis based solely on endoscopic findings, because, in addition to how similar the endoscopic presentations of nonspecific ulcerative colitis (UC) and Crohn's disease (CD) can be, it is also difficult to establish a differential diagnosis with other diseases that cause colon inflammation, such as parasitic, infectious, ischemic and actinic colitis.

It is known that, even in specialized centers, 10 to 20% of the surgical specimens examined by the pathology service do not have a confirmed diagnosis of NS-IBD. Add to that a third manifestation, known as indeterminate colitis.[1]

Colonoscopy appears to be extremely important in the diagnosis of NS-IBD, since, in addition to endoscopic aspect, biopsy material can be collected for microscopic examination and culture, if necessary. Nevertheless, the final diagnosis of these diseases should always be done through the association of clinical history, physical examination, laboratory tests, radiological and endoscopic findings, in addition to histopathology.[2-5]

## INDICATIONS AND CONTRAINDICATIONS

The colonoscopy indications in NS-IBD are described in Box 16.1, while contraindications are shown in Box 16.2.

## ENDOSCOPIC DIAGNOSIS

### Ulcerative colitis (UC)

Active UC presents as continuous and diffuse disease with edema, congestion, friability and mucosal granularity, as well as microulcerations, which may or may not be covered with fibrin (Figure 16.1). These microulcerations can coalesce into larger ulcers, which usually do not exceed 1 cm in their longest axis (Figure 16.2).

The transition area between the mucosa affected by the disease and the normal mucosa is usually clearly identified during colonoscopy. In 95% of cases, the rectum is affected by the disease (Figure 16.1); the terminal ileum, however, is rarely diseased (5%).

Endoscopic alterations found in the ileum are known as backwash ileitis. In the case of pancolitis, they are produced by colonic contents reflux into the terminal ileum through the ileocecal valve, causing a superficial inflammation of the mucosa.

Rectosigmoiditis is the most common form of presentation of the disease (45% of the time), followed by involvement of the descending colon up to the

| Box 16.1 Main indications for colonoscopy in suspected NS-IBD | |
| --- | --- |
| Assessment of disease extension | |
| Differential diagnosis | Between UC and CD |
| | And other specific inflammatory affections |
| Monitoring for disease progression | |
| Evaluation of ileal reservoir | |

| Box 16.2 Main contraindications (absolute and relative) for colonoscopy in suspected NS-IBD |
| --- |
| Toxic megacolon |
| Confirmed or suspected spontaneous perforation |

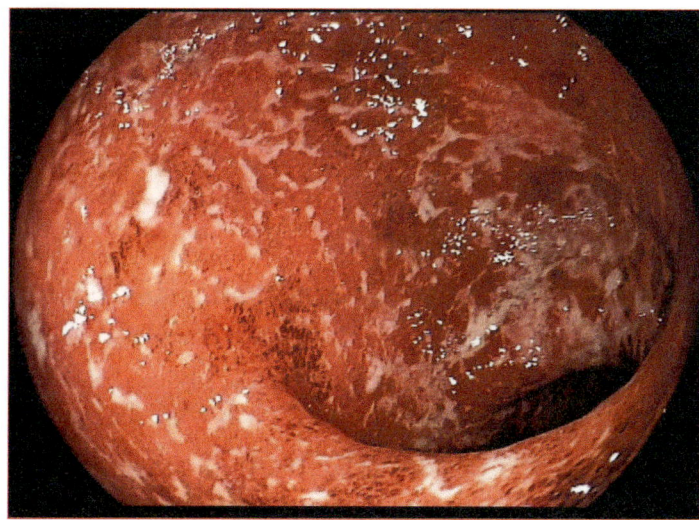

**Figure 16.1** Proctitis (UC). Diffuse and continuous inflammation of the rectal mucosa, with microulcerations covered by fibrin.

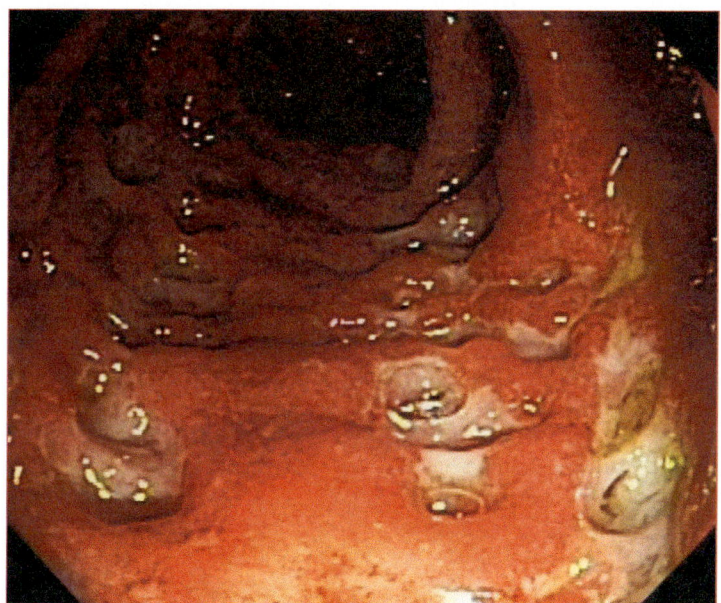

**Figure 16.2** Rectosigmoid Inflammation (UC). This image shows multiple ulcers in mucosa with intense inflammatory process.

splenic angle, or left colitis (40% of the time), and involvement proximal to the splenic angle, called "pancolitis" (15 to 20% of the time).[6]

Another important endoscopic finding is the intensity gradient of UC, which is generally stronger in the distal segments of the organ, and milder in proximal segments.

The submucosal vascular pattern of the normal colon is distorted in UC as a result of mucosal healing process. This endoscopic finding is most commonly observed in patients undergoing colonoscopy in the absence of active disease. In some cases of distal disease, endoscopic changes around the appendiceal orifice alone can be observed, with the remainder of this segment (cecum) showing normal endoscopic aspect.[7-9]

### CROHN'S DISEASE (CD)

It is characterized by areas of normal mucosa interspersed with diseased mucosa, resulting in a pattern called skip lesions, which may be continuous or segmental.[10] Usually, the most typical lesions in CD are aphthous ulcers and/or major ulcers, which can coalesce and/or appear longitudinally to the bowel's lumen and are called linear ulcers (Figure 16.3).

In some cases, infiltration and elevation of the submucosa can be evident, with a cobbled appearance. Unlike UC, the rectum is rarely involved in CD (only 10% of cases) and, when this occurs, there may be concomitant perianal fistula(s). Biopsies performed in the rectum, which has normal endoscopic aspect in CD patients, can reveal up to 14% of lymphoid aggregates (findings that indicate and precede granuloma) and 6% of positive findings for giant cell granuloma.

In 40% of patients with CD, there is ileocolic involvement (Figure 16.4), and in other 25%, the colon is affected exclusively. Thus, 65% of cases can be diagnosed by colonoscopy.

The endoscopic characteristics are described in Table 16.1.

### UC and CD: difficulties in the diagnosis

The more acute and intense clinical manifestation of the disease, the more difficult the differential diagnosis between them, due to the strong and nonspecific inflammation.[11] Twenty percent of patients with prior endoscopic

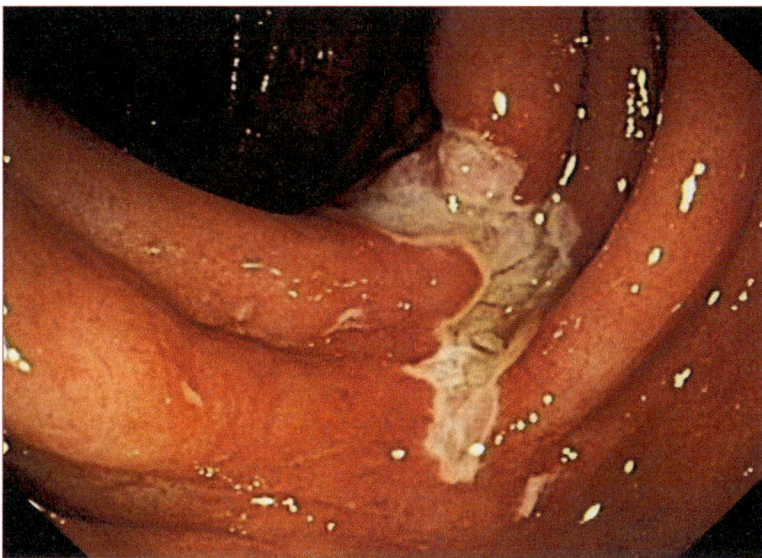

**Figure 16.3** Linear ulcer (CD). A longitudinal ulcered lesion surrounded by normal mucosa is observed.

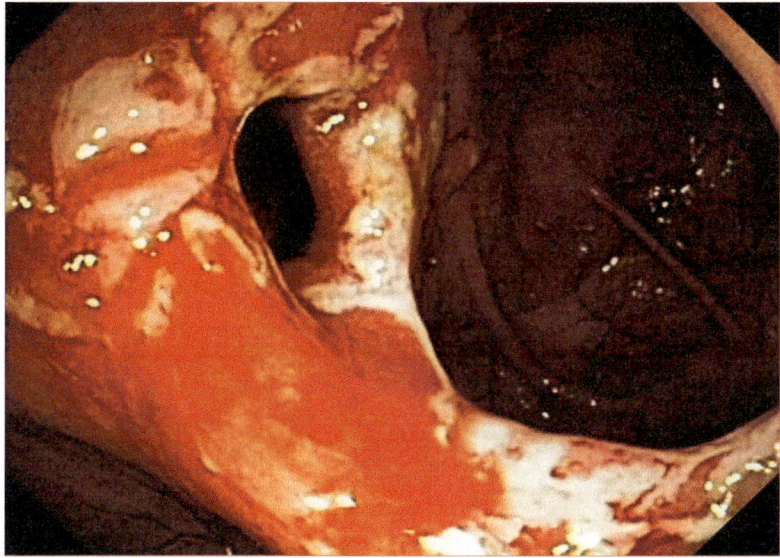

**Figure 16.4** Deformity of the ileocecal valve (CD). Deformities of the ileocecal valve and the presence of ulcers and concomitant inflammatory process in areas of endoscopically normal mucosa in the cecum are observed.

| **Table 16.1** Endoscopic features of UC and CD | |
|---|---|
| **UC** | **CD** |
| Continuous mucosal involvement | Skipped lesions |
| | Cobbled appearance |
| Rectum affected (95%) | Rectum not affected (90%) |
| Distorted submucosal vascular pattern | Linear ulcers |

diagnosis of CD examined in the acute phase of the disease have their diagnosis changed when they are re-examined. Likewise, among patients with previous endoscopic diagnosis of UC who are re-examined, 18% are found to have CD and 10% are patients with other inflammatory diseases, confirmed by pathological and/or laboratory tests.[12,13]

In patients with UC treated with corticosteroids, the endoscopic changes can mimic CD. The use of rectal topical medications such as suppositories or enemas can contribute to a practically normal endoscopic aspect of the rectum in patients with active disease, further complicating the differential diagnosis. Therefore, the predominant endoscopic pattern should be considered, as opposed to isolated lesions, to confirm the suspected diagnosis.

There are some endoscopic findings which may arise both in UC and in CD as well as other inflammatory diseases. These include pseudopolyps (or inflammatory polyps) and mucosal bridges.

The pseudopolyps correspond to "islands" of normal mucosa that remain after a previous inflammatory exacerbation and, during the healing process, develop a polypoid appearance. If filamentous, they are more suggestive of UC.

It is always recommended to remove any lesions with size or shape different from the others (especially if larger than 1 cm), due to the risk of developing malignancy.

The mucosal bridges are perforations caused by ulceration from mucosa to mucosa or adjoining submucosa, resulting in a double *lumen* colon. This finding may also be found in colitis caused by cytomegalovirus.[14]

## Biopsies

Integration between the endoscopist and the pathologist is of utmost importance, with representative samples (size and quantity) being provided, favorable to the final interpretation by the pathologist. Aphthous ulcers should be biopsied to include submucosal tissue in its depth (deep biopsies), since the granuloma is most often found in these lesions (up to 25% of cases). In the case of major ulcers, the edges should be biopsied in order to locate an etiological agent (differential diagnosis against parasitic or infectious colitis), since the bottom of these lesions is generally formed by necrotic or fibrin-leukocyte material. Performing biopsies in endoscopically normal areas may also be important, aiming at comparing possible histopathological changes (Box 16.3).

## DIFFERENTIAL DIAGNOSIS WITH OTHER DISEASES

### Infectious and parasitic diseases

Several infectious and parasitic diseases can mimic the endoscopic signs of a NS-IBD. Usually, diagnosis is made by examination of feces or fecal culture: parasitological or toxin investigation (in the case of pseudomembranous colitis). Those that can become chronic resemble more CD, while more acute and aggressive findings are indicative of UC (Table 16.2).

### Ischemic colitis

This disease usually affects patients with other conditions that compromise the microcirculation, such as diabetes, hypertension, connective tissue disorders, cardiac arrhythmia and atherosclerosis. Women on hormone replacement and users of illegal drugs (e.g., cocaine) should also be considered.

Lesions are segmental and affect mainly the splenic flexure or rectosigmoid transition (terminal circulation). It can present different endoscopic manifestations, such as areas of hyperemia, linear (ischemic) ulcers or even necrosis and sloughing of the mucous layer.

### Actinic colopathy

Actinic colopathy develops from the damage caused in the pelvis following external or internal radiation therapy of this part of the body. Therefore, it is almost always secondary to the treatment of some cancers using this

**Table 16.2** Endoscopic similarity between IBD and other infectious and parasitic diseases[15]

| Condition | UC | CD |
|---|---|---|
| Amebiasis | - | + |
| Schistosomiasis | + (A) | - |
| C. jejuni | + (A) | - |
| Y. enterocolitica | + (A) | - |
| Shigella sp | + (A) | - |
| Salmonella sp | + (A) | + (C) |
| Pseudomembranous colitis | + (A) | - |
| Tuberculosis | - | + |
| Viral colitis | + (A) | + (C) |
| Histoplasmosis | - | + |

(A): acute; (C): chronic.

therapeutic modality, such as the rectum, cervix or prostate. Generally, it appears as proctopathy but can compromise other mobile segments of the colon occupying the pelvis.

### MONITORING FOR DISEASE PROGRESSION

Because of the recurrent or chronic and persistent inflammatory process, these conditions are more likely to develop cancer and, therefore, require colonoscopic follow-up surveillance (Table 16.3).

### EVALUATION OF ILEAL RESERVOIR

Pelvic ileal pouch with sphincter preservation, performed in the surgical treatment of UC (and rarely in CD), can produce over time, in half of the cases, an inflammatory complication called pouchitis.[16] An inflammatory process of varying intensity occurs with possible microulcerations covered with fibrin, or major ulcers, usually as a result from increased or modified gut flora. Endoscopy of the reservoir, with biopsy, is mandatory to guide appropriate treatment.

Last, in patients in whom the distal 2-3 cm of the anal canal mucosa were kept in order to preserve the proper sensitivity and resulting continence, anoscopy should be routinely performed, due to the risk of neoplasia in the remaining mucosa.[17]

| **Box 16.3** When and how to perform biopsies in suspected NS-IBD |
| --- |
| Collect always a sufficient number of specimens |
| Always biopsy aphthous ulcers (deep biopsies) |
| Biopsy the edges of major ulcers |
| Always biopsy any suspicious lesions |
| Even when normal on endoscopy, always biopsy the terminal ileum |

| **Table 16.3** Monitoring of disease progression in NS-IBD | | |
| --- | --- | --- |
| | **Onset of disease** | **Frequency** |
| Distal UC or CD | > 15 years | Every 3 years |
| UC (pancolitis) | > 8 years | Every 1 or 2 years |

Note: Serial colon biopsies (3 to 4 every 10 cm in the case of pancolitis).

## REFERENCES

1. Farmer M, Petras RE, Hunt LE, Janosky JE, Galandiuk S. The importance of diagnostic accuracy in colonic inflammatory bowel disease. Am J Gastroenterol 2000; 95(11):3184-8.

2. Averbach M, Corrêa PAFP. Doenças inflamatórias intestinais. In: Ferrari Jr. AP (ed.). Atlas de endoscopia digestiva. Rio de Janeiro: Rúbio, 2009.

3. Cutait R, Corrêa PAFP, Averbach M. Colonoscopia nas doenças inflamatórias. In: SOBED. Endoscopia digestiva. Rio de Janeiro: Medsi – Médica Científica, 2000.

4. Loftus Jr. EV. Clinical epidemiology of inflammatory bowel disease: Incidence, prevalence, and environmental influences. Gastroenterology 2004; 126:1504-17.

5. Sartor MC, D'Assunção MA. Doenças inflamatórias intestinais. In: Averbach M, Corrêa P (eds.). Colonoscopia. São Paulo: Santos, 2010.

6. Su S, Lichtenstein GR. Ulcerative colitis. In: Feldman M et al. Sleisenger & Fordtran's gastrointestinal and liver disease. 8.ed. Filadélfia: Saunders, 2006.

7. D'Haens G, Geboes K, Peeters M, Baert F, Ectors N, Rutgeerts P. Patchy cecal inflammation associated with distal ulcerative colitis: a prospective endoscopic study. Am J Gastroenterol 1997; 92:1275-9.

8.  Matsumoto T, Nakamura S, Shimizu M, Iida M. Significance of appendiceal involvement in patients with ulcerative colitis. Gastrointest Endosc 2002; 55:180-5.

9.  Yang SK, Jung HY, Kang GH, Kim YM, Myung SJ, Shim KN et al. Appendiceal orifice inflammation as a skip lesion in ulcerative colitis: an analysis in relation to medical therapy and disease extent. Gastrointest Endosc 1999; 49(6):743-7.

10. Lee SD, Cohen RD. Endoscopy in inflammatory bowel disease. Gastroenterol Clin North Am 2002; 31(1):119-32.

11. Bouhnik Y, Lémann M, Maunoury V et al. Doenças inflamatórias intestinais. In: Classen M, Tytgat GNJ, Lightdale CJ. Endoscopia gastrointestinal. Rio de Janeiro: Revinter, 2006.

12. Moum B, Ekbom A, Vatn MH, Aadland E, Sauar J, Lygren I et al. Inflammatory bowel disease: re-evaluation of the diagnosis in a prospective population-based study in southeastern Norway. Gut 1997; 40(3):328-32.

13. Munkholm P. Crohn's disease-occurrence, course and prognosis: an epidemiologic cohort-study. Dan Med Bull 1997; 44:287-93.

14. Marques Jr. O, Averbach M, Zanoni EC, Corrêa PA, Paccos JL, Cutait R. Cytomegaloviral colits in HIV positive patients: endoscopic findings. Arq Gastroenterol 2007; 44:315-9.

15. Corrêa PAFP, Cutait R. Colites específicas. In: Quilici FA. Endoscopia digestiva diagnóstica e terapêutica. Rio de Janeiro: Revinter, 2005.

16. Lohmuller L. Pouchitis and extraintestinal manifestations of IBD after IPAA. Ann Surg 1990; 209:620-8.

17. Corrêa P, Averbach M. Doenças inflamatórias do cólon. In: Averbach M et al. (eds.). Atlas de endoscopia digestiva da SOBED. Rio de Janeiro: Revinter, 2011.

# DIAGNOSIS OF INFLAMMATORY BOWEL DISEASE BY VIDEO CAPSULE ENDOSCOPY

ARTUR A. PARADA

PAULA B. POLETTI

THIAGO FESTA SECCHI

## INTRODUCTION

Inflammatory bowel diseases (IBD) are characterized for being chronic diseases, lasting a lifetime, probably resulting from the interaction between genetic and environmental factors, whose incidence and prevalence has been increasing in recent years. Due to the wide variation in clinical presentation, both in terms of symptoms and clinical signs, the intensity of such, and also the extent and site of involvement of the disease, diagnosis is performed based on a combination of clinical, biological, radiological, endoscopic and histological data.[1-3] As clinical presentations are marked by several episodes of recurrence, frequent repetitions of laboratory, radiological, endoscopic and histological tests are required, not only for definitive diagnosis, but also for monitoring.[1,4]

Despite the differences in presentation and involvement of the digestive tract, the change in diagnosis between Crohn's disease (CD) and ulcerative colitis (UC) during the first year of disease progression may occur in about 10-15% of cases; in another 10%, the involvement is restricted to the colon. In such cases, even with all the investigation, it is not possible to characterize CD or UC, receiving the denomination of IBD unclassified, resulting in delayed institution of suitable treatment for this group of patients.[1]

Despite the large number of diagnostic tests available, early diagnosis of IBD remains a challenge, reflecting on continuity and progression of inflammatory activity, which can result in irreversible damage already established at the time of diagnosis.[5]

Until recently, the evaluation of the involvement of the small intestine in patients with CD and in patients with the unclassified form of IBD was performed using radiological examinations or, partially, through endoscopic exams that allowed visualization of merely the duodenum, proximal jejunum and distal ileum.[1,6] With the introduction of new technologies such as capsule endoscopy and enteroscopes assisted by accessories (balloon-guided or spiral), the endoscopic evaluation of the entire surface of the small intestine became a reality in clinical practice, allowing for more accurate diagnoses, diagnoses of earlier forms, and more appropriate classifications, especially in relation to disease extension, thus enabling better clinical management.[1,4-7]

## ENDOSCOPIC CAPSULE

The introduction of the endoscopic capsule enabled the last endoscopic frontier of the digestive tract to be surpassed, allowing endoscopic access to the entire extension of the small intestine, which had remained accessible only to intraoperative enteroscopy, reserved only for extreme cases due to the characteristics and morbidity inherent in the method.

The development of the endoscopic capsule began in the 1980s. After overcoming numerous technological challenges, in May 2000 in Digestive Disease Week (DDW), Swain presented the results of initial studies on the prototype of the endoscopic capsule system. In 2001, the satisfactory results of clinical studies were demonstrated, and the system gained the approval of the Food and Drug Administration (FDA) and CE Mark Certification for utilization in humans in the search for obscure bleeding.[8,9]

On July 2, 2003, the FDA analyzed 32 studies, totaling 691 patients, which compared the diagnostic accuracy of the endoscopic capsule for diseases of the small intestine (71%) with the accuracy of other tests in use at the time for assessment of the small intestine (intestinal transit, push, enteroscopy, abdominal computed tomography [CT], scintigraphy and intraoperative enteroscopy) (41%). Based on the analysis of these studies, the FDA determined that the endoscopic capsule was to become the first-line diagnostic method for the evaluation and detection of diseases of the small intestine (Figure 17.1).[9]

## ENDOSCOPIC CAPSULE SYSTEM

The components of the endoscopic capsule system are described below.

### Capsule

The capsule has a cylindrical shape measuring $11 \times 27$ to $11 \times 31$ mm, weighing about 3.7 g depending on the brand and model, covered with biocompatible material that is resistant to the action of digestive secretion and is not absorbable.

It is composed of an optical system in convex format, which prevents reflection of light, and one or two spherical lenses, which capture the images; a LED lighting system that provides white light to obtain images; a system

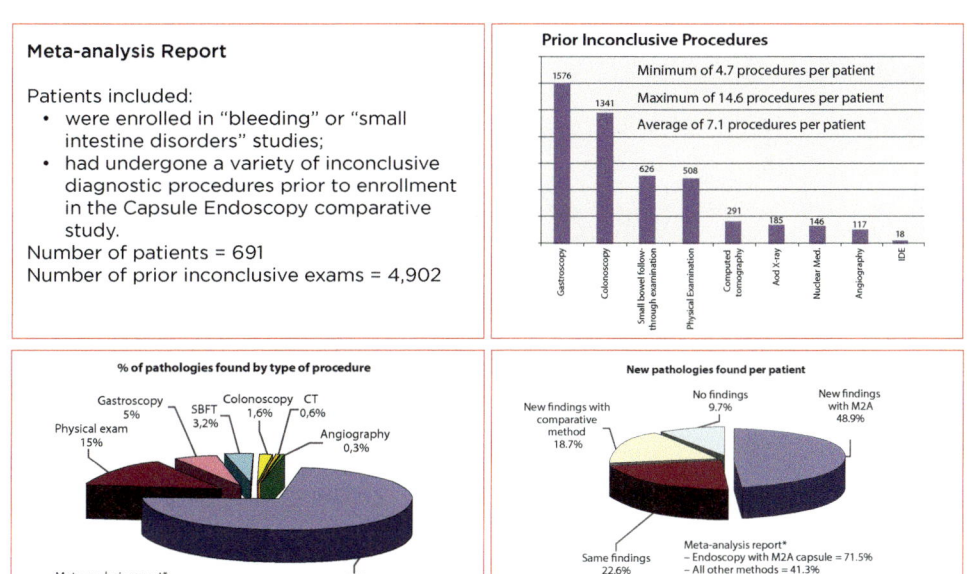

**Figure 17.1** Meta-analysis evaluated by the FDA.[9]

composed of 2 silver oxide batteries, which provide power to the entire system for about 9 to 10 hours; a CMOS (complementary metal oxide semiconductor) or CCD (charged coupled device) image capturing system; and an ASIC (ultra-high frequency VHF radio transmitting telemetry) transmission system consisting of an antenna that emits signals and transmits them by radio frequency to the sensors or HBC (human body communication) system, which transmits the images through the tissues of the human body. The images obtained by the capsule have a visual field of 140 to 160 degrees, with magnification of 1:8, depth ranging from 1 to 30 mm and ability to detect lesions with a size equal to or greater than 1 mm in diameter (Figure 17.2).

### Enteric capsule

Size: 11 × 26.5 mm; weight: 3.7 g; optical system with field of vision: 140 to 156°; image magnification: 1:8; battery duration time: 8 to 11 hours. Captures around two frames per second and about 50,000 to 70,000 images during the examination (Table 17.1 and Figure 17.3).

### Sensors

Adjusted to the patient's abdomen, the sensors pick up radio frequency signals, or they are transmitted to the HBC system by the capsule, and transferred to the recorder.

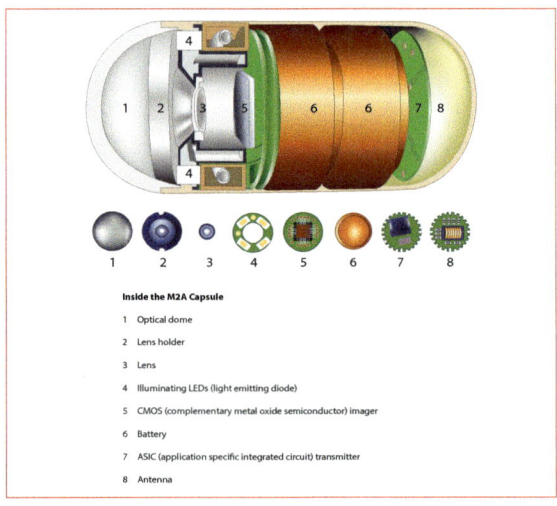

**Inside the M2A Capsule**

1  Optical dome
2  Lens holder
3  Lens
4  Illuminating LEDs (light emitting diode)
5  CMOS (complementary metal oxide semiconductor) imager
6  Battery
7  ASIC (application specific integrated circuit) transmitter
8  Antenna

**Figura 17.2** Capsule.

**Table 17.1** Characteristics of different enteric capsules[10]

|  | Pill cam SB2 | EndoCapsule | MiroCam | OMOM capsule |
|---|---|---|---|---|
| Length (mm) | 26 | 26 | 24 | 27.9 |
| Diameter (mm) | 11 | 11 | 11 | 13 |
| Weight (g) | 3.4 | 3.8 | 3.4 | 6 |
| Frame rate (frames/ second) | 2 | 2 | 3 | 0.5-2 |
| Image sensor | CMOS | CCD | CCD | CCD |
| Field of view | 156º | 145º | 150º | 140º |
| Illumination | 6 LEDs (white) | 6 LEDs (white) | 6 LEDs (white) | NA |
| Antennas (body leads) (n) | 8 | 8 | 9 | 14 |
| Real time (RT) viewing | RT viewer | VE-1 viewer | Miro-Viewer | RT monitoring |
| Recording time (hours) | 8 | 9 | 11 | 7-9 |

CMOS: complementary metal-oxide semiconductor; CCD: charge-coupled device; LED: light emitting diode; NA: not applicable.

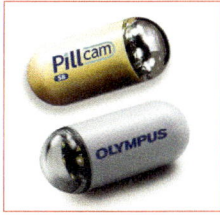  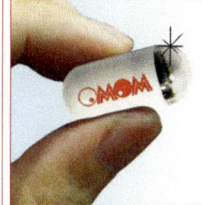 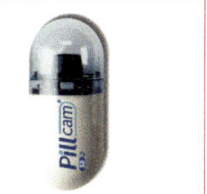

**Figure 17.3** Enteric capsules.

## Recorder

The recorder is a microcomputer with hardware, attached to the belt, which receives the signals of the images captured by the capsule and stores then. Some models of recorder have a system to allow the visualization of the image that is being captured by the capsule in real time, thereby ensuring that the capsule has reached the small intestine.

## WORK STATION

Computer and program that process the images obtained by the capsule and transmitted to the recorder, and transform them into a film to be analyzed. These programs count on various resources that help in the analysis of the images obtained by the capsule (Figures 17.4 and 17.5).

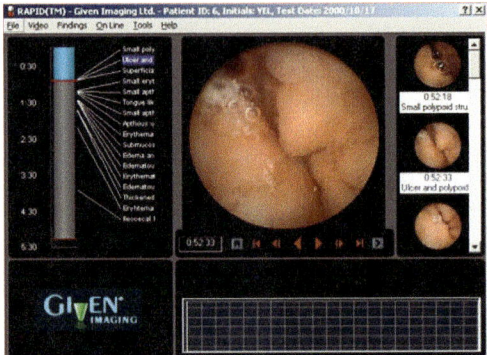

**Figure 17.4** Work station.

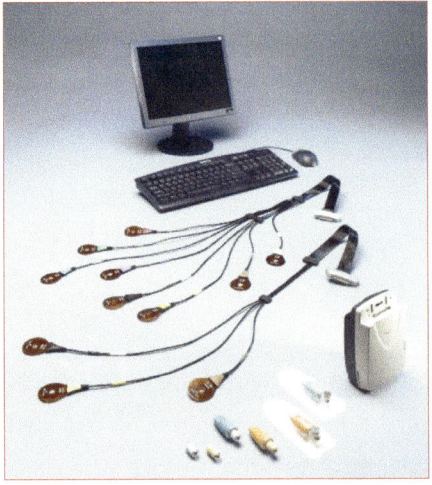

**Figure 17.5** Recorder and work station.

## PREPARATION OF THE EXAMINATION

To date, there is no consensus on the ideal preparation for the realization of enteric capsule examinations. The only recommendation for the examination of the small intestine is an 8 hour fast. Some studies assessed the utilization of a formulation with purgative solutions, such as polyethylene glycol and sodium phosphate, but these do not show conclusive results when compared to a diet with clear liquids when assessing the rate of complete exams, as well as in relation to gastric emptying time and intestinal transit time, though they appear to improve visualization of the mucosa. The use of prokinetics and simethicone has not been shown to be significantly superior, either.[10,11]

## EXAMINATION TECHNIQUE

After installation of the sensors on the surface of the patient's abdomen or chest and connection to the recorder, the endoscopic capsule is swallowed with the aid of a glass of water. It is recommended that a few minutes before the start of the exam (ingestion of the capsule), the patient take a few drops of a surfactant for the elimination of bubbles in the gastrointestinal secretions, although several randomized studies have not shown that this practice is effective in improving visualization of the small bowel mucosa.[11]

Once the capsule is removed from its protective container, it starts capturing images at 2 to 6 frames per second, until the end of the capacity of its batteries, that is, 8 to 12 hours depending on the capsule's model, providing about 50,000 to 260,000 images acquired during its passage through the gut.

For the evaluation of the small intestine after ingestion of the capsule, the patient is told to maintain their usual activities, and may ingest clear liquids after 2 hours and eat a light diet, after 4 hours. After 8 to 12 hours, the patient returns to remove the recorder. As the average gastric emptying time varies from 10 to 319 minutes (63 minutes on average) and the transit time of the small intestine ranges from 70 to 322 minutes (194 minutes on average), the capsule reaches the colon before the end of the batteries in over 90% of cases, providing complete visualization of the small intestine.[8,11]

The analysis of the images is conducted after transmission of the data from the recorder to the work station, which processes and transforms it into a film lasting around 2 hours, which is analyzed by the physician.

The capsule is eliminated in evacuations, which is unnoticed by the patient in the majority of cases, with no need to recover it.[1,8,10,11]

## EXAMINATION ROUTINE

For the realization of the endoscopic capsule examination, after installation of the sensors, the patient merely ingests the capsule with a glass of water and returns after 8 to 12 hours to remove the recorder. Two hours after the start of the examination, they may ingest clear fluids and, after 4 hours, begin a light diet.

## CONTRAINDICATIONS TO CAPSULE ENDOSCOPY EXAMINATIONS

- Absolute: obstructive symptoms, intestinal subocclusion and pregnancy.
- Relative: intestinal motility alterations (gastroparesis), suspected adhesions or fistulas, presence of a pacemaker or implanted defibrillator, large or numerous diverticula of small bowel, Zenker's diverticulum, swallowing disorders and extensive small bowel CD with symptoms suggestive of sub-stenosis, and pregnancy.[10,11]

## INDICATIONS OF CAPSULE ENDOSCOPY IN IBD

The evaluation of the mucosal surface of the small intestine is necessary in various and different times of IBD:

- establishing the definitive diagnosis of CD;
- assessing the extent and seriousness of lesions in the small intestine;
- monitoring the treatment response and healing of the mucosa;
- monitoring post-operative recurrence;
- investigating unexplainable symptoms;
- helping the differential diagnosis of the forms of IBD unclassified and difficult-to-treat UC.[1,2,4,6,10,12-14]

### Suspected CD diagnosis

To date, no single test is considered the gold standard for the diagnosis of CD. This is undertaken through the clinical history data, physical examination, laboratory, radiological, endoscopic and histopathological tests.[1,2,4,12-14]

The involvement of the small intestine in CD is present in about 75-80% of cases, occurring in isolation in approximately 30%, with an average of 36 months between the onset of symptoms and the presence of identifiable abnormalities on radiological investigations of the small intestine, as early lesions, limited to the mucosa, are not detectable on radiological exams, delaying the diagnosis and treatment of these patients.[1,6,13,15-17] It is known that in the natural history of CD, there is involvement of the jejunum in more than 50% of patients with diseases of the small intestine, and the presence of jejunal lesions is related to increased risk of recurrence of the disease and a worse prognosis. These lesions are also more difficult to detect using radiological examinations.[18]

Many studies have been published in recent years in an attempt to determine the actual role of the endoscopic capsule in the diagnosis of small bowel CD, given that despite being the most accurate method in the detection of lesions to the mucosa of this organ, 13% of the normal and asymptomatic population presents small erosions to the mucosa of the small intestine without pathological significance.[1,2,16] Added to this is the fact that many of the mucosal lesions found in CD are not specific and can occur in enteropathy from other etiologies, such as lymphoma of the small intestine, actinic enteropathy, enteropathy by non-steroidal anti-inflammatory drugs (NSAIDs), ischemic enteropathy, opportunistic enteropathies of human immunodeficiency virus (HIV), intestinal tuberculosis and Behcet's disease.[1,2,4,6,12-14,16] Among these, the differential diagnosis most frequently imposed is that of lesions induced by NSAIDs, as ulcers and erosions may be present in the small intestine after a short period of utilization of NSAIDs and the incidence of these lesions in chronic users can reach 70%.[6]

The range of lesions observed by the endoscopic capsule in patients with CD is similar to the lesions of other conventional endoscopy examinations, varying according to the degree of activity and the extent of involvement of the disease. The mucosal alterations found in CD are: erythema, edema, denuded areas with loss of villi, aphthous erosions, fissures, linear ulcers, irregular and confluent ulcers across the circumference of the organ, sub-stenosis and cobblestoning of the mucosa.[13]

The endoscopic criteria for the diagnosis of CD of the small intestine with the endoscopic capsule have been discussed. The most commonly used criterion was proposed by Mow and consists in finding at least three ulcers in the absence of ingestion of NSAIDs for 2 weeks, providing sensitivity of 77%, specificity of 89%, positive predictive value of 55% and negative predictive value of 96%.[1,12] Despite lower sensitivity, other authors argue that the presence of 10 or more aphthous erosions in the same segment or distributed in different segments would be enough to suggest the endoscopic diagnosis of small intestinal CD.[12]

According to endoscopic findings, two severity scores have been proposed: the Lewis score (Table 17.2), which considers the affected portion of the small intestine and endoscopic findings suggestive of inflammatory process, building a score in accordance with these. Thus, a score below 135 corresponds to insignificant inflammatory changes to the mucosa; between 135 and 790 indicates mild inflammatory activity; and greater than 790 indicates moderate to severe inflammation. The Lewis score was evaluated in a study by Rosa et al., displaying a positive predictive value of 82.6%, negative predictive value of 87.9%, sensitivity of 82.6% and specificity of 87.9% in the diagnosis and CD activity in the small intestine.[19]

Another score proposed is the CECDAI (Capsule Endoscopy Crohn's Disease Activity Index) (Table 17.3) or Niv score, which considers the involvement of the small intestine in two segments, proximal and distal, scoring in each of them for the presence of inflammation of the mucosa (A), extent of the disease (B) and the presence of sub-stenosis (C). Thus, patients with score of zero correspond to absence of inflammatory activity, while a score of 26 points reflects intense inflammation.[1,6,17,20] The Niv or CECDAI score was validated by a prospective, multi-center study to follow-up patients with CD of the small intestine.[20]

An interesting fact shown in some studies is that despite the lack of a consensus on the endoscopic criteria for the diagnosis of CD, the absence of lesions or alterations of the mucosa during the assessment of the small intestine through endoscopic capsule examination in patients with suggestive symptoms virtually excludes the diagnosis of CD.[1,21,22]

**Table 17.2** Lewis Score

| Parameters | Number | Longitudinal extension | Descriptors |
|---|---|---|---|
| Villous appearance (worst-affected tertile) | Normal – 0 | Short segment – 8 | Single – 1 |
| | Edematous – 1 | Long segment – 12 | Patchy – 14 |
| | | Whole tertile – 20 | Diffuse – 17 |
| Ulcer (worst-affected tertile) | None – 0 | Short segment – 5 | < 1/4 to 9 |
| | Single – 3 | Long segment – 10 | 1/4 to 1/2 to 12 |
| | Few – 5 | Total tertile – 15 | > 1/2 to 18 |
| | Multiple – 10 | | |
| Stricture (whole study) | None – 0 | Ulcerated – 24 | Transverse – 7 |
| | Single – 14 | Non-ulcerated – 2 | Not transverse – 10 |
| | Multiple – 20 | | |

Lewis score = Score of the worst-affected tertile [(villous parameter x extension x descriptor) + (number of ulcers x extension x size)] + stricture score (number x ulcerated x transverse).

Number of ulcers: single: 1; few: 2 to 7; multiple: > 8.

The ulcer descriptor (size) is determined by the extension of the capsule frame covered by the largest ulcer.

**Table 17.3** CECDAI Score*

| |
|---|
| **A. Inflammation score** |
| 0 = None |
| 1 = Mild to moderate edema/hyperemia/denudation |
| 2 = Severe edema/hyperemia/denudation |
| 3 = Bleeding, exudate, sores, erosion, small ulcer (< 0.5 cm) |
| 4 = Moderate ulcer (0.5 to 2 cm), pseudopolyp |
| 5 = Large ulcer (> 2 cm) |
| **B. Score of the extent of disease** |
| 0 = No disease |
| 1 = Focal disease (single segment) |
| 2 = Patchy disease (2 to 3 segments) |
| 3 = Diffuse disease (more than 3 segments) |
| **C. Stricture score** |
| 0 = None |
| 1 = Single-passed |
| 2 = Multiple-passed |
| 3 = Obstruction (non passage) |

* CECDAI = proximal ([A1 x B1] + C1) + distal ([A2 x B2] + C2).

Various studies and meta-analyses have been published comparing the sensitivity, specificity and accuracy of capsule endoscopy with other endoscopic and radiological methods in an attempt to establish an algorithm to guide the diagnostic investigation of these patients. A meta-analysis of 12 studies comparing the accuracy of capsule endoscopy for the diagnosis of small bowel CD, as well as the evaluation of impairment of the small intestine in patients with a diagnosis of CD showed that the capsule was superior to intestinal transit, computed tomography enterography, colonoscopy with retrograde ileoscopy and magnetic resonance imaging in the evaluation of suspected CD of the small intestine, and more effective in the assessment of involvement of the small bowel in patients with CD compared to intestinal transit assessments, computed tomography enterography and push enteroscopy, but with inferior results to magnetic resonance imaging in this population.[16]

In a prospective study comparing the sensitivity and specificity of capsule endoscopy, computed tomography and magnetic resonance imaging in the evaluation of terminal ileum CD, the capsule showed a sensitivity of 100% and specificity of 91%, while these were 81 and 86% for computed tomography, and 76 and 85% for magnetic resonance imaging, demonstrating a statistically significant difference in the sensitivity of the capsule in relation to tomography, while the difference with the magnetic resonance was not significant. Specificity did not differ among the three methods.[23]

In order to guide the request for assessment of the small intestine through the new endoscopic methods currently available, recommendations from the Societies of Endoscopy and Gastroenterology have been disclosed, as well as some consensuses. These include the consensus held by the World Endoscopy Organization (WEO) and the European Crohn's and Colitis Organization (ECCO), establishing guidelines regarding the use of capsule endoscopy in the diagnostic evaluation of small intestine CD, as shown ahead.[1,22]

**WEO and ECCO recommendations: capsule endoscopy in the investigation of CD**

1. Ileocolonoscopy should be performed before the capsule endoscopy in the investigation of CD.

2. A radiological examination should usually precede the capsule examination in the investigation of CD; the choice of examination will depend on the availability and local knowledge.

3. There is no evidence to support the preparation for the capsule examination in patients for the investigation for CD.

4. Capsule endoscopy is capable of identifying lesions of the mucosa compatible with CD not identifiable using other endoscopic and radiological tests.

5. Like all imaging tests, the diagnosis of CD should not be instituted with the data obtained via the capsule examination alone.

6. A normal capsule endoscopy examination has a high negative predictive value in the diagnostic exclusion of active CD in the small intestine.

7. At present there are no validated endoscopic criteria for the diagnosis of CD through capsule endoscopy.

8. In the diagnosis of lesions of the mucosa of the small intestine compatible with CD, capsule endoscopy appears to be superior to contrasted radiography of the small intestine, magnetic resonance imaging, computed tomography and computed tomography enteroclysis.

9. Guided enteroscopy can be used to assess the small intestine in the investigation of CD and has the advantage of allowing biopsies to be performed. However, there is no data to suggest that the histological findings obtained result in changes in conduct and management of the patient.

10. The decision whether to evaluate the small intestine using capsule endoscopy or guided enteroscopy will depend on the intestinal segment that is suspected of being affected, as well as the availability and local knowledge.

11. Preferably, the capsule should precede enteroscopy in the investigation of CD in the small intestine because it is less invasive and because it can guide the enteroscopy, if necessary.

Recently, the European Consensus on Inflammatory Bowel Disease established that patients with suspected diagnoses of CD without changes in the ileocolonoscopy should undergo endoscopic evaluation of the small intestine via capsule endoscopy in the absence of obstructive signs or symptoms. In patients with suspected sub-stenosis, imaging tests such as CT, enterography and MRI should be preferred.[22]

Despite not providing the performance of biopsies to obtain material for histological study, suspected diagnosis of small intestinal CD is currently the second most important indication for the use of capsule endoscopy in adults and the main indication in the age range of 10 to 18 years, being more cost-effective when, in addition to the clinical symptoms, there is laboratory data suggestive of inflammatory activity, such as anemia, thrombocytosis, serological and/or positive fecal inflammation markers.[10,12,24,25]

### Evaluation of patients with an established diagnosis of CD

Endoscopic examinations play an important role in the evaluation and monitoring of CD, assessing the degree of activity of the disease and allowing therapeutic interventions which were, until recently, limited to the reach of traditional endoscopes, that is, unable to assess the medial and distal jejunum, as well as the proximal, medial and distal part of the ileum. The introduction of capsule endoscopy and guided enteroscopes has enabled these segments to be approached endoscopically. However, their role in the evaluation and monitoring of CD is still being established.[1,4,10,12,14,22]

Capsule endoscopy has demonstrated diagnostic accuracy in patients with diagnosis of CD for detection of lesions and alterations suggestive of disease activity ranging from 78 to 93%, while intestinal transit resulted in an accuracy of 32%, CT enterography/enteroclysis at 38%, and MRI enterography at 79%. Capsule endoscopy and MRI have shown a good correlation in the detection of inflammatory activity, as well as its localization. However, the MRI assessment also allows transmural evaluation and detection of extraintestinal disease activity, making it the first-line imaging examination for monitoring of disease activity.[1,4,10,22]

An interesting aspect to be considered is that, compared with CT and MRI, capsule endoscopy has shown higher capacity for the detection lesions of the proximal small intestine, diagnosing lesions in the jejunum in more than 50% of patients with an ileal CD diagnosis. However, the clinical significance of this finding still needs to be clarified.[22] Given this scenario, the capsule should currently be reserved for patients with an established diagnosis of

**Table 17.4** Endoscopic capsule *vs.* other diagnostic methods

| Modality | Ref. | No. of patients | Diagnostic yield of VCE | Diagnostic yield of the modality compared | IY | p-value |
|---|---|---|---|---|---|---|
| CTE | Eliakim et al. | 35 | 77% | 20% | 47% | <0,05 |
| | Hara et al. | 17 | 71% | 53% | 18% | ND |
| | Voderholzer et al. | 41 | 61% | 49% (CT enteroclysis) | 12% | <0,04 |
| | Solem et al. | 40 | 83% | 83% | 0 | NS |
| MRE | Albert et al. | 27 | 93% | 78% | 15% | NS |
| | Crook et al. | 19 | 93% | 71% | 18% | NS |
| | Jensen et al. | 93 | 100% | 86% | 14% | NS |
| Ileocolo-noscopy | Hara et al. | 17 | 71% | 65% | 6% | NS |
| | Solem et al. | 40 | 83% | 74% | 9% | NS |
| | Leighton et al. | 80 | 55% | 25% | 30% | NA |

VCE: Videocapsule endoscopy; CTE: Computed tomography enterography; MRE: magnetic resonance enterography; IY: incremental yield; NA: not available; NS: not significant.

CD, the investigation of unexplained symptoms such as abdominal pain, diarrhea, flatulence, iron deficiency anemia and bleeding, and investigation of postoperative recurrence when ileocolonoscopy is contraindicated or not possible due to technical difficulties.[1,4,6,12,22]

The potential role of capsule endoscopy in monitoring the healing of the mucosa and thus the response to therapeutic drugs has yet to be established.[1,22] Mucosal healing is defined by the absence of macroscopically visible alterations of inflammatory activity and is an important marker for the efficiency of treatment, and associated reduction in the risk of long term complications.[6] Some important studies have shown that the improvement of clinical symptoms is not correlated, in all cases, with healing or improvement of the lesions of the mucosa. It is currently known that the symptoms of the disease do not occur in the absence of changes to the mucosa, but not all mucosal lesions are associated with symptoms and, therefore, monitoring of mucosal healing seems to be important in order to guide drug treatment.[1,4,6,22]

**WEO and ECCO recommendations: capsule endoscopy in diagnosed CD**

1. Patients presenting symptoms unexplainable on investigation using other methods.[1,4,6,12,22]

2. Patients with postoperative recurrence where it is not possible to conduct an ileocolonoscopy.[1,4,6,12,22]

3. For assessment of healing when necessary.[1,4,6,12,22]

4. The performance of capsule endoscopy must be preceded by MRI enterography or CT enterography, because they enable the identification of obstructive lesions and evaluate the distribution and the transmural and extraintestinal involvement of the disease.[1,4,6,12,22]

## DIFFERENTIAL DIAGNOSIS OF INFLAMMATORY BOWEL DISEASE UNCLASSIFIED

Population studies have shown that about 4 to 10% of adult patients present colonic IBD whose presentation does not allow a differential diagnosis between CD and nonspecific UC, receiving the diagnosis of IBD unclassified. The inability to establish a definitive diagnosis has implications in drug treatment and the clinical evolution of these patients.[1,6,22] It is known that about 30% of them will have the diagnosis of CD during the course of their disease, usually through the identification of lesions in the small intestine.[1,6,22]

Some small studies have evaluated the role of capsule endoscopy in the diagnostic investigation of this group of patients with a diagnosis of CD, through the detection of lesions of the small bowel mucosa, ranging from 17 to 70% in different series. However, the criteria used for establishing the diagnosis were arbitrary.[1,6,22]

It is important to note that the absence of lesions in the small bowel mucosa during capsule examination in this group of patients does not allow us to exclude a future diagnosis of CD.[1,4,6,14,22]

**WEO and ECCO recommendations: capsule endoscopy in the investigation of CD**

1. In patients with a diagnosis of IBD unclassified, the evaluation of the small intestine using capsule endoscopy can help define the diagnosis in the case of identification of lesions suggestive of CD.[1,4,22]

2. The absence of detectable lesions in the capsule endoscopy examination in the small intestine of patients with IBD unclassified does not exclude the possibility of a CD diagnosis in the future.[1,4,22]

3. In patients with IBD unclassified, capsule endoscopy has greater diagnostic accuracy for lesions of the small intestine than intestinal transit. There is no comparative data with other radiological methods.[1,22]

### Differential diagnosis of difficult-to-treat UC

The diagnosis of nonspecific UC is conducted based on typical clinical, laboratory, endoscopic and histological data, without the need to study the small intestine. However, about 10% of these patients will be reclassified as having CD during disease progression. Capsule endoscopy is capable of detecting lesions to the mucosa of the small intestine compatible with CD, thereby allowing the therapy that the patient should receive to be guided.[1,22] In a retrospective study, about 10% of patients with atypical symptoms, 9% of those resistant to drug therapy and 33% of those with recurrence of symptoms after colectomy were reclassified as having CD with the evidence of three or more ulcers in the small intestine via capsule endoscopy.[1]

### WEO and ECCO recommendations: capsule endoscopy in the investigation of nonspecific UC

1. The diagnosis of nonspecific UC does not require the study of the small intestine.[1,22]

2. The evaluation of the small intestine via capsule endoscopy may be indicated in patients with unexplainable symptoms to conventional investigation.[1,22]

3. The capsule is able to identify lesions in the small intestine of patients with ulcerative colitis, especially those with atypical symptoms and refractory to drug therapy. However, the real significance of these lesions still requires clarification.[1,22]

### COMPLICATIONS OF CAPSULE ENDOSCOPY

Capsule retention is the main complication of this new endoscopic method, defined as the proven presence of the capsule by means of a plain abdominal radiograph

2 weeks after ingestion.[26] This 2 week period was established because in up to 20% of cases, there may be incomplete examinations as a result of slow intestinal transit.[26] The capsule retention rates vary according to the indication of the examination: in healthy volunteers, it did not occur (0%); in patients with suspected CD of the small intestine it occurred in 1%; in patients with CD, it occurred in 4 to 5%; in those investigated for obscure bleeding, retention occurred in up to 1.5%; and in suspected cases of subocclusion, in up to 21%.[27] Chronic users of NSAIDs, those undergoing abdominal radiation therapy, individuals with a history of abdominal surgery and enteric anastomosis are also more susceptible to this complication.

At present, there is no diagnostic method that can ensure, in 100% of cases, that there was no retention or impaction of the capsule.[28,29] The suggestion is for realization of radiological examinations with contrast orally, in an attempt to exclude patients with sub-clinical sub-stenosis.[27]

Riccioni et al. have demonstrated good results in the prevention of impaction or capsule retention with the use of the Agile Patency System.[28,29] However, although preliminary studies have presented positive predictive values of 100%, negative predictive values still require further study.

The Agile Patency System consists of a capsule with equal dimensions to the enteric capsule (11 × 26 mm), consisting of biodegradable material, which disintegrates after 30 hours of contact with the digestive fluids. The Agile test capsule includes a marker that emits a radio frequency, thereby enabling verification of whether the test capsule has been expelled from the digestive tract. It is also radiopaque, allowing localization via plain abdominal radiography. The capsule being eliminated within these 30 hours and, therefore, without evidence of disintegration, means that the examination has been safely performed.[28,29]

Patients who have pain and/or abdominal distension during the evaluation with the permeable capsule should not be subjected to examinations with endoscopic capsules.[10,11,28,29]

Carey et al. and Hollerbach et al. were the first to report on the success of the guided introduction of the capsule using endoscopy with different techniques (use of overtube and foreign body graspers) in patients with a history of gastric surgery, esophageal, gastric or pyloric sub-stenosis, dysphagia and gastroparesis. Currently, there are accessories specifically designed for this purpose, called capsule introducers, which make this practice easier, and should be reserved for people with anatomical abnormalities that can make it difficult to swallow the capsule or pass it into the small intestine (e.g. gastrectomy or gastroplasty patients).[10,11,30]

The aspiration of the capsule into the bronchial tree is a complication, fortunately very rare, which is more frequent in elderly male patients. In patients with swallowing dysfunctions, it is advisable to have the passage of the capsule performed endoscopically.[10,11,30]

### WEO and ECCO recommendations: complications of capsule endoscopy in the investigation and monitoring of IBD

1. In patients being investigated for CD, the risk of capsule retention is low and comparable to the risk of indication for research into obscure bleeding.[1,22]

2. In patients with established CD, the risk of retention is higher, and may reach 13%. A normal radiological study does not exclude the risk of capsule retention.[1,22]

3. The permeability capsule reduces the risk of retention and should be considered whenever there are indications or suspicion of sub-stenosis.[1,22]

4. There is no evidence that the capsule causes complications or interference with pacemakers or implantable defibrillators and vice-versa.[1]

### CAPSULE ENDOSCOPY IMAGES: CD

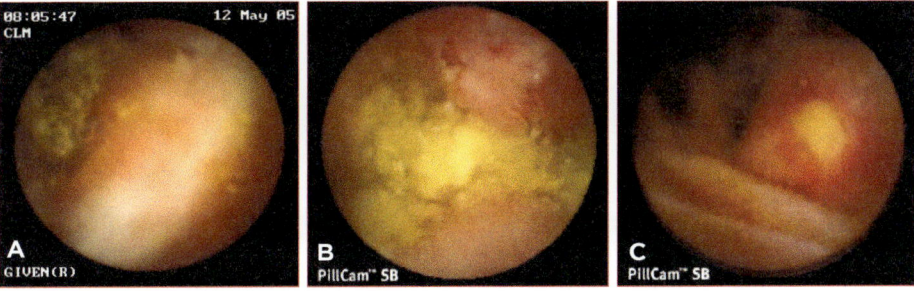

**Figure 17.7** Scarring sub-stenosis (A); longitudinal ulcer (B); ulcer (C).

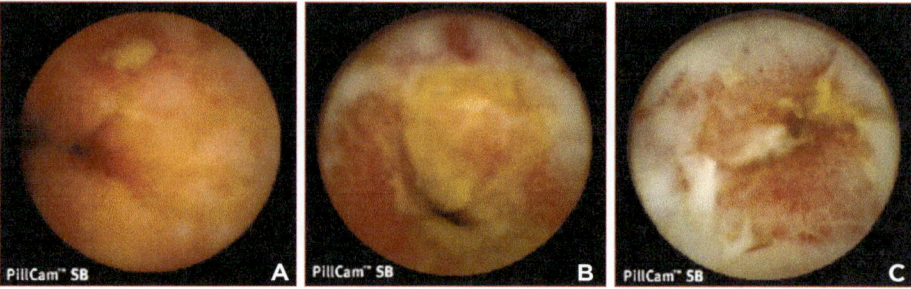

**Figure 17.8** Ulcer (A); confluent ulcers (B and C).

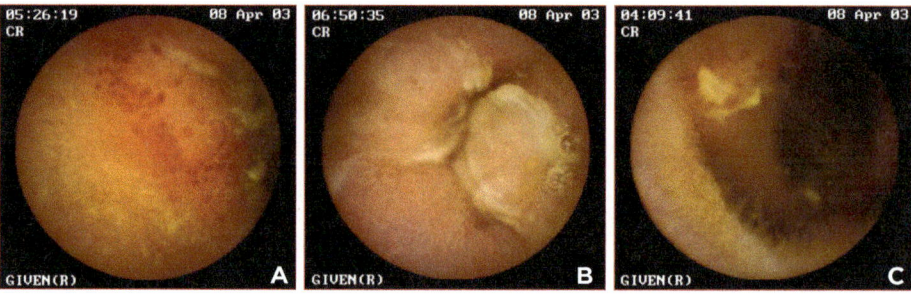

**Figure 17.9** Linear ulcer (A); confluent ulcers (B); ulcer (C).

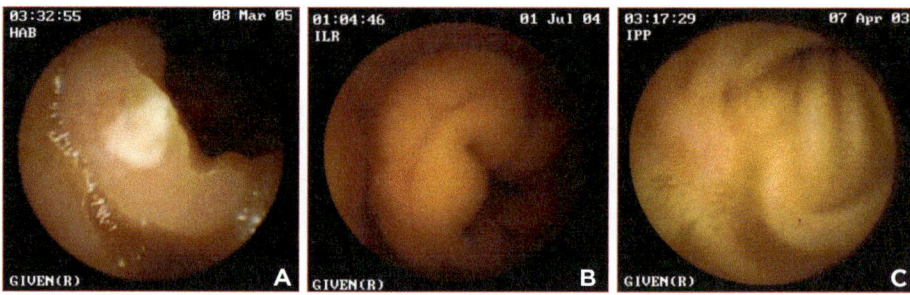

**Figure 17.10** Ulcer (A); fissure (B); aphthous erosion (C).

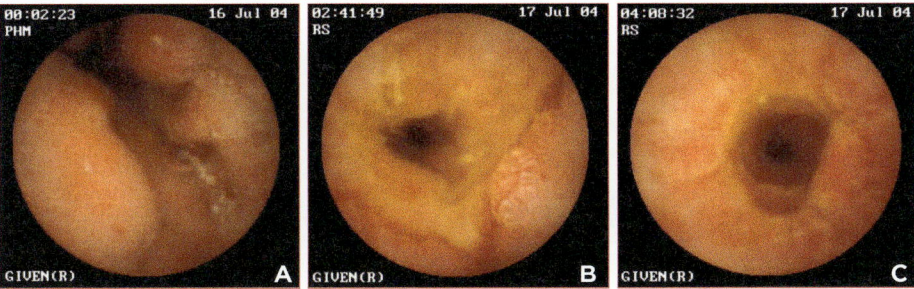

**Figure 17.11** Aphthous erosions (A); circumferential ulcers (B and C).

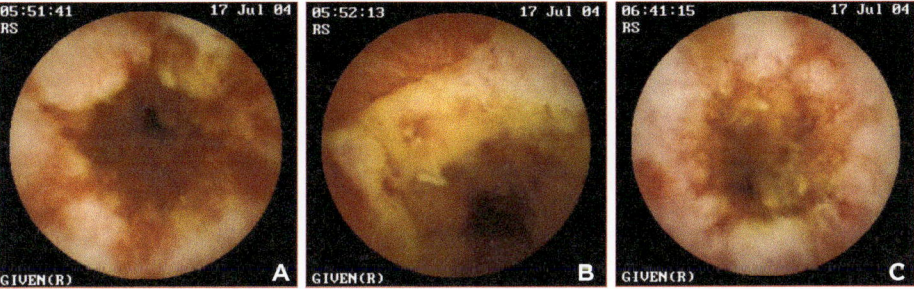

**Figure 17.12** Circumferential ulcers (A, B and C).

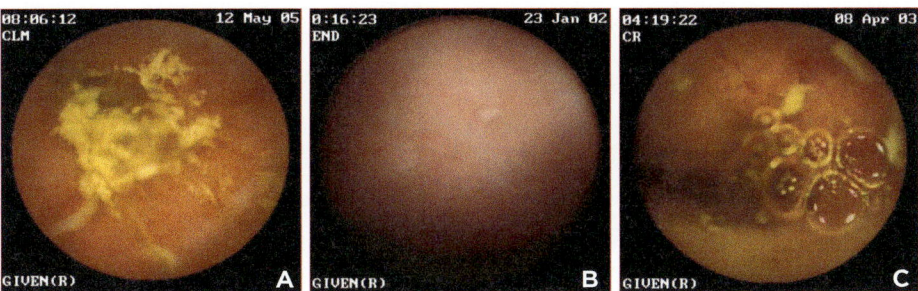

**Figure 17.13** Linear ulcer (A); aphthous erosion (B); erosions and small ulcers (C).

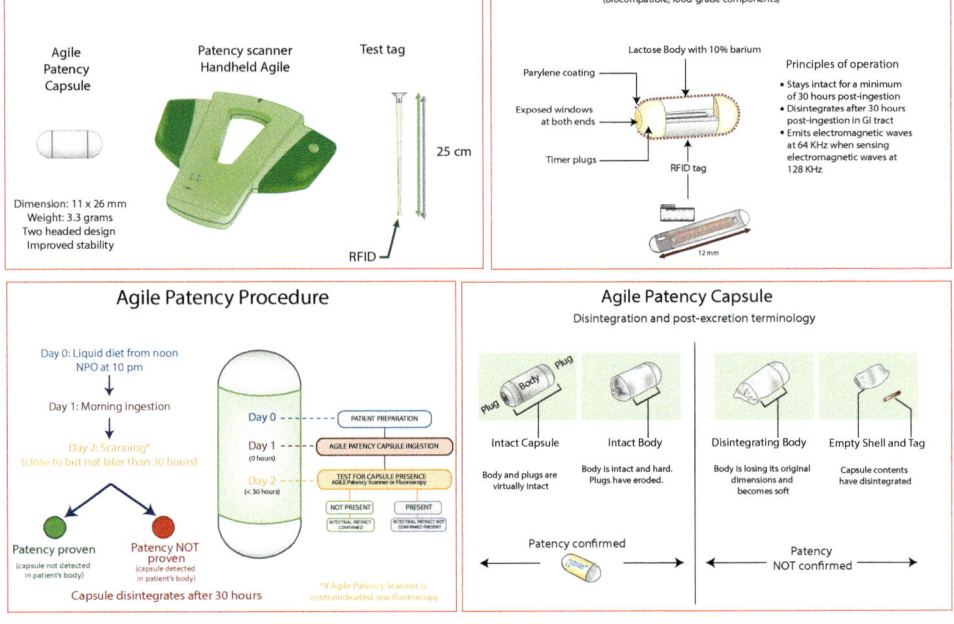

**Figure 17.6** Agile Patency System.

Source: Adapted from Cave DR, 2006.

## REFERENCES

1.  Bourreille A, Ignjatovic A, Aabakken L, Loftus EV Jr, Eliakim R, Pennazio M et al. Role of small-bowel endoscopy in IBD: international OMED–ECCO consensus. Endoscopy 2009; 41:618-37.

2.  Gert Van A et al. for the European Crohn's and Colitis Organisation (ECCO). The second European evidence-based consensus on the diagnosis and management of Crohn's disease: definitions and diagnosis. J Crohns Colitis 2010; 4:7-27.

3.  Dignass A, Eliakim R, Magro F, Maaser C, Chowers Y, Geboes K et al. Second european evidence-based consensus on the diagnosis and management of ulcerative colitis part 1: definitions and diagnosis. J Crohns Colitis 2012; 6(10):965-90.

4.  Peyrin-Biroulet L, Bonnaud G, Bourreille A, Chevaux JB, Faure P, Filippi J et al. Endoscopy in inflammatory bowel disease: recommendations from the IBD Committee of the French Society of Digestive Endoscopy (SFED) Endoscopy 2013; 45:936-43.

5. Panaccione R, Hibi T, Peyrin-Biroulet L, Schreiber S. Implementing changes in clinical practice to improve the management of Crohn's disease. J Crohns Colitis 2012; 6:S235-42.

6. Kopylov U et al. Capsule endoscopy in inflammatory bowel disease. World J Gastroenterol 2014; 20(5):1155-64.

7. Tharian B et al. Enteroscopy in Crohn's disease. World J Gastrointest Endosc 2013; 5(10):476-86.

8. ASGE Technology Evaluation Report. Gastrointestinal Endoscopy 2002; 56:621-4.

9. Internal data at Given Imaging Ltd. Reviewed by the FDA (2001).

10. Ladas SD, Triantafyllou K, Spada C, Riccioni ME, Rey JF, Niv Y et al. ESGE recommendations on VCE in investigation of small-bowel, esophageal, and colonic diseases. Endoscopy 2010; 42:220

11. ASGE Technology Status Evaluation Report: wireless capsule endoscopy. Gastrointestinal endoscopy 2006; 63(4):539-45.

12. Luján-Sanchis M, Sanchis-Artero L, Suárez-Callol P, Medina-Chuliá E. Indications of capsule endoscopy in Crohn's disease. Rev Esp Enferm Dig 2014; 106(1):37-45.

13. Arguelles-Arias F, Rodríguez-Oballe J, Duarte-Chang C, Castro-Laria L, García-Montes JM, Caunedo-Álvarez A et al. Capsule endoscopy in the small bowel Crohn's disease. Gastroenterology Research and Practice Volume 2014, Article ID 529136.

14. Yamagami H, Watanabe K, Kamata N, Sogawa M, Arakawa T. Small bowel endoscopy in Inflammatory Bowel Disease. Clin Endosc 2013; 46:321-26.

15. Leighton JA, Gralnek IM, Cohen SA, Toth E, Cave DR, Wolf DC et al. Capsule endoscopy is superior to small-bowel follow-through and equivalent to ileocolonoscopy in suspected Crohn's disease. Clinical Gastroenterology and Hepatology 2014; 12:609-15.

16. Dionisio PM, Gurudu SR, Leighton JA, Leontiadis GI, Fleischer DE, Hara AK et al. Capsule endoscopy has a significantly higher diagnostic yield in patients with suspected and established small bowel Crohn's disease: A meta-analysis. Am J Gastroenterol 2010; 105:1240-8.

17. Niv Y. Capsule endoscopy in the diagnosis of Crohn's disease. Medical Devices Evidence and Research 2013; 6:85-9.

18. Lazarev M, Huang C, Bitton A, Cho JH, Duerr RH, McGovern DP et al. Relationship between proximal Crohn's disease location and disease behavior and surgery: a

cross-sectional study of the IBD genetics consortium. Am J Gastroenterol 2013; 108(1):106-12.

19. Rosa B, Moreira MJ, Rebelo A, Cotter J. Lewis Score: a useful clinical tool for patients with suspected Crohn's disease submitted to capsule endoscopy. Journal of Crohn's and Colitis 2012; 6(6):692-7.

20. Niv Y, Ilani S, Levi Z, Hershkowitz M, Niv E, Fireman Z et al. Validation of the Capsule Endoscopy Crohn's Disease Activity Index (CECDAI or Niv score): a multicenter prospective study. Endoscopy 2012; 44(1):21-6.

21. Hall B, Holleran G, Costigan D, McNamara D. Capsule endoscopy: high negative predictive value in the long term despite a low diagnostic yield in patients with suspected Crohn's disease. United European Gastroenterology Journal 2013; 1(6):461-6.

22. Vito Annese et al. on behalf for ECCO European evidence based consensus for endoscopy in inflammatory bowel disease. Journal of Crohn's and Colitis 2013; 7:982-1018.

23. Jensen MD, Nathan T, Rafaelsen SR, Kjeldsen J. Diagnostic accuracy of capsule endoscopy for small bowel Crohn's disease is superior to that of MR enterography or CT enterography. Clin Gastroenterol Hepatol 2011; 9:124-9.

24. ASGE guideline: endoscopy in the diagnosis and treatment of inflammatory bowel disease. Gastrointestinal Endoscopy 2006; 63(4):558-66.

25. ASGE Technology Status Evaluation Report. Wireless capsule endoscopy. Gastrointestinal Endoscopy 2013; 78(6):805-16.

26. Cave D, Legnani P, de Francis R, Lewis BS. ICCE Consensus for Capsule retention. Endoscopy 2005; 37:1065-7.

27. Barkim JS, O'Loughlin C. Capsule endoscopy contraindications and how to avoid their occurrence. Gastrointest Endosc Clin N Am 2004; 14:61-5.

28. Herrerias JM, Leighton JA, Costamagna G, Infantolino A, Eliakim R, Fischer D et al. Agile patency system eliminates risk of capsule retention in patients with known intestinal strictures who undergo capsule endoscopy. Gastrointest Endosc 2008; 67:902-9.

29. Riccioni ME, Hasaj O, Spada C, Tringali A, Petruzziello L, Mutignani M et al. "M2A patency capsule" to detect intestinal stictures: preliminary results. Program and abstracts of the Second Conference on Capsule Endoscopy. Berlin, March 2003; 169:23-5.

**30.** Koulaouzidis A, Rondonotti E, Karargyris A. Small-bowel capsule endoscopy: a ten-point contemporary review. World J Gastroenterol 2013; 19(24):3726-46.

# DIFFERENTIAL DIAGNOSIS OF ILEITIS

ELOÁ MARUSSI MORSOLETTO

## INTRODUCTION

The ileum, whose name derives from the Greek word *eileos* and means twisted, comprises roughly 3/5 of distal small intestine. It is located in the pelvic region and right iliac fossa, and remains fixed to the posterior abdominal wall by the mesentery. It is responsible for digestion and absorption of foods, and is also an endocrine (production of peptides) and immunological (Peyer's patches and immunoglobulin production) organ.[1]

In the terminal ileum, there may be bacterial or viral toxic substances resulting from food digestion, which makes the relationship with the lymphoid tissue fundamental for the protection of this segment. The presence of lymphocytes, macrophages and mast cells as a response to luminal antigens can be considered physiological. The ileal mucosa consists of villi measuring 0.5 to 1.5 mm, which appear as finger-like projections perpendicular to the muscularis mucosa; the crypts, which are tubular structures, are found below these two layers. The lining of the villi and crypts is formed by a layer of epithelial cells, with mature (enterocytes and goblet cells) and immature cells.[1]

Due to the immunological characteristics of the terminal ileum, the boundary between a normal ileum and mild ileitis (controlled inflammation) is inaccurate and indicates subjective criteria, which points to the need to establish objective criteria for the microscopic evaluation of the ileum. The diagnosis of diseases that affect this segment should be performed based on the association between clinical, laboratory, endoscopic and histological data.

The examination of the terminal ileum quickly gained interest in the international literature, the first report being published in 1972 by Nagasako. Over the last decades, the technical improvements of endoscopic equipment and refinement of knowledge have simplified intubation and the examination of the last centimeters of the terminal ileum. These advances have been confirmed in several articles, with a success rate for intubation of the terminal ileum during colonoscopy of 74-100%.[2] In experienced hands, it adds three minutes in procedure time and contributes significantly to quality assurance and the diagnostic efficiency.[3] Besides the possibility of macroscopic analysis, the biggest advantage of ileoscopy is that it allows biopsy for histopathological examination (Table 18.1).

Ileitis, or ileal inflammation, is often caused by Crohn's disease (CD) (Figure 18.1). However, a wide variety of diseases can be associated with ileitis, including infectious diseases, spondyloarthropathies, vasculitis, tumors, some drugs, infiltrative disease and other conditions (Table 18.2). The correct diagnosis of the specific cause of ileitis is of paramount importance because an incorrect diagnosis can result in delays or errors in patient management.[4]

## INFECTIOUS ILEITIS

*Yersinia enterocolitica* (or *Yersinia pseudotuberculosis*) is acquired by ingestion of contaminated food or water. Less often, contamination can occur by direct contact with infected domestic or wild animals. Enterocolitis is the most common clinical manifestation, characterized by diarrhea, mild fever and abdominal pain lasting 1-3 weeks. Vomiting occurs in 15 to 40% of cases. Clinical illness results from the organism penetrating the mucosa and invading the underlying intestinal lymphoid tissue, particularly Peyer's patches.[4] Diagnosis is by stool culture. On radiographs, a thickened and nodular mucosal pattern in the terminal ileum is observed. Endoscopic findings include aphthous lesions in the cecum and terminal ileum, with round or oval elevations and ulcerations (Figure 18.2). The ulcers are usually more uniform in size and shape than seen in CD.[4,5]

| Table 18.1 Indications for ileoscopy | |
|---|---|
| **Absolute indications for ileoscopy** | **Absolute indications for ileoscopy and biopsies** |
| Chronic diarrhea | Suspected Crohn's disease |
| Abdominal pain in RLQ | Chronic diarrhea |
| Abnormal radiological findings | Patients with acquired immunodeficiency syndrome (AIDS) |
| Suspected inflammatory bowel disease | Suspected tuberculosis |
| Family history of Crohn's disease | Ileal mucosa lesions on ileoscopy |

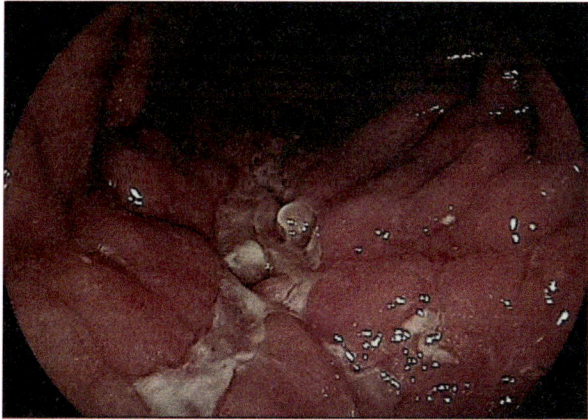

**Figure 18.1** Crohn's disease in the terminal ileum, with characteristic deep and longitudinal ulceration.

| **Table 18.2** Differential diagnosis of ileitis in clinical practice |
| --- |
| Infectious ileitis |
|     *Actinomyces israelli* |
|     *Anisakis simplex* |
|     *Clostridium difficile* |
|     Citomegalovírus |
|     Histoplasma *capsulatum* |
|     *Mycobacterium avium-intracellulare* complex |
|     *Mycobacterium tuberculosis* |
|     *Salmonella spp* |
|     Typhlitis |
|     *Yersinia enterocolitica* and *Yersinia pseudotuberculosis* |
| Ileitis associated with spondyloarthropathy |
| Lymphoid nodular hyperplasia |
| Vasculitis |
| Neoplasia |
|     Lymphoma; carcinoid tumor; adenocarcinoma of the ileum and cecum; leiomyosarcoma |
| Drug-related ileitis |
|     Nonsteroidal anti-inflammatory drugs; potassium chloride |
| Infiltrative causes of ileitis |
|     Eosinophilic gastroenteritis; sarcoidosis; amyloidosis; endometriosis |
| Isolated ulcerations or erosions of terminal ileum |
| Other causes |
|     Backwash ileitis |

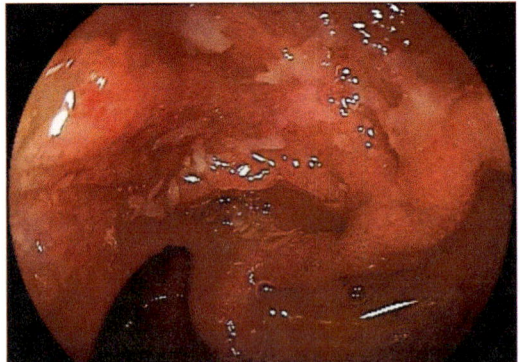

**Figure 18.2** Ileitis caused by Yersinia enterocolitica.

Infection by *Yersinia* can also lead to mesenteric adenitis with terminal ileitis mimicking acute appendicitis, which occurs more commonly in children and young adults. Ultrasound (or computed tomography) examinations can define the diagnosis, thickening of the intestinal mucosa with enlarged mesenteric lymph nodes. In CD, usually there is transmural inflammation with noncompressible fat surrounding the ileum.[4]

Infection by nontyphoidal *Salmonella* occurs after ingestion of contaminated food products of animal origin. Unlike nontyphoidal *Salmonella*, enteric fever (*Salmonella typhi* and *Salmonella paratyphi*) is transmitted person-to-person. *Salmonella* infections generally cause self-limited acute gastroenteritis but may cause bacteremia, vascular infections, and a state of chronic carrier. Because it can affect the regional mesenteric lymph nodes, adenitis and terminal ileitis may occur, mimicking acute appendicitis. Abnormal findings in CT and ileoscopy can be difficult to differentiate from those of other causes, including CD. Biopsies are useful, showing acute ileitis.[4] Cheung et al. reported case of a patient with ulcerations in the cecum and terminal ileum, with histology showing rich granulomatous changes in histiocytes, with necrotic core areas, such as that of an unusual presentation of typhoid fever.[6] Definitive diagnosis is made by stool culture.

*Clostridium difficile* typically causes antibiotic-associated colitis. Small-bowel infections are rare, but there are some reports. Terminal ileum infection by *Clostridium* is caused by virulent BI/NAP1/027 strains. Diagnosis is done with the presence of pseudomembranes on ileal mucosa and/or stool studies showing toxins.[4,5]

Typhlitis, from the Greek *typhlon*, or cecum, is a potentially fatal inflammatory condition of the cecum and ascending colon, which can also affect the terminal ileum. It occurs more often in immunocompromised patients. The exact pathogenesis is unknown but probably involves mucosal damage from chemotherapy, radiotherapy and/or leukemic infiltration, strong neutropenia, decreased host defenses and possible ischemia. Clinically, it presents as pain in the right lower quadrant, fever, nausea, vomiting, bloody diarrhea and/or evidence of peritoneal inflammation. Diagnosis is suggested by ileocecal involvement, with thickened bowel wall in immunosuppressed or neutropenic patients. Endoscopy during pancytopenia is usually contraindicated.[4]

*Mycobacterium tuberculosis* is the causative agent of extrapulmonary tuberculosis, and accounts for 20% of cases in immunocompetent patients and 50% of cases in HIV-positive individuals. Intestinal tuberculosis is the sixth most prevalent form of extrapulmonary tuberculosis. The ileum is the site most commonly affected, alone or with the involvement of the adjacent areas, especially the cecum. The reason for such preferred involvement area is the high density of lymphoid tissue, environment at neutral pH, physiologic stasis and absorptive transport mechanisms that allow the ingested *Mycobacterium* to be absorbed. Studies show that ileocecal region has been involved in about 90% of the cases of extraintestinal tuberculosis.[4,5] Intestinal lesions can be ulcerative (more commonly), hypertrophic or fibrous. With chronic inflammation, the ileal wall may become stenotic or fibrotic, with stricture formation, or may form masses (tuberculomas) leading to intestinal obstruction or perforation.

Symptoms include fever, night sweats, abdominal pain, a palpable mass and altered bowel habits, with or without bleeding. Symptoms are nonspecific and about 70% of patients have normal chest radiographs; thus, many times, differential diagnosis is difficult, especially in the case of CD. Also, the two of them can coexist. Findings that favor a diagnosis of intestinal tuberculosis are high fever in the absence of an intra-abdominal abscess, lack of perianal disease, and shorter duration of symptoms. Thickening of the intestinal wall, lesions forming masses, and regional lymphadenopathy are easily detected on computed tomography. On ileoscopy, longitudinal ulcers, aphthous ulcers, and a cobblestone appearance are more common in CD, while transverse ulcers and pseudopolyps are more common in intestinal tuberculosis, although not always present[4,5] (Figure 18.3).

Cappel et al. described a case of a patient with clinical, endoscopic and histological findings suggestive of CD. He showed limited response to prednisone and azathioprine, with several hospitalizations during the following year. Chest radiography was normal and sputum culture was negative. Anti-TNF therapy was initiated without a PPD (*purified-protein-derivative*) test and/or IGRA (*interferon-gamma-release assay*). After 6 weeks, emergency laparotomy revealed numerous miliary deposits in the intestinal and mesenteric serosa as well as hard enlarged intraperitoneal lymph nodes.

Pathology showed necrotizing granulomas with acid-resistant bacilli and no evidence of CD.[7] This case demonstrates the importance of excluding latent tuberculosis or intestinal tuberculosis mimicking CD before starting biological therapy.

*Mycobacterium avium-intracellulare complex* (MAC) occurs predominantly in persons with acquired immunodeficiency syndrome (AIDS) or other immunosuppressant conditions. It can affect the terminal ileum and mimic CD. Diagnosis can be established with the isolation of MAC from the blood or another sterile site (e.g., lymph node, bone marrow, liver, spleen). MAC detected in the stool is not diagnostic, because the patient may be a colonizer.

Cytomegalovirus (CMV) affects the small bowel in only 4% of the cases of gastrointestinal disease. Diagnosis is done by demonstration of CMV typical inclusion on routine histologic examination, culture, staining for CMV antigen, or DNA. In a large series of 31 immunohistochemically proven CMV infection, gastrointestinal involvement was seen in 22 patients and only two patients presented isolated small bowel involvement with intestinal perforation, probably due to CMV vasculitis, suggested by the presence of CMV inclusions in the endothelial cells.[8]

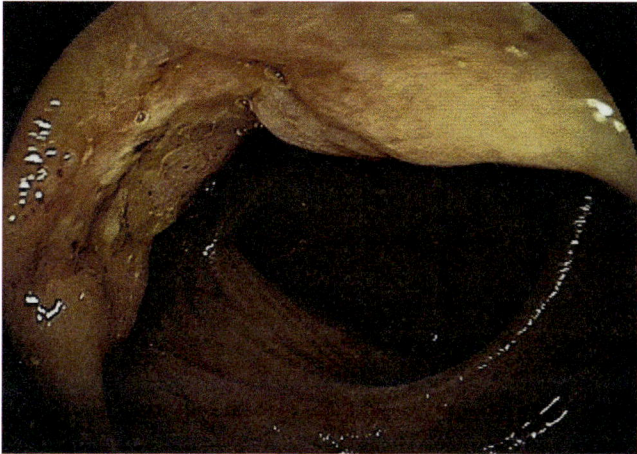

**Figure 18.3** Tuberculosis of the terminal ileum.

*Histoplasma capsulatum* dissemination to the gastrointestinal tract occurs via the reticuloendothelial system by tissue macrophages that accumulate in lymphoid aggregates of Peyer's patches. This probably explains why the terminal ileum is commonly affected. Hepatomegaly and splenomegaly occur in 30 to 100% of cases. On endoscopy, lesions range from mucosal edema and erythema with superficial ulceration to deep ulcers with or without perforation. Diagnosis is made by positive fungal culture or tissue biopsy examination showing diffuse infiltrative lymphohistiocytosis in the mucosa and submucosa.

*Anisakis simplex* (herring worm) and *Pseudoterranova decipiens* (cod worm) are very common in Japan, where there is a habit of ingesting raw marine fish. There are reports of cod contaminated samples in Brazil.[9] Chronic infection can provoke eosinophilic granulomatous ileitis, causing masses and/or obstruction, and nonspecific abdominal symptoms. Endoscopy can occasionally reveal a string-like worm penetrating the gastrointestinal wall.

*Actinomyces israelii* is a species of bacteria present in the normal microflora of the gastrointestinal tract. An injury (from trauma, surgery, or foreign body ingestion) to the mucosa can cause disease. The ileocecal region is the one most commonly involved, probably because of physiologic stasis. Symptoms present indolent course with right lower abdominal pain, abdominal mass and/or fever. Symptoms and signs are nonspecific, and thus the diagnosis is usually delayed with only 10% of cases diagnosed preoperatively.

## ILEITIS ASSOCIATED WITH SPONDYLOARTHROPATHIES

Spondyloarthropathies include ankylosing spondylitis, reactive arthritis, arthritis associated with CD and psoriasis, and undifferentiated spondyloarthropathy.

Distinguishing ileitis associated with spondyloarthropathy and CD can be difficult. HLA-B27 testing may be helpful and is positive in more than 80% of cases of ankylosing spondylitis and in 10 to 35% of CD patients. In ankylosing spondylitis, the gastrointestinal lesions are associated with more severe acroiliitis, spondylitis, and peripheral arthritis. Remission of articular inflammation correlates with disappearance of gut inflammation.

## LYMPHOID NODULAR HYPERPLASIA (LNH)

Focal lymphoid hyperplasia (also known as lymphoid follicular hyperplasia or lymphoid nodular hyperplasia) of the gastrointestinal tract is a rare cause of terminal ileitis. In 1996 the case of a 13-year-old boy was described with a stricture of the gastrointestinal tract diagnosed as CD. There was no response to steroid therapy and surgical resection was required. Histology findings showed focal lymphoid hyperplasia and not CD.[5]

There are no established criteria to differentiate normal lymphoid tissue from hyperplasia, and the latter from pathologic tissue. It is characterized by lymphoid follicle hyperplasia and germinal centers with mitotic activity.[1] The pathogenesis of lymphoid focal hyperplasia is not yet fully clarified. To compensate for the inadequate functioning of the lymphoid tissue, possibly resulting from recurrent infections with *Giardia lamblia*, lymphocyte maturation defects occur.[1] In children, lymphoid hyperplasia can be associated with viral infection.[5] Approximately 20% of adults with common variable immunodeficiency disease (CVID) have lymphoid nodular hyperplasia. Criado and Aguado reported the case of a 32-year-old patient diagnosed with CVID six years before, who was complaining of diarrhea lasting for 2 weeks. Intestinal transit time test showed accelerate intestinal transit and several nodules in the terminal ileum. Ileoscopy confirmed lymphoid follicular hyperplasia. In enteropathy associated with CVID, there is moderate increase of intestinal intraepithelial lymphocytes, which may be the cause of chronic diarrhea in these patients. Differential diagnosis includes lymphoproliferative disease and inflammatory bowel disease (IBD).[10] Regular, flat, yellowish-white nodules measuring about 2 mm in diameter are seen on endoscopy, and may have small apical erosions (Figures 18.4 and 18.5). They are seen more often in children and are not associated with clinical symptoms; thus, LNH is considered physiological, but it can also be associated with viral, bacterial or parasitic infections, as well as immunodeficiency state. It is, therefore, both a physiological and pathological lesion.[11]

## NEOPLASIA

The mucosa of the small intestine encompasses around 90% of the luminal surface area of the digestive system, but accounts for only 2% of GI malignant

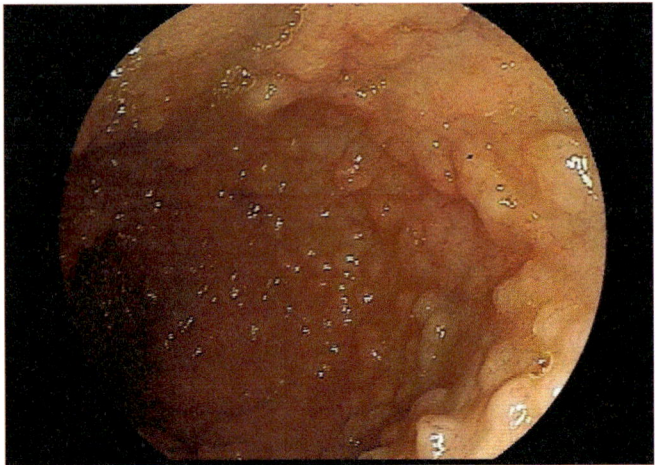

**Figure 18.4** Lymphoid nodular hyperplasia: typical endoscopic appearance with small smooth nodules and normal mucosa staining.

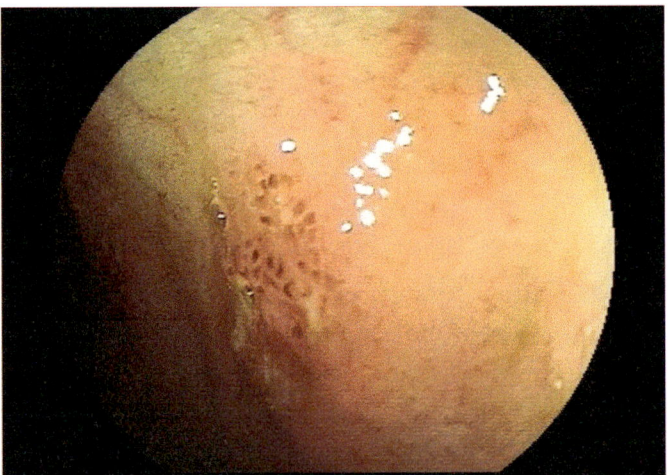

**Figure 18.5** Lymphoid nodular hyperplasia with erosions.

neoplasms, Neoplastic processes with a propensity for the ileum are lymphoma, carcinoid tumor and adenocarcinoma.[4]

Small bowel lymphoma originates in the lymphoid follicle of the submucosa. It can be solitary or diffuse. The solitary form tends to encircle the bowel, narrow the lumen and mimic CD both on radiography and endoscopy.

Regarding carcinoid tumors, 90% originate from the terminal ileum or appendix[5] (Figure 18.6). Diagnosis is usually delayed, on account of the unusual occurrence and nonspecific symptoms.

A review of the literature from 1956 to 1999 showed 131 published cases of small bowel adenocarcinoma in patients with CD. These patients were predominantly male, with an average age of 49 years at diagnosis of cancer and with an average duration of CD of 18 years.[12] Although rare, the small bowel adenocarcinoma is a feared complication of CD, and early diagnosis (pre-operative) is still a challenge.

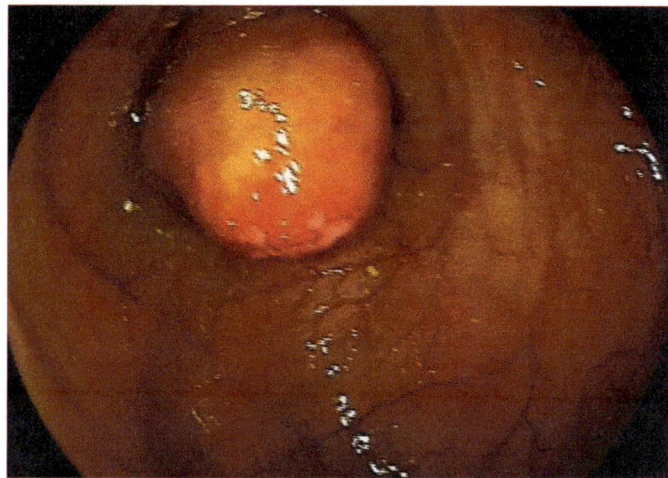

**Figure 18.6** Carcinoid tumor of the terminal ileum.

## VASCULITIS

This can be primary to the intestine or secondary to systemic vasculitis, representing a rare cause of ileitis. The most common form of vasculitis includes systemic lupus erythematosus (SLE) and polyarteritis nodosa (involvement of medium-sized vessels), Henoch-Schonlein purpura and Behcet's disease (involvement of small-sized vessels).[4]

Yavuz et al. described the case of an 18 year old male patient with findings of acute abdomen that underwent urgent surgery. At surgical exploration, 15 cm of the serosa layer of the ileum was hyperemic. The bowel wall was thickened, with obstruction of its *lumen*. With these morphologic findings, a pre-diagnosis of CD was made, with limited bowel resection. At the post-operative 7th day, the patient had maculopapular rashes on both hands and legs. Skin and ileal histology showed leucocytoclastic vasculitis, with a definitive diagnosis of Henoch-Schonlein *purpura*.

This disease's etiology is unknown and it is usually seen 2 to 3 weeks after an upper respiratory tract infection. Although gastrointestinal manifestations are common, they normally occur after the appearance of skin lesions, which makes diagnosis easier.[13]

Aside from vasculitis, ischemic ileitis can have other causes. An important cause is nonocclusive mesenteric ischemia, which is a result of splanchnic hypoperfusion and vasospasm observed predominantly in elderly patients with atherosclerotic vascular disease in the setting of low-flow states (shock, heart failure, drugs, such as cocaine). Because the ileocolic branches are the longest branches of the superior mesenteric artery, the ileocecal region is the most susceptible to ischemia caused by poor perfusion.[4]

## DRUG-RELATED ILEITIS

Although the effects of nonsteroidal anti-inflammatory drugs (NSAID) in the stomach and duodenum are well established, only recently were NSAIDs shown to cause small-bowel injury, which it is now widely recognized as the area most affected by the adverse effects of NSAID (NSAID enteropathy). The pathogenesis and clinical implications of side effects in these areas have many similarities, but they differ in some important aspects. Particularly, a common clinical problem is to differentiate ileitis associated with spondyloarthropathy from NSAID enteropathy and CD.[14]

Although NSAID enteropathy rarely has clinical consequences, in some patients it can lead to erosions or ulcerations, protein-losing enteropathy, and occasionally bleeding, perforation, strictures or obstruction. A type of stricture known as diaphragm disease is characteristic. These diaphragms are numerous, thin-walled, concentric, like mucosal projections that narrow the intestinal lumen. Histology is characterized by prominent submucosal fibrosis without evidence of vascular involvement. The mucosa adjacent to the diaphragms is usually normal.

Differentiating NSAID enteropathy from CD can be difficult, because the two entities can coexist, and NSAIDs can trigger the inflammatory activity of CD. Helpful distinguishing features are that CD classically causes long, thick strictures, while ulcers are often deep, longitudinal and more irregular, unlike NSAID enteropathy, which presents thin fibrotic diaphragms and more sharply demarcated ulcers (Figure 18.7).[4]

There are a few reports of other drugs causing ileitis, such as enteric-coated potassium chloride.[15] After studies published on intestinal ulceration related to this drug, it was withdrawn from the market in 1965. With the development of slow-release formulations, incidence was greatly diminished, but not eliminated.

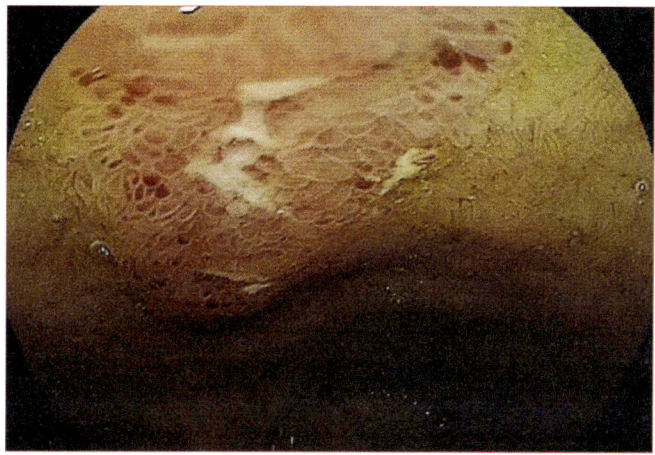

**Figure 18.7** Ileitis by NSAIDs.

Parenteral gold therapy has also been associated with enterocolitis, with reports of edema and ulceration limited to the ileum. Carvalho et al. reported the case of a male patient aged 50 years with abdominal pain and bloody diarrhea that had received kidney transplant seven years before and was treated with tacrolimus, mycophenolate mofetil and prednisone for immunosuppression. Complementary tests revealed fistulizing ileitis, probably secondary to gastrointestinal toxicity of mycophenolate mofetil in patients with renal transplantation. With the cessation of this medication, enteric lesions disappeared and the patient became asymptomatic.[16]

Oral contraceptives, ergotamine, digoxin, and enteric-coated hydrochlorothiazide with potassium are also causes of ileitis.[4]

### INFILTRATIVE CAUSES OF ILEITIS

Infiltrative causes of ileitis include eosinophilic gastroenteritis, systemic mastocytosis, sarcoidosis, amyloidosis and endometriosis. Eosinophilic gastroenteritis commonly involves the stomach and small intestine, but it can causes diffuse colonic involvement. Clinical manifestations depend on the layer and extent of bowel involved with eosinophilic infiltration (mucosa, muscle and/or subserosa).[4] The endoscopic findings vary from normal mucosa to mild erythema, nodularity, and ulceration. Diffuse enteritis with complete loss of villi, submucosal edema, and fibrosis may be present. Multiple biopsies are required because of the patchy nature of the disease.[5] Eosinophilic ileocolitis may be more common than it is thought and the counting of high-power microscopic fields for eosinophils may be a useful way of distinguishing eosinophilic ileocolitis from CD ileocolitis. Rarely, CD is associated with peripheral eosinophilia and/or an eosinophil-rich tissue infiltrate.[5]

In patients with sarcoidosis and suspected ileitis, ileocolonoscopy with biopsy that reveals noncaseating granulomas containing multinucleated giant cells is practically diagnostic. However, proper histologic interpretation is important, since mycobacterial infections, histoplasmosis, CD and lymphoma are granulomatous conditions that can mimic sarcoidosis.

Amyloidosis is characterized by the extracellular deposition of protein in an abnormal fibrillar form. Amyloid deposition in the gastrointestinal tract is most noticeable in the small bowel and results from either mucosal or neuromuscular infiltration. Endoscopic findings include granular appearance, polypoid protrusions, erosions, ulceration, mucosal friability and thickening of the wall. Biopsies of the affected organ reveal amyloid deposits that produce, after staining with Congo red, a pathognomonic red-green birefringence under cross-polarized light microscopy.[4]

Endometriosis is defined by the presence of endometrium outside the womb. Although not very frequent, endometriosis is an important differential diagnosis of CD in young females. Although the rectum and sigmoid colon are the most common sites (85% of cases), the ileum is involved in 1 to 7% of patients. On tissue biopsy, if only the mucosa is sampled, the lesions may be missed as the typical ones are located in the muscularis mucosa. In this regard, radiographic imaging can be helpful. In indeterminate cases, diagnostic laparoscopy is recommended.[17]

## OTHER CAUSES

### Backwash ileitis

Inflammation with damage of the distal ileum in ulcerative colitis (UC) was recognized and was named backwash ileitis during the last half of the nineteenth century. Ileitis is caused by reflux of colonic contents. Most authors agree that backwash ileitis has a morphological pattern of inflammation and injury of the mucosa similar to UC and does not display the typical features of CD.[18]

In general, the severity of ileal inflammation parallels the severity of colonic activity, being more common with pancolitis and cecal involvement. Associations with an aggressive disease course, primary sclerosing cholangitis, and subtotal colectomy for UC have been reported. Definite diagnostic criteria for backwash ileitis have not been determined, but it should be restricted to an active enteritis that involves the ileum for a few centimeters, in a contiguous pattern from the cecum, and with a similar or greater degree of inflammation.[4] The endoscopic appearance should be similar to that of UC, that is, mucosal edema and erythema, superficial erosions and friability (Figure 18.8). When the changes involve long segments, with deep ulcers interspersed with areas of normal mucosa, investigation of the entire small bowel is mandatory to establish or reject the possibility of CD.

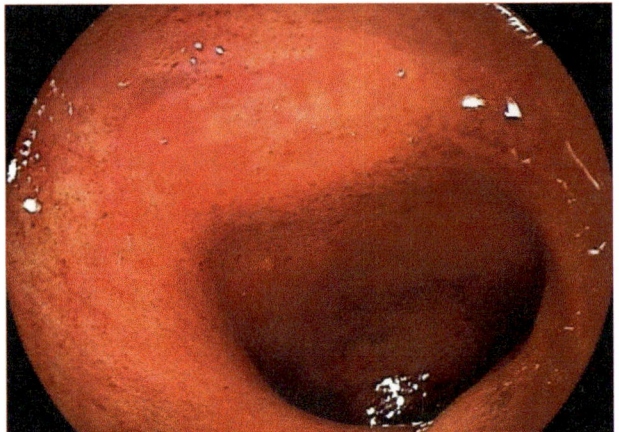

**Figure 18.8** Backwash ileitis.

## Isolated erosions or ulcerations of the terminal ileum

Some asymptomatic individuals may present with aphthous or small ulcerations in the terminal ileum, with no lesions in the colon or ileocecal valve. Most of the ulcerations are surrounded by normal mucosa, regenerating epithelium or edematous mucosa covered by exudates. Patients with isolated terminal ileum ulcerations have nonspecific histological findings and a tendency to resolution without treatment. In most cases, patients do not have CD, but follow up can be needed to determine if that is a case of initial inflammatory disease.[5,19]

Goldstein reports a study including 28 patients with similar mild clinical complaints and ileocolonoscopy showing normal colonic mucosa and small isolated aphthous erosions in the terminal ileum. All 28 lesions were morphologically similar, with focal *lamina propria* edema, mild active inflammation, and crypt disarray. Most had a lymphoid aggregate within the region of edema. Erosion was identified histologically in 21 cases. Mean follow-up duration was 5.8 years. Clinical symptoms resolved in all 28 patients. Typical CD developed in 8 patients (29%) after a mean interval of 3.6 years.[20]

A minority of patients with early CD with minimal inflammatory activity can be asymptomatic, and disease progression to a stage of structural damage sufficient to produce detectable clinical signs and symptoms of complicated CD may take years.

## ENDOSCOPIC DIAGNOSTIC METHODS FOR THE SMALL INTESTINE

Although CD has originally been described as a regional ileitis, a chronic inflammatory condition limited to the terminal ileum, new assessments estimate that 40 to 55% of patients have ileocecal involvement, 15 to 25% have colonic involvement, and 25 to 40% have small bowel involvement. Some patients with CD have inflammatory changes in the mucosa of the proximal small intestine, and conventional ileocolonoscopy cannot detect this involvement.[21] Small bowel transit time has been useful to detect late lesions and complications such as strictures and fistulas but it has low sensitivity to identify early mucosal changes. Also, current imaging methods with multiple slices (computed tomography enterography and magnetic resonance enterography) are useful to demonstrate transmural inflammation and fistulas, but they have low sensitivity to identify subtle early lesions of the disease (Table 18.3).

New endoscopic methods have been used to improve mucosal analysis, and are described below.

**Table 18.3** Advantages and disadvantages of the methods used to evaluate the small bowel[21]

|  | Advantages | Disadvantages |
| --- | --- | --- |
| Endoscopic capsule | Less invasive<br>Does not require sedation | Risk of retention |
| Balloon-assisted enteroscopy | Can collect biopsy samples<br>Therapeutic capacity, such as balloon dilation | Invasive<br>Requires sedation |
| Computed tomography enterography | Non invasive<br>Possibility to show selective contrast radiography<br>Intramural and extraluminal pathology detection | Not able to detect subtle lesions<br>Radiation exposure |
| Magnetic resonance enterography | Non invasive<br>Intramural and extraluminal pathology detection<br>More sensitive to detect fibrosis than computed tomography enterography | Not able to detect subtle lesions |

### Endoscopic capsule (EC)

The endoscopic capsule (EC) was introduced in 2001 as a novel diagnostic method for exploration of the mucosa in the small intestine. The most important feature of an EC is the possibility to capture and transmit 50 to 80.000 images of the small intestine without any connection cable. In addition, the capsule is easily swallowed with a simple sip of water by people of all ages (over 8 years).[2] The advantages of the endoscopic capsule are: noninvasive method, requires no sedation and can be performed easily even in outpatients. However, EC cannot be used to clear the mucosa or take biopsy specimens, and its movement inside the intestine cannot be controlled. EC retention is a major complication of the method, that is, the presence of the capsule in the digestive tract for a minimum of 2 weeks. The rate of capsule retention in patients is close to 1 to 2.5%. Once retained, endoscopic or surgical removal will be necessary.[21] The most common indications for EC are: obscure gastrointestinal bleeding, suspected CD, small bowel tumors and celiac disease.

### Balloon-assisted enteroscopy

Double-balloon endoscopy uses two balloons; one is attached to the tip of the endoscope and the other to the distal end of an overtube which facilitates the insertion of the device through the small intestine. The handles can be easily reduced by slow withdrawal of enteroscope while the balloons are inflated. With this technique, the enteroscope can move being pushed through the overtube, allowing the visualization of the mucosa throughout the small intestine. This method allows deep intubation of the small intestine with air insufflation and washing, thus improving the examination of the mucosa[22] (Figure 18.9), as well as the single-balloon endoscopy, which is based on principles similar to those of double-balloon endoscopy. These two types of enteroscope constitute the balloon-assisted enteroscopy (BAE) and were developed as new techniques for visualization and intervention in the small bowel.

Unlike EC, BAE has the advantage of enabling biopsy fragment collection for histopathological examination (Table 18.4). Diagnostic balloon-assisted enteroscopy shows complication in less than 1% of cases.

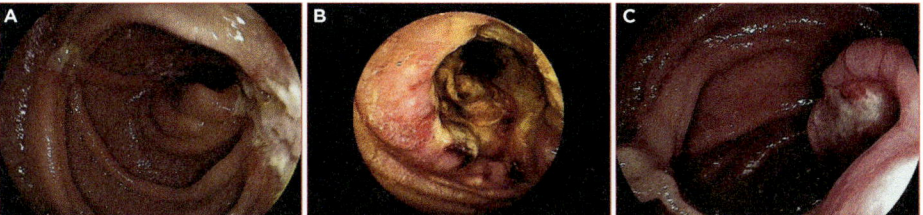

**Figure 18.9** Double-balloon enteroscopy images. A: jejunal CD; B: large-cell lymphoma, jejunal centroblastic pattern; C: jejunal GIST.

| **Table 18.4** Indications for balloon-assisted enteroscopy[2] |
|---|
| After changes demonstrated by EC or radiological images: the most common indication is obscure gastrointestinal bleeding in patients who need more diagnostic tests, and whenever endoscopic therapy is indicated |
| Medium gastrointestinal bleeding: patients with known medium gastrointestinal bleeding and requiring endoscopic hemostasis |
| Diagnosis and treatment of suspected or defined strictures (e.g., Crohn's disease) |
| Mass lesions: endoscopic diagnosis and histopathological confirmation of tumors or masses detected by other imaging methods |
| Pre-surgical tattooing (e.g., small bowel resection area) |
| Foreign body removal from the small intestine (e.g., retained EC) |
| Endoscopic access in patients with anatomy altered by surgery, including cholangiopancreatography after Billroth II or Roux-en Y |
| In difficult colonoscopies |

## Spiral endoscopy

Spiral endoscopy is the new per-oral procedure for visualizing the small bowel. In this procedure, an overtube with a raised helix tip is used to pleat the small bowel. The endoscope coupled to the overtube is advanced into the small bowel by using gentle clockwise rotation of the overtube. Several studies have reported that spiral endoscopy is a safe and effective method for diagnosis and treatment in the small bowel; however, there are no studies comparing the usefulness of spiral endoscopy with that of other modalities in patients with CD, or which have clarified the reason why spiral endoscopy can be relatively traumatic to the mucosa in patients with CD.[21]

## REFERENCES

1. De Melo MM, Gomes Netinho J. Endoscopic aspects in the diagnosis of terminal ileum diseases. Rev Col Bras Cir 2010; 37(3):234-9.

2. Trecca A, Gaj F, Serafini S, Marinozzi G, Silano M. What are the correct indications for ileoscopy? In: Trecca A (ed.). Ileoscopy: technique, diagnosis, and clinical applications. Rome: Springer, 2012.

3. Cherian S, Singh P. Is routine ileoscopy useful? An observational study of procedure times, diagnostic yield, and learning curve. Am J Gastroenterol 2004; 99(12):2324-9.

4. DiLauro S, Crum-Cianflone NF. Ileitis: when it is not Crohn's disease. Curr Gastroenterol Rep 2010; 12(4):249-58.

5. Bojic D, Markovic S. Terminal ileitis is not always Crohn's disease. Ann Gastroenterol 2011; 24(4):271-5.

6. Cheung C, Merkeley H, Srigley JA, Salh B, Webber D, Voyer S. Ileocecal ulceration and granulomatous ileitis as an unusual presentation of typhoid fever. CMAJ 2012; 184(16):1808-10.

7. Cappell MS, Saad A, Bortman JS. Ileocolonic tuberculosis clinically, endoscopically, and radiologically mimicking Crohn's disease: disseminated infection after treatment with infliximab. J Crohns Colitis 2014; 8:560-2.

8. Navaneethan U, Venkatesh PG, Wang J. Cytomegalovirus ileitis in a patient after liver transplantation-differentiating from de novo IBD. J Crohns Colitis 2011; 5(4):354-9.

9. Prado SPT, Capuano DM. Report of nematodes of the Anisakidae family in codfish commercialized in Ribeirão Preto, SP. Rev Soc Bras Med Trop 2006; 39(6):580-1.

10. Said-Criado I, Gil-Aguado A. Nodular lymphoid hyperplasia in common variable immunodeficiency. The Lancet 2014; 383(9911):E2.

11. Mansueto P, Iacono G, Seidita A, D'Alcamo A, Sprini D, Carroccio A. Review article: intestinal lymphoid nodular hyperplasia in children – the relationship to food hypersensitivity. Aliment Pharmacol Ther 2012; 35(9):1000-9.

12. Baars JE, Thijs JC, Bac DJ, Ter Borg PC, Kuipers EJ, van der Woude CJ. Small bowel carcinoma mimicking a relapse of Crohn's disease: a case series. J Crohns Colitis 2011; 5:152-6.

**13.** Yavuz A, Yildiz M, Aydin A, Yıldırım AC, Bulus H, Köklü S. Henoch Schönlein purpura mimicking Crohn's ileitis. J Crohns Colitis 2011; 5(3):271-2.

**14.** Smale S, Tibble J, Sigthorsson G, Bjarnason I. Epidemiology and differential diagnosis of NSAID-induced injury to the mucosa of the small intestine. Best Pract Res Clin Gastroenterol 2001; 15(5):723-38.

**15.** Levin MS, Gyavali CP. Miscellaneous diseases of small intestine. In: Yamada T, Alpers DH, Kallo AN, Kaplowitz N et al. (eds.). Textbook of gastroenterology. 5.ed. Oxford: Wiley-Blackwell, 2011.

**16.** Carvalho R, Almeida N, Portela F, Gomes D, Gregório C, Gouveia H et al. Terminal ileitis in a renal transplanted patient: could it be infectious ileitis, Crohn's disease, or pharmacological toxicity? Inflamm Bowel Dis 2011; 17(6):E52.

**17.** Capell MS, Friedman D, Mikhail N. Endometriosis of the terminal ileum simulating the clinical, roentgenographic, and surgical findings in Crohn's disease. Am Gastroenterol 1991; 86:1057-62.

**18.** Goldstein N, Dulai M. Contemporary morphologic definition of backwash ileitis in ulcerative colitis and features that distinguish it from Crohn disease. Am J Clin Pathol 2006; 126(3):365-76.

**19.** Kwon SO, Kim YS, Oh MK, Kim SY, Cha IH, Jeong SY et al. Clinical significance of erosive or ulcerative lesions isolated in terminal ileum. Intest Res 2012; 10(4):350-6.

**20.** Goldstein NS. Isolated ileal erosions in patients with mildly altered bowel habits. Am J Clin Pathol 2006; 125(6):838-46.

**21.** Yamagami H, Watanabe K, Kamata N, Sogawa M, Arakawa T. Small bowel endoscopy in inflammatory bowel disease. Clin Endosc 2013; 46(4):321-6.

**22.** Safatle-Ribeiro AV, Kuga R, Ishida R, Furuya C, Ribeiro U Jr, Cecconello I et al. Is double-balloon enteroscopy an accurate method to diagnose small-bowel disorders? Surg Endosc 2007; 21(12):2231-6.

# RADIOLOGICAL DIAGNOSIS OF INFLAMMATORY BOWEL DISEASE

DARIO ARIEL TIFERES

ANGELA HISSAE MOTOYAMA CAIADO

MARCOS CARDOSO RESENDE

RENATA EMY OGAWA

## INTRODUCTION

The cross-sectional imaging techniques, computed tomography (CT) and magnetic resonance imaging (MRI), are gradually replacing the conventional radiological studies in the evaluation of patients with inflammatory bowel disease (IBD), because they allow a comprehensive evaluation of the abdomen. Nevertheless, conventional radiological exams still have their importance; they are available in most radiology services, at lower costs. Each method has advantages and disadvantages, and their indication depends on the patient's condition, the availability of the method, and the expertise of the radiologists who interpret the images.

## PLAIN ABDOMINAL RADIOGRAPHS

Plain radiography of the abdomen is still a very useful method in the evaluation of patients with acute abdomen, particularly when perforation and obstruction are suspected.

The basic projections include abdominal radiography in supine position, in the standing position, and a posteroanterior of the chest in standing position. In most cases of acute abdomen, the radiological diagnosis depends on the gas pattern distribution (i.e., gas distribution in dilated and non-dilated loops, and the presence of gas in the peritoneum).

Plain radiography of the abdomen is important in colonic dilatation assessment associated with toxic megacolon and pneumoperitoneum investigation. Toxic megacolon can occur in patients with acute exacerbation of ulcerative colitis (UC), and is a fulminant colitis with transmural inflammation, extensive and deep ulceration, and neuromuscular degeneration.[1] Toxic megacolon may result in perforation, which is more common in the sigmoid colon.[1] Free perforations result in pneumoperitoneum and can be diagnosed by plain abdominal radiographs, unlike blocked perforations.

The most important radiological sign of toxic megacolon is dilation (Figure 19.1), considering the following values for acute exacerbation of colitis: Transverse colon diameter greater than 6 cm or cecal diameter larger than 9 cm.[2]

In chronic UC, a plain abdominal radiograph may show a tubular, ahaustral segment of colon.[2]

Plain radiographs of the abdomen are usually the first examination to evaluate intestinal obstruction in Crohn's disease (CD). Usually, the cause of the obstruction is not identified in this test. Thus, the main function of plain abdominal radiography is to diagnose and quantify the degree of severity of the obstruction.

Obstruction of the small bowel usually leads to dilation of small intestinal loops with gas and liquid accumulation, and decrease in colonic contents and caliper (Figure 19.2). The amount of gas in the colon depends on the duration and degree of obstruction (complete or not).[1] Changes will only be noticeable on abdominal radiographs after 3-5 hours of onset of symptoms, becoming quite evident after 12 hours.[3]

## INTESTINAL TRANSIT

### Radiological anatomy of the small intestine

The small intestine, also called mesenteric (jejunum and ileum), does not have a fixed position in the peritoneal space, and there may be considerable variations in the positioning of each segment.[5] The Kerkrin valves, mucosal folds in the small intestine, usually exhibit a width of 1 to 2 mm,[6,7] while in the

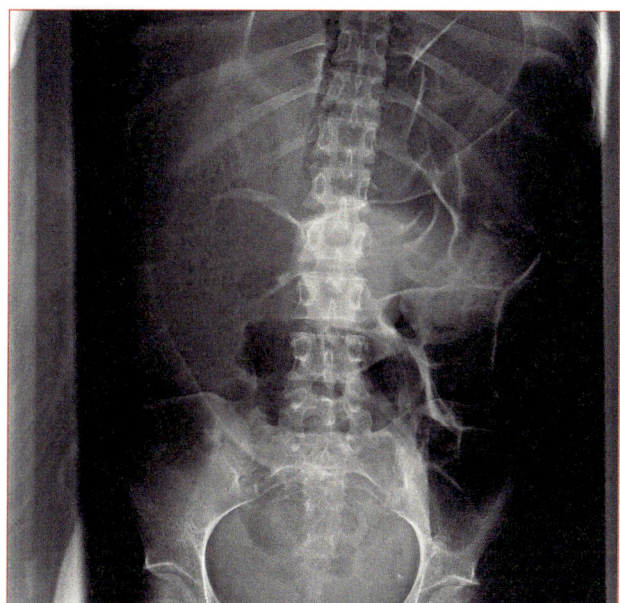

**Figure 19.1** Female patient, 47 years old, with previous diagnosis of UC, progressing to severe diarrhea, pain and progressive abdominal distension. The plain abdominal radiography shows significant distension of the colon suggestive of toxic megacolon.
Source: Caiado et al., 2011.[4]

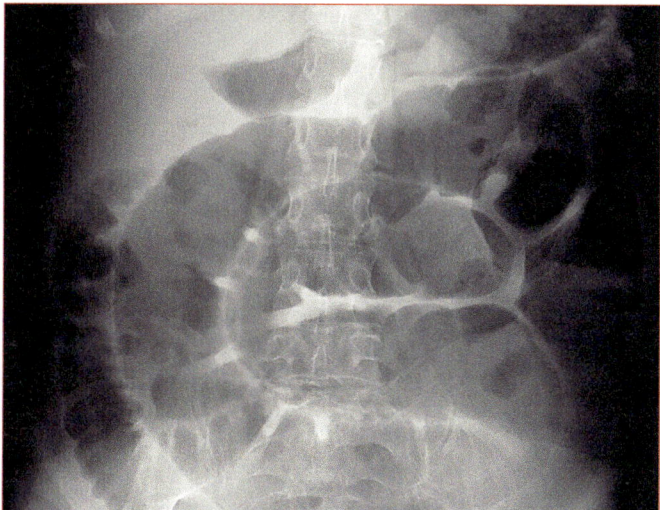

**Figure 19.2** Plain abdominal radiograph demonstrating diffuse large distension of small bowel loops due to stricture of the terminal ileum in a patient with CD.
Source: Caiado et al., 2011.[4]

ileum, the interval is less regular and their number decreases to 4-7 folds/cm in the jejunum and 3-5 folds/cm in the ileum.

On intestinal transit examinations, the jejunum has an average maximum caliper of 4 cm, and the ileum, up to 3 cm.[1]

### Intestinal preparation

In the 48 hours prior to the study, ingesting a low-fiber diet and drinking at least 1-2 L of fluid is recommended.[1] Fasting should be at least 6 hours for adults and for children 3 hours, including fluid restriction.

Some medications such as tranquilizers, antispasmodics and codeine, should be suspended 24 to 48 hours before the test.[1]

### Examination methods

#### Barium study

Typically, the patient ingests between 480 and 600 mL of medium-density barium (50 to 60% weight/volume) and panoramic radiographs are taken every 15 to 30 min until the contrast medium passes into the colon. In addition, a fluoroscopic evaluation is performed with compression of suspicious areas, observing the mobility of loops and the presence of possible hernias, adhesions or masses.

#### Enteroclysis

This technique includes a nasointestinal tube whose end must be positioned approximately 2 to 4 cm after the ligament of Treitz in order to prevent reflux of barium in the duodenum and stomach.[6] Typically, 300 mL of barium are infused at a rate of approximately 100 mL/min.[3] Then, methylcellulose is introduced (1,500 mL at 0.5% concentration)[8] to obtain a double-contrast effect. If there is reflux of methylcellulose, the patient may experience nausea and vomiting. Each intestinal segment filled with contrast is evaluated by fluoroscopy. Alternatively, enteroclysis may be performed with carbon dioxide and an automated pump instead of methylcellulose.

While this technique allows better assessment of the mucosa, it presents disadvantages such as greater patient discomfort and increased exposure to radiation.[1,3]

## Water-soluble contrast

When there is suspicion of acute intestinal perforation into the peritoneal cavity, the use of water-soluble contrast (iodinated) is indicated. In chronic blocked fistulas and in those not suspected of having penetrated the abdominal cavity (i.e., entero-enteral, enterocolic, rectovaginal, colocolic, or perianal fistulas), barium can be used. On the other hand, in fistulas penetrating the urinary tract, iodine is preferably used to avoid the formation of barium "stones".

Iodinated contrast normally promotes weaker bowel opacification, worsened by the medium's osmotic effect that can cause greater dilution. In addition, it has weak adherence to the mucosa, limiting the study of mucosal folds.

The iodinated contrast medium can be absorbed and cause allergic reaction in susceptible patients. Therefore, it should be avoided in patients allergic to iodine.

## Small bowel transit findings in patients with IBD

CD is characterized by transmural and segmental inflammation of the intestinal walls. Radiological evaluation of these patients usually includes intestinal transit examination, since more than 70% of patients have involvement of the small intestine.[9] The terminal ileum is the most frequently involved segment (30 to 40% of cases) (Figure 19.3).[1]

The most common radiological findings include:

- aphthous ulcerations;
- nodules due to the presence of edema or lymphoid hyperplasia;
- granularity due to edema of the mucosa and submucosa;
- mucosal thickening and irregularity, when the edema becomes more pronounced;
- ulceronodular pattern, longitudinal/transverse (when both types are present, with skip areas of normal mucosa, the pattern found is called cobblestoning);
- segmental strictures (Figure 19.4) and fistula formation.[1]

CD findings can be classified as active inflammatory disease (no fistulas or strictures), both penetrating and fibrostricturing, noting that patients may exhibit characteristics of more than one subtype of disease involvement.

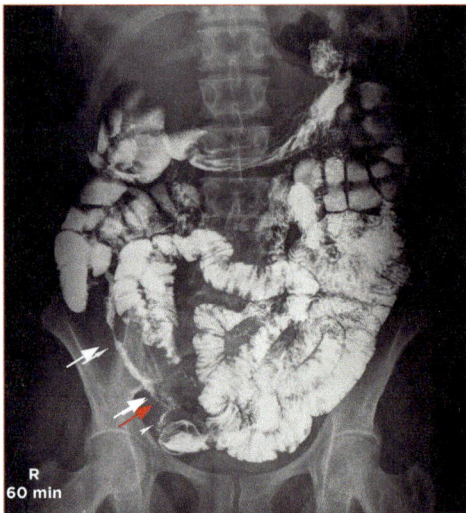

**Figure 19.3** Intestinal transit reveals extensive involvement of the terminal ileum with reduced luminal caliber and ulceronodular appearance of the mucosa (arrow) in a patient with CD. Image of a fistula between ileal loops (red arrow) and other skip areas of disease in the small intestine.

Source: Caiado et al., 2011.[4]

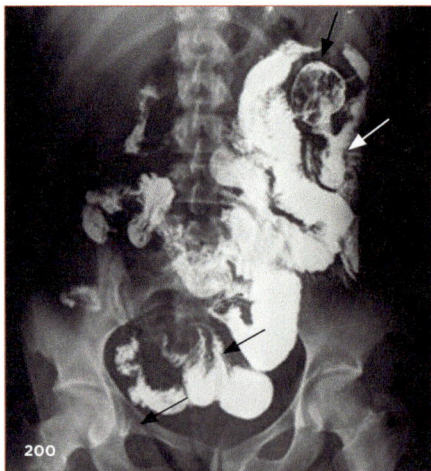

**Figure 19.4** 200-min radiography showing fibrotic scarring disease of the small intestine in the hypogastrium with retractable mesentery component determining subocclusive condition. There are many affected segments (arrows), showing proximal dilation. The colonic loops only began to be opaque after six hours of contrast medium intake.

The active disease is characterized by inflammatory changes in the walls of intestinal loops and mesentery but without strictures or fistulas. In penetrating disease, the formation of deep ulcers may cause transmural involvement from the mucous plan until the deeper layers, progressing to the formation of fistulas and abscesses (Figure 19.5). In fibrostricturing disease, chronic inflammation and the formation of wall fibrosis can reduce the luminal caliber and progress with intestinal subocclusion/occlusion (Figures 19.6 and 19.7).[10]

The small intestinal transit test generally shows no changes in UC. Some interpretation difficulties may occur in situations in which the terminal ileum

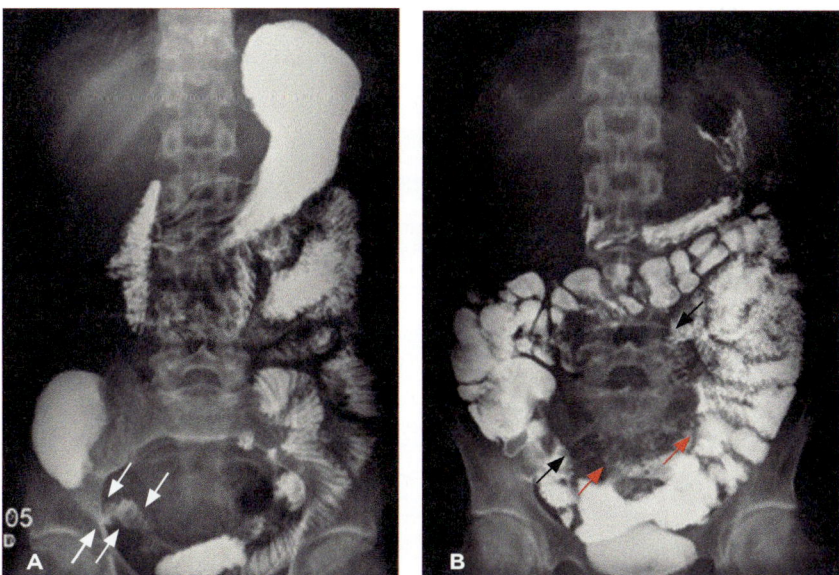

**Figure 19.5** Intestinal transit test in patient with CD and previous right ileocolectomy. A: 5-min radiography in supine position showing opacification of jejunal loops and one small bowel loop with thickened walls and nodular pattern of mucosa in the right iliac fossa (arrows). B: 30-min radiography in supine position showing mass effect in the hypogastrium (arrows), pushing the small bowel loops, with opacification of thin fistulas in the midst (red arrows), no collections. The findings indicate activity in the bowel and inflammatory mass in the mesentery, blocking fistulas.

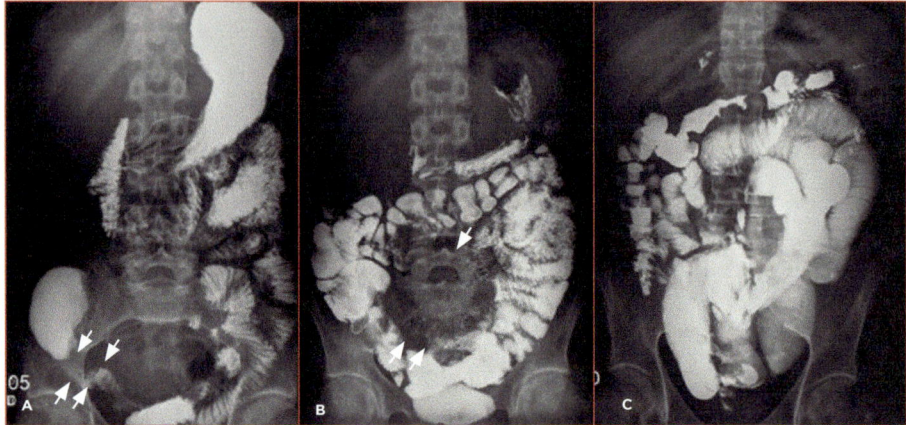

**Figure 19.6** Intestinal transit test in patient with CD. A: 30-min radiography showing sub-stenotic small intestinal loops in the right flank and epigastrium, which maintained fixed positions and caliper on (B) 90 and (C) 180 min images. Retraction and fixation of small bowel loop in the hypogastrium without mucosal injury were already evident, secondary to the mesentery's retraction process. The retraction process also involved the terminal ileum, opacified only in the 180 min (C) image, which displayed reduced caliper, with irregular mucosal folds, partially obscured by the distended loop. Significant distension with fluids is seen in the distal ileum proximal to the terminal ileum (fibrostricturing stage of disease). The contrast medium was diluted in the distended loops by the presence of stasis fluid.

exhibits changes caused by lymph node hyperplasia in young individuals or in cases of backwash ileitis.[10]

## BARIUM ENEMA
### Radiological anatomy of the colon
Both the sigmoid as the transverse colon are mobile and can vary in position because of longer mesenteries. If the cecum presents a long mesentery, it can also vary in position.[5] The hepatic and splenic flexures have a relatively constant position, so that the displacements are generally determined by the

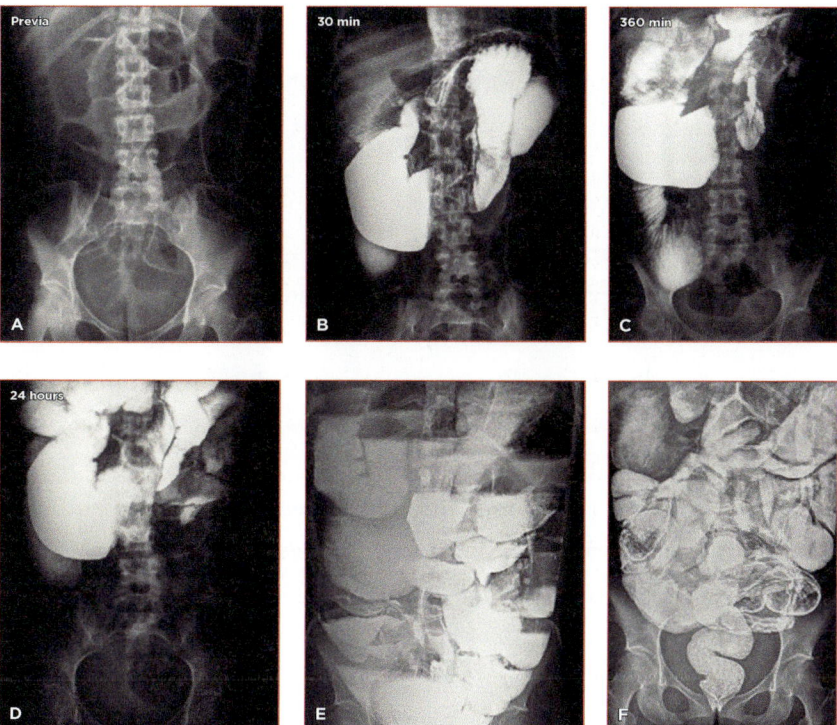

**Figure 19.7** Patient with CD and severe obstruction, with significant bowel distension throughout the abdomen, as shown in the non-contrasted image (A). The image also shows that this is a very thin patient, with little muscle mass and almost no subcu-taneous fat. The 30 (B) and 360 min (C) images reveal severe compression of the mesenteric root over the third portion of the duodenum, which may be a congenital or "acquired" abnormality on account of the intense loss of weight that reduced the mesentery's fat and the space between the upper mesenteric artery and the aorta (D). The intestinal transit test is only capable in these situations to determine the ap-proximate location of the stricture and the time to arrival of the contrast in the colon. In this case, the stricture is in the terminal ileum, as shown in the radiograph perfor-med in orthostasis 3 days after contrast intake (E). The image showing the opacified rectum was taken after 1 week (F).

increase of parenchymal adjacent viscera. The most common anatomical abnormality is the cecum's failure to descend, which can stay in sub-hepatic position.

## Examination methods

### Double-contrast barium enema

The 100% barium sulphate solution is administered through a rectal tube. Then, air is introduced to obtain the double contrast effect. Glucagon or scopolamine butylbromide (Buscopan®) IV or IM can be used to induce hypotonia of the digestive tract. The structures are analyzed by fluoroscopy and recorded on radiographic film.

The double contrast technique was introduced in the 1950s,[1] and showed better results for the evaluation of the colon, especially in the case of mucosal changes, compared with techniques that use single contrast.

A period of 7 to 10 days after conducting deep biopsies and polypectomy is advisable.[3] Moreover, this examination should be avoided during intense activity of UC and CD and suspected toxic megacolon or perforation.

### Water-soluble contrast studies

They are conducted to check the patency and integrity of recent surgical anastomoses.

## Findings of barium enema in patients with IBD

The contrasted examination of the colon and rectum (barium enema) is used to evaluate the extent and complications of IBD. In addition, it can aid in the differential diagnosis between UC and CD. However, radiological findings in both diseases can be overlapping, with identical appearance in some cases.[11] Thus, the final diagnosis should take into consideration clinical, radiological, endoscopic and histological data.

Involvement in UC may be restricted to the rectum, or affect the colon to different extents, typically continuously.[11,12] The most common radiological findings of barium enemas are listed in Table 19.1 (Figure 19.8).

**Table 19.1** Radiological changes in barium enema examination[10]

| Acute | Mucosal granularity |
| --- | --- |
| | Ulcers reaching the submucosa |
| | Loss or thickening of haustra |
| | Inflammatory polyps |
| Chronic | Loss of haustra |
| | Luminal narrowing |
| | Loss of rectal valves |
| | Widening of presacral space |
| | Backwash ileitis |
| | Post-inflammatory pseudopolyps |

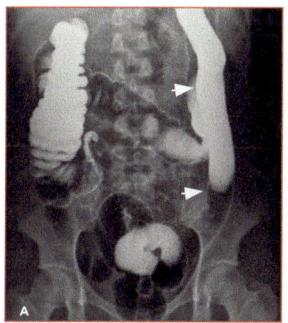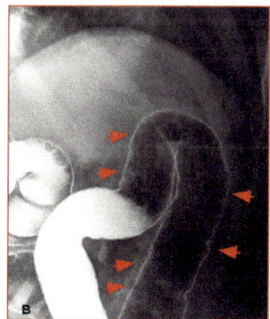

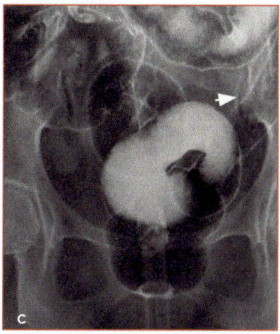

**Figure 19.8** Double-contrast barium enema in patient with UC. Panoramic (A) anteroposterior (AP) projection in supine position; (B) localized projection oblique to splenic flexure; and (C) localized AP projection of rectum and sigmoid colon images showing loss of haustra in the transverse and left colons (arrows) and granularity of mucosal surface observed in segments visualized in the double-contrast (red arrows in B). There is little adhesion of the contrast medium to the mucosal surface, possibly due to hypersecretion.

In CD, barium enema examination should be avoided during intense inflammation phase, due to the high risk of perforation. In the chronic phase, the examination may be normal or show characteristically diseased patches separated by areas of apparently normal mucosa (skip disease pattern, Figure 19.9). The findings are listed in Table 19.2.

After total proctocolectomy with reconstruction of intestinal transit, barium enema examination can be performed to evaluate the ileal pouch (Figure 19.10).

## COMPUTED TOMOGRAPHY ENTEROGRAPHY AND MAGNETIC RESONANCE ENTEROGRAPHY

CD is essentially a chronic condition and can affect any segment of the gastrointestinal tract, but preferably the distal small intestine and proximal colon.

Endoscopic methods are commonly used to assess the stomach, the esophagus, the colon and the distal portion of the small intestine. However, to assess the remainder of the small intestine, as well as CD complications, imaging methods are the main tool.

In the last decade, technical advances in CT and MRI determined a substantial increase in the use of these methods for the evaluation of the small intestine. Regarding the CT scan, for example, the multislice (or multidetector) technology provides millimeter images of the entire abdomen, allowing a multiplanar evaluation of the intestinal loops. The introduction of neutral oral contrast, with attenuation similar to water,[11] and new contrast administration techniques, increasing the distention of the loops, are also important factors that improve the evaluation of the small intestine.[13]

CT enterography differs from conventional CT of the abdomen because of the use of large volumes of oral contrast and fine slices, with multiplanar reconstruction, acquired by multislice CT.[13-15] Currently, its main indication is the evaluation of patients with suspected disease, or follow already diagnosed with CD. Other indications include the evaluation of obscure gastrointestinal bleeding and investigation for intestinal cancer.[13]

The objectives of CT enterography include discriminating the intestinal loop, distending the lumen, displaying the intestinal wall, identifying

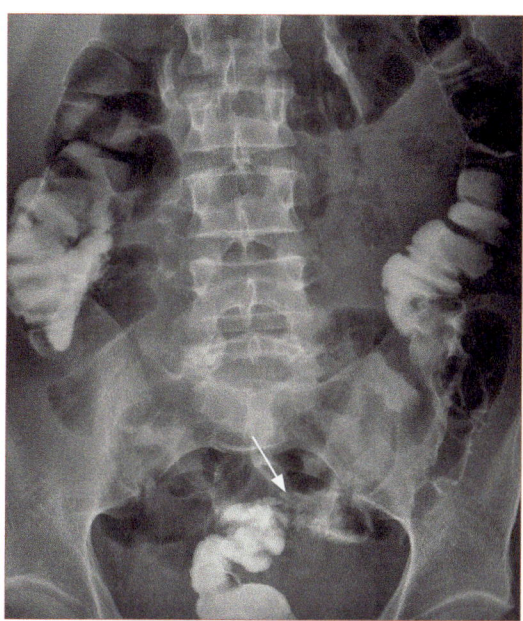

**Figure 19.9** Double-contrast barium enema revealing luminal narrowing of the sigmoid colon and irregularity of the mucosa (arrow) in a patient with CD.

Source: Caiado et al., 2011.[4]

| Table 19.2 Radiological changes on barium enema[10] | |
|---|---|
| Early | Lymphoid nodular hyperplasia |
| | Aphthous ulcers |
| | Deep ulcers |
| | Confluent ulcers |
| | Cobblestone appearance |
| | Asymmetrical involvement |
| | Inflammatory pseudopolyps |
| | Patchy distribution |
| | Skip lesions |
| Late | Fissures |
| | Fistulas |
| | Loss of haustra |
| | Sacculations |
| | Post-inflammatory pseudopolyps |
| | Strictures |

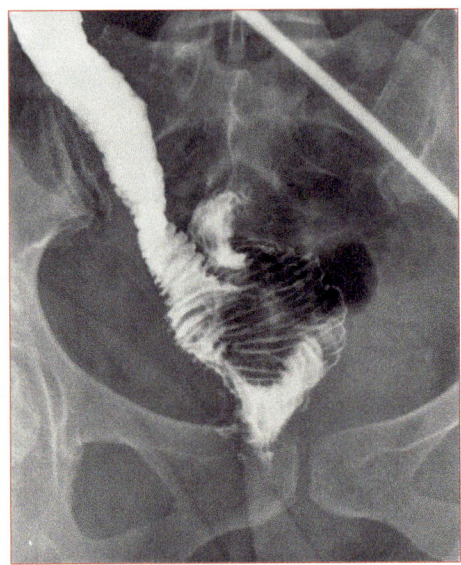

**Figure 19.10** Double-contrast barium enema in a patient with UC after total proctoco-lectomy and reconstruction of intestinal transit through ileal pouch.
Source: Caiado et al., 2011.[4]

the feeding vessels, and evaluating the mesentery (Figure 19.11). Oral gastrointestinal contrast is therefore essential. Currently, in most cases, the use of neutral oral contrast media (i.e., with attenuation similar to water) is recommended. These oral contrast media, associated with the use of intravenous iodinated contrast agent, allow improved demarcation of the intestinal wall and assessment of bowel segments with increased wall enhancement, as well as hypervascular masses and other inflammatory and vascular processes.[14,15]

Different low-attenuation contrast media may be used,[11,13,15,16] the most common being water itself, water with methylcellulose, 0.1% barium solution with sorbitol (Volumen®), not yet available in Brazil, and polyethylene glycol (PEG) solution.[11,13,16-18] Typically, such solutions are administered in large volumes (between 1.5 and 2 L) in an interval ranging between 40 and 60 minutes, in order to maximize loop distension.

The use of intravenous contrast is of capital importance in CT enterography,[13] since, without it, the enhancement of bowel wall cannot be assessed, and

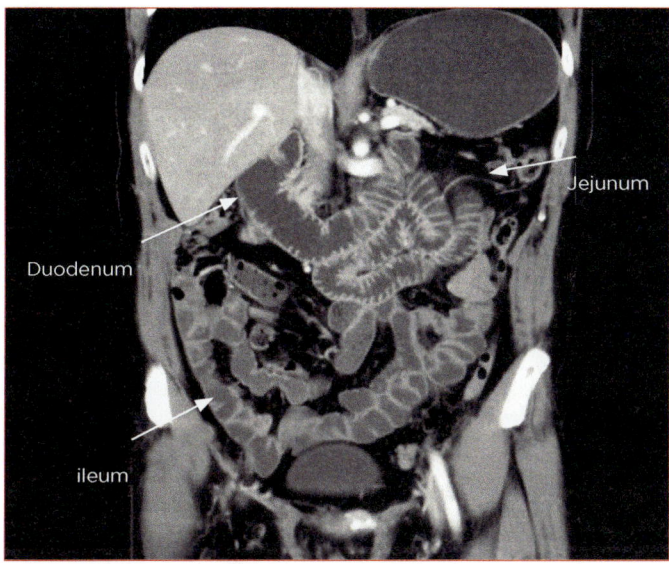

**Figure 19.11** CT enterography, coronal image revealing well-distended small bowel loops with normal morphology.

intraluminal masses and gastrointestinal bleeding may go unnoticed. When using intravenous iodinated contrast medium is not possible, positive oral contrast media, such as solutions based on barium and iodine, should be used. They can also be used in cases of suspected abscesses or fistulas, favoring the delimitation of extraluminal collections.

CT enterography has been increasingly used as the first imaging test in patients with suspected CD. The main findings in the acute phase of the disease include increased mucosal enhancement, stratification and parietal thickening, mesenteric fat densification and engorgement of the *vasa recta* (comb sign) (Figures 19.12 and 19.13).[19-21]

Mucosal enhancement is the most sensitive finding in active CD, correlating significantly with histological[20,22] and clinical[23] findings of disease activity (Figures 19.14 and 19.15).

Wall stratification is the view of intestinal wall layers after the administration of intravenous contrast. Generally, there is an internal layer of mucosal enhancement, an external layer of serosal and muscular[24] enhancement,

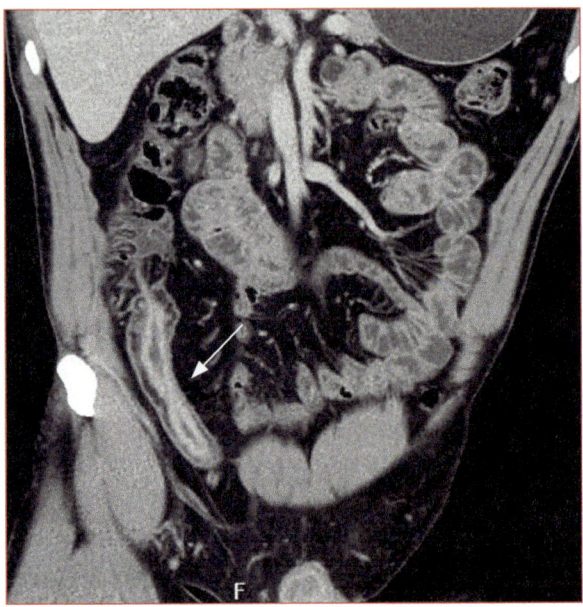

**Figure 19.12** CT enterography, coronal image revealing enteritis in terminal ileum (arrow) with parietal thickening and layer stratification, and increased mucosal enhancement.

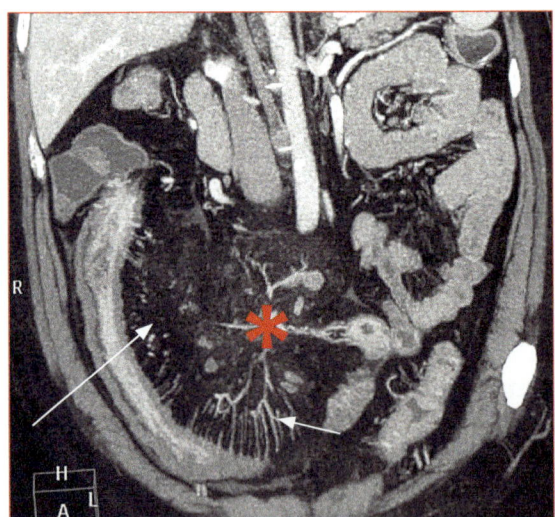

**Figure 19.13** CT enterography, coronal image revealing enteritis in a long segment of distal terminal ileum (large arrow), with parietal thickening and layer stratification, and increased mucosal enhancement. Engorgement of the vascular arch (small arrow), fat proliferation and densification of the mesentery (asterisk) are associated.

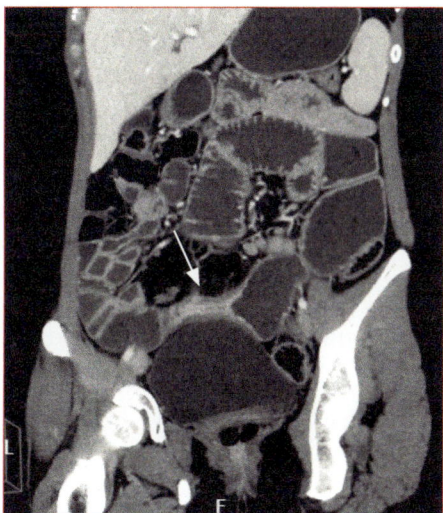
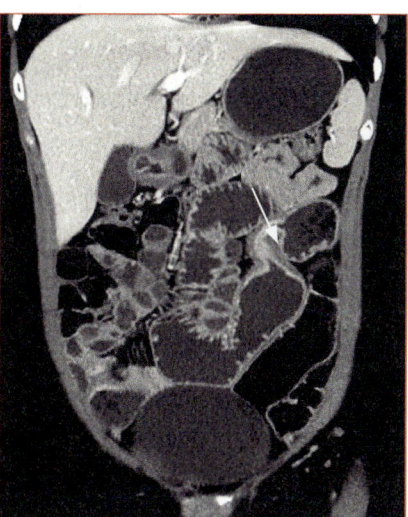

**Figure 19.14** CT enterography, coronal images revealing two segments of ileum with chronic inflammatory involvement (arrows), and luminal stricture causing dilated small bowel segments proximally. There are signs of inflammatory activity characterized by an increase in the degree of enhancement of the mucosa.

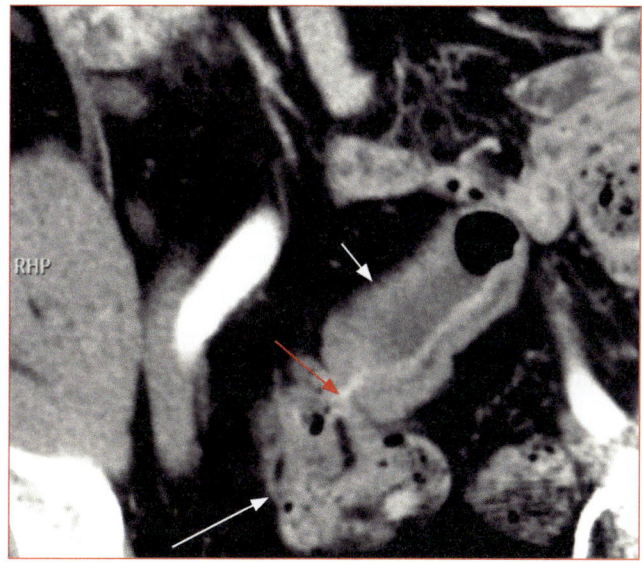

**Figure 19.15** CT enterography, coronal image localized to the right iliac fossa, revealing enterocolic fistula (red arrow) between distal ileal segment (small arrow) and cecum (large arrow).

and an intramural layer that can have various degrees of attenuation. The presence of intramural edema (water attenuation) is indicative of active inflammation, while the presence of fat represents previous or chronic inflammatory process. Intramural attenuation of soft tissues can also be observed, representing inflammatory infiltration and resulting in a bilaminar appearance of intestinal loop.[14]

The most common extraenteric finding in active CD is fibroadipose proliferation of the mesentery, due to edema and engorgement of the *vasa recta*. Engorged prominent *vasa recta* penetrating the intestinal loop perpendicularly to the lumen, also called comb sign, is a very specific sign of active CD (Figure 19.13).[6,21,25] The main findings of CT enterography in the chronic phase of CD are: diffuse wall thickening, submucosal fat deposition, sacculations, fibroadipose proliferation and strictures.[19]

In CD, involvement occurs preferably in the mesenteric border of the intestinal loop, which may result in wall fibrosis and retraction. The presence of asymmetric fibrosis of the intestinal loop, combined with the steady increase in intraluminal pressure during bowel movements, leads to antimesenteric border sacculations.[26]

Fibroadipose proliferation occurs along the mesenteric border of affected segments, partially covering the inflamed loop.[24] Although often found in active disease, fibroadipose proliferation can remain latent during the disease.[14]

Strictures can occur in active disease due to inflammation and spasms. However, they are more often related to the chronic phase of the disease, with fibrotic process, when diffuse parietal thickening with little enhancement, absence of stratification of layers[27] and luminal narrowing of the loop are observed, which may progress to bowel obstruction and the need for surgical resection (see Figure 19.14).[24]

Other common complications of CD include fistulas and abscesses.[24] The cumulative risk of developing fistulas in patients with CD is 33% after 10 years and 50% after 20 years,[27] perianal being the most common type, in which case they can be better assessed by MRI.

Enteroenteric and enterocolic fistulas are usually passageways enhanced with intravenous contrast materials (see Figure 19.15), but they can be difficult to characterize in some cases. In this scenario, the use of a positive oral contrast medium can be very useful to better characterize the fistula.[24] Abscesses often communicate with the intestinal loops, and are seen more frequently in the retroperitoneum and mesentery.[14]

As with CT, enterography can also be performed by MRI, with similar techniques regarding bowel distention, oral contrast media and the use of intravenous contrast medium (Figure 19.16). The findings seen on MRI are basically the same as those found by CT in the different stages of the disease (Figure 19.17).[3,28-30]

The biggest advantage of MRI is that it does not use ionizing radiation. Given the chronic and recurrent nature of CD, several tests are often required over time, particularly in young patients. In this context, the use of methods that do not use ionizing radiation is desirable.

However, compared to CT, MRI also has lower spatial resolution for assessment of intestinal loops and higher incidence of artifacts, which means lower sensitivity for detection of incipient inflammatory changes. Other disadvantages of MRI include lower availability, higher cost, longer examination time[3,28-30] and greater variability in the quality of exams. The choice between the two methods also depends, in addition to the local availability, on the radiologist's experience performing each technique. A strategy that has proven to be interesting is to start the patient's image investigation in cases of suspected CD with CT enterography, and proceed with due monitoring of the changes found using MRI enterography.

## MRI IN THE EVALUATION OF PERIANAL FISTULAS

Perianal complications are frequent in patients with CD and include anal fistulas and fissures, perianal fistulas, abscesses, anorectal strictures and carcinoma.[31] In the evaluation of perianal region, MRI appears to be superior to CT and fistulography due to high contrast resolution, allowing accurate visualization of the anatomy of the area.[32]

MRI and very useful in surgical planning of these complications, since it defines the location and extent of fistulas (Figure 19.18), and also detects abscesses that could go unnoticed on clinical examination. As the presence of post-contrast enhancement indicates active inflammation, the method

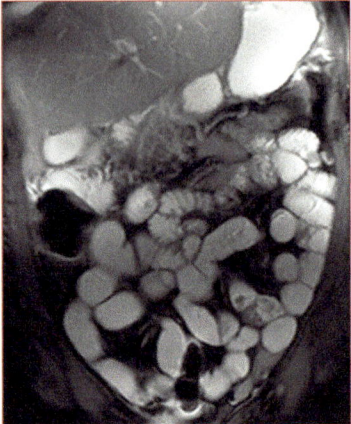

**Figure 19.16** MRI enterography, T2-weighted coronal image showing well-distended small bowel loops, with normal morphology.

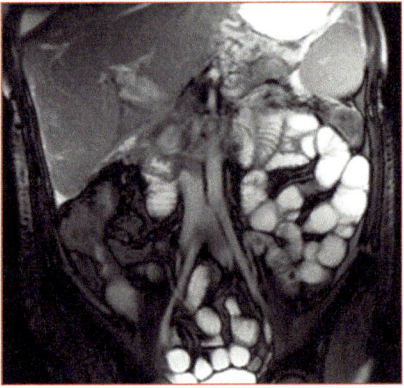

**Figure 19.17** MRI enterography, T2-weighted coronal image showing parietal thickening of terminal ileum, with discreet stenosis (arrow). The remaining segments of small bowel have normal morphology.

may further assist in determining the degree of perianal disease activity. In addition to the usual protocol for the evaluation of the pelvis, the examination includes high resolution sequences, directed towards the analysis of the anal canal.

The report must describe the fistula's trajectory, openings and location – using clock-dial positions based on the patient in lithotomy position –, as well as detailed information on the involved anatomical structures, for example its relation with the levator ani muscle and anal sphincters. In case of multiple trajectories, the report specifies any communications. It also points out the presence of fistulas ending in a cul-de-sac, and the location and volume of abscesses.

## CONCLUSIONS

CT and MRI enterography protocols allow evaluation of the intestinal wall of the mesentery, with detection of extra-luminal extent of disease. For these reasons, such methods are gradually replacing the intestinal transit as an image exam of the small intestine. In addition, they make it possible to evaluate segments that cannot be achieved endoscopically, and are quick options, with low complication rates and good tolerability for patients. Nevertheless, the usefulness of the joint use of the various exam modalities, with specific indications for each clinical situation, is unquestionable, and this is a very important step to deliver the best possible care to the patient.

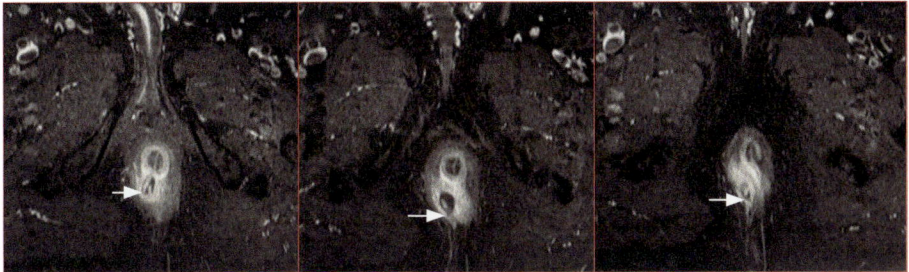

**Figure 19.18** MRI directed to the evaluation of the perianal region. The post-contrast T1 sequence shows perianal trans-sphincteric fistula originating from approximately 2.5 cm of the anal border, at 7 o'clock position, with downward trajectory and cutaneous exteriorization in the right intergluteal cleft (arrows). Post-contrast enhancement of the fistula's wall and adjacent sphincteric muscles revealing inflammatory activity.

## REFERENCES

1.  Sutton D. Textbook of radiology and imaging. 7.ed. Philadelphia: Churchill Livingstone, 2003.

2.  Engstrom PF. Diagnóstico e tratamento das doenças do intestino: doença inflamatória do intestino. Rio de Janeiro: Editora de Publicações Científicas Ltda., 2002.

3.  Lin MF, Narra V. Developing role of magnetic resonance imaging in Crohn's disease. Curr Opin Gastroenterol 2008; 24(2):135-40.

4.  Caiado AHM, Savino ASS, Hashimoto CL. Doenças inflamatórias intestinais – Retocolite ulcerativa e doença de Crohn. Rio de Janeiro: Rubio, 2011.

5.  Juhl JH, Crummy AB, Kuhlman JE. Paul and Juhl's essentials of radiologic imaging. 7.ed. Philadelphia: Lippincot-Raven Publishers, 1998.

6.  Lee SS, Ha HK, Yang SK, Kim AY, Kim TK, Kim PN et al. CT of prominent pericolic or perienteric vasculature in patients with Crohn's disease: correlation with clinical disease activity and findings on barium studies. AJR Am J Roentgenol 2002; 179(4):1029-36.

7.  Vogel J, da Luz Moreira A, Baker M, Hammel J, Einstein D, Stocchi L et al. CT enterography for Crohn's disease: accurate preoperative diagnostic imaging. Dis Colon Rectum 2007; 50(11):1761-9.

8.  Sailer J, Peloschek P, Schober E, Schima W, Reinisch W, Vogelsang H et al. Diagnostic value of CT enteroclysis compared with conventional enteroclysis in patients with Crohn's disease. AJR Am J Roentgenol 2005; 185:1575-81.

9.  Travis SP, Stange EF, Lemann M, Oresland T, Chowers Y, Forbes A et al. European evidence based consensus on the diagnosis and management of Crohn's disease: current management. Gut 2006; 55(Suppl.1):i16-35.

10. Haddad MT. Doença inflamatória intestinal. Rio de Janeiro: Lemos, 2005.

11. Megibow AJ, Babb IS, Hecht EM, Cho JJ, Houston C, Boruch MM et al. Evaluation of bowel distention and bowel wall appearance by using neutral oral contrast agent for multi-detector row CT. Radiology 2006; 238(1):87-95.

12. Koo CW, Shah-Patel LR, Baer JW, Frager DH. Cost-effectiveness and patient tolerance of low-attenuation oral contrast material: milk versus VoLumen. AJR Am J Roentgenol 2008; 190(5):1307-13.

13. Macari M, Megibow AJ, Balthazar EJ. A pattern approach to the abnormal small bowel: Observations at MDCT and CT enterography. Am J Roentgenol 2007; 188(5):1344-55.

14. Paulsen SR, Huprich JE, Fletcher JG, Booya F, Young BM, Fidler JL et al. CT enterography as a diagnostic tool in evaluating small bowel disorders: review of clinical experience with over 700 cases. Radiographics 2006; 26(3):641-57; discussion 657-62.

15. Young BM, Fletcher JG, Booya F, Paulsen S, Fidler J, Johnson CD et al. Head-to-head comparison of oral contrast agents for cross-sectional enterography: small bowel distention, timing, and side effects. J Comput Assist Tomogr 2008; 32(1):32-8.

16. Sood RR, Joubert I, Franklin H, Doyle T, Lomas DF. Small bowel MRI: comparison of a polyethylene glycol preparation and water as oral contrast media. J Magn Reson Imaging 2002; 15(4):401-8.

17. Hebert JJ, Taylor AJ, Winter TC, Reichelderfer M, Weichert JP. Low-attenuation oral GI contrast agents in abdominal-pelvic computed tomography. Abdom Imaging 2006; 31(1):48-53.

18. Maglinte DD, Sandrasegaran K, Lappas JC. CT enteroclysis: techniques and applications. Radiol Clin North Am 2007; 45(2):289-301.

19. Tochetto S, Yaghmai V. CT enterography: concept, technique, and interpretation. Radiol Clin North Am 2009; 47(1):117-32.

20. Booya F, Fletcher JG, Huprich JE, Barlow JM, Johnson CD, Fidler JL et al. Active Crohn's disease: CT findings and interobserver agreement for enteric phase CT enterography. Radiology 2006; 241(3):787-95.

21. Meyers MA, McGuire PV. Spiral CT demonstration of hypervascularity in Crohn's disease: "vascular jejunization of the ileum" or the "comb sign". Abdom Imaging 1995; 20(4):327-32.

22. Bodily KD, Fletcher JG, Solem CA, Johnson CD, Fidler JL, Barlow JM et al. Crohn's disease: mural attenuation and thickness at contrast-enhanced CT enterography – correlation with endoscopic and histologic findings of inflammation. Radiology 2006; 238(2):505-16.

23. Del Campo L, Arribas I, Valbuena M, Maté J, Moreno-Otero R. Spiral CT findings in active and remission phases in patients with Crohn's disease. J Comput Assist Tomogr 2001; 25(5):792-7.

24. Zamboni GA, Raptopoulos V. CT enterography. Gastrointest Endosc Clin N Am 2010; 20(2):347-66.

25. Colombel JF, Solem CA, Sandborn WJ, Booya F, Loftus Jr EV, Harmsen WS et al. Quantitative measurement and visual assessment of ileal Crohn's disease activity by computed tomography enterography: correlation with endoscopic severity and C reactive protein. Gut 2006; 55(11):1561-7.

26. Paulsen SR, Huprich JE, Hara AK. CT enterography: noninvasive evaluation of Crohn's disease and obscure gastrointestinal bleed. Radiol Clin North Am 2007; 45(2):303-15.

27. Schwartz DA, Loftus Jr EV, Tremaine WJ, Panaccione R, Harmsen WS, Zinsmeister AR et al. The natural history of fistulizing Crohn's disease in Olmsted County, Minnesota. Gastroenterology 2002; 122(4):875-80.

28. Fidler JL, Guimaraes L, Einstein DM. MR imaging of the small bowel. Radiographics 2009; 29(6):1811-25.

29. Siddiki H, Fidler J. MR imaging of the small bowel in Crohn's disease. Eur J Radiol 2009; 69(3):409-17.

30. Laghi A, Paolantonio P, Passariello R. Small bowel. Magn Reson Imaging Clin N Am 2005; 13(2):331-48.

31. Ruffolo C, Citton M, Scarpa M et al. Perianal Crohn's disease: is there something new? World J Gastroenterol 2011; 17:1939-46.

32. de Miguel Criado J, del Salto LG, Rivas PF et al. MR imaging evaluation of perianal fistulas: spectrum of imaging features. Radiographics 2012; 32:175-94.

## BIBLIOGRAPHY

1. Gore RM, Levine MS. Textbook of gastrointestinal radiology. 3.ed. Philadelphia: Saunders, 2008.

2. Thompson SE, Raptopoulos V, Sheiman RL, McNicholas MM, Prassopoulos P. Abdominal helical CT: milk as a low-attenuation oral contrast agent. Radiology 1999; 211(3):870-5.

3. Wold PB, Fletcher JG, Johnson CD, Sandborn WJ. Assessment of small bowel Crohn's disease: noninvasive peroral CT enterography compared with other imaging methods and endoscopy  feasibility study. Radiology 2003; 229(1):275-81.

# HISTOPATHOLOGICAL DIAGNOSIS IN INFLAMMATORY BOWEL DISEASE

HUMBERTO SETSUO KISHI

ROBERTO EL IBRAHIM

## INTRODUCTION

Inflammatory bowel disease (IBD) includes two major disease groups: ulcerative colitis (UC) and Crohn's disease (CD). The pathogenesis of these diseases is still unclear, but the approaches currently point to the importance of deregulation of the innate and adaptive immune response directed against luminal bacteria and their products found in the intestinal lumen, in addition to specific genetic and molecular factors.[1,2] The diagnosis of each of these entities is based on the correlation between clinical, endoscopic, anatomical and pathological data. There are no pathognomonic findings that alone allow the specific diagnosis of UC or CD.

The interpretation of biopsies from the large bowel requires knowledge on normal, healthy colon, according to topography in which biopsies are taken. Even though the presence of plasma cells is constant in the colorectal mucosa, cellularity of the lamina propria is variable, so that the cecum and the right colon are more cellular than the other portions of intestine, with a progressive decrease toward the left colon. Another significant regional feature is the irregularity in the crypts format found in distal rectum, as compared to the remainder of the colonic mucosa. Also, the number of intraepithelial lymphocytes is usually higher in the cecum and right colon. Under these conditions, the specimens obtained during colonoscopy must be identified separately so that a normal mucosa is not improperly interpreted as abnormal.

The main microscopic markers of chronicity in colitis are basal lymphoplasmacytosis and crypt architectural distortion (Figure 20.1). Even in the early stages of IBD, the infiltrate of lymphocytes and plasma cells within the crypts and in the most superficial portion of the muscularis mucosa is already present, compared to other types of nonspecific colitis. The Paneth cells, present in the right colon, also characterize lesion chronicity in the colonic mucosa if found beyond the hepatic flexure.[3,4]

## ULCERATIVE COLITIS

### Clinical features

UC is a chronic IBD confined to the colon. White and young people are the ones who most often suffer from this disease, but other age ranges and racial groups may be affected. Its incidence has an inverse correlation with smoking. It is manifested by diarrhea, urgency to evacuate, bloody diarrhea, abdominal

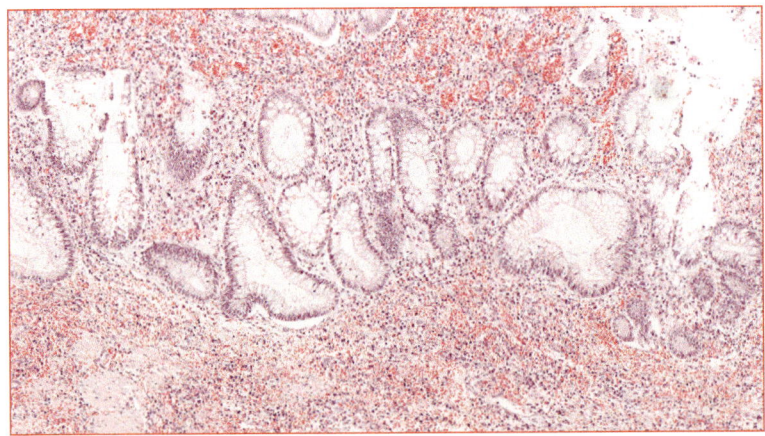

**Figure 20.1** Hematoxylin and Eosin stain: 200-times magnification. Examples of crypt architectural changes.

and perianal pain. The most characteristic symptom is the presence of blood and mucus in the stools.

Abdominal pain correlates with the topography of the lesions, so that disease in the left colon causes hypogastric pain while pancolitis causes diffuse abdominal pain. The bleeding originates from diffuse ulceration of the mucosa on telangiectatic vessels within the lamina propria. Factors associated with disease progression include: extent of lesions at diagnosis, joint symptoms, young age at diagnosis and severe bleeding. Infection by cytomegalovirus, *Salmonella*, *Clostridium*, medications and ischemia can complicate or worsen the symptoms.

## Pathological features

### Macroscopy

In UC, the involvement of the distal segments of the large intestine is most commonly observed. Since the inflammatory process is limited to the mucosa, external examination of the resected specimens is normal, except in cases presenting toxic megacolon or associated carcinoma.

In active colitis, there is blood leakage from the mucosa in the opening of the intestine. Typically, the lesions are present in the rectum with continuous proximal extension in varying dimensions. Transition between the ulcers and normal mucosa is abrupt. The use of enemas with corticosteroids can lead to a normal appearance in the rectum.

In inactive colitis, the appearance of the mucosa is diffusely hemorrhagic, and uniformly granular and erythematous. Ulcers, polyps and pseudopolyps are common (Figure 20.2). Often, post-inflammatory polyps may persist and indicate previous episodes of colitis. They are more common in the colon than in the rectum and may coalesce, creating the aspect of mucosal bridges. Large areas of ulceration with mucopurulent exudate and partial or total loss of the mucosa may be observed. The ulcers have a linear pattern of distribution, following the path of colonic tenia, and can reach up to the muscularis propria in fulminant forms of UC and toxic megacolon. In the quiescent phases, the mucosa is granular, with no hemorrhagic component, with or without post-inflammatory polyps. When in remission, the mucosa exhibits normal or

**Figure 20.2** UC: ulcerations and pseudopolyps with a pattern of continuous involvement.

atrophic appearance with loss of mucosal folds. The intestine may present shorter than usual, with loss of haustra.

### Microscopy

UC is a disease which exhibits inflammation primarily restricted to mucosa, with occasional extension to the submucosa. The signs of chronicity are similar to those of CD, with variable degrees of active inflammation. Activity is determined by the presence of neutrophils, and its level is proportional to that of the lymphoplasmacytic infiltrate in the lamina propria. The term chronic active colitis denotes acute inflammatory infiltrate in a context of chronic injuries. In these situations, the glandular epithelium is attacked by neutrophils, and mucin depletion and ulceration of the surface occur. The activity can be graded as mild, moderate or severe. The stages of formation of ulcers can be summarized as follows:

- cryptitis: aggression of the crypts by neutrophils;
- crypt abscesses: neutrophilic exudate, with cellular debris in the lumen of the glands (Figure 20.3);
- ulcers: the inflammatory infiltrate spreads laterally within the lamina propria, beneath the lining epithelium, detaching it and causing the ulcer.

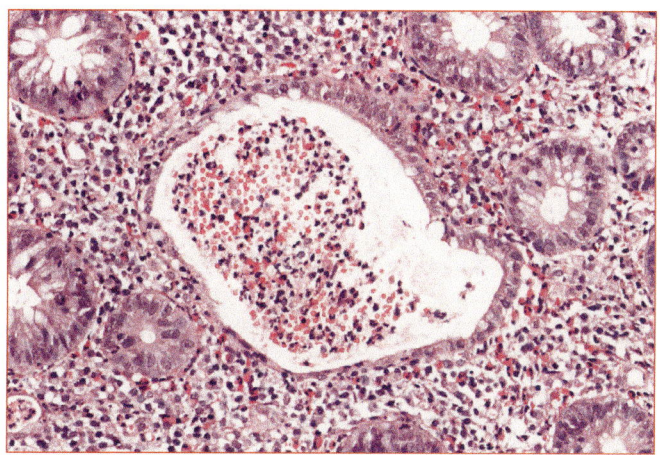

**Figure 20.3** Hematoxylin and Eosin stain: 400-times magnification. Crypt microabscess.

They tend to be small and superficial, with or without formation of mucin granulomas adjacent to the base of the crypts.

Other microscopic findings include lymphoid aggregates, especially in the rectum, lymphocytic infiltration and occasional eosinophils.

In the resolution phase of colitis, there is reduced activity and lesions to the crypts, regeneration and remodeling of crypts, and, last, the disappearance of the basal lymphoplasmacytic infiltrate.

When the disease is quiescent, the mucosa may present with normal appearance or diffusely atrophic with loss of parallelism between crypts, damage, ramifications and shortening of the crypts.

Patients with pancolitis may present with ileal lesions due to reflux of cecal content, which is known as backwash ileitis.[5] Morphological changes include villous atrophy, crypt regeneration without evident inflammation, neutrophilic and mononuclear inflammatory infiltrates in the lamina propria, foci of cryptitis and crypt abscesses, in addition to superficial and focal erosions. In general, the level of ileal lesions reflects the severity of the disease in the colon.[5]

## CROHN'S DISEASE

CD is an idiopathic disease that mainly affects the ileum and cecum, although any area of the gastrointestinal tract may be affected. It is characterized by presenting diseased patches interspersed with endoscopically normal areas. Its highest peak of incidence occurs between the second and third decades of life, with a lower secondary peak between the fourth and fifth decades.

### Clinical features

The first symptoms of CD can be initially mild, leading to late diagnosis in months to years. Any segment of the gastrointestinal tract can be affected, with approximately 30 to 40% of patients with lesions confined to the small intestine, 30 to 40% suffering ileocolonic involvement, and only 10 to 20% with exclusive involvement of the large bowel.[4]

The location and extent of the injuries are critical to the clinical manifestations. The progressive transmural inflammation with deep ulcers and scars can lead to symptoms associated with intestinal obstruction, perforation, bleeding and fistulas. Lower gastrointestinal bleeding, for example, is the result of fistulas or ulcers, while obstructive symptoms are usually complications from strictures in the distal ileum.

### Pathological features

#### Macroscopy

Patients with more severe disease are prime candidates for intestinal resection. In such cases, the pathologist examines surgical specimens in advanced stages, with segmental mucosal lesions, and with areas of partial or transmural involvement of the bowel. Although not specific, aphthous ulcers are the first sign of illness. Next, a hemorrhagic contour is formed, which facilitates visualization. Aphthous ulcers eventually progress to discontinuous, serpiginous or linear ulcers. At this stage, the adjacent mucosa shows hyperemia and edema. Islands of nonulcerated mucosa interspersed with ulcers give a typical cobblestone appearance (Figure 20.4). As the progression of disease occurs, the entire thickness of the bowel changes with fibrosis of all layers and mucosal atrophy.

Other findings include pseudopolyps and inflammatory polyps with variable sizes and forms, including worm-like filiform polyps. Fistulas, adhesions and strictures are more common in the distal ileum and may occur spontaneously or in post-surgical states with residual disease. Perforations are unusual, since the inflammatory process is slow, so that adhesions between loops with plastron formation are protective. Externally, there may be hyperemia, serositis, dense fibrous adhesions between loops, other abdominal and pelvic organs or the abdominal wall, as well as retraction of pericolic adipose tissue in the affected areas.

### Microscopy

In biopsies, the classical histological features of DC are non-specific, particularly when only the mucosa and the submucosa are represented in the samples. Nevertheless, the pattern and distribution of the lesions can contribute to the diagnosis, if they are correlated with clinical and endoscopic findings. In surgical specimens, on the other hand, all the lesions may be contemplated.

CD has areas of segmental colitis alternating with areas that are not affected. Since the lesions are multifocal, a single microscope slide can display normal mucosa and acute, chronic and regenerative lesions. The architectural changes present in active disease include irregularly-shaped glands, with ramifications, shortened or with variable diameters. Pyloric metaplasia and Paneth cell hyperplasia are also evident.

Heterogeneity in density and distribution of lymphoplasmacytic infiltration are striking characteristics of this disease. Lymphoplasmacytic groupings are often separated by paucicellular areas with edema. The lymphoid aggregates of CD contain central crypts, while common lymphoid follicles have their crypts shifted to the periphery. Lymphoid follicles with germinal centers in mucosa-submucosa junction are very suggestive of CD when associated with fibrosis or edema of submucosa, without significant mucosal lesions (Figure 20.5).

The granulomas of CD are small, compact and with no associated necrosis. Usually, they are accompanied by a halo of lymphocytes in the surroundings, they can have giant cells (Figure 20.6) or not, but are not activity indicators.

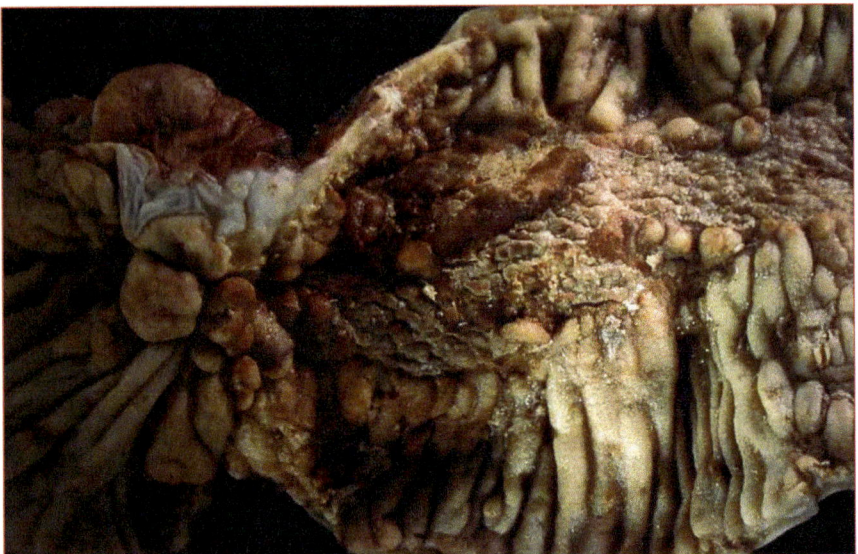

**Figure 20.4** Crohn's disease: longitudinal serpiginous ulcers delimiting areas of preserved mucosa with a focus of stenosis.

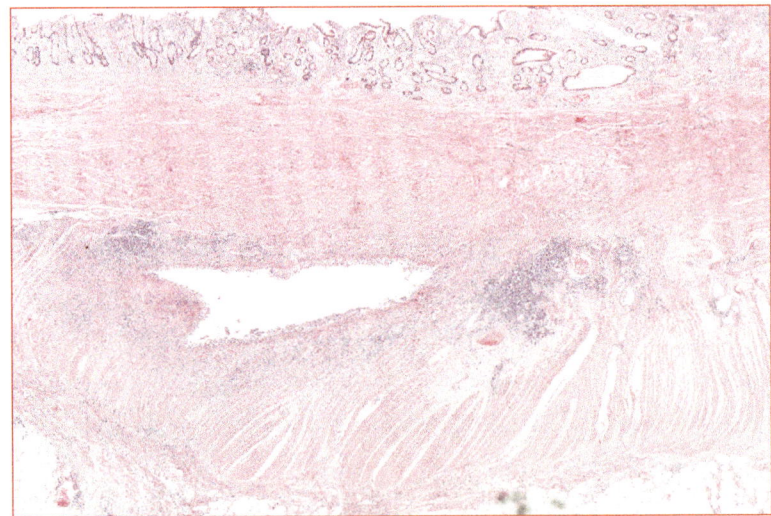

**Figure 20.5** Hematoxylin and Eosin stain: 2.5-times magnification. Crohn's Disease. Transmural inflammation with fistulas and lymphoid aggregates in the submucosa--muscularis propria limit is seen. The mucosa shows chronic inflammatory changes.

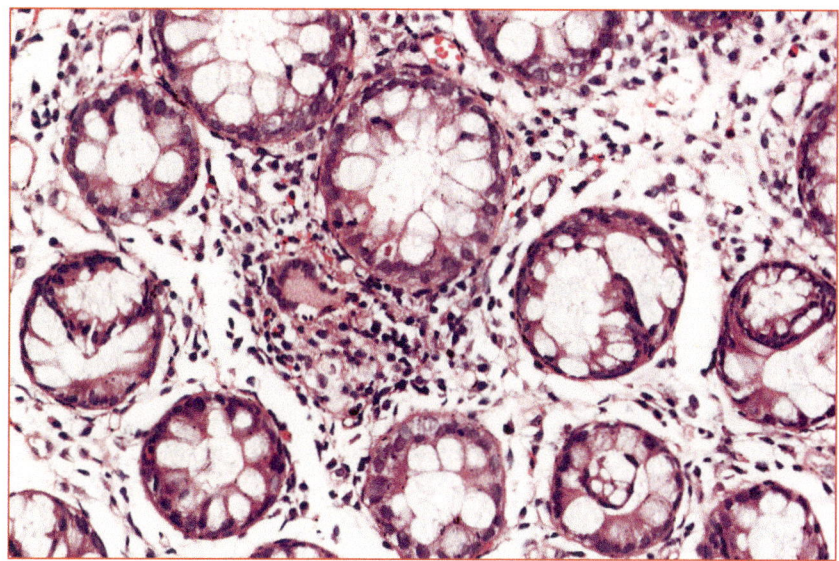

**Figure 20.6** Hematoxylin and Eosin stain: 400-times magnification. Granuloma in the lamina propria with multinucleated giant cell.

If there is necrosis or suppuration, or if they are associated with fissures, they are not considered specific to the disease. Pericryptic mucin granulomas with foamy macrophages can also occur in UC. Investigation for specific infectious agents should be performed.

Ulcers arise from increased activity and superficial epithelial damage, as well as crypt damage. Aphthous ulcers arise associated with lymphoid aggregates and are small. Ulcers associated with fissures are deep, narrow and perpendicular to the mucosa. They can affect the entire wall and are accompanied by acute inflammatory infiltrate with granulation tissue and prominent histiocytes.

Other common histopathological features include obliterating vascular lesions, nerve hypertrophy and deep enteritis/cystic colitis. Differential diagnoses take into account UC (Table 20.1), forms of infectious enterocolitis and granulomatous gastroenterocolites, especially infectious. The same changes of the colon can be observed in the terminal ileum and, particularly, pyloric metaplasia, which is a chronicity marker.

**Table 20.1** Histopathologic differences between CD and UC

| Appearance | CD | UC |
|---|---|---|
| Distribution of inflammation | Multifocal, transmural | Diffuse, mucosa and submucosa, transmural in toxic megacolon |
| Crypt distortion | Minimal | Pronounced |
| Paneth cell metaplasia | Can occur | Common |
| Cytoplasmic mucin | Slightly decreased | Depleted |
| Vascular telangiectasia | Occasional | Prominent |
| Edema | Pronounced | Minimal |
| Lymphoid hyperplasia | Common, separated from the muscularis mucosa, transmural and pericolic tissues, associated with submucosal edema and fibrosis | Rare, mucosa and submucosa, not associated with edema and fibrosis |
| Crypt abscesses | Present, in small number | Common |
| Granulomas (sarcoid) | Common | Absent |
| Aphthous ulcers | Common | Rare |
| Fissures | Common | Absent |
| Submucosa | Normal, inflamed or with reduced thickness | Normal or with reduced thickness |
| Lymphoid aggregates in the submucosa | If present, suggest CD; especially if deep | Generally absent |
| Nerve hypertrophy | Common | Rare |
| Inflammatory pseudopolyps | Less common | Common |
| Filiform polyposis, giant polyps, post-inflammatory polyps | Occur | Occur |
| Ileal inflammation | Common | Minimal, generally less than 10 cm affected |
| Anal involvement | Granulomas | Nonspecific |
| Lymph nodes | Granulomas | Reactive hyperplasia |

## DYSPLASIA IN INFLAMMATORY BOWEL DISEASE

Intraepithelial neoplasia (dysplasia) is a glandular neoplastic proliferation that can occur in IBD patients, but with macroscopic and microscopic features that distinguish it from an adenoma.[6] The risk of developing adenocarcinoma increases after 8 to 10 years, and is higher in patients with early onset and extensive disease (pancolitis). Considering that the adenocarcinoma invasion can take place associated with intraepithelial neoplasia with relatively discreet morphological changes, high-grade intraepithelial neoplasia is diagnosed in colitis based on less severe abnormalities than the criteria of intraepithelial neoplasia of adenomas.[7] These dysplasias can be flat or raised (dysplasia-associated lesion or mass – DALM). DALM and flat high-grade dysplasia usually lead to total colectomy.

## INDETERMINATE COLITIS AND INFLAMMATORY BOWEL DISEASE UNCLASSIFIED

The injuries described in the previous sections are not always clearly present in biopsies and surgical specimens. The main clinical conditions in which lesions are not specific include the early stages of the disease, inflammatory diarrhea in children, patients with concomitant liver disease, patients under treatment and those with fulminant disease. In these situations, the differential diagnosis between CD and UC in biopsies cannot be established. The term IBD unclassified is suggested, thus, especially when there is mild focal basal lymphoplasmacytic infiltrate, with or without relevant architectural changes.[8] Follow-up biopsies and review of previous biopsies can help establish a more accurate diagnosis.

The diagnosis of CD or UC cannot be clearly established in approximately 10 to 15% of surgical specimens collected from colectomies. In these situations, the term indeterminate colitis should be employed.[9]

## REFERENCES

1.  Xavier RJ, Podolsky DK. Unravelling the pathogenesis of inflammatory bowel disease. Nature 2007; 448:427.

2.  Cho JH. The genetics and immunopathogenesis of inflammatory bowel disease. Nat Rev Immunol 2008; 8:458.

3.  Noffsinger A, Fenoglio-Preiser C, Maru D, Gilinksy N. Inflammatory bowel disease in gastrointestinal diseases: Atlas of nontumor pathology. Washington: American Registry of Pathology in collaboration with the Armed Forces Institute of Pathology, 2007.

4.  Greenson JK, Odze RD. Inflammatory disorders of the large intestine. In: Surgical pathology of the GI tract, liver, biliary tract and pancreas. 2.ed. Philadelphia: Saunders Elsevier, 2009.

5.  Haskell H, Andrews Jr. CW, Reddy SI, Dendrinos K, Farraye FA, Stucchi AF et al. Pathological features and clinical significance of "backwash" ileitis in ulcerative colitis. Am J Surg Pathol 2005; 29:1472-81.

6.  Hamilton SR, Vogelstein B, Kudo S. Carcinoma of the colon and rectum. In: Pathology and genetics of tumors of the digestive system. World Health Organization Classification of Tumors. Lyon: IARC Press, 2000.

7.  Svrcek M, Cosnes J, Beaugerie L, Parc R, Bennis M, Tiret E et al. Colorectal neoplasia in Crohn's colitis: a retrospective comparative study with ulcerative colitis. Histopathology 2007; 50:574-83.

8.  Guindi M, Riddell RH. Indeterminate colitis. J Clin Pathol 2004; 57:1233-44.

9.  Geboes K, van Eyken P. Inflammatory bowel disease unclassified and indeterminate colitis: the role of the pathologist. J Clin Pathol 2009; 62:201-5.

# DIFFERENTIAL DIAGNOSIS OF INFLAMMATORY BOWEL DISEASE

SENDER JANKIEL MISZPUTEN

## INTRODUCTION

The approach to differential diagnoses in the so called inflammatory bowel diseases (IBD), namely Crohn's disease (CD) and nonspecific ulcerative colitis (UC), should take into consideration, on one hand, aspects that allow distinguishing them from each other, and, on the other hand, differentiating them from acute inflammation or chronic recurring inflammatory processes, but with established etiology,[1] which can affect the same anatomical regions or present with similar clinical signs and symptoms.

There is no additional procedure, a gold standard that can diagnose IBD directly and safely. Thus, it would be desirable, at first, to seek differentiating complaints from a specific process, with a known origin, such as celiac disease, irritable bowel syndrome, infectious colitis,[2] of the nonspecific inflammations – which is seldom an easy task – to analyze later what the best hypothesis can be, CD or UC, if the disease is located in segments common to both. This fact makes diagnosis a real challenge for the doctor, even when experienced. The diagnosis must be supported by data from the clinical history, as detailed as possible, physical examination, some laboratory tests (unfortunately relative importance), and especially the analysis of radiological, endoscopic and histological exams. All this suggests the involvement of a multidisciplinary team, including a gastroenterologist, a radiologist, an endoscopist and a pathologist to confirm the disease.[3] Despite the availability of increasingly sophisticated methods for differentiation, IBD limited to the colon quite often is characterized after a period of disease progression, when changes in clinical signs or results of laboratory tests allow a definition. At this stage, it is called indeterminate or unclassified colitis.

Often the diagnosis of these diseases is delayed because they rarely occur and even specialists sometimes forget about this possibility, but also because symptoms vary considerably, and complementing technical resources are not always accessible. These aspects result in the initial acceptance of the complaints stemming from common gastroenteric infectious processes, both by the physician and the patient, even if they have been repeated in several episodes.

When there is a chance of CD or UC, in most cases, the diagnosis can be done easily, with high probability of success, after conventional investigation. Although similar in some respects, they are distinct diseases in many morphological and clinical aspects, requiring different approaches for confirmation. Less frequently, but still involving 10 to 15% of cases,[4,5] patients should undergo non-routine imaging and laboratory procedures in order to define the nature of the inflammation, which may require greater time and cost.

## DIFFERENTIAL DIAGNOSIS OF INFLAMMATORY ENTEROCOLITIS
### Clinical features

The usually distinct anatomical distribution between UC and CD may prove to be an important aspect for the observed differences in terms of clinical manifestations of these two inflammatory processes and, consequently, their diagnosis.

### Ulcerative colitis

UC exclusively affects the colon; in adults, more often on the left side,[6] and in almost all cases, the rectum; sometimes, inflammation may spread beyond into the terminal ileum, in the case of pancolitis. For all these facts, the most commonly reported clinical complaints are those of change in bowel frequency with the presence of diarrhea, characterized by multiple urgent evacuations, due to repeated stimuli (tenesmus), with stool of variable consistency, containing mucus and blood, sometimes with elimination of mucus and blood with no feces, occurring during day and night, usually preceded of diffuse abdominal pain or in the left lower quadrant, or colicky pain, with temporary relief after each passage of colorectal content. Systemic symptoms such as low fever and weight loss are associated to intestinal symptoms in a manner compatible with the inflammatory activity, all signs and symptoms disappearing during remission stage. Recurrence is expected. Extraintestinal, joint, ophthalmic or dermatological manifestations are observed in approximately 25% of patients,[7] even preceding the intestinal complaints, and may be a reference point for hypothesizing such inflammatory process.

Physical examination reveals changes depending on the disease development time, the severity of symptoms, and the frequency with which they appear. Signs of malnutrition are occasionally present, based on several reasons: anorexia; restrictive diet that the patient sometimes adopts assuming that this can ease the discomfort, or as a medical suggestion, not always well-indicated; abdominal pain motivated by diet and derived from loss of protein produced by inflammatory lesions. Loss of blood in stool, iron depleted by inflammation, diets with a low content of this element or intestinal malabsorption explain the pale color of the skin and mucosa seen in most cases during the active stages of disease.[8] Fever and tachycardia usually accompany the systemic response to inflammation. On abdominal palpation, tenderness in the region of colon segments, associated with increased number of bowel sounds, is observed. Complementing the physical examination with anal inspection and digital rectal examination is essential, due to the potential to find anorectal and perianal lesions, leading to the diagnosis of other inflammatory disease, not UC.

### Crohn's Disease

Inflammation that affects the terminal ileum and colon is mainly diagnosed as CD and obviously requires differentiation from other inflammatory diseases involving these segments. Among the chronic processes of the terminal ileum, the most important differential diagnosis is intestinal tuberculosis, especially in countries where prevalence is significant, such as Brazil.[9] Personal and family history for this infection should be carefully considered during anamnesis.

When the colon is the primary site of inflammation, intestinal and general manifestations are not different from those observed in UC, especially if the process reaches distal portions of the segment. This means that, in this location, based on clinical data, distinguishing the two diseases, at first, would not be possible. However, complications such as perianal and perineal lesions, fissures, large plicomas and fistulas are observed more often in this model of inflammation, even when they are limited to the small intestine, and may even precede the intestinal complaints.

A second complication of Crohn's enterocolitis is related to its tendency to stenosis due to retraction of the scar in diseased areas, which would manifest as complaints that resemble subocclusion (more persistent pain, reduced distension and caliper of stool in the case of anorectal strictures).

Diarrhea associated with abdominal pain, tending to persist for longer periods and continues with the same intensity even after the passage of intestinal contents, and weight loss are the most frequently reported complaints to this inflammation in any anatomical location.[2,10] Weight loss affects about 60% of patients, a rate higher than that found in UC,[3] and their similarity also comes from changes in appetite, spontaneous or suggested dietary changes due to an expectation to relieve symptoms, malabsorption by reduction the absorptive area as a result of inflammation and fistulas, and bacterial overgrowth caused by sub-occlusions and intestinal resections.[11] Bleeding in stool occurs in 40 to 50% of cases, fewer than in UC, and is more common in the distal presentations of Crohn's colitis. In the active phase of the disease, mild fever and signs of anemia are usually present on physical examination. Because of its invasive nature, CD tends to be associated with fistulas, most often directed to the perianal area.

On physical examination, in addition to signs of anemia, weight loss and fever present in periods of inflammatory activity, the transmural character of CD allows the identification of a palpable mass in the right lower quadrant of the abdomen, a fact that differentiates it from UC, resulting from the expansion of inflammation to structures adjacent to the ileocecal segment, which is the predominant anatomical area in this condition. This symptom, although strongly indicative of CD, does not exclude other inflammatory diseases in the region. However, the joining of fistulous orifices, perianal or enterocutaneous, anal fissures and plicomas point to CD as the first diagnosis.

Extraintestinal manifestations are more common in the presentation of colonic CD.

### Microscopic colitis

Microscopic colitis comprises collagenous colitis and lymphocytic colitis. Clinically, both are characterized by chronic diarrhea without blood or mucus. Its peak incidence occurs from 60 years of age, with changes in bowel frequency accompanied by weight loss, abdominal pain, daytime and nighttime evacuations and fecal urgency.[12,13] Diagnosis is based on histological investigation biopsies, since the macroscopic appearance of colonoscopy is normal.

### Gastroenteric infections

Gastroenteric infections by bacteria, viruses, parasites and fungi resemble the symptoms of acute exacerbation of IBD, causing diarrhea with mucus and blood, abdominal pain and fever. The most frequent causes stem from contamination by *Salmonella, Shiguella, Campylobacter jejuni*, enterohemorrhagic *Escherichia coli, Yersinia enterocolitica, Clostridium difficile* and *Entamoeba histolytica*. Infectious proctitis mimics nonspecific ulcerative proctitis, stemming from sexually transmitted diseases caused by *Neisseria gonorrhoeae, Chlamydia trachomatis*, herpes simplex or *Treponema pallidum*.[14] Clearly, information on medical history and epidemiology, and similar symptoms in others reinforce the hypothesis of infection, as well as the patient's sexual behavior. The current or recent use of antimicrobial or chemotherapeutic agents allows strong suspicion in cases of diarrhea caused by *Clostridium difficile*, which leads to a form of acute ulcerative colitis, namely pseudomembranous colitis, manifested by diarrhea, with fecal elimination of mucus, blood and pseudomembranes.

In patients with IBD treated with corticosteroids or immunomodulators, the reappearance of intestinal symptoms may be the expression of their association with some acute infectious process by opportunistic microorganisms,[15] having in *Clostridium difficile* its main etiological suspect. Even in the absence of immunosuppressive therapy, recurrence of clinical manifestations of UC or CD must place this infection as a differential diagnosis with decompensated inflammatory activity, because of its higher incidence in this group of individuals with increased risk of acquiring it in the community, regardless of the medication being used.[16,17]

Infection by cytomegalovirus (CMV) is common in humans and can behave like a flare agent of IBD, promoting lesions in the colon such as ulcers and erosions, which overlap those recognized in UC. Its activation as an opportunistic infectious agent occurs mostly during periods of decreased immune defenses. There seems to be differences in the clinical significance of this infection according to the IBD: it is more frequent in severe forms of UC and refractory to steroids and does not interfere in the progression of CD. It is unclear whether the virus is simply an adjunct or a factor of disease exacerbation.[18] A recent publication recognized the persistence of the virus in the colon after remission of an episode of relapse, suggesting thereby that it is not responsible for recurrence.[19] This viral disease must be considered as a differential diagnosis when there is decompensation of colitis, and even more in cases of treatment with immunosuppressants. It has been described in corticosteroid-naive patients.[20]

The possibility of intestinal and general symptoms as an expression of acquired immune deficiency (AIDS), an obligatory differential diagnosis for chronic consumptive diarrhea, is also important.

Extraintestinal manifestations such as arthritis, erythema nodosum and oral aphthous ulcers are described with some strains of *Chlamydia trachomatis*, which also induce the formation of perianal fistulas and resemble CD lesions.

### Ischemic infections

This is a medical condition that is prevalent in older age groups, but may occur in young people. The reduced blood flow that promotes the formation of colic or enteric ulcers is mostly caused by hypotension events (sepsis, loss of volume by bleeding, dehydration, and left heart failure). Medications such as nonsteroidal anti-inflammatory drugs (NSAIDs), antihypertensive drugs and contraceptives, as well as hypercoagulable states, vasculitis and atheromatosis also justify the development of ischemia in the mesenteric microcirculation. Diarrhea with blood and mucus in geriatric patients should therefore include, among other possibilities, a diagnosis of ischemic colitis.[21,22]

## Actinic infections

The same vascular behavior is observed in radiation-induced colitis (actinic), particularly in the first months post-treatment with ionizing substances. It is a progressive and dose-dependent vasculitis, which causes thrombosis of small vessels in the region affected by radiation. The main symptoms of actinic colitis are diarrhea with mucus and blood and abdominal pain, especially in the distal segments of the colon and rectum. It is easily differentiated from the IBDs based on history information about past pelvic radiotherapy.[23]

## Neoplastic infections

Colon tumors have, in many patients, prolonged evolution. Its symptoms repeat those seen in IBDs, such as diarrhea, weight loss and anemia. Since they also affect young individuals, investigation must be performed as soon as possible for proper differentiation of these disease entities. Family history of either disease, if present, is a good base for a first diagnostic supposition.

## LABORATORY AND IMAGING ASPECTS

### Laboratory

General laboratory procedures have relative value for the investigation of inflammation in the alimentary canal. In the active phase of UC and CD, these allow the assessment of states of anemia (with dosages of hemoglobin, iron and ferritin) and hypoalbuminemia, and of the intensity of the process, indirectly, by measuring inflammatory parameters of acute phase, erythrocyte sedimentation and C-reactive protein (CRP) levels, if there are no genetic defects for their production,[24] as well as platelet count. CRP is a protein produced in small quantities by hepatocytes, which quickly rises in the blood after the initiation of an inflammatory process, stimulated by interleukin-6, tumor necrosis factor (TNF-alpha) and interleukin-1 beta. It has a short half-life when compared with other acute phase proteins.[25] Real-time dosage ends up being the best serological indicator for the analysis of presence and intensity of inflammation, both in the exacerbation period and during recovery.

These parameters, however, not only cannot always define accurately the presence of intestinal inflammation but also do not characterize the type of disease. Laboratory differentiation between UC and CD using blood tests can be summarized as serological measurement of perinuclear anti-neutrophil cytoplasmic antibody (pANCA) and anti-*Saccharomyces cerevisiae* antibodies (ASCA), detected in a number of patients with UC and CD, respectively,[5,26] with the possibility that none of them reveal positive results due to low specificity.

Other markers are described in an attempt to confirm CD and establish a prognosis for disease progression, particularly the family of anti-glycans, which include ASCA, sugars attached to proteins found in various human cells,[27] but not used in routine laboratory tests.

Stool examination should be amplified with various investigations: parasitological, culture, leukocytes, and occult blood if the clinical history has not confirmed the presence of blood in stool. Coproculture turns out to be fundamental in determining infectious processes, either in the onset of symptoms or during disease progression. Dosage of fecal leukocyte markers calprotectin and lactoferrin,[28-30] which advantageously replace serological markers since they are more specific, helps differentiate the diarrheal states of inflammatory nature from those resulting from functional disturbances. In patients with IBD, it is possible to evaluate the intensity of the process and predict, with reasonable notice, the recurrence of inflammation.[7,31] These are proteins of neutrophils, which reach high levels in the presence of infectious or inflammatory processes. Some Brazilian centers offer fecal calprotectin dosage, which allows the evaluation of IBD from the diagnosis and during the progression of the disease without the need for repeated use of invasive methods for follow-up. Stool evaluation for fecal leukocytes, an easily accessible, although less sensitive procedure, is the simplest way to suggest inflammatory disease. However, it is noteworthy that none of these fecal methods is specific enough to identify the model of disease.

Special attention should be given to intestinal infection conditions associated with IBD. The highest incidence of *Clostridium difficile* infection in these patients must be considered in cases of clinic decompensation, especially if there is no adequate response to treatment in this stage of disease progression. The search for toxins from the microorganism in stool is paramount, and can guide the therapy indicated for this complication, if positivity is confirmed.

## Imaging

Imaging methods are the best alternative for the differential diagnosis of the two IBDs because they allow identification of the nature, location and extent of disease, its presentation format and possible complications.

Plain radiographs of the abdomen can be very useful, but only in case of complications of IBD, particularly in suspected subocclusion, both organic, as in CD strictures, and functional, as in toxic megacolon, a severe event that can lead to differentiation between these two forms of inflammation. It is also indicated for suspected perforation.

At the stage of diagnosis, barium enema using double contrast technique is recommended to assess ulcerative disease of the colon, allowing recognition of disease extent and characteristics, ranging from superficial ulcers in the case of UC to the deep ulcers of CD, as well as their complications, fistulas and strictures. When the progression of the colonoscope is hindered in the areas of narrowed colon, this radiological method is recommended for analysis of intestinal segments not studied by endoscopy, thus increasing the security in guiding medical management.

In CD, diagnostic imaging classically involves performing the conventional intestinal transit test, using compression technique for better assessment of the inflamed area, which facilitates the recognition of strictures or fistulas with high sensitivity and specificity, especially in cases of ileal disease. Keep in mind that this inflammatory process can affect several segments of the alimentary canal simultaneously, so that a detailed study of the entire gut is recommended. The method is flawed for small lesions or isolated superficial ulcers, leaving therefore some cases without proper diagnosis, in addition to not providing information about signs of disease activity.

Transabdominal ultrasound is a method used in many centers to diagnose the inflammatory process, and the changes that accompany inflammation activity, being a noninvasive and nonionizing method. When used with high frequency technique, it can identify the thickening of the wall in the diseased intestinal loop, its stratification, and the conditions of mesenteric fat. Complemented with color Doppler to study blood flow through the vessels of the thickened segment, it allows to differentiate between inflammatory activity state or remission.[32,33] While in CD this method shows good correlation with the indices of clinical and endoscopic activity, both in adults and children,[34] its clinical importance in UC is not well established.

Ultrasonography enhanced by contrast, used in patients with CD, has also attracted the attention of experts because of its high sensitivity and specificity to detect inflammatory activity and its strong correlation with the CD activity index (*Crohn's Disease Activity Index* – CDAI),[35] an important factor for decision making in both therapy planning and follow-up of outcomes. Anticipating the need for future controls, it is worth remembering that this is a non-invasive method of study that has no exposure to radiation.

To assess lesions in the perianal region, frequent site of fistulas in CD, endoanal or transperineal ultrasound[36,37] and pelvic magnetic resonance imaging (MRI) are the methods with best accuracy indices.[38,39]

In recent years, the study of bowel loops, previously done with conventional barium methods - transit and enema - began to be performed with computed tomography (CT) and MRI enterography, used for diagnostics of suspected CD and its complications, strictures, abscesses and fistulas. While colonoscopy explains the conditions of the mucosa, CT enterography is focused beyond the mucosa, that is, a transmural view of the bowel wall,[40] with the possibility of analyzing the enhanced images, thickening and stratification of the layers, the microcirculation, engorgement of the *vasa recta*, peri-enteric and peri-colonic mesenteric inflammation, increased number and size of lymph nodes, all signs of disease activity, and recognizes extra-intestinal complications. It is able to differentiate lesions of CD and intestinal tuberculosis.[41] In cases of strictures, it can clarify its nature, inflammatory or fibrotic. The major limitation of CT is the patient's exposure to radiation.

MRI enterography has in its favor the fact that it does not use X-rays, unlike CT. Such advantage is relevant in the case of a generally young population, which expects a need to repeat the imaging tests with some frequency over time of disease progression. On the other hand, the images are not as detailed as those obtained by CT, acquisition takes longer (responsible for the longer procedure duration) and there is a difficulty of acceptance by claustrophobic patients. Anyway, the recognition of parietal, vascular and extra-intestinal lesions is identical to that of CT enterography.

CT or MRI colonography has the same goals of enterographies: to evaluate the hyper-enhancement, edema and thickening of the colon wall, presence of ulcers, lymphadenopathy, and mesenteric inflammatory changes.[42] Its initial use is limited to cases where colonoscopy has not been completed and there is intolerance by the patient to the endoscopic procedure, but its non-invasive nature allows indication for the control of IBD in the colon. One possible benefit of MRI colonography is to distinguish UC from Crohn's colitis in patients with indeterminate forms of the inflammatory process, if signs of transmural inflammation or extraluminal complications such as fistulas and abscesses are not identified.[39]

Anyway, new techniques have been proposed to minimize the radiation produced in CT enterography, reducing its risk and, in MRI enterography, to speed up image acquisition and shorten the time of the procedure.

The endoscopic capsule, another minimally invasive method, should be indicated for patients with strong suspicion of CD in the small intestine and whose conventional, radiological and endoscopic exams have not clarified the diagnosis, as well as in the control of disease progression.[43,44] The important collaboration offered by the capsule is the investigation of the entire small intestine with photographic information of possible endoluminal lesions in different segments, the only obstacle to the procedure being suspected or confirmed stenotic disease. Its specificity is debatable, since about 10% of healthy individuals have erosions in the enteric mucosa, which shows that such a finding is insufficient to establish the diagnosis of CD.[3] Impossibility to perform biopsies and extraluminal assessment of possible complications is another limitation of the method. It is still an expensive procedure and only found in big cities.

More recently, the endoscopic capsule for colon became available in some centers, representing another option for cases that cannot be diagnosed by classical colonoscopy: non-performance due to contraindications, incomplete procedure or patient's refusal to undergo endoscopic exploration.[45] Few controlled studies have been published, but a comparative study against colonoscopy revealed high level of agreement with the results of endoscopy.[46]

Enteroscopy in its various techniques is not a first-line method for the diagnosis of diseases of the small intestine, but has in its favor the capacity of identifying small lesions in any segment of the small intestine, as well as obtaining biopsies from all of the diseased areas. Another possibility is its therapeutic potential in the management of strictures by means of maneuvers of dilation.[47] Moreover, the indication of such endoscopy should be careful, because it requires prolonged sedation and has greater risk for perforation (adhesions and kinking of intestinal loops, mucosal lesions that will undergo distension, etc.).

Undoubtedly, the best method to be employed in the differentiation of colitis and ileocolonoscopy (gold standard), since it allows the analysis of characteristics of the inflammation (continuous in UC, skip lesions in CD), extent and degree of intensity (mild, moderate or severe). The macroscopic normality does not invalidate the existence of inflammation (lymphocytic or collagenous colitis). Serial biopsies in all segments, used for improved discrimination of the inflammation model and of presence of dysplasia, have been implemented with a focus on suspect areas, recognized by use of chromoscopy and image magnification. Histology usually does not differentiate the two most common inflammatory diseases, unless the samples are deep enough to show the limits of the inflammatory process: up to the *muscularis mucosa* in UC, and below it in CD. Granulomas with epithelioid cells, which could facilitate the interpretation of the origin of the disease, are identified on a small number of patients, even after the analysis of surgical specimens.

Mucosal healing is an important goal to be achieved with therapy for IBD. Real-time assessment is obtained by confocal laser endomicroscopy, which allows analyzing the residual inflammation and distortion of the vascular and crypt architecture, defining the level of histological healing.[48]

In collagenous colitis, in addition to the lymphocytic infiltration in the lamina *propria* and presence of intraepithelial lymphocytes, there is an expansion of subepithelial collagen. In lymphocytic colitis, the histological changes are similar, except for the increased thickness of the collagen layer, absent in this disease.[49,50]

Although the difference between microscopic colitis and those of IBD is clear, it is possible in small number of cases that both exhibit similar pathological features with erosions or ulcers, distorted crypt architecture, and cryptic abscesses, which does not prevent the morphological pattern usually labeled as specific to colitis occasionally to derive from a microscopic colitis. The possibility of being part of the spectrum of classic IBD is also considered.[51]

Monitoring the progress of colic inflammation by colonoscopy and biopsies is recommended, due to the increased risk that these diseases pose to the development of cancer.[52]

Dysplasia is found less frequently in inflammations of the small intestine compared to those of the colon and, thus, presents lower risk for cancer.

## REFERENCES

1. Sands BE. From symptom to diagnosis: clinical distinctions among various forms of intestinal inflammation. Gastroenterology 2004; 126(6):1518-32.

2. Quintana C, Galleguillos L, Benavides E, Quintana JC, Zúñiga A, Duarte I et al. Clinical diagnostic clues in Crohn's disease: a 41-year experience. ISRN Gastroenterol 2012; Article ID 285475:1-6.

3. van Assche G, Dignass A, Panes J, Beaugerie L, Karagiannis J, Allez M et al. The second evidence-based Consensus on the diagnosis and management of Crohn's disease: definitions and diagnosis. J Crohns Colitis 2010; 4:7-27.

4. Jung SA. Differential diagnosis of inflammatory bowel disease: what is the role of colonoscopy? Clin Endosc 2012; 45(3):254-62.

5. Zisman TL, Rubin DT. Novel diagnostic and prognostic modalities in inflammatory bowel disease. Gastroenterol Clin N Am 2009; 38:729-52.

6. Magro F, Langner C, Driessen A, Ensari A, Geboes K, Mantzaris GJ et al. European consensus on the histopathology of inflammatory bowel disease. J Crohns Colitis 2013; 7(10):827-51.

7. Iskandar HN, Ciorba MA. Biomarkers in inflammatory bowel disease: current practices and recent advances. Transl Res 2012; 159(4):313-25.

8. Goldberg ND. Iron deficiency anemia in patients with inflammatory bowel disease. Clin Exp Gastroenterol 2013; 6:6:1-70.

9. Pulimood AB, Amarapurkar DN, Ghoshal U, Phillip M, Pai CG, Reddy DN et al. Differentiation of Crohn's disease from intestinal tuberculosis in India in 2010. World J Gastroenterol 2011; 17(4):433-43.

10. Wilkins T, Jarvis K, Patel J. Diagnosis and management of Crohn's disease. Am Fam Physician 2011; 84(12):1365-75.

11. Hartman C, Eliakim R, Shamir R. Nutritional status and nutritional therapy in inflammatory bowel diseases. World J Gastroenterol 2009; 15(21):2570-8.

12. Nyhlin N, Bohr J, Eriksson S, Tysk C. Microscopic colitis: a common and an easily overlooked cause of chronic diarrhoea. Eur J Intern Med 2008; 19(3):181-6.

13. Storr MA. Microscopic colitis: epidemiology, pathophysiology, diagnosis and current management-an update 2013. ISRN Gastroenterol 2013; 352718.

14. Gallegos M, Bradly D, Jakate S, Keshavarzian A. Lymphogranuloma venereum proctosigmoiditis is a mimicker of inflammatory bowel disease. World J Gastroenterol 2012; 18(25):3317-21.

15. Dave M, Purohit T, Razonable R, Loftus Jr EV. Opportunistic infections due to inflammatory bowel disease therapy. Inflamm Bowel Dis 2014; 20(1):196-212.

16. Nitzan O, Elias M, Chazan B, Raz R, Saliba W. Clostridium difficile and inflammatory bowel disease: Role in pathogenesis and implications in treatment. World J Gastroenterol 2013; 19(43): 7577-85.

17. Reddy SS, Brandt LJ. Clostridium difficile infection and inflammatory bowel disease. J Clin Gastroenterol 2013; 47(8):666-71.

18. Garrido E, Carrera E, Manzano R, Lopez-Sanroman A. Clinical significance of cytomegalovirus infection in patients with inflammatory bowel disease. World J Gastroenterol 2013; 19(1):17-25.

19. Criscuoli V, Rizzuto MR, Montalbano L, Gallo E, Cottone M. Natural history of cytomegalovirus infection in a series of patients diagnosed with moderate-severe ulcerative colitis. World J Gastroenterol 2011; 17(5):633-8.

20. Inoue K, Wakabayashi N, Fukumoto K, Yamada S, Bito N, Yoshida N et al. Toxic megacolon associated with cytomegalovirus infection in a patient with steroid-naïve ulcerative colitis. Intern Med 2012; 51(19):2739-43.

21. Theodoropoulou A, Koutroubakis IE. Ischemic colitis: clinical practice in diagnosis and treatment. World J Gastroenterol 2008; 14:7302-8.

22. Tortora A, Purchiaroni F, Scarpellini E, Ojetti V, Gabrielli M, Vitale G et al. Colitides. Eur Rev Med Pharmacol Sci 2012; 16(13):1795-805.

23. Kennedy GD, Heise CP. Radiation colitis and proctitis. Clin Colon Rectal Surg 2007; 20(1):64-72.

24. Carlson CS, Aldred SF, Lee PK, Tracy RP, Schwartz SM, Rieder M et al. Polymorphisms within the C-reactive protein (CRP) promoter region are associated with plasma CRP levels. Am J Hum Genet 2005; 77(1):64-77.

25. Dubinsky MD. Serologic and laboratory markers in prediction of the disease course in inflammatory bowel disease. World J Gastroenterol 2010; 16(21):2064-8.

26. Nisihara RM, de Carvalho WB, Utiyama SR, Amarante H, Baptista ML. Diagnostic role and clinical association of ASCA and ANCA in Brazilian patients with inflammatory bowel disease. Dig Dis Sci 2010; 55(8):2309-15.

27. Dotan I. New serologic markers for inflammatory bowel disease diagnosis. Dig Dis 2010; 28(3):418-23.

28. D'Incà R, Dal Pont E, Di Leo V, Ferronato A, Fries W, Vettorato MG et al. Calprotectin and lactoferrin in the assessment of intestinal inflammation and organic disease. Int J Colorectal Dis 2007; 22(4):429-37.

29. Langhorst J, Elsenbruch S, Koelzer J, Rueffer A, Michalsen A, Dobos GJ. Noninvasive markers in the assessment of intestinal inflammation in inflammatory bowel diseases: performance of fecal lactoferrin, calprotectin, and PMN-elastase, CRP, and clinical indices. Am J Gastroenterol 2008; 103(1):162-9.

30. Sutherland AD, Gearry RB, Frizelle FA. Review of fecal biomarkers in inflammatory bowel disease. Dis Colon Rectum 2008; 51:1283-91.

31. Gisbert JP, Bermejo F, Pérez-Calle JL, Taxonera C, Vera I, McNicholl AG et al. Fecal calprotectin and lactoferrin for the prediction of inflammatory bowel disease relapse. Inflamm Bowel Dis 2009; 15(8):1190-8.

32. Strobel D, Goertz RS, Bernatik T. Diagnostics in inflammatory bowel disease: Ultrasound. World J Gastroenterol 2011; 17(27):3192-7.

33. Kralik R, Trnovsky P, Kopáčová M. Transabdominal ultrasonography of the small bowel. Gastroenterol Res Pract 2013; 2013 Article ID:896704.

34. Drews BH, Barth TF, Hänle MM, Akinli AS, Mason RA, Muche R et al. Comparison of sonographically measured bowel wall vascularity, histology, and disease activity in Crohn's disease. Eur Radiol 2009; 19:1379-86.

35. Migaleddu V, Scanu AM, Quaia E, Rocca PC, Dore MP, Scanu D et al. Contrast-enhanced ultrasonographic evaluation of inflammatory activity in Crohn's disease. Gastroenterology 2009; 137(1):43-52.

36. de la Portilla F, León-Jiménez E, Cisneros N, Rada R, Flikier B, Vega J et al. Use of anorectal ultrasounds in perianal Crohn's disease: consistency with clinical data. Rev Esp Enferm Dig 2006; 98(10):747-54.

37. Kim Y, Park YJ. Three-dimensional endoanal ultrasonographic assessment of an anal fistula with and without H(2)O(2) enhancement. World J Gastroenterol 2009; 15(38):4810-5.

**38.** Hvas CL, Dahlerup JF, Jacobsen BA, Ljungmann K, Qvist N, Staun M et al. Diagnosis and treatment of fistulising Crohn's disease. Dan Med Bull 2011; 58(10):C4338. Review.

**39.** Gee MS, Harisinghani MG. MRI in patients with inflammatory bowel disease. J Magn Reson Imaging 2011; 33(3):527-34.

**40.** Loftus EV. Using CT and MR enterography to diagnose and monitor IBD. Gastroenterol Hepatol (NY) 2010; 6(12):754-6.

**41.** Park MJ, Lim JS. Computed tomography enterography for evaluation of inflammatory bowel disease. Clin Endosc 2013; 46(4):327-36.

**42.** Rimola J, Rodríguez S, García-Bosch O, Ricart E, Pagès M, Pellisé M et al. Role of 3.0-T MR colonography in the evaluation of inflammatory bowel disease. Radiographics 2009; 29(3):701-19.

**43.** Kopylov U, Seidman EG. Clinical applications of small bowel capsule endoscopy. Clin Exp Gastroenterol 2013; 6:129-37.

**44.** Niv Y. Capsule endoscopy in the diagnosis of Crohn's disease. Med Devices (Auckl) 2013; 6:85-9.

**45.** Riccioni ME, Urgesi R, Cianci R, Bizzotto A, Spada C, Costamagna G. Colon capsule endoscopy: advantages, limitations and expectations. Which novelties? World J Gastrointest Endosc 2012; 4(4):99-107.

**46.** Herrerías-Gutiérrez JM, Argüelles-Arias F, Caunedo-Álvarez A, San-Juan-Acosta M, Romero-Vázquez J, García-Montes JM et al. PillCamColon Capsule for the study of colonic pathology in clinical practice. Study of agreement with colonoscopy. Rev Esp Enferm Dig 2011; 103(2):69-75.

**47.** Jeon SR, Kim JO. Deep Enteroscopy: which technique will survive? Clin Endosc 2013; 46(5):480-5.

**48.** Gheorghe C, Cotruta B, Iacob R, Becheanu G, Dumbrava M, Gheorghe L. Endomicroscopy for assessing mucosal healing in patients with ulcerative colitis. J Gastrointestin Liver Dis 2011; 20(4): 423-6.

**49.** Ianiro G, Cammarota G, Valerio L, Annicchiarico BE, Milani A, Siciliano M et al. Microscopic colitis. World J Gastroenterol 2012; 18(43):6206-15.

**50.** Freeman HJ. Long-term natural history and complications of collagenous colitis. Can J Gastroenterol 2012: 627-30.

51. Jegadeesan R, Liu X, Pagadala MR, Gutierrez N, Butt M, Navaneethan U. Microscopic colitis: is it a spectrum of inflammatory bowel disease? World J Gastroenterol 2013; 19(26):4252-6.

52. Kim YG, Jang BI. The role of colonoscopy in inflammatory bowel disease. Clin Endosc 2013; 46(4):317-20.

# CONVENTIONAL MEDICAL THERAPY IN ULCERATIVE COLITIS

WILSON ROBERTO CATAPANI

## INTRODUCTION

In this chapter, directed to physicians, nurses, nutritionists and other health professionals, the conventional treatment of ulcerative colitis (UC) will be addressed, not including biological therapy, probiotics and nutritional therapy, which will be described in Chapters 25 and 31, respectively.

According to the precepts of evidence-based medicine, the best models for decision-making are properly conducted controlled randomized clinical trials and meta-analyses; the worst level of evidence comes from opinions based on the personal experience of experts. The reader of this chapter will notice that some older references are listed; these are, however, still valid as best evidence available to date. Often, clinical studies suffer major deficiencies, such as small sample size and poorly defined outcomes. On the other hand, consensuses prepared by experts, whose guidelines are made through rigorous and thorough assessment of the literature, and not the experience of specialists, have been very useful in many clinical situations. The European Crohn's and Colitis Association (ECCO)[1] consensus follows this rule and will be the basis for what will be presented in this chapter.

The treatment of UC, or simply ulcerative colitis, generally seeks to induce the remission of active disease, maintain remission, prevent complications and provide quality of life to the patient.

To achieve these goals, clinical, biochemical and endoscopic criteria are used to guide therapy, seeking an accurate evaluation of the severity and extent of disease, characteristics that assist in the choice of route of administration and dosage of the drugs used, indication for surgery or hospitalization and other aspects related to treatment. Thus, the first measure to be taken regarding a patient with UC is the evaluation of the severity and extension of the inflammatory process, before starting any treatment.

The extent of disease is more accurately assessed by colonoscopy, which is also very useful to estimate the severity of the inflammatory process. For the assessment of disease severity, clinical and laboratory parameters are used, and since their details are beyond the scope of this chapter, they will be addressed in other sections of this book. However, we present here most commonly used reference for this purpose, which is the adapted Truelove-Witts index found in Table 22.1. According to the ECCO consensus,[1] severe UC is properly defined by this criterion. Patients presenting more than 6 bloody stools daily and signs of toxicity (tachycardia greater than 90 bpm, fever greater than 37.8°C (100.04°F), hemoglobin concentration below 10.5 g/dL or erythrocyte sedimentation rate [ESR] greater than 30 mm/h) should be hospitalized.

According to the 2010 guidelines for the treatment of UC dictated by the World Gastroenterology Organization,[2] effective drugs in therapy are aminosalicylates, corticosteroids, immunosuppressants and antibiotics.

Aminosalicylates available in Brazil are mesalazine and sulfasalazine. They are useful in the treatment of acute disease attacks, as well as for maintenance. They are available in tablets (sulfasalazine 500 mg and mesalazine 400, 500 and 800 mg), 3 g enemas, and mesalazine 250 mg and 1 g suppositories. Recently, mesalazine sachets of 1 and 2 g, and mesalazine MMX, which releases the drug at pH 7 in the terminal ileum, providing treatment in a single dose, were made available on the market.

Corticosteroids are useful in the acute phase of the disease, but should be avoided as maintenance treatment, because of their side effects. They are available in Brazil in intravenous (hydrocortisone and methylprednisolone) and oral (prednisone, prednisolone) forms. Budesonide is also available in Brazil in the form of enemas and tablets, but has limited use due to its high cost.

The most commonly used immunosuppressants are thiopurines (6-mercaptopurine and azathioprine) and methotrexate. They are slow-acting and are not indicated for the treatment of acute phase of the disease; however, these drugs are very useful for maintenance in patients who respond poorly to aminosalicylates or in corticosteroid-dependent individuals.

The most commonly used antibiotics are metronidazole and ciprofloxacin. They are best used in fulminant colitis, quite empirically, since there is no consistent data on effectiveness. Antibiotics are not useful for the maintenance treatment of UC.

As already discussed, the general principles of treatment include the severity and extent of the inflammatory process. Furthermore, disease behavior, frequency of relapses, response to previous treatments, and side effects of drugs already used are also evaluated.

**Table 22.1** UC Classification according to severity of acute disease (Truelove and Witts)

|  | Mild | Moderate | Severe |
|---|---|---|---|
| 1. Number of bowel movements/day | ≤ 4 | 4 to 6 | > 6 |
| 2. Blood in the stool | ± | + | ++ |
| 3. Body temperature | Normal | Intermediate values | Average body temperature at night >37.5°C (99.5°F) or >37.8°C (100.04°F) in 2 out of 4 days |
| 4. Pulse | Normal | Intermediate | > 90 bpm |
| 5. Hemoglobin (g/dL) | > 10,5 | Intermediate | < 10,5 |
| 6. ESR (mm/first hour) | < 30 | Intermediate | > 30 mm, first hour |

ESR: erythrocyte sedimentation rate.

## INDUCTION OF REMISSION IN ACTIVE DISEASE

In clinical practice, the most important is to distinguish patients with severe activity outbreaks, who should be hospitalized, from those in mild or moderate activity, who can be treated as outpatients. Therefore, among various disease activity indices, the Truelove-Witts index is still the most validated and used.[1] Obviously, in the case of a patient experiencing the first acute outbreak of his or her life, the most important differential diagnoses are infectious diseases (*Entamoeba histolytica* or invasive bacteria of the intestinal wall), cytomegalovirus in immunocompromised patients or HIV positive, and Crohn's disease (CD) of the colon.

For patients with mild to moderate activity, treatment of proctitis should initially be topical, with mesalazine suppositories.[1] In a meta-analysis of 778 patients, topical mesalazine induced remission in 31 to 80% of patients compared to 7 to 11% of patients using placebo.[3] 1 g suppositories are very effective, with clinical and endoscopic remission in about 64% after 2 weeks.[4] Topical mesalazine is more effective than oral mesalazine for the treatment of proctitis, and is also more effective than topical corticosteroids. In cases of poor response can be used combined therapy mesalazine 2g + beclomethasone dipropionate 3 mL;[5] however, in Brazil, this corticosteroid is not available in soluble form. If the topical treatment does not lead to a satisfactory response, oral mesalazine, oral prednisone or prednisolone can be added according to the disease activity. This subject will be discussed later in this book.

In the treatment of left colitis or pancolitis, combined therapy (oral mesalazine plus topical) is more effective than oral or topical therapy alone.[1-6] Patients in mild to moderate activity can be initially treated with sulfasalazine or mesalazine; in general, the doses indicated to induce remission are equivalent to 4 g/day for both drugs. Response to treatment can be relatively slow and progressive, reaching up to 8 weeks for maximum effectiveness; however, if no response is observed until the fourth week, there is usually no improvement after this period. Patients who do not respond to this scheme should be treated with corticosteroids in the acute phase, typically started at a dose of 40 to 60 mg/day of prednisone.[1] The optimal duration of therapy, including a loading dose and subsequent weaning, is unknown, and various schemes are possible. However, it is reasonable to assume that it should be 8 to 16 weeks from the start of therapy until complete weaning has occurred. The concomitant use of corticosteroids and aminosalicylates is not necessary while the patient is receiving high or moderate doses of corticosteroids.

In the treatment of severe or fulminant active colitis with conventional therapy, systemic effects of the inflammatory process should also be considered. Measurements of hemoglobin, hematocrit, albumin and electrolytes are required to estimate the need for blood and fluid replacement. Hypokalemia is particularly common in these patients, especially those receiving corticosteroids. Fulminant colitis can be complicated by toxic megacolon when the transverse colon has a caliper equal to or larger than 7 cm, associated with signs of toxicity such as fever, hypotension and tachycardia. Supportive measures should be carried out as soon as possible, while the patient is hospitalized, starting with hydration and correction of anemia and avoiding the use of opioid analgesics and antidiarrheal drugs such as loperamide and diphenoxylate, due to risk of inducing toxic megacolon.

The use of oral aminosalicylates is not recommended in severe colitis,[7] and these drugs should be suspended if they are being used by the patient. Methylprednisolone at a dose of 0.75 to 1 mg/kg/day or hydrocortisone 400 mg/day is the most common treatment schedule.[1] In this situation, the ideal patient care requires the joint efforts of both clinical gastroenterologist and surgeon. Surgery should be discussed as an option at this stage, compared with alternative treatment using cyclosporine or infliximab.[1]

The use of cyclosporin as continuous intravenous infusion at 4 mg/kg in severe colitis is effective.[8] However, its use is avoided on account of side effects such as kidney failure, hypertension, hypomagnesemia, hyperkalemia, seizures, and more. Toxicity appears to be greater with the concomitant use of corticosteroids, and the cessation of this drug is recommended if cyclosporine is introduced. Increased toxicity is also associated with higher doses in the range of 4 mg/kg. The decision to use cyclosporin should be compared with the decision to refer the patient to colectomy, since multiple case series suggest that colectomy rate within 6 months to 1 year in patients who initially responded to oral cyclosporin is between 45 and 70%.[9]

As an adjuvant therapy, altered intestinal microbial flora with antibiotics can be helpful, although there is no consistent data from studies in humans. When there is no sepsis, the use of systemic antibiotics is not necessary, but non-absorbable oral agents such as vancomycin can be useful. Systemic antibiotics such as metronidazole and ciprofloxacin, added as therapy to patients treated with intravenous corticosteroids, did not increase the effectiveness of treatment.[10,11]

Nutritional therapy is discussed separately in Chapter 31; however, except in cases of ileus secondary to severe colonic inflammation, patients generally tolerate mild or enteral feedings, and parenteral nutrition, in principle, should not be used. Nevertheless, it should be borne in mind that, if indicated, parenteral nutrition does not have the purpose of inducing remission, as this is not its effect; it aims to maintain the nutritional status of the patient.

## MAINTENANCE THERAPY

The natural history of UC shows that when kept without any treatment after the acute phase, 75 to 80% of patients relapse within 1 year after diagnosis. This is the rationale for maintenance therapy, in addition to improved quality of life and prevention of dysplasia and cancer, which, according to the ECCO consensus, aims to maintain remission without corticosteroids, defined both clinically and endoscopically. Maintenance therapy is recommended for all patients; intermittent therapy is acceptable only in a small number of patients with limited disease extension. The choice of drug used for maintenance is determined by the extent of disease, frequency of relapses, ineffectiveness of previous treatments, severity of the latest exacerbation, drug safety, and prevention of cancer.[1]

In patients who responded to induction of remission with aminosalicylates and corticosteroids, mesalazine is the first-line drug to maintain remission.[12] In proctitis and left colitis, the combined regimen (oral and topical) can be used. The minimal effective dose is around 1 g/day. If rectal topical treatment is chosen, the total dosage of 3 g/week in divided doses is usually sufficient. Although sulfasalazine is also effective, its toxicity is higher. The option of using suppositories should be discussed with the patient, since the long-term tolerance for intrarectal treatment is highly variable from one individual to another. At present, there is insufficient evidence to recommend the preferential use of one type of mesalazine presentation over another for maintenance treatment.[1] However, formulations that reduce the number of daily doses of the drug significantly improve treatment compliance.[13,14]

Thiopurines, azathioprine and 6-mercaptopurine are recommended for patients with early or frequent recurrence being treated with mesalazine or sulfasalazine at appropriate doses, patients intolerant to such drugs, or those dependent on corticosteroids to maintain remission.[1] In a meta-analysis of the Cochrane Collaboration, there was no clear evidence of a dose-response effect for azathioprine. Adverse events occurred in 11 of 127 patients, including pancreatitis and bone marrow aplasia.[15] In a randomized controlled study comparing mesalazine 3.2 g/day and azathioprine 2 mg/kg/day in patients dependent on corticosteroids, clinical and endoscopic remission occurred in 53% of patients with azathioprine and 21% of patients with mesalazine (predictive value 4.78, 95% confidence interval, 1.57-14.5).[16]

Regarding the use of ciprofloxacin, metronidazole and methotrexate for maintaining remission in UC, the ECCO consensus believes that the existing studies are not satisfactory, and there is no evidence so far to support this recommendation.[1]

## REFERENCES

1.  Travis SPL, Stange EF, Lémann M, Øresland T, Bemelman WA, Chowers Y et al. European evidence-based consensus on the management of ulcerative colitis: current management. J Crohns Colitis 2008; 2:24-62.

2.  Bernstein CN, Fried M, Krabshuis JH, Cohen H, Eliakim R, Fedail S et al. World Gastroenterology Organization Practice Guidelines for the diagnosis and management of IBD in 2010. Inflamm Bowel Dis 2010; 16(1):112-24.

3.  Marshall JK, Irvine EJ. Rectal aminosalicylate therapy for distal ulcerative colitis: a meta-analysis. Aliment Pharmacol Ther 1995; 9:293-300.

4.  Gionchetti P, Rissole F, Ventura A, Brignola C, Ferretti M, Peruzzo S et al. Comparison of mesalazine suppositories in proctitis and distal proctosigmoiditis. Aliment Pharmacol Ther 1997; 11:1053-7.

5.  Mulder CJ, Fockens P, Meijer JWR, van der Heide H, Wiltinik EH, Tytgat GN. Beclomethasone dipropionate (3mg) vs. 5-aminosalicylic acid (2g) vs. the combination of both (3 mg/2 g) as retention enemas in active ulcerative proctitis. Eur J Gastroenterol Hepatol 1996; 8:549-53.

6. Safdi M, DeMicco M, Sninsky C, Banks P, Wruble L, Deren J et al. A double-blind comparison of oral vs. rectal mesalamine vs. combination therapy in the treatment of distal ulcerative colitis. Am J Gastroenterol 1997; 92:1867-71.

7. Shanahan F, Targan S. Sulfasalazine and salicylate-induced exacerbation of ulcerative colitis. N Engl J Med 1987; 317(7):455.

8. Lichtiger S, Present DH, Kornbluth A, Gelernt I, Bauer J, Galler G et al. Cyclosporine in severe ulcerative colitis refractory to steroid therapy. N Engl J Med 1994; 330(26):1841-5.

9. Haslam N, Hearing SD, Probert CS. Audit of cyclosporin use in inflammatory bowel disease: limited benefits, numerous side-effects. Eur J Gastroenterol Hepatol 2000; 12(6):657-60.

10. Chapman RW, Selby WS, Jewell DP. Controlled trial of intravenous metronidazole as an adjunct to corticosteroids in severe ulcerative colitis. Gut 1986; 27(10):1210-2.

11. Mantzaris GJ, Hatzis A, Kontogiannis P, Triadaphyllou G. Intravenous tobramycin and metronidazole as an adjunct to corticosteroids in severe ulcerative colitis. Am J Gastroenterol 1994; 89(1):43-6.

12. Sutherland L, Macdonald JK. Oral 5-aminosalicylic acid for maintenance of remission in ulcerative colitis. Cochrane Database Syst Rev 2006; 19:CD000544.

13. Kane S, Huo D, Magnanti K. A pilot feasibility study of once daily vs. conventional dosing mesalamine for maintenance of ulcerative colitis. Clin Gastroenterol Hepatol 2003; 1:170-3.

14. Kruis W, Gorelov A, Kiudelis G et al. Once daily dosing of 3 g mesalamine (Salofalk® granules) is therapeutic equivalent to a three-times daily dosing of 1 g mesalamine for the treatment of active ulcerative colitis. Gastroenterology 2007; 132(Suppl.4):A-130.

15. Timmer A, McDonald JW, Macdonald JK. Azathioprine and 6-mercaptopurine for maintenance of remission in ulcerative colitis. Cochrane Database Syst Rev 2007; 24:CD000478.

16. Ardizzone S, Maconi G, Russo A, Imbesi V, Colombo E, Bianchi Pono G. Randomised controlled trial of azathioprine and 5-aminosalicylic acid for treatment of steroid dependent ulcerative colitis. Gut 2006; 55:47-53.

# CONVENTIONAL MEDICAL THERAPY IN CROHN'S DISEASE

ADÉRSON OMAR MOURÃO CINTRA DAMIÃO
FLÁVIO FEITOSA
LUCIANE REIS MILANI

## INTRODUCTION

Generally speaking, when it comes to inflammatory bowel disease (IBD), we speak of ulcerative colitis (UC) and Crohn's disease (CD).[1-3] IBD therapy requires pretreatment measures that are critical for success. Diagnosis should be as accurate as possible, although it is not always feasible to determine which of the two diseases the patient has. Moreover, determining the degree of inflammatory activity of UC and CD, disease extent and behavior (CD, pattern of inflammation, stricturing, penetrating or fistulizing) is also relevant.[1-3]

## GENERAL APPROACH TO INFLAMMATORY BOWEL DISEASE

Most patients with IBD can be treated on an outpatient basis. Only patients with severity criteria or infected, at risk for sepsis, should be hospitalized. This criterion includes dehydrated patients with tachycardia or tachypnea, infected, sub-occluded or malnourished patients, and those who develop serious complications of the disease (e.g., toxic megacolon). In most cases, common sense helps the physician in decision making.

The doctor, when admitting a patient to the hospital, must take extra care with hydration and electrolyte replacement of the patient, as well as nutritional support, especially in cases that are being prepared for surgery. Prophylaxis of thromboembolic events is recommended.[3,4]

Before indicating the therapeutic options and guiding practices in IBD, the conventional therapeutic arsenal available for the treatment of CD and UC will be presented. Biological agents will be discussed in Chapters 24 (Biological therapy in Crohn's disease) and 25 (Biological therapy in ulcerative colitis). Medical treatment of IBD is aimed at rapid and sustained remission of symptoms, with the normalization of the disease's activity scores, laboratory tests and endoscopic remission.[1-4]

## SALICYLIC DERIVATIVES

This group of drugs includes the traditional sulfasalazine (SSZ) and mesalazine. When ingested, SSZ is metabolized in the colon by intestinal bacteria to sulfapyridine (largely absorbed) and 5-aminosalicylic acid (5-ASA or mesalazine or mesalamine), which is poorly absorbed. Mesalazine is the active ingredient of the drug, acting topically. The various mechanisms of action of 5-ASA include inhibiting leukotriene synthesis and free radical scavenging capacities.5 Folic acid (2 to 5 mg/day) should be given concurrently with SSZ because of the risk of macrocytic anemia, one of the side effects of SSZ.[1,2]

Although the use of salicylic derivatives appear to confer a protective effect against the development of dysplasia and/or neoplasia in patients with UC,[6] a retrospective study showed that the use of thiopurines presents more pronounced and impressive effects than those of 5-aminosalicylates, also constituting a protective measure against the emergence of dysplasia and/or colorectal cancer in patients with IBD.7 Apparently, the most important factor to be considered in this context is maintaining the mucosa free of inflammation.

Side effects with SSZ have been reported in up to 45% of patients.[1-3,5] Generally, these effects are dose-dependent, related to high serum levels of sulfapyridine, especially occurring in individuals with low genetic capacity of hepatic drug acetylation (slow acetylators) and include: abdominal pain, nausea, vomiting, anorexia, headache, hemolysis, male infertility, and more.

Less frequently, the side effects of treatment with SSZ can be hypersensitivity (allergy or idiosyncratic) with fever, rash, lymphadenopathy, Stevens-Johnson syndrome, agranulocytosis, hepatitis, pancreatitis and exacerbation of diarrhea. Thus, due to side effects of SSZ, release strategies for 5-ASA (mesalamine in the United States and mesalazine in Europe) in the digestive tract have been developed.[1-3,5]

Most patients allergic or intolerant to SSZ (80 to 90%) tolerate 5-ASA well. However, some (10 to 20%) reproduce the side effects of SSZ when using 5-ASA, which corroborates the fact that some of the side effects of SSZ are caused by 5-ASA, and not by sulfapyridine.[5]

Mesalazine may also be used as enema or suppository, 1 to 4 g/day. It is indicated for distal active UC (proctitis and proctossigmoiditis), with rates of clinical, endoscopic and histologic improvement of 60-90% in the first weeks of treatment.[1,2,4,5]

Although the use of salicylic derivatives is effective in the treatment of UC (active phase and remission),[8] its use in CD is questionable due to little improvement compared to placebo.[9,10] However, SSZ can be used in mild cases of active CD with colonic involvement.[3,5,11]

## CORTICOSTEROIDS

Traditional corticosteroids (such as prednisone, prednisolone and hydrocortisone) are effective drugs in moderate and severe cases of IBD during the active phase.[12] In active CD, treatment with oral prednisone at a dose of 1 mg/kg/day led to clinical remission in 92% of patients after 7 weeks. However, of those in clinical remission, only 29% also had endoscopic remission.[13] In addition, after one year, dependency and refractoriness indices of corticosteroids can reach up to 70%.[14]

In general, in active moderate to severe UC and CD, oral prednisone (0.75 to 1 mg/kg/day, not exceeding 60 mg/day) can be used until clinical remission. Then, corticosteroid dosage is reduced (10 mg/week, up to 0.5 mg/kg/day and, later, 5 mg/week until complete withdrawal).[15] If during "weaning" there is relapse of the disease, the dose of corticosteroid may be increased to the penultimate dose preceding that in which relapse occurred in order to retry drug withdrawal. Currently, however, going "back and forth" with corticosteroids is not accepted and therefore as soon as the patient begins to show dependency or refractoriness to this drug, it is best to introduce immunosuppressive medication (e.g., azathioprine, 6-mercaptopurine ) to allow weaning.[11,15,16]

In severe cases of hospitalized patients, intravenous hydrocortisone can be administered, 100 mg every 6 or 8 hours, replaced by oral prednisone as soon as the patient's condition allows.[11,15-17]

The side effects of traditional corticosteroids are well known, particularly when used for prolonged periods, even at low doses: increased appetite and weight, edema, insomnia, emotional lability, psychosis, acne, Cushing, osteoporosis, osteonecrosis, delayed growth, suppression of the hypothalamic-hypophyseal-adrenal axis, infections, myopathy, cataracts, skin atrophy, striae, ecchymoses, fatty liver, diabetes, hypertension, glaucoma and acute pancreatitis.[11,15-17]

Because of the side effects of traditional corticosteroids, new corticosteroids have been developed in an attempt to reduce such effects. The most studied has been budesonide, which is rapidly metabolized (approximately 90%) into inactive products after its first passage through the liver. At an oral dose of 9 mg/day, it was better than placebo to induce remission in patients with moderate CD located in the ileum, ileum and cecum, and ascending colon (51% *versus* 20%, p< 0.001), with no increased side effects in the group treated with budesonide. It is commercially available as enema (2 mg/100 mL) or tablets (3 mg). Therefore, budesonide is recommended for induction of remission in patients with mild/moderate CD in the terminal ileum and/or right colon, as an alternative to conventional therapy with corticosteroids.[11,15-17]

## ANTIBIOTICS

Patients with CD have quantitative and qualitative changes in gut microbiota (dysbiosis).[18,19] This change is involved in the pathogenesis of the disease, inducing an abnormal immune response in genetically susceptible patients.[20] The composition of the intestinal microflora in patients with CD has an increased concentration of invasive bacteria, especially *Escherichia coli*, and a decrease in the number of protective bacteria such as *Bifidobacterium*, *Lactobacillus* and *Faecalibacterium prausnitzii*, which have anti-inflammatory properties.[21]

Although there is no doubt that antibiotics are useful in certain situations that can complicate IBD, such as fistulas, abscesses, sepsis, infections in general and toxic megacolon, its use as primary or adjunctive treatment in uncomplicated CD or UC is controversial.[22] Studies with antibiotics are usually not controlled and with a small number of patients, which prevents more definitive conclusions. Furthermore, there is also a concern about infection by *Clostridium difficile*, which can occur with the use of antibiotics, especially ciprofloxacin - one of the most widely used antibiotics in IBD.[23,24]

The mechanisms of action of antibiotics in IBD are: reduction of the concentration of luminal bacteria and bacteria attached to the mucosa; selective elimination of more aggressive species; and decreased bacterial translocation and invasion of tissue. Some antibiotics also have the potential to immunosuppressive action.[22]

Metronidazole, a nitroimidazole compound active against anaerobic bacteria and some parasites, and ciprofloxacin, especially active against *E. coli* and *Enterobacteriaceae*, are the most studied and used in IBD. Several randomized controlled studies have been conducted in order to evaluate the efficacy of these drugs for induction of remission in CD. Studies indicate that the use of metronidazole is more effective in patients with active CD than in those with disease limited to the small intestine, probably because of the higher concentration of bacteria in the colon.25 Prantera et al.26 compared the association between oral ciprofloxacin (500 mg every 12 hours) and oral metronidazole (250 mg every 6 hours) with oral corticosteroids (methylprednisolone, 0.7 to 1 mg/kg/day) in patients with active CD for 12 weeks. At the end of the study, clinical remission (CD activity index [CDAI] ≤ 150) was achieved by 45.5% of patients in the "antibiotics" group versus 63% in the "corticosteroid" group (P = not significant). Cessation of the drug caused by side effects occurred in 27.3% of patients using antibiotics and in 10.6% of the group treated with methylprednisolone.26 The authors concluded that the combination of metronidazole and ciprofloxacin could be an alternative to treatment with corticosteroids in patients with active CD. Arnold et al.,27 in turn, evaluated the effect of oral ciprofloxacin for 6 months, 500 mg every 12 hours, added to the treatment regimen in patients with moderately active CD considered resistant. At the end of study, the mean

CDAI was 112 in the group receiving additional ciprofloxacin versus 205 in the placebo group (p <0.001).[27]

The side effects of metronidazole, especially when used for more than 4 months and at high doses (20 mg/kg/day) are nausea, metallic taste, gastrointestinal intolerance and peripheral neuropathy – sometimes irreversible –, characterized by paresthesia/hypoesthesia and burning sensation in the upper and/or lower limbs. Headache, nausea, diarrhea (including that caused by *Clostridium difficile*) and *rash* are side effects reported with the use of ciprofloxacin. Spontaneous rupture of the Achilles tendon has also been reported with the use of ciprofloxacin.[28]

Studies with rifaximin, an antibiotic active against Gram-positive and Gram-negative bacteria, have shown promising results in the treatment of active CD. At a dose of 800 mg, 2 times/day, it was significantly better than placebo in inducing remission (defined in the study as CDAI <150) after 12 weeks of therapy (62% versus 43%). This effect was maintained for an additional 12 weeks of follow-up in 65% of treated patients. It was also observed that patients with colonic CD showed better response to treatment.[29]

Treatment with antimicrobial agents active against atypical mycobacteria (*Mycobacterium avium* subspecies *paratuberculosis*) aiming at inducing remission showed controversial results.[22,30]

Metronidazole has also been tested in patients who underwent ileal/ileocolonic resection due to CD. A dose of 20 mg/kg/day was given to patients for 3 months *versus* placebo. Endoscopic recurrence at 1 year was lower in the metronidazole group (p=0.02). Clinical recurrence was also statistically lower after one year, but the effect was not sustained after 2-3 years.[31] Side effects were more common in the metronidazole group, which led the authors to conduct a similar study on ornidazole (1 g/day orally), for 1 year.[23] Again, the results showed benefit with ornidazole in reducing the frequency of endoscopic and clinic relapses, but the side effects were also more prevalent in this group. D'Haens et al. evaluated the association of metronidazole with azathioprine versus metronidazole and placebo postoperatively. After 12 months, patients on combination therapy (metronidazole and azathioprine) showed lower endoscopic recurrence (43.7% *versus* 69%, p=0.048).[32]

In anal/perianal CD, uncontrolled studies with metronidazole (20 mg/kg/day) showed improvement and/or closure of fistulas in 50 to 60% of treated cases. The results usually appear after 2 months of therapy. Discontinuation of medication generates high levels of symptomatic recurrence, reaching up to 78% recurrence after 4 months of drug cessation.[23,28] Similar results were obtained with ciprofloxacin (1 to 1.5 g/day) for 3 to 12 months. The combination of ciprofloxacin (1 to 1.5 g/day) and metronidazole (500 to 1.500 mg/day) also showed benefit in the treatment of perianal fistulas in uncontrolled studies (up to 80% reduction in fistula drainage or closure after 3 months treatment). However, patients relapsed after discontinuing the medication.[23,28] The use of antibiotics has also been evaluated in perianal disease as an adjunct to therapy with immunosuppressant or infliximab. In fact, in preliminary studies, the combination of ciprofloxacin (1 g/day) and infliximab appears more effective than infliximab alone, and the use of ciprofloxacin (500 to 1.000 mg/day) and/or metronidazole (1 to 1.5 g/day) increased the activity of azathioprine in patients with perianal fistulas.[23,28]

In UC, there are fewer studies on the effects of antibiotics. Preliminary studies have shown some benefit in the short term (combined with corticosteroid therapy) in patients with moderate/severe UC using tobramycin (120 mg orally, every 8 hours, for 7 days). Ciprofloxacin (500 to 750 mg, two times/day, for 6 months) also yielded positive results, particularly in the short term.[23] Ohkusa et al.[33] observed that the combination of amoxicillin (1.5 g/day), tetracycline (1.5 g/day) and metronidazole (750 mg/day) for 2 weeks in order to neutralize *Fusobacterium varium* was better than placebo to induce and maintain clinical and endoscopic improvement in patients with active UC.

A recent meta-analysis evaluated the efficacy of antibiotic treatment in IBD.[34] In general, antibiotics were beneficial compared to placebo (for active CD, confidence interval [CI] 0.73-0.99; for perianal CD, reducing fistula drainage, CI 0.66-0.98; for quiescent CD [antimycobacterial drugs only], CI 0.46-0.84; for active UC, CI 0.43-0.96). The authors point out, however, that various types and combinations of antibiotics were used, which makes it difficult to draw definitive conclusions.

Regarding pouchitis, studies, mostly uncontrolled, have shown the beneficial effects of the following antibiotics: ciprofloxacin (1 g/day), metronidazole (800 mg to 1.2 g/day), rifaximin (2 g/day), and combinations of rifaximin + ciprofloxacin, metronidazole + ciprofloxacin, rifaximin + ciprofloxacin and ciprofloxacin + tinidazole.[23]

### IMMUNOMODULATORS (OR IMMUNOSUPPRESSANTS)

This group of drugs commonly includes azathioprine (AZA), 6-mercaptopurine (6-MP), chloroquine, cyclosporine and methotrexate. More recently, tacrolimus (FK506) and mycophenolate mofetil have been tested. Below, the most studied and used drugs, AZA and 6-MP, methotrexate and cyclosporine, will be discussed.

Undoubtedly the most studied immunomodulators in IBD, and with which there is considerable accumulated experience, are AZA and 6-MP. After absorption, AZA is quickly converted into 6-MP in erythrocytes, producing active metabolites of the group of 6-thioguanine nucleotides (6-TGN). AZA and 6-MP are potent immunosuppressants that inhibit the activity of T and B lymphocytes, in addition to natural killer (NK) cells. They also induce apoptosis, which is beneficial for patients with IBD – especially those with CD – whose lymphocytes and monocytes have reduced apoptosis. At high doses, AZA also inhibits prostaglandin synthesis.[5]

In IBD, AZA and 6-MP have been used in doses of 2 to 3 mg/kg/day (2.5 mg/kg/day on average) and 1 to 1.5 mg/kg/day, respectively. Both are delayed-action drugs, and thus require at least 3 to 4 months of treatment before considering the intervention as therapeutic failure.[4,11,15-17]

Well-controlled studies and recent meta-analyzes, however, revealed that the thiopurines (AZA and 6-MP) do not appear to be as effective in inducing remission in patients with CD recently diagnosed and at high risk of progression to a disabling condition (presence of at least 2 criteria: age <40 years, active perianal lesion and use of corticosteroids in the previous 3 months).[35-37] In the AZTEC study, Panes et al.[35] compared early use of AZA (2.5 mg/kg/day) *versus* placebo in patients with recent diagnosis (< 8 weeks) of CD during 76 weeks. AZA was no more effective than placebo in inducing corticosteroid-free medical remission (44.1% of the AZA group *versus* 36.5% placebo, p=0.48). It was not superior to placebo in prevention of relapse (defined in the study as CDAI> 175), either. However, AZA was superior to placebo in preventing relapse in a subset of patients with more severe disease (CDAI> 220). Another study by the French group GETAID (RAPID study)36 compared early use of AZA (defined as use up to 6 months after diagnosis) with conventional treatment (step-up) during the first 3 years after diagnosis. The study failed to demonstrate that early use of AZA was more effective than conventional treatment to increase the time of corticosteroid-free remission in the first 3 years (67% early AZA versus 56% conventional treatment group, p=0.69). Patients in the early AZA group showed proportions similar to those of conventional treatment in rates of relapse, hospitalization, intestinal surgery and indication for anti-TNF use. The only apparent advantage of using early AZA was that it statistically reduces the development of active perianal lesions (14% in the early treatment group versus 27% in the conventional therapy group, p=0.049), thus yielding less indication for perianal surgery (3% early-treatment group versus 13% conventional group, p=0.04).

Therefore, the indications for primary treatment with AZA and 6-MP are: corticosteroid-resistant (or refractory) and corticosteroid-dependent disease (facilitates dose reduction/cessation of corticosteroids [steroid-sparing effect], promoting maintenance of remission), as part of combination therapy (AZA/6-MP + anti-TNF) for severe and refractory cases of CD, in penetrating (fistulizing) CD, and postoperatively to avoid relapses.[11,15-17,38]

Side effects of AZA and 6-MP occur in approximately 15% of patients, and may be allergic, including fever, rash, malaise, nausea, vomiting, abdominal pain, diarrhea, hepatitis and pancreatitis, or non-allergic, including bone marrow depression (leukopenia, neutropenia, thrombocytopenia, and anemia), infection, liver enzyme abnormalities and cancer.[11,15-17]

The frequency of infections (7%) and neoplasms (3%) (e.g., colon cancer, breast cancer, testicular cancer, melanoma and leukemia) is similar to that expected in patients with IBD who are not treated with AZA or 6-MP. However, some authors have shown 4 to 5 times higher risk of developing non-Hodgkin lymphoma in patients treated with AZA or 6-MP, particularly in patients over 65 years of age.39,40 But the risk of neoplasia with AZA or 6-MP in doses commonly used in IBD should be contrasted with the complications, limitations, disabilities and deleterious effects of IBD activity.11,15-17 Recently, increased risk of skin cancer (non-melanoma) has been reported in patients using thiopurines, which justifies the use of sunscreen.41 AZA and 6-MP can be used during pregnancy at the discretion of the treating physician.[11,15-17]

Cyclosporine is a peptide extracted from the fungus *Tolypocladium inflatum*, which undoubtedly revolutionized organ transplants and the treatment of autoimmune diseases. Its main mechanism of action is reduction in the production of interleukin 2 (IL-2) by T-helper cells. In IBD, the drug was effective to treat severe UC unresponsive after 7-10 days of therapy with corticosteroids, and refractory and fistulizing CD.[11,15-17] Doses normally used are 2 to 4 mg/kg/day, infused intravenously at a constant rate, for 1 to 2 weeks, followed by oral administration at a 6 to 8 mg/kg/day dosage. Results in the short term are favorable and vary between 60 and 85%, especially in UC. However, in the medium and long term, the drug does not lead to good results, unless an AZA or 6-MP immunosuppressant is added. During the weaning of oral cyclosporine, there is a period in which the patient uses corticosteroids, cyclosporine and AZA or 6-MP. In this situation, the prophylaxis of *Pneumocystis jiroveci* (formerly *carinii*) with sulfamethoxazole/trimethoprim is indicated. Major obstacles to cyclosporine therapy are its high cost, the need for strict monitoring of serum levels (ideally maintained between 150 and 350 ng/mL), interaction with other drugs, and toxicity.[4,11,15-17]

Side effects are relatively frequent, reaching 50%. In general, they are dose-related, and in most cases subside upon dose reduction or discontinuation. They are, in order of frequency: paresthesia, hypertension, hypertrichosis, kidney failure, headache, opportunistic infections, gingival hyperplasia, dizziness and anaphylaxis. Grand mal seizures can occur in patients with low serum cholesterol (<120 mg/dL) or hypomagnesemia. Finally, there is the rare possibility of colitis or jejunitis and even lymphoma caused by cyclosporine. Therefore, the use of cyclosporine should be limited to centers experienced in the management of the drug, and with infrastructure to monitor the patient and treat complications.[4,11,15-17] In a recent controlled prospective study, cyclosporine was compared to infliximab in severe UC refractory to corticosteroids. The results showed that both drugs were effective in the short term, with a similar profile of adverse effects.[42] Similarly, a meta-analysis comparing cyclosporine *versus* infliximab in severe UC refractory to corticosteroids found no differences between the two drugs in relation to colectomy rate at 3 and 12 months, adverse effects and postoperative complications.[43] But an Australian retrospective study showed superiority of infliximab compared to cyclosporine.[44] Differences in method and characteristics of the recruited population may explain these differences.

Methotrexate is a folate antagonist and interferes with DNA synthesis. It acts on the activity of cytokines and inflammatory mediators, blocking the binding of IL-1 to its receptor and reducing the synthesis of IL-2, IL-6, IL-8, interferon-gamma and leukotriene B4.[11,15-17]

It is the main substitute for AZA or 6-MP, in cases of intolerance or adverse effect. At a weekly dose of 15 to 25 mg, intramuscularly or subcutaneously (usually, 25 mg/week), methotrexate promoted remission in nearly 60% of patients with refractory CD after 12 to 16 weeks of treatment.[24] It can also be used as a maintenance drug (15 mg/week).[21] Adverse reactions occur in 10 to 25% of patients and include: nausea, diarrhea, stomatitis, leukopenia, hair loss, transaminase elevation, hypersensitivity pneumonia, liver fibrosis or cirrhosis. Concomitant administration of folic acid (1 to 2 mg/day orally) helps to prevent stomatitis, diarrhea and bone marrow toxicity. When methotrexate is given in conjunction with sulfamethoxazole/trimethoprim, AZA or 6-MP, the risk of severe leukopenia becomes even higher. It is teratogenic and can cause abortion; therefore it is contraindicated for pregnant women, or those who wish to become pregnant. The risk of lymphoma associated with methotrexate therapy appears to be very low.[45] In UC, uncontrolled studies suggest some benefit from methotrexate in an oral dose of about 20 mg/week.[46]

## THERAPEUTIC APPROACH TO ULCERATIVE COLITIS

In a patient diagnosed with UC, the degree of activity (mild, moderate or severe) and the extent of disease (distal UC, left hemicolon or pancolitis) should be determined. Colonoscopy is the recommended method for assessing disease extent, but should be avoided in very severe flare-ups.[4,15,16]

Initially, general measures are recommended, such as clarification about the disease, including information about its chronic nature, the need for medical returns and maintenance of a good doctor-patient relationship. Antidiarrheal and antispasmodic drugs should be used with caution because of the risk of d eveloping toxic megacolon. Tranquilizers and antidepressants may be prescribed if necessary. Attention should be paid to the patient's nutritional status, and the use of enteral and/or parenteral nutrition is recommended to correct malnutrition and prepare for surgery. Bloody diarrhea can trigger dehydration, anemia and electrolyte disturbances, imbalances that must be properly corrected. Antibiotics (e.g., ciprofloxacin, 500 mg every 12 hours, in combination with metronidazole, 250 to 500 mg every 8 hours, both orally or intravenously) are indicated, at the physician's discretion, in severe cases or with proven infection. Thromboembolism prophylaxis with subcutaneous heparin is recommended in severe cases.[3,4]

Medical treatment of UC follows the traditional scheme called step-up, a designation that corresponds to the initial use of drugs with low potential for side-effects and, as required, progresses to more powerful alternatives from a therapeutic point of view, but more likely to produce side effects (Figure 23.1).[15,16]

Thus, in patients with mild/moderate UC, it is recommended first to use salicylic derivatives (SSZ 3 to 4 g/day or mesalazine, 2 to 4 g/day without exceeding 4.8 g/day) orally. Combination with topical treatment (mesalazine enema) favors the therapeutic response, regardless of the extent of UC.[15,16,47] If the patient does not respond to this treatment, corticosteroids (e.g., prednisone, 0.75 to 1 mg/kg/day, orally, without exceeding 60 mg/day) can be added and slowly withdrawn (about 5 to 10 mg/week), as soon as the patient reaches clinical remission. Salicylic derivatives should be maintained indefinitely to reduce the chance of relapse (Table 23.1).[4,15,16]

Corticosteroid-dependent patients requiring doses of corticosteroids, even if low, in order to remain oligo- or asymptomatic, and those refractory to steroid treatment, who do not respond after 4 to 6 weeks of therapy at appropriate doses, should start using oral immunosuppressants – AZA, 2 to 3 mg/kg/day, or 6-MP, 1 to 1.5 mg/kg/day. Prolonged courses of oral corticosteroids and their frequent reintroduction are not currently accepted,

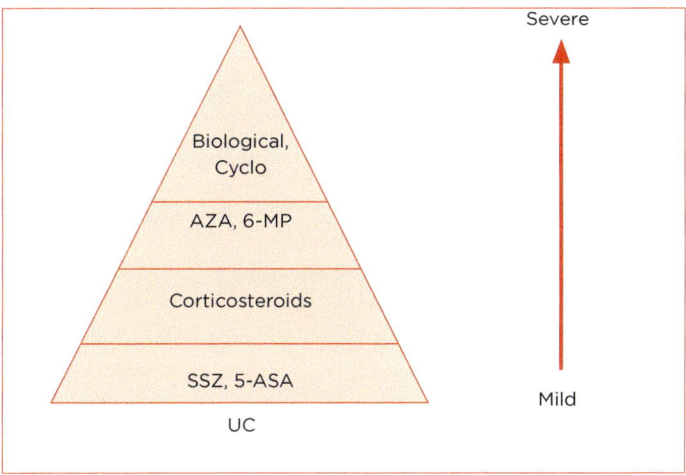

**Figure 23.1** Step-up strategy for the treatment of UC.

SSZ: sulfasalazine; 5-ASA: mesalazine; AZA: azathioprine; 6-MP: 6-mercaptopurine; Cyclo: Cyclosporine.

| Table 23.1 Salicylic derivatives in the treatment of UC | | | |
|---|---|---|---|
| **Drug** | **Commercial brand** | **Doses** | |
| | | **Active UC** | **UC – remission** |
| Topical mesalazine | Asalit (suppository 250 mg; enema 3 g)* | 1 to 4 g/day | 1 g/day or 1 to 4 g every 2 or 3 days |
| | Rowasa, Salofalk, Mesasal (enema 4 g) | | |
| | Pentasa (suppository and enema 1 g)* | | |
| | CD-ASA 5 (enema 2 and 3 g)* | | |
| | Mesacol (suppository 250 and 500 mg)* | | |
| Oral mesalazine | Mesacol (400 and 800 mg)* | 3 to 4.8 g/day | 2 to 3 g/day |
| | Asalit (400 mg)* | | |
| | Pentasa (500 mg)* | | |
| | Mesacol MMX (1.2 g)* | | |
| | Pentasa sachet (1 and 2 g)* | | |
| Sulfasalazine (oral) | Azulfin (500 mg)* | 2 to 3 g/day | 2 to 4 g/day |
| Olsalazine (oral) | Dipentum (500 mg) | 2 a 3 g/dia | 1 to 3 g/day |
| Balsalazide (oral) | Colazal, Colazide (750 mg) | 2.25 to 6.75 g/day | 2.25 to 6.75 g/day |

*Products sold in Brazil.

and its use in maintenance therapy should also be avoided. It is suggested to start with AZA or 6-MP, 50 mg/day, and then depending on the periodic blood tests (blood cell count to assess leukopenia, aminotransferase, amylase, etc.) progress to the optimal dose. Nonresponsive patients are candidates for biological therapy (e.g., anti-tumor necrosis [anti-TNF]), such as infliximab and adalimumab. Cyclosporine is also an option.[4,15,16]

In patients with moderate/severe UC, the use of corticosteroids is recommended since the beginning, even intravenously, if necessary (e.g., hydrocortisone, 100 mg every 6 or 8 hours). If the patient does not respond, cyclosporine or biological agent is indicated. In all these more severe or unresponsive situations, surgical treatment should be considered as a possible option. If the patient responds medically, maintaining treatment with an immunosuppressant or salicylic derivatives is recommended. The highest dosage of mesalamine per tablet or sachet is an interesting alternative for patients with IBD treated with salicylate because it allows the total dose to be taken once a day, which can improve patient compliance with the treatment (see Table 23.1 ).[48-51]

### THERAPEUTIC APPROACH TO CROHN'S DISEASE

Just as in UC, general measures should be taken. Similarly, it is important to recognize the degree of CD activity, its extent and behavior (inflammatory, stricturing and penetrating/fistulizing). Disease extent is evaluated by means of endoscopy and imaging, such as CT and MRI.[3,11,15,17]

Nutritional therapy is recommended for malnourished patients with CD and those who will undergo surgery. Nutritional supplementation with vitamins and minerals is often required. In children and adolescents, nutritional therapy (enteral probe or orally if tolerated) can be used as the sole primary measure to replace corticosteroids with clear advantages in terms of speed of growth.[15,52]

In addition to the traditional strategy (step-up), there is also an earlier more potent strategy in CD, called *top-down* (Figure 23.2).[53] In the step-up approach, treatment begins with salicylic derivatives (e.g., sulfasalazine) in mild cases of CD affecting the colon, and budesonide in mild/moderate cases of CD in the ileum and cecum and/or ascending colon.[11,15,17] In patients with moderate/severe disease or those unresponsive to initial clinical treatment, prednisone can be employed. If the patient is refractory or becomes dependent on corticosteroids, AZA or 6-MP is indicated.[11,15,17] Methotrexate is also an option (25 mg/week, intramuscularly or subcutaneously, for 12 weeks followed by maintenance with 15 mg/week).[15,17,24] Non-response to these measures brings into focus biological therapy (e.g., infliximab, adalimumab, certolizumab).[11,15,17,24] Exclusive nutrition therapy, usually enteral (oligomeric or polymeric diets), can also be attempted at that stage. In children and adolescents, nutrition therapy exclusively for 6 to 8 weeks is a primary effective measure in most cases (clinical response 70 to 80%). Maintenance with AZA or 6-MP is effective, even postoperatively.[15] The same applies to maintenance therapy with biological agents.[54,55] Corticosteroids and salicylic derivatives are not useful drugs to maintain remission in CD. The step-up strategy, although widespread, does not seem to affect the natural course of the disease; however, it improves quality of life indices.[3]

In top-down strategy, treatment is started with biological agents and immunosuppressants (e.g., AZA, 6-MP or methotrexate), avoiding the use of corticosteroids (Figure 23.2B). This strategy proved to be more effective than step-up in patients with moderate/severe CD. In fact, endoscopic remission in two years with the *top-down* strategy was much higher than that obtained with a step-up approach (73.1% top-down *versus* 30.4% step-up, p < 0.002) (Figure 23.3).[53] There was also a reduction in hospitalization rates and surgery with the use of biological agents, at least in the short to medium term (up to 5 years).[56] In immunosuppressant-naive patients with moderate/severe CD, corticosteroid-dependent or refractory to standard treatment, the combination of a biological agent (infliximabe) and an oral immunosuppressor (AZA) was more effective than AZA or infliximabe alone (Sonic study).[57] Thus, in selected cases, such as young patients with perianal CD or CD severe enough to receive corticosteroids at high doses early in therapy, the top-down strategy seems to be useful, with considerable endoscopic remission rates, and potential to impact the natural course of the disease.[3]

More recently, special attention has been given to postoperative recurrence in CD and the possibility of preventing clinical and/or endoscopic

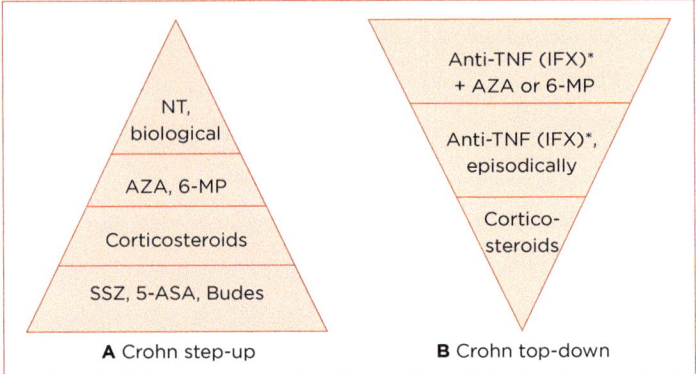

**Figure 23.2** (A) Traditional approach (step-up) to Crohn's disease. (B) Top-down strategy.

SSZ: sulfasalazine; 5-ASA: mesalazine; AZA: azathioprine; 6-MP: 6-mercaptopurine; NT: nutritional therapy; Budes:

budesonide.

*Anti-TNF (in this case, infliximab [IFX] was the biological agent tested; it was used after the induction episodically).

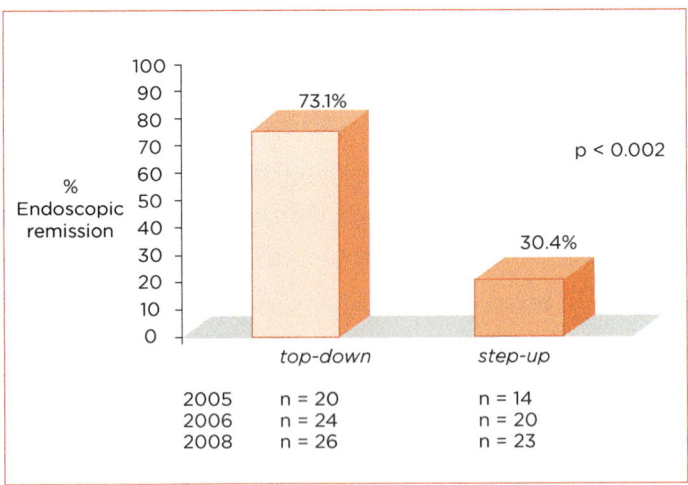

**Figure 23.3** Endoscopic remission with top-down and step-up strategies.

recurrence medically.[32,54,55,58] D'Haens et al.[32] evaluated patients with CD who underwent ileal or ileocecal resection with ileocolonic anastomosis. All patients considered at high risk for relapse (with at least one of the following: young patient – less than 30 years old – smoker, treated with corticosteroids in the last three months, previous intestinal resection, penetrating disease) received metronidazole (750 mg/day) for 3 months after surgery (within 15 days). One group also received AZA (100 mg/day if weight <60 kg and 150 mg/day if weight >60 kg) concomitantly for 12 months, and the other received placebo. After 1 year, endoscopic recurrence was lower in the metronidazole + AZA group. Side effects were equally distributed in both groups. The authors concluded that the use of metronidazole + AZA prevents postoperative recurrence in patients at high risk for recurrence. Reinisch et al.[58] compared mesalazine (4 g/day) and AZA (2 to 2.5 mg/kg/day) in patients with postoperative endoscopic recurrence (assessed 6 and 24 months after surgery). After 1 year, the group treated with mesalazine showed higher clinical recurrence compared to the group treated with AZA and endoscopic improvement was most evident in the AZA group. However, unlike the study by D'Haens et al.,[32] there were more side effects in the AZA group, perhaps due to the different route of delivery. Nevertheless, the authors preferred the use of AZA in patients with severe endoscopic recurrence (i2a, i3 and i4 indices in the modified Rutgeerts score), preferably evaluated 6-12 months after the surgical event.[58] Infliximab has also been evaluated in the prevention of postoperative recurrence in CD with good results.[54,55,59-62] Regueiro et al.,[59] in 2009, observed a significantly lower endoscopic recurrence rate in patients treated with infliximab compared to placebo after 1 year of follow-up (9.1% infliximab group versus 84.6% placebo, p=0.0006). Clinical recurrence was also lower in the group treated with infliximab, although with no statistically significant difference (20% infliximab versus 46.2%, p=0.38). Another study conducted in 2013 compared the use of infliximab (after induction maintenance) versus AZA (2.5 mg/kg/day) in patients after ileocolic surgery. After one year of follow-up, 40% of patients treated with AZA showed endoscopic recurrence compared to 9% in the infliximab group (p=0.14); 80% of patients treated with AZA had severe histological activity *versus* 18% in the infliximab group (p=0.008). Regarding clinical symptoms, there was no statistically significant difference between groups.[60] More recently, an article was published comparing the use of adalimumab (ADA), AZA and mesalazine in endoscopic recurrence after 2 years of treatment. Endoscopic recurrence rate was significantly lower in the ADA group (6.3%), compared to the AZA (64.7%) and mesalazine (83.3%) groups; the same was seen for the rate of clinical recurrence: ADA (12.5%), AZA (64.7%) and mesalazine (50%).[62] More studies are needed to confirm this therapeutic advantage and assess the economic implications of biological therapy in this group of patients.

## REFERENCES

1. Danese S, Fiocchi C. Ulcerative colitis. N Engl J Med 2011; 365:1713-25.

2. Ordás I, Eckmann L, Talamini M, Baumgart DC, Sandborn WJ. Ulcerative colitis. Lancet 2012; 380:1606-19.

3. Baumgart DC, Asndborn WJ. Crohn's disease. Lancet 2012; 380:1590-605.

4. Dignass A, Lindsay JO, Sturm A, Windsor A, Colombel JF, Allez M et al. Second European evidence-based consensus on the diagnosis and management of ulcerative colitis part 2: current management. J Crohns Colitis 2012; 6:991-1030.

5. Damião AOMC. Doença inflamatória intestinal. In: Silva P (ed.). Farmacologia. 8.ed. Rio de Janeiro: Guanabara Koogan, 2010; p.991-1030.

6. Munkholm P, Loftus Jr. EV, Reinacher-Schick A, Kornbluth A, Mittmann U, Esendal B et al. Prevention of colorectal cancer in inflammatory bowel disease: value of screening and 5-aminosalicylates. Digestion 2006; 73:11-9.

7. van Schaik FD, van Oijen MG, Smeets HM, van der Heijden GJ, Siersema PD, Oldenburg B et al. Thiopurines prevent advanced colorectal neoplasia in patients with inflammatory bowel disease. Gut 2012; 61:235-40.

8. Ford AC, Achkar JP, Khan KJ, Kane SV, Talley NJ, Marshall JK et al. Efficacy of 5-aminosalicylates in ulcerative colitis: systematic review and meta-analysis. Am J Gastroenterol 2011; 106:601-16.

9. Ford FC, Khan KJ, Talley NJ, Moayyedi P. 5-aminosalicylates prevent relapse of Crohn's disease after surgically induced remission: systematic review and meta-analysis. Am J Gastroenterol 2011; 106:413-20.

10. Ford FC, Kane SV, Khan KJ, Achkar JP, Talley NJ, Marshall JK et al. Efficacy of 5-aminosalicylates in Crohn's disease: systematic review and meta-analysis. Am J Gastroenterol 2011; 106:617-29.

11. Dignass A, Van Assche G, Lindsay JO, Lémann M, Söderholm J, Colombel JF et al. The second European evidence-based Consensus on the diagnosis and management of Crohn's disease: current management. J Crohns Colitis 2010; 4:28-62.

12. Ford AC, Bernstein CN, Khan KJ, Abreu MT, Marshall JK, Talley NJ et al. Glucocorticosteroid therapy in inflammatory bowel disease: systematic review and meta-analysis. Am J Gastroenterol 2011; 106:590-9.

13. Modigliani R, Mary J, Simon J. Clinical, biological and endoscopic picture of attacks of Crohn's disease. Evolution on prednisolone. Gastroenterology 1990; 98:811-8.

14. Munkholm P, Langholz E, Davidsen M, Binder V. Frequency of glucocorticoid resistance and dependency in Crohn's disease. Gut 1994; 35:360-2.

15. Brazilian Study Group of Inflammatory Bowel Diseases. Consensus Guidelines for the management of inflammatory bowel disease. Arq Gastroenterol 2010; 47:313-25.

16. Kornbluth A, Sachar DB. Ulcerative colitis practice guidelines in adults: American College of Gastroenterology, Practice Parameters Committee. Am J Gastroenterol 2010; 105:501-23.

17. Lichtenstein GR, Hanauer SB, Sandborn WJ, The Practice Parameters Committee of the American College of Gastroenterology. Management of Crohn's disease in adults. Am J Gastroenterol 2009; 104:465-83.

18. Chassaing B, Darfeuille-Michaud A. The commensal microbiota and enteropathogens in the pathogenesis of inflammatory bowel diseases. Gastroenterology 2011; 140:1720-28.

19. Damião AOMC, Milani LR, Feitosa FC. Probióticos na doença inflamatória intestinal: qual a sua importância? In: Quilici FA, Miszputen SJ (eds.). Doença inflamatória intestinal. Grupo de Estudos da Doença Inflamatória Intestinal do Brasil (GEDIIB). São Paulo: Elsevier, 2010. p.97-108.

20. Podolsky DK. Inflammatory bowel disease. N Engl J Med 2002; 347:417-29.

21. Swidsinski A, Loening-Baucke V, Herber A. Mucosal flora in Crohn's disease and ulcerative colitis – an overview. J Physiol Pharmacol 2009; 60 (Suppl 6):61-71.

**22.** Scribano ML, Prantera C. Use of antibiotics in the treatment of Crohn's disease. World J Gastroenterol 2013; 19:648-53.

**23.** Keohane J, Shanahan F. Therapeutic manipulation of the microbiota in inflammatory bowel disease: antibiotics and probiotics. In: Targan SR, Shanahan F, Karp LC (eds.). Inflammatory bowel disease: translating basic science into clinical practice. London: Wiley-Blackwell, 2010.

**24.** Feagan B, McDonald JWD. Crohn's disease. In: McDonald JWD, Burroughs AK, Feagan BG, Fennerty MB (eds.). Evidence-based gastroenterology & hepatology. 3.ed. Oxford: Wiley-Blackwell, 2010.

**25.** Sutherland L, Singleton J, Sessions J, Hanauer S, Krawitt E, Rankin G et al. Double blind, placebo controlled trial of metronidazole in Crohn's disease. Gut 1991; 32:1071-5.

**26.** Prantera C, Zannoni F, Scribano ML, Berto E, Andreoli A, Kohn A et al. An antibiotic regimen for the treatment of active Crohn's disease: a randomized, controlled clinical trial of metronidazole plus ciprofloxacin. Am J Gastroenterol 1996; 91:328-32.

**27.** Arnold GL, Beaves MR, Pryjdun VO, Mook WJ. Preliminary study of ciprofloxacin in active Crohn's disease. Inflamm Bowel Dis 2002; 8:10-5.

**28.** Bressler B, Sands BE. Review article: medical therapy for fistulizing Crohn's disease. Aliment Pharmacol Ther 2006; 24:1283-93.

**29.** Prantera C, Lochs H, Grimaldi M, Danese S, Scribano ML, Gionchetti P et al. Rifaximin-extended intestinal release induces remission in patients with moderately active Crohn's disease. Gastroenterology 2012; 142:473-81.

**30.** Peyrin-Biroulet L, Neut C, Colombel JF. Antimycobacterial therapy in Crohn's disease: game over? Gastroenterology 2007; 132:2594-98.

**31.** Rutgeerts P, Hiele M, Geboes K, Peeters M, Penninckx F, Aerts R et al. Controlled trial of metronidazole treatment for prevention of Crohn's recurrence after ileal resection. Gastroenterology 1995; 108:1617-21.

**32.** D'Haens GR, Vermeire S, Van Assche G, Noman M, Aerden I, Van Olmen G et al. Therapy of metronidazole with azathioprine to prevent postoperative recurrence of Crohn's disease: a controlled randomized trial. Gastroenterology 2008; 135:1123-9.

**33.** Ohkusa T, Kato K, Terao S, Chiba T, Mabe K, Murakami K et al. Newly developed antibiotic combination therapy for ulcerative colitis: a double-blind placebo-controlled multicenter trial. Am J Gastroenterol 2010; 105:1820-9.

**34.** Khan KJ, Ullman TA, Ford AC, Abreu MT, Abadir A, Marshall JK et al. Antibiotic therapy in inflammatory bowel disease: a systematic review and meta-analysis. Am J Gastroenterol 2011; 106:661-73.

**35.** Panés J, Lópes-SanRomán, Bermejo F, García-Sánchez V, Esteve M, Torres Y et al. Early azathioprine therapy is no more effective than placebo for newly diagnosed Crohn's disease. Gastroenterology 2013; 145:766-74.

**36.** Cosnes J, Bourrier A, Laharie D, Nahon S, Bouhnik Y, Carbonnel F et al. Early administration of azathioprine vs conventional management of Crohn's disease: a randomized controlled trial. Gastroenterology 2013; 145:758-65.

**37.** Chande N, Tsoulis DJ, MacDonald JK. Azathioprine or 6-mercaptopurine for induction of remission in Crohn's disease. Cochrane Database Syst Rev 2013; 4:CD000545.

**38.** Khan KJ, Dubinsky MC, Ford AC, Ullman TA, Talley NJ, Moayyedi P. Efficacy of immunosupressive therapy for inflammatory bowel disease: a systematic review and meta-analysis. Am J Gastroenterol 2011; 106:630-42.

**39.** Beaugerie L, Brousse N, Bouvier AM, Colombel JF, Lémann M, Cosnes J et al. Lymphoproliferative disorders in patients receiving thiopurines for inflammatory bowel disease: a prospective observational cohort study. Lancet 2009; 374:1617-25.

**40.** Herrinton LI, Liu L, Weng X, Lewis JD, Hutfless S, Allison JE et al. Role of thiopurine and anti-TNF therapy in lymphoma in inflammatory bowel disease. Am J Gastroenterol 2011; 106:2146-53.

**41.** Ariyaratnam J, Subramanian V. Association between thiopurine use and nonmelanoma skin cancers in patients with inflammatory bowel disease: a meta-analysis. Am J Gastroenterol 2014; 109:163-9.

**42.** Laharie D, Bourreille A, Branche J, Allez M, Bouhnik Y, Filippi J et al. Ciclosporin versus infliximab in patients with severe ulcerative colitis refractory to intravenous steroids: a parallel, open-label randomised controlled trial. Lancet 2012; 380:1909-15.

**43.** Chang KH, Burke JP, Coffey JC. Infliximab versus cyclosporine as rescue therapy in acute severe steroid-refractory ulcerative colitis: a systematic review and meta-analysis. Int J Colorectal Dis 2013; 28:287-93.

**44.** Croft A, Walsh A, Doecke J, Cooley R, Howlett M, Radford-Smith G. Outcome of salvage therapy for steroid-refractory acute severe ulcerative colitis: ciclosporin vs. infliximab. Aliment Pharmacol Ther 2013; 38:294-302.

**45.** Mariette X, Cazals-Hatern D, Warszawki J, Liote F, Balandraud N, Sibilia J et al. Lymphomas in rheumatoid arthritis patients treated with methotrexate: a 3-year prospective study in France. Blood 2002; 99:3909-15.

**46.** Cummings JRF, Herrlinger KR, Travis SPL, Gorard DA, McIntyre AS, Jewell DP. Oral methotrexate in ulcerative colitis. Aliment Pharmacol Ther 2005; 21:385-9.

**47.** Ford AC, Khan KJ, Achkar JP, Moayyedi P. Efficacy of oral vs. topical, or combined oral and topical 5-aminosalicylates, in ulcerative colitis: systematic review and meta-analysis. Am J Gastroenterol 2012; 107:167-76.

**48.** Ford AC, Khan KJ, Sandborn WJ, Kane SV, Moayyedi P. Once-daily dosing vs. conventional dosing schedule of mesalamine and relapse of quiescent ulcerative colitis: systematic review and meta-analysis. Am J Gastroenterol 2011; 106:2070-7.

**49.** Kane SV, Robinson A. Review article: understanding adherence to medication in ulcerative colitis – innovative thinking and evolving concepts. Aliment Pharmacol Ther 2010; 32:1051-8.

**50.** Kruis W, Kiudelis G, Rácz I, Gorelov IA, Pokrotnieks J, Horynski M et al. Once daily versus three times daily mesalazine granules in active ulcerative colitis: a double-blind, double-dummy, randomised, non-inferiority trial. Gut 2009; 58:233-40.

**51.** Ng SC, Kamm MA. Review article: new drug formulations, chemical entities and therapeutic approaches for the management of ulcerative colitis. Aliment Pharmacol Ther 2008; 28:815-29.

**52.** Damião AOMC, Amarante D, Pinheiro-César M. Terapêutica nutricional nas doenças inflamatórias intestinais. In: Cury DB, Moss AC (eds.). Doenças inflamatórias intestinais: retocolite ulcerativa e doença de Crohn. Rio de Janeiro: Rubio, 2011. p.151-70.

**53.** D'Haens G, Baert F, Van Assche G, Caenepeel P, Vergauwe P, Tuynman H et al. Early combined immunosupression or conventional management in patients with newly diagnosed Crohn's disease: an open randomised trial. Lancet 2008; 371:660-7.

**54.** Regueiro M. Management and prevention of postoperative Crohn's disease. Inflamm Bowel Dis 2009; 15:1583-90.

55. Sorrentino D, Paviotti A, Terrosu G, Avellini C, Geraci M, Zarifi D. Low-dose maintenance therapy with infliximab prevents postsurgical recurrence of Crohn's disease. Clin Gastroenterol Hepatol 2010; 8:591-9.

56. Schnitzler F, Fidder H, Ferrante M, Noman M, Arijs I, Van Assche G et al. Mucosal healing predicts long-term outcome of maintenance therapy with infliximab in Crohn's disease. Inflamm Bowel Dis 2009; 15:1295-301.

57. Colombel JF, Sandborn WJ, Reinisch W, Mantzaris GJ, Kornbluth A, Rachmilewitz D et al. Infliximab, azathioprine, or combination therapy for Crohn's disease. N Engl J Med 2010; 362:1383-95.

58. Reinisch W, Angelberger S, Petritsch W, Shonova O, Lukas M, Bar-Meir S et al. Azathioprine versus mesalazine for prevention of postoperative clinical recurrence in patients with Crohn's disease with endoscopic recurrence: efficacy and safety results of a randomised, double-blind, double-dummy, multicentre trial. Gut 2010; 59:752-9.

59. Regueiro M, Schraut W, Baidoo L, Kip KE, Sepulveda AR, Pesci M et al. Infliximab prevents Crohn's disease recurrence after ileal resection. Gastroenterology 2009; 136:441-50.

60. Armuzzi A, Felice C, Papa A, Marzo M, Pugliese D, Andrisani G et al. Prevention of posoperative recurrence with azathioprine or infliximab in patients with Crohn's disease: an open-label pilot study. J Crohns Colitis 2013; 15:623-9.

61. Yamamoto T, Watanabe T. Strategies for the prevention of posoperative recurrence of Crohn's disease. Colorectal Dis 2013; 15:1471-80.

62. Savarino E, Bodini G, Dulbecco P, Assandri L, Bruzzone L, Mazza F et al. Adalimumab is more effective than azathioprine and mesalamine at preventing postoperative recurrence of Crohn's disease: a randomized controlled trial. Am J Gastroenterol 2013; 108:1731-42.

# BIOLOGICAL THERAPY IN CROHN'S DISEASE

RUSSELL D. COHEN

## INTRODUCTION

Biological therapies have revolutionized the treatment of Crohn's disease (CD) ever since their introduction in 1998. Therapies targeting tumor necrosis factor- -alpha (TNF-$\alpha$) have dominated the biological use in these patients, with multiple agents showing efficacy in the treatment of CD. The anti-integrin antibodies have recently expanded to include two such agents; their use has been limited due to safety considerations with the first agent (natalizumab), but is expected to greatly increase with the introduction of vedolizumab in mid-2014. Biological agents don't only successfully meet the "traditional" endpoints of inducing and maintaining clinical response and remission, and improve patient quality of life, but their efficacy has also allowed for the improvement or normalization of mucosal appearance and prevention of disease relapse following CD surgery. The widespread use of biological agents in CD is limited by concerns of safety (largely unfounded) as well as their high medication cost. Economic analyses have supported their use in patients with CD.

Given the low prevalence of CD worldwide, many of the initial clinical trials for the biologics were smaller than might be seen in other diseases;

as a result, important information has been discovered since these drugs have been released on the market that was not as well appreciated during the clinical investigations. Infusion or injection reactions, the development and implications of anti-drug antibodies, the role of concomitant immunomodulators, incidence of medication-induced lupus-like reaction, and issues regarding risk of neoplasm have been the subject of many investigations worldwide. Given the epidemiology of CD, the use of biological agents in children and pregnant and lactating women is also an important topic that needs to be addressed outside of the traditional clinical trials. Lessons learned from the biological drugs that "made it" to release for CD, as well as those that didn't, also may provide deeper insight into specific characteristics of disease, which may impact future therapeutic approaches.

## BIOLOGICAL AGENTS USED TO TREAT CROHN'S DISEASE

There are two "families" of biological agents that are currently used to treat patients with CD: agents directed at blocking TNF-$\alpha$ (Table 24.1) and agents directed at blocking specific integrins (Table 24.2). We will discuss each of these families separately, as they are quite different and are typically not considered as inter-changeable agents.

**Table 24.1** Anti-TNF therapies used in Crohn's disease

| Agent | Mode of Delivery | Induction Protocol | Maintenance Protocol |
|---|---|---|---|
| Infliximab | Intravenous | 5 mg/kg in weeks 0,2,6 | 5 mg/kg every 8 weeks |
| Adalimumab | Subcutaneous | 160 mg in week 0 | 40 mg every 2 weeks |
| | | 80 mg in week 2 | |
| | | 40 mg in week 4 | |
| Certolizumab | Subcutaneous | 400 mg in weeks 0,2,4 | 400 mg every 4 weeks |

**Table 24.2** Anti-integrin therapies used in Crohn's disease

| Agent | Mode of Delivery | Induction Protocol | Maintenance Protocol |
|---|---|---|---|
| Natalizumab | Intravenous | 300 mg in week 0 | 300 mg every 4 weeks |
| Vedolizumab | Intravenous | 300 mg in weeks 0,2,6 | 300 mg every 8 weeks |

## BIOLOGICAL AGENTS THAT TARGET TUMOR NECROSIS FACTOR-ALPHA

### Infliximab

#### Luminal Crohn's disease

Infliximab is a chimeric $IgG_1$ monoclonal antibody given by intravenous infusion. It was the first biological therapy released for the treatment of CD. Its approval in 1998 was based on clinical trials including only 203 patients. A dose-ranging single infusion study in 108 patients with moderate to severely active CD received a single infusion of infliximab at 5mg/kg, 10mg/kg, or placebo. Week 4 response and remission rates were 64% and 33% for the patients receiving infliximab, and only 17% (p<0.001) and 4% (p=0.005) in those who received placebo, respectively.[1] Subsequently, the 73 patients who initially responded to therapy entered a further clinical trial including 4 additional infusions of either infliximab 10mg/kg or placebo at 8 week intervals.[2] The week 44 remission rates were 53% in those who received active drug, *versus* only 20% in those who received placebo. In the safety analysis, there was one B-cell duodenal lymphoma in a patient who was receiving placebo-maintenance retreatment after a single infliximab infusion 9 ½ months earlier.

The maintenance benefit of infliximab was subsequently supported by the much larger ACCENT-I trial, including 573 patients with moderately to severely active luminal CD.[3] In this trial, responders to an initial 5mg/kg dose of infliximab were randomly assigned to receive additional blinded infusions of either drug (at 5mg/kg or 10mg/kg) or placebo at weeks 2 and 6, and then every 8 weeks thereafter until week 46. Week 30 remission rates were 39% in the 5mg/kg group and 45% in the 10mg/kg group *versus* 21% in those who received placebo. By week 54, the rates were 28%, 38%, and 14%, respectively. Tumors were found in 2 patients who had received placebo (one epithelial-cell skin cancer; one natural killer cell lymphoma) and in 4 patients who had received infliximab maintenance (one basal cell skin cancer, one breast cancer, one hypernephroma, and one bladder cancer). Two patients developed sepsis (one died), one had myocardial infarction, and one presented tuberculosis; these were all patients allocated to infliximab-treatment arms.

### Fistulous Crohn's Disease

Infliximab is the only biological therapy to date with large clinical trials specifically designed to evaluate the response of CD related fistulas. In the initial trial with 94 CD patients who suffered from either abdominal enterocutaneous or perianal fistulas, 3 infusions of infliximab 5mg/kg or 10mg/kg at weeks 0, 2 and 6 were successful in closing at least 50% of draining fistulas in 68% and 56%, respectively, compared to only 26% in those who received placebo.[4] Complete closure of all fistulas was seen in 55%, 38%, and 13%, respectively. The median duration of fistula closure was 3 months.

The much larger ACCENT II trial gave all patients an initial three-dose load (infliximab 5mg/kg at week 0, 2, and 6) after which the responders (defined as at least a 50% decrease in the number of draining fistulas) were randomized to receive five additional doses of drug at 5mg/kg or placebo every 8 weeks.[5] A response was maintained at week 54 in 46% of those who continued to receive infliximab, *versus* 23% who were withdrawn from therapy and received placebo. Complete response (an absence of draining fistulas) was maintained in 36% *versus* 19%, respectively. There was one case of cutaneous nocardiasis and one cytomegalovirus infection, both during induction with infliximab. New fistula-related abscesses occurred during the maintenance phase in 12% of patients receiving infliximab and 17% of those receiving placebo.

### Use of Concomitant therapy

Although the use of concomitant therapy with the immunosuppressants azathioprine, 6-mercaptopurine, or methotrexate to did not seem to impact the clinical results reported in the fore-mentioned trials, it soon became apparent that the dosing of patients who receive infliximab along with an immunosuppressant was important to achieve optimal clinical results, and to minimize the formation of anti-infliximab antibodies, which accounted for some infusion-related adverse events and neutralization of the infliximab. The rates of anti-drug antibodies in the ACCENT I trial were 8.2% in those who received combination therapy, compared to 17.5% in patients treated with infliximab alone (i.e. "monotherapy").[3] The rate of infusion reactions was 16.5% in those patients with anti-infliximab antibodies, as compared to only 8.4% in patients who did not form such antibodies. Similarly, the anti-infliximab antibody rates in the ACCENT II trial were 9% on combination therapy versus 24.4% on infliximab monotherapy, with infusion reactions seen in 29.5% of those with, and only 16.3% in those without, anti-infliximab antibodies.[5]

In one of the earlier analyses from Belgium, 125 CD patients received induction doses of infliximab (single infusion if luminal disease, 3 infusions at weeks 0, 2, and 6 if fistulous disease), followed by repeat infusions only upon relapse.[6] They showed that patients on concomitant immunomodulators had a lower incidence of anti-infliximab antibodies (40% vs. 75%, p<0.01), lower titers of anti-infliximab antibodies (p<0.001), and higher infliximab concentrations (p<0.001). In a separate study, researchers from Boston conducted a placebo-controlled blinded prospective study providing premedication with 200mg hydrocortisone (or placebo) prior to infliximab infusions in 80 CD patients.[7] There were lower rates of anti-infliximab antibodies (26% vs. 42%, p=0.06) and lower titers of antibodies (1.6 vs. 3.4 mcg/mL; p=0.02) seen in the patients who received the hydrocortisone premedication, with no difference in infection or neoplasm rates.

The benefit of combination therapy was subsequently proven in the SONIC trial, which compared monotherapy with either azathioprine (2.5mg/kg), infliximab, or a combination of both agents.[8] Corticosteroid-free remission rates (week 26) were 30.0%, 44.4%, and 56.8%, respectively (p<0.001). Mucosal healing rates (week 26) were 16.5%, 30.1%, and 43.9%, respectively (p<0.001). Anti-infliximab antibodies were found in 14.6% on infliximab monotherapy and only 0.9% of patients on combination therapy. This impacted trough infliximab levels, which were 1.6 mcg/mL vs. 3.5 mcg/mL, respectively (p<0.001). The rate of infusion reaction was 5.6% of the azathioprine group (they received placebo infusions), 16.6% of the infliximab monotherapy patients, and 5.0% of the combination therapy group. Serious infections developed in 5.6% of those in the azathioprine group, 4.9% of those in the infliximab group, and 3.9% of patients in the combination-therapy group. Colon cancer developed in two patients receiving azathioprine alone; one patient on azathioprine (alone) died from sepsis following a colectomy.

Despite the obvious advantage to the use of combination therapy in CD patients, there has been hesitancy due to safety concerns, despite the lack of concerning safety outcomes with combination therapy in any of the studies discussed above. Database studies have suggested that the risk for infections does increase as one adds corticosteroids and/or immunomodulators to monotherapy with an anti-TNF agent; importantly, the absolute risk remains very low. The risk of lymphomas has been demonstrated to be higher in IBD patients receiving thiopurines (hazard ratio 5.28 (95% CI 2.01-13.9)[9] but this has not been seen in patients on anti-TNF monotherapy. It is also unclear whether combination therapy increases risk for lymphoma compared to thiopurine alone in IBD patients (SIR 1.7, 95% CI 0.5-7.1).[10] The exceeding low incidence rates of hepatosplenic T-cell lymphoma prevent accurate statistical evaluation of any increase in risk with combination therapy as compared to thiopurines alone.

### Infliximab Levels and Antibodies

The ability to check anti-infliximab antibodies and drug levels has existed for quite some time, but the current cost of the most accurate commercial assay prevents the wide-spread use of this technology. Tested at the time of drug trough, higher infliximab levels have been correlated to more durable clinical responses, lower rates of relapse, and are inversely related to the levels of anti-infliximab antibodies.[11,12]

The paradigm of checking anti-drug antibodies suggests that patients who have developed such antibodies should be switched to an alternative anti-TNF agent, rather than increasing the dose of the anti-TNF agent.[13]

### Summary of Infliximab Use

Given the proven induction and maintenance benefit in these trials, it is standard practice to initially dose infliximab at 5mg/kg for induction, given at weeks 0, 2, and 6, followed by maintenance infusions every 8 weeks (see Table 24.1). Patients who respond to infliximab but subsequently relapse prior to the 8-week reinfusion time typically have their reinfusions performed earlier (i.e. every 6 or 7 weeks); this may also be overcome by increasing the dose of infliximab. Those who lose response often can be "recaptured" by increasing the dose to 7.5mg/kg or up to 10mg/kg. The best results are when infliximab is used in combination with an immunosuppressant (azathioprine, 6-mercaptopurine, or methotrexate); the safety risk is increased only minimally. An alternative to combination therapy is one dose of intravenous hydrocortisone (200mg) prior to every infliximab infusion. Checking trough infliximab levels may be helpful when assessing for cause of loss of response; patients with low drug levels and no anti-infliximab antibodies can be treated with dose escalation; those who have developed antibodies should be switched to a different biological agent. Screening for tuberculosis, hepatitis B, and other opportunistic infections specific to the locality is required prior to initiation of therapy (Table 24.3).

| Table 24.3 Infectious screening requirements prior to initiation of therapy* | | | |
|---|---|---|---|
| Agent | Tuberculosis | Hepatitis B | JC virus |
| Infliximab | + | + | − |
| Adalimumab | + | + | − |
| Certolizumab | + | + | − |
| Natalizumab | − | − | + |
| Vedolizumab | − | − | − |

* General recommendations; individual patients may require screening for infections, based upon clinical condition, previous history of infection, suspected exposures, and/or geographic location.

## Adalimumab

### Luminal Crohn's disease

Adalimumab is a human $IgG_1$ monoclonal antibody given by subcutaneous injection. Adalimumab use in CD was first studied only after the drug was released for rheumatologic disease. Early success in patients who had previously lost response or were intolerant to infliximab led[14] to subsequent clinical trials to determine the optimal dosing strategy for this agent. The CLASSIC-I trial demonstrated that, unlike in rheumatoid arthritis, a quadruple-loading dose of 160mg, followed by a double dose of 80mg two weeks later, was required for induction of remission in CD.[15] In the CLASSIC-II trial, 55 patients had received 2 blinded doses in the CLASSIC-I trial then received two open-label doses of adalimumab 40mg at weeks 0 and 2; responders were randomized to receive either adalimumab 40mg dosed weekly or every two weeks, or placebo up to 56 weeks.[16] Remission rates were maintained in 83% (dosed weekly) and 79% (dosed every 2 weeks) with adalimumab as compared to just 44% in those who had received placebo ($p<0.05$).

In the CHARM trial, 499 patients who had responded to an initial open-label induction of adalimumab (80 mg at week 0, 40 mg at week 2) were randomized to receive either adalimumab 40mg dosed weekly or every two weeks or placebo.[17] Remission maintenance rates at 26 weeks were 47%, 40%, and 17%, respectively (p<0.001). At week 56, the rates were 41%, 36%, and 12%, respectively (p<0.001).

The safety data from the patients who received active drug in these 1-year studies revealed only a few serious events. In the CLASSIC-II trial, there were 6 abdominal abscesses, one patient who developed pneumonia and sepsis, another patient had parvovirus and nocardiasis, while a different patient had viral meningitis. In the CHARM study, there were 8 abscesses in the patients who received adalimumab, and 5 in those who received placebo. There was 1 case of pneumonia in an adalimumab patient, and no cancers or other concerning serious events. During the post-randomization open-label trial that followed, there were two cases of tuberculosis, both in patients who had normal PPD and chest x-rays at baseline.

Adalimumab was also shown to be safe and effective in CD patients who had previously responded to infliximab, but subsequently stopped therapy due to loss of response or intolerance. In the GAIN trial, 325 such patients with moderate to severely active CD were randomized to receive two injections of either placebo or adalimumab 160 mg at week 0, followed by 80 mg at week 2.[18] Week 4 remission rates were 7% vs. 21%, respectively. Remission rates were similar in patients who had lost response to infliximab (8% vs. 20%) and those who had a previous intolerance to infliximab (5% vs. 22%). Serious adverse events were more common in the placebo group (3 abscesses, 1 Staphylococcal sepsis, 2 Crohn's flares, 2 cases of severe abdominal pain) than in those who received adalimumab (2 cases of dehydration).

### Fistulous Crohn's Disease

There have been no dedicated placebo-controlled trials of CD fistulas with adalimumab. In the CLASSIC-I trial, only 32 of the patients had draining fistulas, and there was no apparent treatment benefit.[15] The only large placebo-controlled experience was a subgroup of the CHARM trial; 117 of the randomized patients had active enterocutaneous or perianal fistulas. Complete fistula closure was achieved in 30% of the adalimumab-treated patients *versus* 13% for placebo at week 26 (p=0.043); at week 56, the rates were 33% and 13%, respectively (p=0.016). All of the patients who had achieved fistula closure at week 26 continued to have complete closure at week 52. A subsequent analysis of these patients who subsequently received open-label adalimumab showed that 24% had maintained fistula closure four years later.[19]

### Use of Concomitant therapy; Adalimumab Levels and Antibodies

The incidence of anti-drug antibodies was low in the 276 patient CLASSIC-II adalimumab study, none of whom developed anti-adalimumab antibodies if they were also on concomitant immunomodulators, while 3.8% developed antibodies if on monotherapy.[16] A breakdown of injection reaction rates by anti-adalimumab status was not provided. The rates of anti-adalimumab antibodies are higher (comparable to those seen with infliximab) in other disease states, and the impact of concomitant immunosuppression (typically with methotrexate) results in lower antibody rates, higher drug levels, and better clinical outcomes.[13,20]

### Summary of Adalimumab Use

It is important to give the correct loading regimen with adalimumab when treating CD: 160mg at week 0, 80mg at week 2, and then 40mg every 2 weeks thereafter (see Table 24.1). Patients who respond to adalimumab but subsequently relapse prior to the 2-week reinjection time should have their dose adjusted to 40mg every week;[21] some patients who relapse may benefit from 80mg every 2 weeks. There is growing evidence that the best results are when adalimumab is used in combination with methotrexate. This may also be true if adalimumab is used with azathioprine or 6-mercaptopurine; the safety risk is increased only minimally. Checking trough adalimumab levels may be helpful when assessing for cause of loss of response; patients with low drug levels and no anti-adalimumab antibodies can be treated with dose escalation; those who have developed antibodies should be switched to a different biological agent. Screening for tuberculosis, hepatitis B, and other opportunistic infections specific to the locality is required prior to initiation of therapy (see Table 24.3).

## Certolizumab

### Luminal Crohn's disease

Certolizumab pegol is a pegylated humanized Fab' fragment that is given by subcutaneous injection. Induction trials have unfortunately fallen short of achieving remission. In the PRECiSE 1 trial, 662 adults with moderate-to-severe CD received either 400 mg of certolizumab pegol or placebo subcutaneously at weeks 0, 2, and 4 and then every 4 weeks.[22] While response rates were higher with the drug than placebo at week 6 (37% vs. 26%, p=0.04), and at both weeks 6 and 26 (22% vs.12%; p=0.05), the rates of remission in the two groups did not differ significantly (p=0.17). In a subsequent induction study in Crohn's patients naïve to anti-TNF therapies, the remission rate with certolizumab pegol 400mg given at weeks 0, 2, and 4 was 32%, as compared to 25% for placebo (p=0.174),[23] although a larger benefit of the drug over placebo was seen in patients who had elevated CRP levels.

Certolizumab has proven effective in a large open-label induction trial followed by a blinded maintenance arm in the initial responders. In the PRECiSE 2 trial 23, 428 CD patients who had responded to an initial open-label induction of certolizumab pegol (400 mg at weeks 0, 2, and 4) were randomized to receive either drug or placebo out to 26 weeks.[24] Clinical response was maintained in 63% versus 36% receiving placebo as week 26 (p<0.001); remission was maintained in 48% with certolizumab versus 29% in the placebo group (p<0.001).

The subsequent 539 patient WELCOME trial in CD patients (all of whom had either lost response or developed reactions to infliximab) dosed all patients with open-label induction of certolizumab pegol (400 mg at weeks 0, 2, and 4).[25] The 62% who initially responded (39% had achieved remission) then received maintenance certolizumab 400mg dosed every 2 or 4 weeks. Week 26 response rates were 40% for those dosed every 4 weeks and 37% in those dosed every 2 weeks (p=0.55); the remission rates were 29% and 30%, respectively (p=0.81).

The safety data from the trials was not alarming. In the PRECiSE 1 study, serious infections were seen in 2% of patients on certolizumab and less than 1% on placebo (mostly perianal abscesses).[22] In the certolizumab group, 1 patient developed metastatic lung cancer (previously treated with infliximab, methotrexate, azathioprine, and prednisone) and 1 patient developed rectal cancer; in the placebo group, 1 patient developed cervical cancer and there was one Hodgkin's disease. In the PRECiSE 2 study, serious infections were seen in 3% of patients receiving certolizumab (including one case of pulmonary tuberculosis) vs. <1% of those on placebo; there were no neoplasms seen.[24] In the WELCOME study, the serious infection rate was 3.2% (no tuberculosis), and there were no malignancies.[25]

### Fistulous Crohn's Disease

There have been no dedicated placebo-controlled trials of CD fistulas with certolizumab. Of the 58 responders to open-label induction with certolizumab in the PRECiSE 2 trial who also had draining fistulas, fistula closure at week 26 did not differ between those who received maintenance certolizumab *versus* placebo (54% vs. 43%; p=0.069) although more patients who received drug had complete closure of all fistulas (36% vs. 17%; p=0.038).[26]

### Use of Concomitant Therapy; Certolizumab Levels and Antibodies

At the time of this writing, there were no commercially available assays to measure certolizumab levels or antibodies to certolizumab. In the PRECiSE 1 trial, anti-drug antibodies were found in 8% of the certolizumab-treated patients; the rate was only 4% in patients on concomitant immunomodulators *versus* 10% in patients on monotherapy.[22] Similarly, in the PRECiSE 2 trial, anti-drug antibodies were found in 8% of the certolizumab-treated patients; the rate was only 2% in patients on concomitant immunomodulators *versus* 12% in patients on monotherapy.[24]

### Summary of Certolizumab Use

Certolizumab pegol is viscous; each 400 mg dose consists of 2 injections (200mg each). The loading dose is 400 mg at weeks 0, 2, and 4, followed by 400mg every 4 weeks (see Table 24.1). In the PRECiSE 4 clinical trial, patients who lost response were given a single "booster" dose of 400mg two weeks after the last dose;[27] in practice, this does not seem to be sufficient to maintain remission, and 400mg every 2 weeks may be required.[28] When possible, combination therapy with azathioprine, 6-mercaptopurine, or methotrexate is recommended. When assays for certolizumab levels become available, checking trough levels may be helpful when assessing for cause of loss of response; patients with low drug levels and no anti-certolizumab antibodies should be considered for dose escalation to 400mg every 2 weeks; those who have developed antibodies should be switched to a different biological agent. Screening for tuberculosis, hepatitis B, and other opportunistic infections specific to the locality is required prior to initiation of therapy (see Table 24.3).

## BIOLOGICAL AGENTS THAT TARGET SPECIFIC INTEGRINS

### Natalizumab

#### Luminal Crohn's disease

Natalizumab is a humanized $IgG_4$ antibody given by intravenous infusion once a month. Its mechanism of action is by binding to the 4 integrin, blocking the adhesion and subsequent migration of leukocytes into the gut. It's efficacy in patients with moderate to severe Crohn's disease was the subject of multiple large trials. In the ENACT-1 induction trial, 905 patients received either monthly infusions of natalizumab 300mg or placebo.[29] Week 10 response rates were 56% vs. 49% (p=0.05); remission rates were 37% vs. 30% (p=0.12). Better results were seen when only patients with elevated CRP levels at entry were evaluated; week 10 response rates in the elevated CRP subgroup were 58% vs. 45% for placebo and remission rates were 40% vs. 28% (p<0.05 for both).

Responders were then re-randomized to maintenance infusions of natalizumab or placebo monthly through week 56 (ENACT-2).[29] Here, the efficacy was more convincing, with week 36 sustained response rates of 61% vs. 28% for placebo (p<0.001); sustained remission rates were 44% vs. 26% for placebo (p=0.003). Analysis at the week 60 time point showed sustained response rates of 54% vs. 20% for placebo, and sustained remissions of 39% vs. 15% for placebo.

Subsequently, the 509 patient ENCORE trial was conducted only in patients with moderate to severely active CD who had elevated CRP levels.[30] After getting 3 infusions of natalizumab at weeks 0, 4, and 8, the week 12 sustained response rates were 48% vs. 32% for placebo (p<0.001) and remission rates were 26% vs. 16% (p=0.002).

The safety data from the natalizumab studies showed similar (very low) rates of severe infections between drug and placebo; 2% each for the ENACT-1 induction trial; 2% for placebo and <1% for the ENCORE induction trial; 3% for natalizumab and 2% for placebo in the ENACT-2 maintenance trial. In the ENACT trials, the only cancers seen were a single basal cell carcinoma of the skin found in each the drug arm and the placebo arm. In the ENCORE trial, 1 patient in the natalizumab arm developed a basal cell carcinoma.

The notable safety issue with natalizumab, which was recognized only after release of the drug for the treatment of multiple sclerosis, was the rare risk of debilitating or fatal progressive multifocal leukoencephalopathy, which subsequently has been attributed to the JC virus.[31] The risk was highest in patients who had been on therapy for 2 or more years, and on previous or concomitant immunosuppressants. The subsequent availability of blood testing for antibodies to the JC virus has allowed this therapy to continue to be used in patients with moderate to severe CD who do not have evidence of prior JC virus exposure. Current recommendations are to retest every 6 or 12 months, and to avoid concomitant immunosuppressants.

### Fistulous Crohn's Disease

There have been no dedicated placebo-controlled trials of CD fistulas with natalizumab.

### Use of Concomitant Therapy; Natalizumab Levels and Antibodies

Due to the increased risk of progressive multifocal leukoencephalopathy in patients on concomitant immunosuppressants, combination therapy with these agents is not allowed, with the exception of corticosteroids, which should be tapered off within the first 6 months. However, there are data from clinical trials on the impact of concomitant immunosuppressants.

At the time of this writing, there were no commercially available assays to measure natalizumab levels, although there are assays offered to measure anti-natalizumab antibodies. In the ENACT-1 trial, 8% of patients tested positive for anti-natalizumab antibodies; the rate was 14% in patients on natalizumab monotherapy, 6% for those on natalizumab with corticosteroids, and 3% in those on natalizumab with immunosuppressants.[29] Acute infusion reactions were more common in patients with anti-natalizumab antibodies (45%) than in those negative (9%; p<0.001). Week 12 response rates were numerically lower in patients with antibodies (53%) than in those without antibodies (62%; p=0.18).

In the ENACT-2 trial, 9% of patients tested positive for anti-natalizumab antibodies; persistently positive antibody rates were 5% overall, 10% in patients on natalizumab monotherapy, 2% for those on natalizumab with corticosteroids, and none in those on natalizumab with immunosuppressants.[29] Acute infusion reactions were more common in patients with anti-natalizumab antibodies (19%) than in those negative (7%; p=0.02). Week 60 sustained response rates were 0% in patients with persistent anti-natalizumab antibodies as compared to 56% in those without antibodies.

Anti-natalizumab antibodies were found in 9.5% of patients in the ENCORE trial; the rate was 14% in patients on natalizumab monotherapy, 10.4% for those on natalizumab with corticosteroids, and 4.5% in those on natalizumab with immunosuppressants.[30] Acute infusion reactions were more common in patients who tested positive for anti-natalizumab antibodies (22%) than in those negative (8%). Patients who developed anti-natalizumab antibodies also had numerically lower week 12 sustained response and remission rates (40.9% and 22.7%) than those without antibodies (49.5% and 26.1%, respectively, p=NS).

### Summary of Natalizumab Use

Natalizumab 300 mg is infused once every 4 weeks; there is no "loading" regimen; dose escalation is not permitted (see Table 24.2). Concomitant immunosuppressants are not allowed, and corticosteroids should be weaned off within 6 months of initiating therapy. Any suspicion for the development of progressive multifocal leukoencephalopathy must be immediately evaluated and excluded. Screening for antibodies to the JC virus is required prior to initiation of therapy; this blood test should be repeated every 6 or 12 months, and the therapy should be discontinued in patients who test positive. Patients (and their treating physicians) in the US must be enrolled in the "CD-TOUCH" drug safety monitoring program; outside of the US, check the country's specific monitoring policy. Screening for other opportunistic infections is not required prior to initiation of natalizumab, although care should be taken in individual circumstances (see Table 24.3).

## Vedolizumab

### Luminal Crohn's disease

Vedolizumab is a humanized $IgG_4$ antibody given by intravenous infusion once a month. Its mechanism of action is by binding to the 4 7 integrin complex, blocking the mucosal addressin – cell adhesion molecule 1 (MAdCAM-1), selectively preventing the adhesion and subsequent migration of leukocytes into the gut. Since it does not impact leukocyte trafficking into the central nervous system, theoretically (and in practice thus far) it should not increase the risk of progressive multifocal leukoencephalopathy.[32]

The GEMINI 2 trial was a placebo-controlled double-blinded randomized trial that contained both an induction and maintenance protocol.[33] Approximately 50% of the patients had previous exposure to an anti-TNF therapy. In the induction phase, cohort 1 consisted of 368 patients with moderate to severely active CD who were randomized to receive two infusions (week 0 and 2) of either vedolizumab 300mg or placebo. Remission rates at week 6 were 14.5% vs. 6.8% for placebo (p=0.02); response rates were 31.4% vs. 25.7% for placebo (p=0.23). Cohort 2 consisted of 747 patients who received open-label vedolizumab 300mg at weeks 0 and 2; their week 6 remission and response rates were 17.7% and 34.4%, respectively.

In the maintenance trial, responders from the blinded induction (cohort 1) or the open-label induction (cohort 2) were randomized 1:1:1 to receive blinded maintenance infusions of vedolizumab 300mg every 4 weeks or every 8 weeks or placebo (although patients from cohort 1 who responded to placebo infusions continued to receive placebo maintenance). Week 52 remission rates for vedolizumab were 39.0% (every 8 week dosing) and 36.4% (every 4 week dosing) as compared to 21.6% for placebo (p<0.001 and p=0.004, respectively). Glucocorticoid-free remission rates for vedolizumab were 31.7% (every 8 week) and 28.8% (every 4 week) compared to 15.9% for placebo (p=0.02 and p=0.04, respectively). Response rates were 43.5%, 45.5%, and 30.1%, respectively (p=0.01, p=0.005). Cancers were seen in 1/301 placebo patients and 4/814 vedolizumab patients (1 each: basal-cell skin cancer, squamous-cell skin cancer, breast cancer, carcinoid of appendix).

The GEMINI 3 clinical trial looked at induction in 315 moderate to severely active Crohn's patients previously exposed to one or more anti-TNF agents with either an inadequate response, loss of response, or intolerance.[34] Remission rates at week 6 to vedolizumab 300mg infusions at weeks 0 and 2 were 15.2% vs. 12.1% for placebo (p=0.433; response rates were 39.2% vs. 22.3%, p=0.001). Following an additional blinded infusion at week 6, week 10 remission rates were 26.6% vs. 12.1% (p=0.001).

Data in an additional 101 TNF-naïve patients was also reported; week 6 remission rates were 31.4% vs. 12.0% for placebo; week 6 response rates were 39.2% vs. 24.0% for placebo. At week 10, the remission rates were 35.3% vs. 16.0% for placebo, and response rates 51.0% vs. 22.0% for placebo.

Safety analyses from the trials were encouraging; in the GEMINI 2 study, vedolizumab patients had numerically higher rates of nasopharyngitis but lower rates of headaches and abdominal pain than placebo.[33] Serious infection rates were 5.5% (drug) vs. 3.0% (placebo). In the GEMINI 3 trial, serious infections were seen in 2/209 patients who received vedolizumab (anal abscess, urinary tract infection). The rates of nasopharyngitis were similar between patients who received drug and placebo (4% each).

### Fistulous Crohn's Disease

There have been no dedicated placebo-controlled trials of CD fistulas with vedolizumab. Subgroup analysis of the 57 patients with draining fistulae at baseline in the GEMINI 2 trial reported week 52 fistula closure rates in the vedolizumab patients of 41.2% (every 8 week dosing) and 22.7% (every 4 week dosing) as compared to 11.1% in placebo (p=0.03 and p=0.32, respectively).[33]

### Use of Concomitant Therapy; Vedolizumab Levels and Antibodies

At the time of this writing, there were no commercially available assays to measure vedolizumab levels or antibodies to vedolizumab. Concomitant immunosuppressants were not allowed in patients enrolled in GEMINI 2 from sites in the United States; the authors stated that concomitant immunosuppression was associated with decreased immunogenicity.[33] Anti-vedolizumab antibodies were detected in 4.1% of patients.

In the GEMINI 3 study, concomitant therapy with immunosuppressants or with corticosteroids led to higher rates of clinical remission at week 10 for patients who were anti-TNF failures, as well as for the entire population.[34] Anti-drug antibodies were reported in 1% of patients.

### Summary of Vedolizumab Use

Vedolizumab induction should be 300mg given at weeks 0, 2, and 6, followed by maintenance infusions every 8 weeks (see Table 24.2). There is not data yet on pushing up the infusions on patients who relapse earlier; it is likely that those who relapse sooner than eight weeks may benefit from infusions given at four –week intervals. Drug level and antibody testing, when made available, may potentially be helpful for this purpose. The best results are when vedolizumab is used in combination with an immunosuppressant (azathioprine, 6-mercaptopurine, or methotrexate); this may be particularly advisable if the patient has already failed an anti-TNF therapy. Screening for JC virus exposure or other opportunistic infections is not required prior to initiation of vedolizumab, although care should be taken in individual circumstances (see Table 24.3).

### CONCLUSION

The advent of biological therapies has revolutionized the treatment of luminal and fistulous CD. Therapies that target TNF-$\alpha$ work rapidly for induction, especially full antibodies (infliximab and adalimumab). The anti-integrin antibodies, natalizumab and vedolizumab, may have a slower onset of action; bridging with corticosteroids or other acute therapy may be needed in sick patients. All of the therapies are effective in maintaining clinical response and remission. The best results occur when these agents are given together with a concomitant immunosuppressant (not allowed with natalizumab), in part due to lower incidence of anti-drug antibodies and other acute reactions. Safety concerns for opportunistic infections are higher with the anti-TNF agents; this should be considered when evaluating the location and risks of the patient population being treated. Natalizumab should only be used after first verifying an absence of exposure (via blood test) to the JC virus; this test should be repeated every 6-12 months, and therapy discontinued if positive. The future promises alternative biological therapies which target additional pathways (i.e. anti-IL12, anti-IL23, anti-IL6, JAK inhibitors), some of which are already used in other inflammatory diseases.

## REFERENCES

1.  Targan SR, Hanauer SB, van Deventer SJ, Mayer L, Present DH, Braakman T et al. A short-term study of chimeric monoclonal antibody cA2 to tumor necrosis factor alpha for Crohn's disease. Crohn's Disease cA2 Study Group. New England Journal of Medicine 1997;337(15):1029-35.

2.  Rutgeerts P, D'Haens G, Targan S, Vasiliauskas E, Hanauer SB, Present DH et al. Efficacy and safety of retreatment with anti-tumor necrosis factor antibody (infliximab) to maintain remission in Crohn's disease. Gastroenterology 1999;117(4):761-9.

3.  Hanauer SB, Feagan BG, Lichtenstein GR, Mayer LF, Schreiber S, Colombel JF et al. Maintenance infliximab for Crohn's disease: the ACCENT I randomised trial. Lancet 2002;359(9317):1541-9.

4.  Present DH, Rutgeerts P, Targan S, Hanauer SB, Mayer L, van Hogezand RA et al. Infliximab for the treatment of fistulas in patients with Crohn's disease. New England Journal of Medicine 1999;340(18):1398-405.

5.  Sands BE, Anderson FH, Bernstein CN, Chey WY, Feagan BG, Fedorak RN et al. Infliximab maintenance therapy for fistulizing Crohn's disease. The New England Journal of Medicine 2004; 350(9):876-85.

6.  Baert F, Noman M, Vermeire S, Van Assche G, G DH, Carbonez A et al. Influence of immunogenicity on the long-term efficacy of infliximab in Crohn's disease. The New England Journal of Medicine 2003;348(7):601-8.

7.  Farrell RJ, Alsahli M, Jeen YT, Falchuk KR, Peppercorn MA, Michetti P. Intravenous hydrocortisone premedication reduces antibodies to infliximab in Crohn's disease: A randomized controlled trial. Gastroenterology 2003;124(4):917-24.

8.  Colombel JF, Sandborn WJ, Reinisch W, Mantzaris GJ, Kornbluth A, Rachmilewitz D et al. Infliximab, azathioprine, or combination therapy for Crohn's disease. The New England Journal of Medicine 2010; 362(15):1383-95.

9.  Beaugerie L, Brousse N, Bouvier AM, Colombel JF, Lemann M, Cosnes J et al. Lymphoproliferative disorders in patients receiving thiopurines for inflammatory bowel disease: a prospective observational cohort study. Lancet 2009; 374(9701):1617-25.

10. Siegel CA, Marden SM, Persing SM, Larson RJ, Sands BE. Risk of lymphoma associated with combination anti-tumor necrosis factor and immunomodulator therapy for the treatment of Crohn's disease: a meta-analysis. Clin Gastroenterol Hepatol 2009; 7(8):874-81.

11. Lee LY, Sanderson JD, Irving PM. Anti-infliximab antibodies in inflammatory bowel disease: prevalence, infusion reactions, immunosuppression and response, a meta-analysis. Eur J Gastroenterol Hepatol 2012; 24(9):1078-85.

12. Nanda KS, Cheifetz AS, Moss AC. Impact of antibodies to infliximab on clinical outcomes and serum infliximab levels in patients with inflammatory bowel disease (IBD): a meta-analysis. The American Journal of Gastroenterology 2013; 108(1):40-7; quiz 8.

13. Yanai H, Lichtenstein L, Assa A, Mazor Y, Weiss B, Levine A et al. Levels of Drug and Antidrug Antibodies Are Associated With Outcome of Interventions After Loss of Response to Infliximab or Adalimumab. Clin Gastroenterol Hepatol 2014; S1542-3565(14)01074-X.

14. Sandborn WJ, Hanauer S, Loftus Jr. EV, Tremaine WJ, Kane S, Cohen R et al. An Open-Label Study of the Human Anti-TNF Monoclonal Antibody Adalimumab in Subjects with Prior Loss of Response or Intolerance to Infliximab for Crohn's Disease. The American Journal of Gastroenterology 2004; 99(10):1984-9.

15. Hanauer SB, Sandborn WJ, Rutgeerts P, Fedorak RN, Lukas M, MacIntosh D et al. Human anti-tumor necrosis factor monoclonal antibody (adalimumab) in Crohn's disease: the CLASSIC-I trial. Gastroenterology 2006; 130(2):323-33; quiz 591.

16. Sandborn WJ, Hanauer SB, Rutgeerts P, Fedorak RN, Lukas M, MacIntosh DG et al. Adalimumab for maintenance treatment of Crohn's disease: results of the CLASSIC II trial Gut 2007; 56(9):1232-9.

17. Colombel JF, Sandborn WJ, Rutgeerts P, Enns R, Hanauer SB, Panaccione R et al. Adalimumab for maintenance of clinical response and remission in patients with Crohn's disease: the CHARM trial. Gastroenterology 2007; 132(1):52-65.

18. Sandborn WJ, Rutgeerts P, Enns R, Hanauer SB, Colombel JF, Panaccione R et al. Adalimumab induction therapy for Crohn disease previously treated with infliximab: a randomized trial. Annals of Internal Medicine 2007; 19;146(12):829-38.

19. Panaccione R, Colombel JF, Sandborn WJ, D'Haens G, Zhou Q, Pollack PF et al. Adalimumab maintains remission of Crohn's disease after up to 4 years of treatment: data from CHARM and ADHERE. Alimentary Pharmacology & Therapeutics 2013; 38(10):1236-47.

20. Roblin X, Marotte H, Rinaudo M, Del Tedesco E, Moreau A, Phelip JM et al. Association between pharmacokinetics of adalimumab and mucosal healing in patients with inflammatory bowel diseases. Clin Gastroenterol Hepatol 2014; 12(1):80-4 e2.

**21.** Rutgeerts P, Van Assche G, Sandborn WJ, Wolf DC, Geboes K, Colombel JF et al. Adalimumab induces and maintains mucosal healing in patients with Crohn's disease: data from the EXTEND trial. Gastroenterology 2012; 142(5):1102-11 e2.

**22.** Sandborn WJ, Feagan BG, Stoinov S, Honiball PJ, Rutgeerts P, Mason D et al. Certolizumab pegol for the treatment of Crohn's disease. The New England Journal of Medicine 2007; 357(3):228-38.

**23.** Sandborn WJ, Schreiber S, Feagan BG, Rutgeerts P, Younes ZH, Bloomfield R et al. Certolizumab pegol for active Crohn's disease: a placebo-controlled, randomized trial. Clin Gastroenterol Hepatol 2011; 9(8):670-8 e3.

**24.** Schreiber S, Khaliq-Kareemi M, Lawrance IC, Thomsen OO, Hanauer SB, McColm J et al. Maintenance therapy with certolizumab pegol for Crohn's disease. The New England Journal of Medicine 2007; 357(3):239-50.

**25.** Sandborn WJ, Abreu MT, D'Haens G, Colombel JF, Vermeire S, Mitchev K et al. Certolizumab pegol in patients with moderate to severe Crohn's disease and secondary failure to infliximab. Clin Gastroenterol Hepatol 2010; 8(8):688-95 e2.

**26.** Schreiber S, Lawrance IC, Thomsen OO, Hanauer SB, Bloomfield R, Sandborn WJ. Randomised clinical trial: certolizumab pegol for fistulas in Crohn's disease - subgroup results from a placebo-controlled study. Alimentary Pharmacology & Therapeutics 2011; 33(2):185-93.

**27.** Sandborn WJ, Schreiber S, Hanauer SB, Colombel JF, Bloomfield R, Lichtenstein GR. Reinduction with certolizumab pegol in patients with relapsed Crohn's disease: results from the PRECiSE 4 Study. Clin Gastroenterol Hepatol 2010; 8(8):696-702 e1.

**28.** Stein AC, Rubin DT, Hanauer SB, Cohen RD. Incidence and Predictors of Clinical Response, Re-induction Dose, and Maintenance Dose Escalation with Certolizumab Pegol in Crohn's Disease. Inflammatory Bowel Diseases 2014; 20(10):1722-8.

**29.** Sandborn WJ, Colombel JF, Enns R, Feagan BG, Hanauer SB, Lawrance IC et al. Natalizumab induction and maintenance therapy for Crohn's disease. The New England Journal of Medicine 2005; 353(18):1912-25.

**30.** Targan SR, Feagan BG, Fedorak RN, Lashner BA, Panaccione R, Present DH et al. Natalizumab for the treatment of active Crohn's disease: results of the ENCORE Trial. Gastroenterology 2007; 132(5):1672-83.

**31.** Van Assche G, Van Ranst M, Sciot R, Dubois B, Vermeire S, Noman M et al. Progressive multifocal leukoencephalopathy after natalizumab therapy for Crohn's disease. The New England Journal of Medicine 2005; 353(4):362-8.

**32.** Milch C, Wyant T, Xu J, Parikh A, Kent W, Fox I et al. Vedolizumab, a monoclonal antibody to the gut homing alpha4beta7 integrin, does not affect cerebrospinal fluid T-lymphocyte immunophenotype. Journal of Neuroimmunology 2013; 264(1-2):123-6.

**33.** Sandborn WJ, Feagan BG, Rutgeerts P, Hanauer S, Colombel JF, Sands BE et al. Vedolizumab as induction and maintenance therapy for Crohn's disease. The New England Journal of Medicine 2013; 369(8):711-21.

**34.** Sands BE, Feagan BG, Rutgeerts P, Colombel JF, Sandborn WJ, Sy R et al. Effects of vedolizumab induction therapy for patients with Crohn's disease in whom tumor necrosis factor antagonist treatment failed. Gastroenterology 2014; 147(3):618-27 e3.

# BIOLOGICAL THERAPY IN ULCERATIVE COLITIS

FÁBIO VIEIRA TEIXEIRA

ROGÉRIO SAAD HOSSNE

## INTRODUCTION

The term inflammatory bowel disease (IBD) is often used in the literature to define a group of diseases involving the alimentary tract, especially the small and the large intestine. The main IBD are Crohn's disease (CD) and ulcerative colitis (UC).[1-7]

UC is a chronic IBD of unknown nature, with an annual incidence of 0.5 to 24.5 new cases per 100,000 inhabitants, and a prevalence of 7.6 to 246 cases/100,000 inhabitants.[1-4] Unfortunately, in Brazil, both the prevalence and incidence of this disease are still unknown.[2-5]

In UC, inflammation is restricted to the intestinal mucosa and can manifest from a mild form until severe colitis with systemic involvement.[1,2]

Conventional treatment of UC, as suggested by most medical consensuses, should be stepwise and progressive (step-up), according to disease severity.[5-8] Patients with mild disease should be treated with aminosalicylates, and those with manifestations of moderate disease, corticodependents or refractory to salicylates should receive more effective treatment with immunomodulators.[1-3]

The primary goal of treatment of IBD is to induce remission in patients with acute disease. Once clinical remission is achieved, maintenance without corticosteroids and with the lowest possible dose of medication is the goal.[1-5]

Until some time ago, it was believed that the most important goal of treatment was the patient's clinical remission, usually based on scores for signs and symptoms in CD (Crohn's disease activity index – CDAI <150 points) and UC (Mayo score = 2 points). However, after the advent of a new class of drugs known as immunobiological agents, which induces healing of inflamed intestinal mucosa, there was a change in that goal. In addition to the attempt of achieving sustained clinical remission, mucosal healing should be a new goal.[5-18]

Biological therapy, represented by its main class of drugs, anti-tumor necrosis factor (anti-TNF) antibodies, soon became the top of a mountain, whose base is represented by other drugs used for decades to treat IBD. These new drugs interfere directly into the individual's immune response, reducing the activation of T cells and inducing apoptosis in immune cells, controlling, thus, the still unknown and complex mechanism that causes diseases such as UC and CD.[3-30,33-57]

The aim of this chapter is to provide the reader with the history, indications, mechanism of action, side effects, efficacy, and safety of biological drugs currently used for the treatment of UC.

## HISTORY OF BIOLOGICAL THERAPY

The term biological therapy refers to treatment of a disease using natural or biologically modified compounds.[5,51-53] In the pathogenesis of CD and UC, there is the involvement and activation of certain pro-inflammatory cytokines (TNF, interleukins [IL], integrins etc.). As a result, some biological agents are more effective compared to others. Most often, these drugs used to treat IBD are monoclonal antibodies. Antibodies are glycoproteins produced by B lymphocytes from the immune system. An antibody only binds to its corresponding antigen. However, some antigens have several antigenic regions that can bind to an antibody; they are the epitopes. Antibodies may be produced by clones of B lymphocytes targeted to several epitopes, which are called polyclonal antibodies. But it is also possible to produce specific antibodies that bind to a single epitope, known as monoclonal antibodies. This biotechnology can create antibodies directed against specific antigens and thereby inactivate tumor enzymes; stimulate or block receptors; switch physiological functions of the cells on and off; and also block the action of proinflammatory cytokines, among other actions. There are hundreds of monoclonal antibodies used in therapy and their names derive from nomenclature rules established by the World Health Organization (WHO). They are all named using a suffix -mab, from monoclonal antibody, and infixes indicating the type and primary indication. For instance, -o- are murine, -xi- are chimeric, -zu- are humanized and -u- are human. Infixes -tu- or -tum- qualify oncologic drugs, and -li- or -lim-, immunological agents, etc.[5,51-53]

## ANTI-TNF

Long ago, it was observed that some cancer patients showed regression of their tumors when they experienced serious bacterial infections. Infections, especially Gram-negative, induce the production of a substance that is called tumor necrosis factor (TNF). This proinflammatory cytokine is a non-glycosylated protein produced by various cells of the immune system, especially macrophages and lymphocytes. TNF plays a role in many inflammatory diseases, such as rheumatoid arthritis, psoriasis and IBD. TNF blocking neutralizes the production of other proinflammatory cytokines such as IL-1 and IL-8.[51-53]

Several drugs have been created in order to block the action of cytokines or physiological mechanisms related to inflammation: anti-TNF, protease inhibitors, T-cell traffic inhibitors, T cell polarization inhibitors, T cell activation inhibitors, epithelial repair promoters and enhancers of the immune system.[1,3-5,51-53]

In Brazil, two anti-TNF agents were approved by Anvisa for treating UC. Until September 2014, infliximab (IFX) was the only drug approved. In early October 2014, adalimumab (ADA) was approved by Anvisa for the treatment of UC. Recently, health agencies in the United States (Food and Drug Administration – FDA) and the European Community (European Medicines Agency – EMA) also approved the use of ADA and golimumab (GLM) for the treatment of moderate/severe UC.[7,13,17,29,49,51-53]

## INFLIXIMAB (IFX)

Since the mid-1990s, the role played by TNF-alpha began to be known as an important mediator of the inflammatory response in the human intestine. TNF induces immune response; blocking, however, does not completely interrupt such response but prevents it from being effective.

In the US, IFX (Remicade® – Janssen, Brazil) was approved by the FDA to be used in patients with CD in 1998. In Brazil, the drug was approved by ANVISA for CD only in 2000, and for the treatment of UC in 2006.[1,2,21,22] IFX is an IgG1 chimeric monoclonal antibody (75% human protein and 25% mouse protein). The antibody binds to a subunit of the soluble TNF and TNF precursor adhered to the cell membrane. The drug probably blocks the interaction between TNF and its receptors. Soluble TNF, once locked, prevents the chemotaxis of macrophages and T cells, with consequent reduction of the inflammatory process. Conversely, blocking of TNF adhered to the cell membrane induces T cell apoptosis.[4,23,51-53] The drug is administered intravenously. IFX is indicated for patients with moderate to severe CD and UC, refractory or intolerant to conventional therapy with corticosteroids, salicylates and immunomodulators. Dosage is based on body weight; to induce remission, the dose is 5 mg/kg of body weight administered at week 0 (start of treatment), week 2 (2 weeks after the initial dose) and week 6 (6 weeks after the initial dose). To maintain remission, the patient should receive infusions of 5 mg/kg every 8 weeks.[1] With a dose of 5 mg/kg of weight, the average half-life of the drug is from 7.7 to 9.5 days.[4,23]

The efficacy of IFX in inducing and maintaining remission in CD patients was described in two randomized, double-blind studies: Accent 1 and Accent 2.[9,10] However, TNF's role in the pathogenesis of UC was cause for much debate in the recent past. Although the first pivotal studies have demonstrated limited efficacy of IFX in patients with UC, following the publication of two randomized, double-blind and placebo-controlled studies, efficacy of the drug in patients with moderate and severe UC was proven.[2,6,7,24-28,58] The ACT (Active Ulcerative Colitis) 1 and 2 trials demonstrated the therapeutic success of IFX in patients with moderate to severe UC refractory to conventional treatment. Almost 70% (ACT 1=69% and ACT 2=64%) of patients who received IFX induction at the dose of 5 mg (weeks zero, 2 and 6) responded to treatment at week 8 (treatment response defined as reduction of 3 points in the Mayo score) and of these, about 35% (ACT 1 = 39% and ACT 2 = 34%) remained in remission.[6] In addition, most patients showed healing of the colonic mucosa, and reduction or complete cessation of corticosteroids.[6,58]

After the publication in 2005 of the ACT 1 and ACT 2 trials, other sub-analyzes or extensions of this cohort of patients have been reported in the literature. In 2009, Sandborn et al.[52] evaluated colectomy rates, number of hospitalizations and number of operations related to UC in patients treated with IFX, 5 mg and 10 mg, compared to placebo. The number of colectomies after 54 weeks was significantly lower in patients who received IFX 10 mg, compared with those who received placebo ($p = 0.007$).[6,23,28,54,56,58] In addition, the number of hospitalizations and surgeries related to UC was also significantly lower in the group treated with IFX. Moreover, when a multivariate analysis was performed, some factors related colectomy could be identified. High levels of CRP at diagnosis, late disease diagnosis (disease > 3 years), Mayo score between 10 and 12 points, and use of corticosteroids were factors associated with a higher chance of colectomy. In addition to these factors, non-use of IFX was also directly associated with the possibility of colectomy.[58]

### Severe colitis refractory to intravenous corticosteroid therapy

Jarnerot et al. assessed IFX as a rescue therapy in severe UC patients hospitalized and refractory to intravenous steroids. Patients were randomized to receive IFX 5 mg/kg (single dose) or placebo. The authors concluded that IFX can be used as rescue therapy in the treatment of patients with severe colitis to reduce the number of colectomies at least in the short term (3 months).[25] There was a significant reduction in the number of surgeries in patients receiving IFX (colectomy: IFX = 29% *versus* placebo = 67%, p=0.017), in a follow-up of three months.[25] The same Scandinavian group published the results of the following three years in the same cohort of patients.[25] Fifty percent of patients receiving IFX did not require surgery, and most of them remained in remission without using corticosteroids. However, 76% of the placebo group had their colons removed (p=0.012).[26] None of the patients treated with IFX and who presented healing of intestinal mucosa after 3 months of treatment required colectomy. In turn, 50% of placebo recipients who were not in endoscopic remission after 3 months of treatment had their colons removed. The difference was statistically significant (p=0.02).[25,26] It is noteworthy that patients treated with IFX received a single dose of the drug; maintenance was done with salicylates or azathioprine in both groups, placebo and IFX. It is known, however, that the response to biological therapy is more effective when patients receive induction (week 0, week 2 and week 6) associated with maintenance at 5 mg/kg every 8 weeks. Maybe the results with IFX therapy, as recommended in the label, could be even better. However, this is only a supposition.

The findings of these studies support the conclusion that the benefit of IFX as rescue therapy remains in the long term, and healing of the intestinal mucosa is directly related to a good clinical outcome with lower colectomy rates.[25,26]

Recently, a retrospective study evaluated the long-term response to IFX as rescue therapy in patients with severe colitis, hospitalized and who were refractory to intravenous corticosteroids. Some of these patients were part of previously published randomized trials.[25,26] This study was conducted by the Swedish Organization for the Study of Inflammatory Bowel Disease (SOIBD) and this is certainly the largest cohort of patients with UC who received rescue therapy with IFX published in the literature.[55] As in previous Scandinavian studies, the primary outcome was survival without colectomy in the period from 3 to 12 months. 211 patients from 12 Swedish centers were evaluated from September 1999 to February 2010.[55] As rescue therapy following failure of betamethasone administered for 5 to 10 days, patients received one, two or three infusions of IFX according to the preference of each service. Subsequently, they were treated with salicylates (18%), immunomodulators (54%) or IFX (25%; of these, 58% received combined therapy with azathioprine) to maintain remission. The colectomy rate in the period of 3-12 months was 5% (2/37) in those who received IFX, 10% (8/81) in those treated with azathioprine as maintenance drug and 13% (4/31) in those treated with salicylates. However, there was no statistical difference between the groups, since only 14 colectomies were performed during the period. The lack of significant differences between treatments may be due to a type II error.[55] The authors concluded that with IFX rescue therapy is effective in patients with severe UC refractory to intravenous corticosteroids.[55] The results are similar to those of previously published randomized trials.[25,26]

### Long-term clinical response to IFX therapy

Patients with UC treated with IFX also maintain long-term clinical remission. Ferrante et al., from University of Leuven, Belgium, published in 2008 results of the follow-up of individuals with UC treated as outpatients.[27] Rates of clinical remission, mucosal healing and colectomy were evaluated. Approximately 68% of patients treated with IFX achieved sustained clinical response in a follow-up of 33.4 (17 to 51.1) months. Univariate analysis showed some predictors of clinical response in the long run. As seen in the ACT 1 and ACT 2 trials, decrease in serum levels of C-reactive protein (CRP) after the start of treatment with IFX was associated with better clinical response, clinical remission and mucosal healing. It is clear, therefore, that the more "inflamed" the patient is, the better response of biological therapy can be expected. In addition, in this same group of Belgian patients, those who received IFX combined with an immunomodulatory agent, such as azathioprine, tended to better clinical response in the long term, compared those who received other treatment. Recently, monitoring of this cohort for 65 months on average revealed that one of the main predictors of remission maintenance in the long run was the use of combination therapy: infliximab + azathioprine (OR: 1:47; 95CI: 1.04-2.08).[28]

### Combination therapy in the treatment of UC

Combination therapy seems to actually be more effective to treat UC than monotherapy with immunomodulatory or anti-TNF agents. The study known as SUCCESS, presented at the ECCO congress in Dublin, Ireland, in 2011 and recently published as full-text article, was designed for this purpose.[49,50] UC patients with moderate/severe disease who had never been treated with immunomodulators or anti-TNF agent were recruited. Patients were treated for 16 weeks with 2.5 mg/kg of body weight of azathioprine, IFX 5 mg/kg (loading dose) and maintenance every 8 weeks; or combined therapy: IFX + azathioprine. Patients receiving combination therapy showed better clinical response and clinical remission rates compared to the placebo group. About 63% of patients receiving combination therapy had endoscopic remission *versus* 55% of those receiving IFX monotherapy, or 37% of those treated with azathioprine monotherapy.[45,50] Therefore, the SUCCESS study showed that combination therapy, at least in the short term (16 weeks), is better than monotherapy.[45,50]

### ADALIMUMAB (ADA)

Adalimumab (ADA) (Humira® – Abbvie, Brazil) was approved by Anvisa for treating people with CD in 2007. In February 2012, ADA was approved by EMA for the treatment of UC and, in October 2012, the drug was also approved in the US by the FDA. Recently, in October 2014, the drug was approved by Anvisa for use in patients with UC.

ADA is an anti-TNF, a fully human monoclonal and recombinant IgG1 antibody which binds with great affinity and specificity to soluble TNF.[3,4,6] As in IFX, ADA can induce apoptosis of T cells. The drug is administered subcutaneously and, unlike IFX, its dose is not weight-dependent. Each ADA syringe has 40 mg. The loading dose for remission is based on the Classic 1 and 2 trials:[11,12] 160 mg (4 syringes) at week 0; 80 mg (2 syringes) at week 2 (2 syringes after week 0). Reinish et al.[29] evaluated the use of ADA to induce remission in patients with moderate/severe UC. Patients included had moderate/severe UC (Mayo score $\geq$ 6 and endoscopic subscore > 2) were naive to treatment with other anti-TNF drug, and refractory to conventional therapy (mesalazine, azathioprine and corticosteroids). Originally, the study was designed with two groups: induction with ADA 160 mg at week 0; and 80 mg at week 2; and 40 mg every 2 weeks. As required by the EMA, another group was included with an induction dose of 80 mg at week 0 and 40 mg every 2 weeks, in addition to the placebo group. In 8 weeks, the rate of clinical remission in the group receiving induction with ADA 80/40 mg was similar to placebo (p = 0.833). On the other hand, the group that received induction with ADA 160/80 mg had clinical remission rates at 8 weeks superior to placebo (18.5 *versus* 9.2%: p=0.0031). This study leads to the conclusion that ADA 160/80 mg was superior to the 80/40 mg dose for induction of remission in patients with moderate/severe UC.[29]

Recently, a multicenter study (North America, Europe, Australia, New Zealand and Israel) evaluated the use of ADA to treat moderate/severe UC in the long term. In this study, called *Ulcerative colitis long-term remission and maintenance with adalimumab*, or simply ULTRA 2, patients with moderate/

severe UC (Mayo score ≥ 6 and endoscopic subscore> 2), refractory to conventional therapy (mesalazine, corticosteroids and azathioprine) and intolerant or loss of response to other anti-TNF (IFX) were included.[30] The ULTRA 2 study design was similar to that adopted in the ACT 1 and ACT 2. Perhaps some crucial differences were related to statistical analysis of patients who left the study (missing data), as will be discussed ahead.

Patients received induction with ADA 160 mg/80 mg or placebo. At week 8, 16.5% of the patients treated were in remission *versus* 9.3% in the group that received placebo (p=0.019). At week 52, 17.3% of patients achieved clinical remission *versus* 8.5% in the placebo group (p = 0.004). When endoscopic remission after eight weeks of starting treatment was evaluated, 41.1% of patients receiving ADA and 31.7% of those receiving placebo had healed mucosa (p = 0.032). At week 52, 25% of the group receiving ADA and 15.4% in the placebo group presented healing of the intestinal mucosa (p <0.05). Among the patients naive to anti-TNF who were treated with ADA, 21.3% achieved clinical remission in 8 weeks, compared to 11% in the placebo group (p=0.017). Furthermore, after 52 weeks of treatment, 22% of the ADA group and 12.4% of the placebo group were in remission (p = 0.029). As observed in patients with CD naive to treatment with anti-TNF, clinical remission in patients with UC never exposed to an anti-TNF seems to be higher compared to those who failed a previous treatment.[30]

Therefore, there is evidence regarding the effectiveness of ADA in the treatment of UC refractory to salicylates, corticosteroids and immunosuppressants, and in the event of failure of other agents such as IFX.[30]

The use of ADA in patients with active UC led to a decrease in the number of colectomies after 1 year of follow-up, according to a study by the Spanish group.[31]

A multicenter study by Taxonera et al.[31], with 30 patients with active UC previously treated with IFX, revealed that those patients who achieved clinical response at week 12 avoided colectomy in the long term (60 weeks).[31]

A study conducted by Feagan et al.[32] showed that in patients with moderate to severe UC, the addition of ADA reduced the number of hospitalizations for any cause (UC-related complications or drug) compared with placebo.[32]

## GOLIMUMAB (GLM)

Golimumab (GLM) (Simponi® – Janssen, Brazil) was approved by Anvisa for use in patients with rheumatoid arthritis in April 2013.[21] For UC, GLM was approved in the US by the FDA in May 2013 and, in September of that year, by the EMA for use in countries belonging to the European Community. In Brazil, there is still no information on the product's approval for UC.

GLM is a human monoclonal IgG antibody produced by a murine hybridoma cell line with recombinant DNA technology and anti-TNF function; administration is subcutaneous and applications are monthly. This drug was approved by Anvisa, the FDA and the EMA for use in rheumatoid arthritis, ankylosing spondylitis and psoriatic arthritis, with good results.[35-40] Its application in UC was initially tested in a phase 2 and 3 trial. The purpose of the phase 2 study was to identify the optimal dose of GLM needed to induce remission, while the phase 3 study aimed to confirm the efficacy and safety of the dose.

The PURSUIT-SC (Program of Ulcerative Colitis Research Studies Utilizing an Investigational Treatment – Subcutaneous) study to evaluate the response to induction with GLM was conducted in 217 locations throughout Eastern and Western Europe, North America, Asia and South Africa.[36,41] The publication of the manuscript included the phase 2 and 3 studies. The parameters clinical response and remission were similar to those obtained with IFX and ADA as described in the ACT 1 and 2, and ULTRA 2 trials. With regard to mucosal healing, however, 0 endoscopic subscore was accepted in the PURSUIT study, unlike the studies on IFX and ADA, which accepted Mayo endoscopic subscores 0 and 1. Furthermore, in the PURSUIT study, quality of life data were included using the validated questionnaire IBDQ, which is structured with 32 questions ranging from 1 point (serious problem) to 7 (no problem) points. Thus, the result of the IBDQ ranges from 32 to 224 points, with higher scores related to better quality of life.

In the phase 3 PURSUIT-SC, 761 patients with moderate/severe UC were included, naive to treatment with biological agents, who were randomized into three groups: induction with GLM 400/200 mg=257 patients; 200/100 mg=253 patients; and placebo=251 patients. After 6 weeks, a larger proportion of patients who received induction with GLM 400/200 mg (54.9%) and GLM 200/100 mg (51%) had better clinical response than the placebo group (30.3%), p <0.001. The effectiveness of GLM was also observed in secondary endpoints, such as mucosal healing and improved quality of life.

The study that assessed maintenance of remission with GLM in moderate/severe UC patients was PURSUIT-M.[37,40] In the maintenance phase, only randomized responding patients were analyzed, from the induction studies (n = 464). The primary endpoint was to evaluate the clinical response at week 54 (one year of treatment) and main secondary endpoints were to analyze clinical remission and mucosal healing at weeks 30 and 54, among others. The two phases of this study demonstrated an acceptable safety profile for the groups analyzed, with no major differences between the study groups and the placebo. At week 54, the authors observed that 47 and 49.4% of subjects who received maintenance with GLM showed clinical response to treatment, compared to 31.2% of those receiving placebo, p <0.01 and p <0.001, respectively.[37,40]

## Comparison between the drugs

To date, there are three anti-TNF drugs approved for use in patients with moderate/severe UC refractory to conventional therapy and other anti-TNF. In Brazil, only IFX and ADA are approved for use in UC. GLM still awaits the approval of Anvisa.

Analyzing efficacy results requires detailed attention on the study populations (disease severity, extent, duration of disease), use of medications prior to recruitment (salicylates, immunosuppressants, anti-TNF, etc.) and the criteria used to define the variables (clinical response, clinical remission and mucosal healing). Studies evaluating the use of biological agents in UC have been designed very similarly. However, some differences in the way the authors analyze the efficacy results (clinical response, clinical remission, mucosal healing), which are dichotomous variables, can make a big difference in the final numbers. Dichotomous variables are those that can have two results; for example, clinical remission: yes or no. Statistical analysis should take into account all patients randomized at baseline: intent-to-treat (ITT) analysis. It is very important that the analysis of a dichotomous variable is made with ITT because participants recruited for a clinical trial are analyzed according to the intervention to which they were allocated, despite having received the intervention or not. For example, patients who were excluded during the study, by their own choice or due to side effects or failure to respond, will be included in the final statistical analysis. In addition, some incomplete or missing data may or may not be considered. If the incomplete and/or missing data from a study are analyzed using LOCF (last observation carried forward), that patient who was responsive to therapy in the last assessment but did not completed the study will be included in the analysis as a responder. On the other hand, if an NRI (nonresponder imputation) analysis is used, a patient who left the study, despite having responded to therapy, will be considered as non-responder. Clearly the latter, NRI, is much stricter than the first, LOCF. For example, the efficacy results in ACT 1 and ACT 2 studies used an analysis of LOCF type.[7] On the other hand, the ULTRA study, involving ADA, adopted a NRI type analysis.[36-40] Therefore, among other reasons already described, it would be tricky to analyze the results of the two studies. Comparison of efficacy between a drug and another is not possible, unless a randomized study is designed for this purpose (head-to-head comparison).

### NEW DRUGS

New drugs are being studied for the management of UC and in most of them, the mechanism of action is based on the type of inflammatory immune response of this disease (mediated by T lymphocytes – Th2 response).[33,34,38,41,42,57]

The primary focus is on reducing the effects of this activation through various mechanisms: induction of T cell apoptosis, blocking of costimulation of T cells, inhibition of pro-inflammatory cytokines, or increase of tissue concentration of anti-inflammatory cytokines. The main drugs and their mechanisms of action are described below.

- Tofacitinib: JAK3 inhibitor (kinase).
- Dersalazine: combination of platelet activation antagonist and anti-TNF properties associated with 5-ASA.
- Vedolizumab: blocks alpha-4 beta-7 integrins.
- Etrolizumab: blocks alpha-4 beta-7 integrins.
- rHuMab-beta7: blocks beta-7 integrin.
- Vidofludimus: inhibition of interleukins IL-17A and IL-17F.
- Anrukinzumab: inhibition of interleukin IL-13.

- Enkorten: anti-inflammatory and immunomodulatory action.
- Sotrastaurin: inhibition of protein kinase C.
- Baziliximab: inhibition of interleukin IL-2R.
- Daclizumab: inhibition of interleukin IL-2R.

These drugs are in various stages of clinical research, between phases I and III, and none have yet been released worldwide by regulatory agencies.

### Vedolizumab (VDZ)

VDZ is an inhibitor of alpha-4 beta-7 integrins, whose properties reduce the inflammatory activity, blocking the influx of leukocytes to the endothelium of the intestine. Unlike natalizumab, VDZ has a more selective action in the gut and, thus, less potential for adverse effects. The drug is available in 400 mg vials for intravenous infusion. The randomized phase 3 study called GEMINI 1 was recently presented, with promising results in the treatment of UC.[34,37] Three groups of patients were evaluated in a randomized, double-blinded manner: placebo (n = 126), VDZ every 8 weeks (n = 122) and VDZ every 4 weeks (n = 125). The primary objective was to evaluate clinical remission rates, while secondary objectives were remission and sustained response, remission without corticosteroids and mucosal healing. The results were evaluated after 52 weeks (one year of treatment) and both doses of VDZ were superior to placebo with respect to all variables.[34,37] The safety analysis of this study did not show higher rates of adverse effects in patients treated with the drug or discontinuation of treatment or infections. Therefore, the drug was considered safe, for its selective characteristics in the gut and absence of systemic effect. Important results from the GEMINI 1 study point out VDZ as one of the most promising therapies in UC.

### Tofacitinib (TOFA)

TOFA is a small molecule that is administered orally, whose mechanism of action is inhibition of JAK3 (just another kinase 3) enzyme, present exclusively in the hematopoietic tissue.[42,57] This inhibition results in blockage in a lower level of the inflammatory cascade and consequent inhibition of interleukins 2, 4, 7, 9, 15 and 21.[42,57] These cytokines play an important role in the activation, proliferation and function of lymphocytes, and their inhibition results in reduced inflammatory activity.

A phase 2 study for dose calculation was performed in patients with UC to verify the effectiveness and safety profile of the drug.[42] In this prospective, double-blind placebo-controlled study, 194 adults were randomized into five groups according to the dose of TOFA: 0.5, 3, 10, 15 mg and placebo. The tablets were taken 2 times/day, and the primary outcome was the assessment of clinical response at week 8; evaluation of clinical remission was considered a secondary outcome.

In the US, this drug was approved by the FDA for use in patients with rheumatoid arthritis. The biggest attraction in the use of this molecule, in addition to its significant efficacy, is the fact that it is an oral prescription drug, which can definitely reduce treatment costs compared to injectable drugs with more complex pharmacological characteristics. Surely, previous experience with the drug in patients with rheumatoid arthritis will define more accurately its safety profile.

## SIDE EFFECTS AND SAFETY

Generally, the side effects and risks of anti-TNF are equivalent to those of all the drugs available on the market. Several factors may limit the use of anti-TNF agents in the treatment of IBD. Most patients have a benign course of the disease, and the use of drugs with proven toxicity and immunogenicity may not be justified. In short, the risk does not justify the benefit.

Anti-TNF drugs could make the patient more susceptible to opportunistic infections, since they reduce the immune response. This seems to be true. Several reports of reactivation of latent tuberculosis, reactivation of latent hepatitis B, cytomegalovirus infection, pneumonia caused by *Pneumocystis carinii*, abscesses, cellulitis, histoplasmosis, aspergillosis, and other viral, fungal or bacterial infections, have been reported in the literature after use of anti-TNF.[41-46] Given this problem, there seems to be a consensus on the need for close clinical and serological evaluation of the patient prior to starting treatment with anti-TNF drugs.[1-5] Data published from the TREAT™ study, in turn, showed no increase in cases of serious infection or death in patients who were treated with IFX.[41] Lichtenstein et al. analyzed 6,290 patients with CD; 3,179 were treated with IFX and 82% received at least two infusions. Mortality rates were similar in patients treated and not treated with IFX (53/100 patients *versus* 43/100 patients; relative risk 1.24; 95%CI). In the multivariate analysis, only the use of prednisone was associated with increased mortality (p < 0.016).[43]

Risk of lymphoma is another concern with the use of biological agents. Siegel et al.[43] analyzed data from the National Cancer Institute, US agency that monitors the cases of cancer (Surveillance Epidemiology and End Results – SEER), and found an increased incidence of non-Hodgkin lymphoma in patients treated with IFX: 13 cases of non-Hodgkin lymphoma (6.1/10,000 patients per year), were recorded, compared to the number of cases expected by the SEER (1.9 cases/10,000 patients per year).[45] In most cases, the patients were treated with IFX combined with immunomodulatory drugs. However, once they have received the combined therapy, the relationship between lymphoma and use of IFX cannot be established.[45]

Recently, a case-control and multicenter study conducted in Italy evaluated the risk of cancer (all types) in patients with CD treated or not treated with IFX. The authors concluded that the frequency of neoplasia was similar between the two groups, in a follow-up of 10 years.[46] Thus, it would be necessary to consider the benefit of treatment at an extremely low risk.[43,46] As for ADA in a recent work on the drug's safety, serious adverse event (SAE) risks and the safety of its use were studied in 19.041 patients with rheumatoid arthritis, psoriatic arthritis, ankylosing spondylitis, psoriasis, idiopathic arthritis and CD (2,228 patients). The results in the latter group of patients were: mean age 38.3 years, mean duration of disease of 11.7 years, serious infections in 5.18/100 patients (tuberculosis [0.13], opportunistic infections [0.08], histoplasmosis [0], malignancies [0.46], lymphoma [0.08], demyelinating diseases [0.13] and lupus-like syndrome [0.04]).[48]

Compared with serious infections, these occurred more frequently in patients with rheumatoid arthritis or CD, and indices were lower in patients with the four other evaluated diseases. Possible reasons for these observations include differences inherent to the disease studied regarding the risks of various SAE, differences in the severity and disease duration, comorbidities and use of concomitant medications (e.g., corticosteroids or other immunosuppressants). Data associated with ADA presented in this report support the security for long-term use in patients with six different inflammatory diseases mediated by immune response. Based on the confirmed effectiveness, and the substantial benefits of ADA in these conditions, as well as IFX, the risk of treatment should be weighed against the risk of uncontrolled inflammatory disease and its consequences in the long term.[45,48]

## REFERENCES

1. Consenso sobre tratamento da doença inflamatória intestinal. Grupo de estudos da doença inflamatória do Brasil (GEDIIB). Arg Gastroenteral 2010; 47(3):313-25.

2. Teixeira FV, Regadas FSP. Afecções benignas do colo e reto. In: Saad Jr. R, Salles RAR, de Carvalho WR, Maia AM (eds.). Tratado de cirurgia do CBC. São Paulo: Atheneu, 2009.

3. Teixeira FV, Saad-Hossne R, Sobrado CW et al. Tratamento clínico da retocolite ulcerativa inespecífica. Projeto Diretrizes, Associação Médica Brasileira, 2014 [no prelo].

4. Ferrante M, D'Haens G, Rutgeerts P, Vermeire S, Van Assche G. Optimizing biologic therapies for inflammatory bowel disease (ulcerative colitis and Crohn's disease). Curr Gastroenterol Rep 2009; 11(6):504-8.

5. Ahmadi A, Valentine JF. Biologic therapies in inflammatory bowel disease. In: Handbook of inflammatory bowel disease. Baltimore: Lippincott Williams & Wilkins, 2010.

6. Rutgeerts P, Vermeire S, Van Assche G. Biological therapies for inflammatory bowel diseases. Gastroenterology 2009; 136(4):1182-97.

7. Rutgeerts P, Sandborn WJ, Feagan BG, Reinisch W, Olson A, Johanns J et al. Infliximab for induction and maintenance therapy for ulcerative colitis. N Engl J Med 2005; 353(23):2462-76. Erratum in: N Engl J Med 2006; 354(20):2200.

8. D'Haens G, Baert F, van Assche G, Caenepeel P, Vergauwe P, Tuynman H et al. Belgian Inflammatory Bowel Disease Research Group; North-Holland Gut Club. Early combined immunosuppression or conventional management in patients with newly diagnosed Crohn's disease: an open randomised trial. Lancet 2008; 371(9613):660-7.

9. Hanauer SB, Feagan BG, Lichtenstein GR, Mayer LF, Schreiber S, Colombel JF et al.; ACCENT I Study Group. Maintenance infliximab for Crohn's disease: the ACCENT I randomised trial. Lancet 2002; 359(9317):1541-9.

10. Sands BE, Blank MA, Diamond RH, Barrett JP, Van Deventer SJ. Maintenance infliximab does not result in increased abscess development in fistulizing Crohn's disease: results from the ACCENT II study. Aliment Pharmacol Ther 2006; 23(8):1127-36.

11. Hanauer SB, Sandborn WJ, Rutgeerts P, Fedorak RN, Lukas M, MacIntosh D et al. Human anti-tumor necrosis factor monoclonal antibody (adalimumab) in Crohn's disease: the CLASSIC-I trial. Gastroenterology 2006; 130(2):323-33.

12. Sandborn WJ, Hanauer SB, Rutgeerts P, Fedorak RN, Lukas M, MacIntosh DG et al. Adalimumab for maintenance treatment of Crohn's disease: results of the Classic II trial. Gut 2007; 56(9):1232-9.

13. Colombel JF, Sandborn WJ, Rutgeerts P, Enns R, Hanauer SB, Panaccione R et al. Adalimumab for maintenance of clinical response and remission in patients with Crohn's disease: the Charm trial. Gastroenterology 2007; 132(1):52-65.

14. Sandborn WJ, Rutgeerts P, Enns R, Hanauer SB, Colombel JF, Panaccione R et al. Adalimumab induction therapy for Crohn disease previously treated with infliximab: a randomized trial. Ann Intern Med 2007; 146(12):829-38.

15. Colombel JF, Sandborn WJ, Reinisch W, Mantzaris GJ, Kornbluth A, Rachmilewitz D et al. Sonic Study Group. Infliximab, azathioprine, or combination therapy for Crohn's disease. N Engl J Med 2010; 362(15):1383-95.

16. Baert F, Moortgat L, Van Assche G, Caenepeel P, Vergauwe P, De Vos M et al. Belgian Inflammatory Bowel Disease Research Group; North-Holland Gut Club. Mucosal healing predicts sustained clinical remission in patients with early-stage Crohn's disease. Gastroenterology 2010; 138(2):463-8.

17. Peyrin-Biroulet L. Anti-TNF therapy in inflammatory bowel diseases: a huge review. Minerva Gastroenterol Dietol 2010; 56(2):233-43.

18. Lichtenstein GR, Hanauer SB, Sandborn WJ. Practice Parameters Committee of American College of Gastroenterology. Management of Crohn's disease in adults. Am J Gastroenterol 2009; 104(2):465-83.

19. Schnitzler F, Fidder H, Ferrante M, Noman M, Arijs I, Van Assche G et al. Mucosal healing predicts long-term outcome of maintenance therapy with infliximab in Crohn's disease. Inflamm Bowel Dis 2009; 15(9):1295-301.

20. Sandborn WJ, Feagan BG, Stoinov S, Honiball PJ, Rutgeerts P, Mason D et al. Precise 1 Study Investigators. Certolizumab pegol for the treatment of Crohn's disease. N Engl J Med 2007; 357(3):228-38.

21. Ghosh S, Goldin E, Gordon FH, Malchow HA, Rask-Madsen J, Rutgeerts P et al. Natalizumab Pan-European Study Group. Natalizumab for active Crohn's disease. N Engl J Med 2003; 348(1):24-32.

22. Kotze PG, Albuquerque IC, Moraes AC, Vieira A, Souza F. Cost-minimization analysis with infliximab (IFX) and adalimumab (ADA) for the treatment of Crohn's disease (CD). Rev Bras Coloproct 2009; 29(2):158-68.

23. Teixeira FV, Saad-Hossne R, Carpi MR, Teixeira ACA, Teixeira Jr P. Infliximabe no tratamento inicial da retocolite ulcerativa moderada e grave. Terapia *top down*: relato preliminar de dois casos. Rev Bras Coloproct 2008; 28(3):289-93.

24. Present DH, Rutgeerts P, Targan S, Hanauer SB, Mayer L, van Hogezand RA et al. Infliximab for the treatment of fistulas in patients with Crohn's disease. N Engl J Med 1999; 340(18):1398-405.

25. Jarnerot G, Hertervig E, Friis-Liby I, Blomquist L, Karlen P, Granno C et al. Infliximab as rescue therapy in severe to moderately severe ulcerative colitis: a randomized, placebo-controlled study. Gastroenterology 2005; 128(7):1805-11.

26. Gustavsson A, Järnerot G, Hertervig E, Friis-Liby I, Blomquist L, Karlén P et al. Clinical trial: colectomy after rescue therapy in ulcerative colitis – 3-year follow-up of the Swedish-Danish controlled infliximab study. Aliment Pharmacol Ther 2010; 32(8):984-9.

27. Ferrante M, Vermeire S, Fidder H, Schnitzler F, Noman M, Van Assche G et al. Long-term outcome after infliximab for refractory ulcerative colitis. J Crohns Colitis 2008; 2(3):219-25.

28. Ferrante M, Arias MT, Vermeire S, Noman M, Van Assche AG, Wolthuis A et al. Predictors of long-term relapse-free and colectomy-free survival in patients with ulcerative colitis treated with infliximab. J Crohns Colitis 2013; S171-2.

29. Reinisch W, Sandborn WJ, Hommes DW, D'Haens G, Hanauer S, Schreiber S et al. Adalimumab for induction of clinical remission in moderately to severely active ulcerative colitis: results of a randomised controlled trial. Gut 2011; 60(6):780-7.

30. Sandborn WJ, van Assche G, Reinisch W, Colombel JF, D'Haens G, Wolf DC et al. Adalimumab induces and maintains clinical remission in patients with moderate-to-severeulcerative colitis. Gastroenterology 2012; 142(2):257-65.

31. Taxonera C, Estellés J, Fernández-Blanco I et al. Adalimumab induction and maintenance therapy for patients with ulcerative colitis previously treated with infliximab. Aliment Pharmacol Ther 2011; 33:340-8.

**32.** Feagan BG, Sandborn WJ, Lazar A et al. Adalimumabe therapy is associated with reduced risk of hospitalization in patients with ulcerative colitis. Gastroenterology 2014; 146:110-8.

**33.** Parikh A, Leach T, Wyant T, Scholz C, Sankoh S, Mould DR et al. Vedolizumab for the treatment of active ulcerative colitis: a randomized controlled phase 2 dose-ranging study. Inflamm Bowel Dis 2012; 18(8):1470-9.

**34.** Feagan B, Rutgeerts P, Sands BE, Sandborn WJ, Colombel JF, Hanauer S et al. Induction therapy for ulcerative colitis: results of GEMINI I, a randomized, placebo-controlled, doubleblind, multicenter phase 3 trial. Gastroenterology 2012; 142:S160-1.

**35.** Hutas G. Golimumab, a fully human monoclonal antibody against TNFα. Curr Opin Mol Ther 2008; 10(4):393-406.

**36.** Rutgeerts P, Feagan B, Marano C, Strauss R, Johanns J, Zhang H et al. A phase 2/3 randomized, placebo-controlled, double-blind study to evaluate the safety and efficacy of subcutaneous golimumab induction therapy in patients with moderately to severely active ulcerative colitis - PURSUIT-SC. Gut 2012; 61(Suppl 3)A78. Presented at UEGW, 2012.

**37.** Rutgeerts P, Feagan B, Marano C, Strauss R, Johanns J, Zhang H et al. A phase 3 randomized, placebo-controlled, double-blind study to evaluate the safety and efficacy of subcutaneous golimumab maintenance therapy in patients with moderately to severely active ulcerative colitis - PURSUIT-maintenance. Gut 2012; 61(Suppl 3)A79. Presented at UEGW 2012.

**38.** Danese S. New therapies for inflammatory bowel disease: from bench to the bedside. Gut 2012; 61:918-32.

**39.** Sandborn WJ, Feagan BG, Marano C, Zhang H, Strauss R, Johanns J et al; PURSUIT-SC Study Group. Subcutaneous golimumab induces clinical response and remission in patients with moderate-to-severe ulcerative colitis. Gastroenterology 2014; 146(1):85-95.

**40.** Sandborn WJ, Feagan BG, Marano C, Zhang H, Strauss R, Johanns J et al.; PURSUIT-Maintenance Study Group. Subcutaneous golimumab maintains clinical response in patients with moderate-to-severe ulcerative colitis. Gastroenterology 2014; 146(1):96-109.

**41.** Perrier C, Rutgeerts P. New drug therapies on the horizon for IBD. Dig Dis 2012; 30 (Suppl 1):100-5.

**42.** Sandborn WJ, Ghosh S, Panes J, Vranic I, Su C, Rousell S et al. Tofacitinib, an oral Janus kinase inhibitor, in active ulcerative colitis. N Engl J Med 2012; 367(7):616-24.

**43.** Lichtenstein GR, Feagan BG, Cohen RD, Salzberg BA, Diamond RH, Chen DM et al. Serious infections and mortality in association with therapies for Crohn's disease: Treat registry. Clin Gastroenterol Hepatol 2006; 4(5):621-30.

**44.** Beaugerie L, Seksik P, Nion-Larmurier I, Gendre JP, Cosnes J. Predictors of Crohn's disease. Gastroenterology 2006; 130(3):650-6.

**45.** Siegel CA, Marden SM, Persing SM, Larson RJ, Sands BE. Risk of lymphoma associated with combination anti-tumor necrosis factor and immunomodulator therapy for the treatment of Crohn's disease: a meta-analysis. Clin Gastroenterol Hepatol 2009; 7(8):874-81.

**46.** Lin MV, Blonski W, Lichtenstein GR. What is the optimal therapy for Crohn's disease: step-up or top-down? Expert Rev Gastroenterol Hepatol 2010; 4(2):167-80.

**47.** Burmester GR, Mease P, Dijkmans BA, Gordon K, Lovell D, Panaccione R et al. Adalimumab safety and mortality rates from global clinical trials of six immune-mediated inflammatory diseases. Ann Rheum Dis 2009; 68(12):1863-9.

**48.** Biancone L, Petruzziello C, Orlando A, Kohn A, Ardizzone S, Daperno M et al. Cancer in Crohn's Disease patients treated with infliximab: a long-term multicenter matched pair study. Inflamm Bowel Dis 2011; 17(3):758-66.

**49.** Panaccione R, Ghosh S, Middleton S, Velazquez JRM, Khanlif I, Flint L et al. Infliximab, azathioprine, or infliximab + azathioprine for treatment of moderate to severe ulcerative colitis: The UC Success Trial. Gastroenterology 2011; 140(Suppl. 1): S134.

**50.** Panaccione R, Ghosh S, Middleton S, Márquez JR, Scott BB, Flint L et al. Combination therapy with infliximab and azathioprine is superior to monotherapy with either agent in ulcerative colitis. Gastroenterology 2014; 146(2):392-400.e3.

**51.** WHO (World Health Organization). General policies for monoclonal antibodies. 2009. Available at: http://www.who.int/medicines/services/inn/Generalpoliciesformonoclonalantibodies2009.pdf. Accessed in: 09/2014.

**52.** Aggarwal BB, Gupta SC, Kim JH. Historical perspectives on tumor necrosis factor superfamily: 25 years later, a golden journey. Blood 2012; 119(3):651-64.

**53.** Giambelluca MS, Rollet-Labelle E, Bertheau-Mailhot G, Laflamme C, Pouliot M. Post-transcriptional regulation of tumor necrosis factor alpha biosynthesis: relevance to pathophysiology of rheumatoid arthritis. OA Inflammation 2013; 1(1):1-6.

**54.** Sandborn WJ, Rutgeerts P, Feagan BG, Reinisch W, Olson A, Johanns J et al. Colectomy rate comparison after treatment of ulcerative colitis with placebo or infliximab. Gastroenterology 2009; 137(4):1250-60.

**55.** Sjöberg M, Magnuson A, Björk J, Benoni C, Almer S, Friis-Liby I et al.; Swedish Organization for the Study of Inflammatory Bowel Disease (SOIBD). Infliximab as rescue therapy in hospitalised patients with steroid-refractory acute ulcerative colitis: a long-term follow-up of 211 Swedish patients. Aliment Pharmacol Ther 2013; doi: 10.1111/apt.12387. [Epub ahead of print]

**56.** Teixeira FV, Kotze PG. Novas estratégias no manejo da RCU. Tratamento precoce com biológicos na RCU. In: Teixeira FV, Kotze PG. Retocolite ulcerativa inespecífica: estado atual do tratamento no século XXI. Rio de Janeiro: DOC, 2013. p.199-216.

**57.** Kotze PG, Teixeira FV. Novas estratégias no manejo da RCU. Novas drogas e tratamentos. In: Teixeira FV, Kotze PG. Retocolite ulcerativa inespecífica: estado atual do tratamento no século XXI. Rio de Janeiro: DOC, 2013. p.217-25.

**58.** Teixeira FV, Kotze PG. Tratamento medicamentoso. Biológicos. In: Teixeira FV, Kotze PG. Retocolite ulcerativa inespecífica: estado atual do tratamento no século XXI. Rio de Janeiro: DOC, 2013. p.73-91.

# HEALING OF THE INTESTINAL MUCOSA IN INFLAMMATORY BOWEL DISEASE

ORLANDO AMBROGINI JUNIOR

MARJORIE ARGOLLO

CLÁUDIA UTSCH BRAGA

## INTRODUCTION

Inflammatory bowel disease (IBD) is a term used to refer to two conditions that affect the alimentary canal, represented by nonspecific or idiopathic ulcerative colitis (UC) and Crohn's disease (CD). With a chronic course and unknown etiology, both progress with intestinal inflammation, with episodes of worsening and unpredictable relapses, causing symptoms and occasional complications, interspersed with variable phases of clinical remission. Their mortality rates, worldwide, are considered low, but morbidity remains significant.

It is accepted that IBD occurs in individuals with genetic predisposition, as a result of interaction between environmental and microbial factors, and the intestinal immune system.

To date, there is no full understanding of the etiology of IBD, which hampers medical options to prevent the disease and manage recurrence, as well as the adoption of measures to modify the natural course of the disease, despite existing treatments which produce an expectation of longer remission and less severe inflammatory flares.

The progression of CD tends to be more aggressive than that of UC, due to its local and systemic complications, which increases the indication for surgical treatment, although the latter does not guarantee healing of the process after resection of diseased segments. Highly recurrent, new lesions usually occur in the remaining intestine, in apparently healthy portions.

Far from being a cure, mucosal healing (MH) on the long term, with targeted treatment and occurring at the right time, has the advantage of allowing prevention of debilitating lesions and/or preventing progression of existing lesions.[1]

## GOALS OF TREATMENT: TREATING BEYOND THE SYMPTOMS

The current concept of treatment of IBD is that it should be early and optimized, with well-defined objectives in an attempt to slow the progression to irreversible tissue damage and consequent debilitating disease. Still under debate, these goals are not well established in the literature; however, they suggest stricter control of inflammatory activity in the disease.[2]

IBD is characterized by periods of relapse and remission assessed primarily by signs and symptoms. However, it is known that although they remain asymptomatic for a certain period, patients have subclinical inflammation, based on evidence of disease activity found in laboratory, radiological and/or endoscopic examinations; i.e. there is no clear and strong correlation between clinical and endoscopic activity in IBD. This is even more evident in CD than in UC.[3-5]

In the past, the goal of therapeutic approach in IBD involved improvements in quality of life, aimed at control of symptoms. However, with a deeper understanding of the pathophysiology and long-term progression, it was concluded that uncontrolled inflammation can develop into complications such as stenosis, fistulas, abscesses, toxic megacolon, need for surgery and development of colorectal cancer, with great impact the morbidity and mortality of patients.[2,6]

In other chronic diseases such as diabetes, hypertension and especially rheumatoid arthritis, the strategy used is treat-to-target, meaning treating a disease to a well-defined target level. It is thus possible to extrapolate this strategy for IBD in an attempt to interfere with the natural course of the disease.

In order to interfere with the course of the disease and the degree of inflammation, already well established evaluation criteria are needed in order to guide the conduct of each case, as described in Table 26.1.[3,7]

The therapeutic goal in IBD is not limited to control of symptoms, but includes control and remission of the inflammatory markers described above. According to Panaccione et al., milestones to be achieved in the CD are suggested, which are described in Table 26.2.[2]

## REMISSION (CLINICAL, ENDOSCOPIC, HISTOLOGICAL, DEEP AND SUSTAINED)

Remission is a term used to describe the state of the disease with little risk of progression, therefore implying the absence of evidence of inflammation. Before discussing the concepts surrounding the term remission, we must highlight the importance of assessing the disease according to the time of progression of symptoms, which, as described in the literature, can be classified as early- (up to 2 years after the onset of symptoms) or delayed-onset (after 2 years). The reason to differentiate between the two forms is that the therapeutic goal to be achieved varies because the disease established for longer periods may have irreversible tissue damage and greater difficulty to reverse inflammation.[2,3,6]

**Table 26.1** Investigating inflammatory activity in IBD

| Clinical data | Laboratory data | Radiological data | Endoscopic data | Histological data |
|---|---|---|---|---|
| CDAI (CD) | PCR | CT and MRI (including enterography) | UC: Mayo score | Biopsies |
| Truelove and Witts (UC) | **Fecal biomarkers** | | CD: CDEIS and SES-CD | |
| | Calprotectin and lactoferrin | US Doppler and CEUS | | |

PCR: C-reactive protein; CT: computed tomography; MRI: magnetic resonance imaging; CEUS: contrast enhanced ultrasound; CDEIS: Crohn's Disease Endoscopic Index of Severity; SES-CD: Simple Endoscopic Score for Crohn's Disease.

**Table 26.2** Definitions of remission proposed for patients with CD

| Stage of disease | Biological remission (control of inflammation) | Clinical remission (control of symptoms) | Outcome |
|---|---|---|---|
| Initial stage (up to 2 years) | Mucosal healing; colonoscopy: no ulcers (except for a certain number of aphthous ulcers <5 mm) Improvement in serum and fecal markers of inflammatory activity: PCR <5 mg/L, fecal calprotectin <250 mcg/g | Clinical practice: total absence of symptoms; 1 to 2 formed stools per day without abdominal/colicky pain CDAI <150 points | Total absence of symptoms, no disease progression, no complications, no limitation, normal quality of life |
| Advanced disease (>2 years) | Mucosal healing; colonoscopy: no ulcers (except for a certain number of aphthous ulcers <5 mm) Improvement in serum and fecal markers of inflammatory activity: PCR <5 mg/L, fecal calprotectin <250 mcg/g | Clinical practice: improvement of inflammatory symptoms (may experience residual symptoms of pain or diarrhea due to previous surgical treatment or bowel damage) CDAI 150 to 220 points | Stabilization of non-inflammatory symptoms without progression of structural damage and limitations, improved quality of life |

For some time, it was believed that disease remission would be related with the absence of signs and symptoms; however, it is currently known that there can be active inflammation even in their absence.

Clinical remission does not mean the absence of symptoms, but it can be defined in CD as CDAI < 150 in the initial disease, or between 150 and 220 in diseases with late diagnosis. In UC, defecation frequency ≤ 3 times/day, with no bleeding or fecal urgency, is used in clinical practice.[2,7]

Endoscopic remission, also called mucosal healing (MH), is still a controversial subject in the literature and has few well-defined criteria.

In the context of IBD, the term "mucosal healing" refers to endoscopic assessment according to the degree of inflammation, and is defined as resolution of visible ulceration or erosion. It is divided into: complete healing, characterized by disappearance of lesions; almost complete, with only aphthous ulcers < 5 mm or erosions remaining after treatment of previous deep ulcers; and partial, when there is reduction > 33% of deep ulcers. Persistence or worsening of lesions means absence of healing.[8]

CDEIS is the gold standard for endoscopic activity evaluation in CD, but it is complex, subject to interobserver variation and is based on the presence of ulcers, with scores between 0 and 44 points. The following are considered endoscopic remission parameters: CDEIS < 6, with another response criteria (decrease of more than 5 points in CDEIS); complete endoscopic remission (CDEIS < 3).; and mucosal healing (absence of ulcers).[9]

Mucosal healing is paradoxically more difficult to define in UC because mucosal inflammation is not always associated with the presence of visible ulcers. Some proposals for endoscopic scores, such as modified Baron and Baron, Mayo Clinic, Sutherland indices, Powell-Tuck and Rachmilewitz, are described, although not validated.[10] Mayo Clinic's subscore is the most used and defines MH as a value < 1, characterized by normal mucosa or presenting loss of vascular pattern but without friability.[11,12]

CRP levels < 5 mg/L, after infection is cleared, can predict mucosal healing with sensitivity > 70%, but low specificity, around 40%. In addition, levels below 5 mg/L would be related to a milder course of the disease, with higher rates of spontaneous remission (*STORI Trial*).[2,10,13]

Calprotectin values < 250 mcg/g are more effective in predicting MH with sensitivity of 80% and specificity around 50% in colon diseases. In small bowel diseases, calprotectin values remain undefined for evaluating endoscopic remission.[2,10,13]

Good correlation is described between imaging tests, such as CT and MRI (including enterography), and MH, with the advantage of assessing transmural inflammation and, in cases of CD, graduating inflammatory activity and identifying established tissue damage, in addition to being less invasive with lower complication rates when compared to endoscopic exams.[10,13]

Despite the strategy to access the degree of tissue inflammation at regular intervals after starting treatment, the *ACCENT* study showed no significant difference comparing patients who had complete *versus* partial endoscopic remission in terms of progression to surgery.[14]

The term complete remission can be used to describe patients who achieve clinical and endoscopic remission, although this is still questionable with respect to the long-term progression of the disease.[15]

Histological evaluation is based on criteria that combine changes produced by acute inflammation characterized by epithelial damage, mono and polymorphonuclear infiltrates, erosions and/or ulcers and granulomas, and chronic architectural changes associated with a correction factor which adds points to a score according with the number of affected samples. Values range from 0 (normal) to 12 (intense inflammation in all samples). In addition, a subscore is proposed and calculated based on inflammatory cell infiltrate found in the epithelium and *lamina propria*, ranging between 0 and 7.[16,17]

When remission is maintained for 2 years or more, it is called sustained remission.[2]

## ADVANTAGES OF MUCOSAL HEALING AND SUSTAINED DEEP REMISSION

It is suggested that MH is an important marker of therapeutic efficacy and prognosis in the long term, with consequent impact on the course of the disease, improvement in quality of life, decreased need of corticosteroids, lower rates of hospitalization and surgery, as well as lower cost of treatment.

Significant increase in the incidence of colorectal cancer (CRC) in diseases that affect the colon is described, particularly in UC, resulting from the sequence inflammation-dysplasia-cancer. It has been shown that control of inflammation substantially reduces this complication. The *CESAME* study showed decreased incidence of colorectal cancer (CRC) in extensive colitis lasting more than 10 years, which had endoscopic remission with azathioprine (AZA).[18]

In CD, there is not a direct relation between clinical remission and MH; however, patients with deep and extensive ulcers show a more aggressive clinical course, with increased development of penetrating complications and surgery, regardless of CDAI score. The presence of deep ulcerations predicts a more aggressive disease and MH could progress with favorable outcome, thus reducing the risks of complications. Patients achieving MH (absence of ulcers, SES-CD = 0) in 2 years had higher rates of steroid-free clinical remission for 3 to 4 years.[8]

In UC, it was observed that endoscopic and microscopic changes in the rectum persisted despite the apparent resolution of symptoms. In patients in clinical and endoscopic remission who continued with acute inflammatory changes in histology, there was a higher incidence of relapse in the subsequent year. The *IBSEN* study showed that MH, one year after diagnosis, was associated with lower risk of colectomy in five years.[11]

## WHICH TREATMENTS CAN ACHIEVE THE GOAL OF SUSTAINED DEEP REMISSION?

Drug therapy is not an isolated tool in the attempt to achieve disease remission. Practical aspects to achieve remission involve psychosocial factors such as education and patient knowledge regarding their disease, time of progression and adherence. The doctor-patient relationship plays an important role in achieving this goal.

The physician is also responsible for evaluating and optimizing the established standard therapy, identifying high-risk patients who would benefit from an early aggressive treatment for faster deep remission, and for monitoring regularly the treatments adopted and the need for continued treatment.[3]

Drug treatment involves using drugs with different mechanisms of action that can contribute to achieving the goal of sustained deep remission.

In CD, corticosteroids have little or no effect on the induction of MH; in UC, on the contrary, they seem to be effective. Studies of oral budesonide and corticosteroid enemas caused MH in UC, suggesting that its use favors endoscopic remission regardless of the route of administration in mild disease. This fact is not observed in the presence of deep ulcerations, and there is no response even with intravenous treatment.[6,8]

The *ASCEND I/II* studies found rates of MH defined as endoscopic scores from 0 to 1 in 80 and 64% of patients with moderate UC treated with mesalazine, 4.8 and 2.4 g/day, respectively, after 6 weeks. In CD, no benefit was observed under treatment with aminosalicylates, and its use is thus limited.[6,19]

In contrast to the aforementioned drugs, AZA seems to promote MH in CD. Studies in patients with UC and CD have shown that endoscopic remission is also achieved in most cases that had clinical remission with use of AZA.[6,8,20]

Although studies have demonstrated that methotrexate (MTX) can promote MH in CD, such frequency is lower compared to AZA and tumor necrosis factor-alpha inhibitors (anti-TNF-alpha).[6,8]

Cyclosporine, a calcineurin inhibitor used to treat severe UC unresponsive to corticosteroid therapy, is more effective in inducing MH until the fourth week. Patients with early response after use appear to have less chance of progressing to total colectomy after 1 year.[8]

Anti-TNF-alpha agents showed a major advance in obtaining clinical and endoscopic remission, reaching between 30 and 40% in CD and UC. Its benefit is enhanced when immunosuppressants are combined.[8,21]

The EXTEND trial, first study that has MH as the primary endpoint, showed that at week 12, 27% of patients receiving ADA presented mucosal healing versus 12% in the placebo group (p = 0.056). At week 52, MH was observed in 24% in the ADA group versus 0% in the placebo group, showing clear advantage of the ADA group.[21]

Maintenance therapy with ADA led to higher rates of mucosal healing in patients with early CD (<5 years duration) compared to patients with longer disease progression. This analysis suggests that the greatest impact of ADA occurs early in the disease course, possibly reflecting the progressive nature of CD. It may also be noted that the CDEIS remission rates in both weeks 12 and 52 were significantly higher in the ADA continued-use group as compared to the ADA induction only/placebo group.[21]

Table 26.3 summarizes relevant data regarding drug treatment and the proposed therapeutic goals.[6]

Conclusion is that MH should be recognized as a therapeutic goal in IBD. Rationale is to maintain an intact mucosal barrier, including intestinal epithelial cells. New endoscopic scores, endomicroscopy, serum markers and imaging studies comprise an arsenal of prospects for better evaluation of MH.[8]

**Table 26.3** Evidence of drug therapy in relation to the objectives in the short and long term in IBD

|  | 5-ASA | Corticosteroids | AZA | MTX | Anti-TNF |
|---|---|---|---|---|---|
| **Short-term outcome** | | | | | |
| Clinical remission | RCU | RCU, DC | RCU?, DC? | DC | RCU, DC |
| Corticosteroid-free | ? | ? | ? | ? | ? |
| Deep remission | RCU | RCU | RCU | ? | RCU, DC |
| Treating beyond the symptoms | ? | ? | ? | ? | ? |
| **Long-term outcome** | | | | | |
| Reduction of surgical risk | ? | ? | Conflicting | ? | RCU, DC |
| Reduction of "disability" | ? | ? | ? | ? | ? |
| Reduction of permanent harm | ? | ? | ? | ? | ? |

AZA: azathioprine; MTX: methotrexate.

## REFERENCES

1. Odze R. Diagnostic problems and advances in inflammatory bowel disease. J Mod Pathol 2003; 16(4):347-58.

2. Panaccione R, Colombel JF, Louis E, Peyrin-Biroulet L, Sandborn WJ. Evolving definitions of remission in Crohn's disease. Inflamm Bowel Dis 2013; 19(8):1645-53.

3. Panaccione R, Hibi T, Peyrin-Biroulet L, Schreiber S. Implementing changes in clinical practice to improve the management of Crohn's disease. J Crohns Colitis 2012; 6(suppl 2):S235-42.

4.  Dignass A, Eliakim R, Magro F, Maaser C, Chowers Y, Geboes K et al. Second European evidence-based consensus on the diagnosis and management of ulcerative colitis Part 1: Definitions and diagnosis. J Crohns Colitis 2012; 6:965-90.

5.  Dignass A, Lindsay JO, Sturm A, Windsor A, Colombel JF, Allez M et al. Second European evidence-based consensus on the diagnosis and management of ulcerative colitis Part 2: Current management. J Crohns Colitis 2012; 6:991-1030.

6.  Sandborn WJ, Hanauer S, Van Assche G, Panés J, Wilson S, Petersson J et al. Treating beyond symptoms with a view to improving patient outcomes inflammatory bowel diseases. J Crohns Colitis 2014; 8(9):927-35.

7.  Reenaers C, Louis E, Belaiche J. Current directions of biologic therapies in inflammatory bowel disease. Therap Adv Gastroenterol 2010; 3(2):99-106.

8.  Neurath MF, Travis SPL. Mucosal healing in inflammatory bowel diseases: a systematic review. Gut 2012; 61:1619-35.

9.  Carter DL, Lang A, Eliakim R. Endoscopy in inflammatory bowel disease. Minerva Gastroenterol Dietol 2013; 59(3):273-84.

10. Baert F, Moortgat L, Van Assche G. Mucosal healing predicts sustained clinical remission in patients with early-stage Crohn's disease. Gastroenterol 2010; 138:463-8.

11. Froslie KF, Jahsen J, Moum BA, Vatn MH. Mucosal healing in inflammatory bowel disease: results from a Norwegian population-based cohort. Gastroenterol 2007; 133(2):412-22.

12. Samaan MA, Mosli MH, Sandborn WJ, Feagan BG, D'Haens GR, Dubcenco E et al. A systematic review of the measurement of endoscopic healing in ulcerative colitis clinical trials: recommendations and implications for future research. Inflamm Bowel Dis 2014; 20(8):1465-71.

13. Benitez JM, Meuwis MA, Reenaers C, Van Kemseke C, Meunier P, Louis E. Role of endoscopy, cross-sectional imaging and biomarkers in Crohn's disease monitoring. Gut 2013; 62(12):1806-16.

14. Hanauer SB, Feagan BG, Lichtenstein GR, Mayer LF, Schreiber S, Colombel JF et al. Maintenance infliximab for Crohn's disease: the ACCENT I randomised trial. Lancet 2002; 359:1541-9.

15. Rogler G, Vavricka S, Schoepfer A, Lakatos PL. Mucosal healing and deep remission: What does it mean? World J Gastroenterol 2013; 19(43):7552-60.

16. D'Haens G, Geboes K, Peeters M, Baert F, Penninckx F, Rutgeerts P. Early lesions caused by infusion of intestinal content in excluded ileum in Crohn's disease. Gastroenterology 1998; 114:262-7.

17. D'Haens G, Geboes K, Rutgeerts P. Endoscopic and histologic healing of Crohn's (ileo-) colitis with azathioprine. Gastrointestinal Endoscopy 1999; 50(5):667-71.

18. Beaugerie L, Seksik P, Bouvier A. Thiopurine therapy is associated with a three-fold decrease in the incidence of advanced colorectal neoplasia in IBD patients with longstanding extensive colitis: results from the CESAME cohort. Gastroenterology 2009; 136:A54.

19. Hanauer SB, Sandborn WJ, Kornbluth A, Katz S, Safdi M, Woogen S et al. Delayed-release oral mesalamine at 4.8 g/day (800 mg tablet) for the treatment of moderately active ulcerative colitis: the ASCEND II trial. Am J Gastroenterol 2005; 100:2478.

20. Walker-Smith JA. Mucosal healing in Crohn's disease. Gastroenterology 1998; 114:419-2.

21. Rutgeerts P, Van Assche G, Sandborn WJ, Wolf DC, Geboes K, Colombel JF et al. Adalimumab induces and maintains mucosal healing in patients with Crohn's disease: data from the EXTEND trial. Gastroenterology 2012; 142(5):1102-11.

# PHARMACOTHERAPY IN INFLAMMATORY BOWEL DISEASE

LUCIANA DOS SANTOS

RAQUEL GUERRA DA SILVA

MAYDE SEADI TORRIANI

ELVINO BARROS

## INTRODUCTION

Chronic inflammatory bowel diseases (IBD) represent a wide variety of inflammatory conditions of the bowel, but the term used here refers to Crohn's disease (CD) and rectal colitis or ulcerative colitis, which are diseases characterized by recurrent chronic inflammation of the bowel, without a definite etiology. The two diseases are differentiated from each other through clinical, radiological, endoscopic and histological findings.[1-3]

Ulcerative colitis (UC) is a chronic and recurrent disease, characterized by diffuse mucosal inflammation involving the colon and, invariably, the rectum. The inflammation in UC is typically restricted to the mucosal surface and extends from the rectum along the entire length of the colon. CD is a complex entity involving the small intestine and the colon, and possibly the entire gastrointestinal tract, as well as other organs. It is characterized by transmural inflammation in certain segments of the gastrointestinal tract.[4,5]

Both UC and CD may be associated with numerous extra-intestinal manifestations, including oral ulcers, arthritis, spondylitis, sacroiliitis, uveitis, erythema nodosum, etc. Although sharing some common characteristics, these diseases show differences in genetic predisposition, risk factors and clinical and endoscopic findings. Family history is the most important independent risk factor. Smoking has a different role in the two diseases: it is considered a protective factor in UC and a risk factor in CD.

The pathophysiology of the inflammatory process is complex, multifactorial, and includes genetic, immunological and environmental aspects, with bacterial invasion and the presence of antimicrobial peptides. Tumor necrosis factor (TNF-alpha) is high in blood samples, in feces and the mucosa of patients with UC and is considered an important factor in the pathogenesis of this disease; an exacerbation of T cells can also be observed, causing mucosal injury.

## PHARMACOLOGICAL TREATMENT

The medical management of these patients is complex and involves the use of many classes of drugs. In UC and CD, the same classes of medications are used, although they are distinct diseases. There is no specific and curative treatment for either of the diseases. The most commonly used drugs are derivatives of 5-aminosalicylic acid, corticosteroids, immunomodulators and the most recent medicines - biological agents. Below is a brief summary of pharmacological treatment in UC and CD.

### Ulcerative colitis (UC)

The treatment consists in the use mesalamine, corticosteroids, immunosuppressants and anti-TNF-alpha monoclonal antibodies. The success of treatment depends on many factors, including: correct choice of drugs for induction or maintenance, correct dose and adherence to treatment.[6]

Mesalazine taken orally or locally (suppository) is the drug of choice. The combined oral and topical use leads to higher remission rates compared to individual use of either formulation for treating moderate UC. If there is no response, the use of corticosteroids is indicated. Patients with corticosteroid dependency or those with recurrence of the disease with the use of mesalazine may be treated with azathioprine or mercaptopurine.

Patients with active UC of moderate intensity that are unresponsive to conventional treatment may be treated with infliximab or adalimumab in an isolated form or in combination with azathioprine. Patients with the severe form of the disease should be hospitalized and receive treatment with corticosteroids intravenously. If no improvement occurs within 3 days of intravenous corticosteroids, the use infliximab, cyclosporine or tacrolimus is indicated. Prophylaxis with sulfamethoxazole/trimethoprim for *Pneumocystis carinii* is recommended in patients using corticosteroids, cyclosporine or tacrolimus. Use of probiotics is not effective in the treatment of UC.

### Crohn's disease (CD)

The treatment of such patients includes drugs, nutritional counseling and surgery in order to control the activity of the disease and improve quality of life.[5] Patients who do not respond or who are dependent on corticosteroids need to receive immunomodulators or biological agents.

The administration of azathioprine and mercaptopurine is indicated for induction and remission of the disease in the active phase. Anti-TNF-alpha agents are effective for inducing remission and maintained as supportive treatment after remission. When infliximab shows no therapeutic response, adalimumab can be effective for the induction of clinical remission.

## FUTURE MANAGEMENT

The management of IBD has been changing in recent years and there are good expectations for improving symptoms, quality of life and control of the disease. A large number of new medications are being developed and launched onto the market. The role of biological medication will revolutionize the treatment of these patients in the short and medium term. The complicated issue is always the same: the cost of treatment.

## ASSESSMENT OF THE MAIN DRUGS USED IN IBD

### Aminosalicylates

Aminosalicylates have as their active ingredient 5-amino salicylic acid (5-ASA), which confers anti-inflammatory properties through the presence of an amino group at the 5-position (meta) in the structure of salicylic acid. This therapeutic class is indicated for treating mild to moderate UC and in CD, both in the active phase and in remission. The mechanism of action, especially in products containing 5-ASA, appears to be associated with different proposed mechanisms including inhibition of cytokines, prostaglandins, leukotriene synthesis, elimination of free radicals and immunosuppressive activity.

Among the precautions, attention is recommended in patients with hypersensitivity to salicylates and/or derivatives thereof, those with a history of gastric or duodenal ulcer and with severe impairment of renal and liver function.[7] Adverse effects are uncommon but include nausea, skin rash, diarrhea, pancreatitis and acute interstitial nephritis. There are different formulations for topical and oral use. We can highlight mesalazine and sulfasalazine.

**Mesalazine**

Mesalazine is a formulation of 5-ASA found in different pharmaceutical forms for oral or anorectal use.

Formulations

For oral use, it is available in tablets and in granular form (sachet) with microgranules coated with an extended release ethylcellulose forming a semipermeable membrane and protecting against the acidity of the gastrointestinal tract. It is also available in the form of tablets coated with acrylic, which allows most of the drug to be released only in the terminal ileum and colon; and extended release with MMX (*Multi-Matrix System*), a technology that covers the tablet, allowing its release when it reaches the terminal portion of the small intestine and colon, allowing the mesalazine to be distributed adequately throughout the colon; thus, this formulation allows administration once/day.

Formulations coated with acrylic resin, sensitive to pH, dissolve at pH ≥ 7, releasing 10 to 15% of the 5-ASA in the terminal ileum and the remainder in the proximal colon. Extended release formulations with ethylcellulose coating release 30 to 40% of the 5-ASA in the duodenum and continue releasing throughout the colon, depending on the pH and the contact time of the drug.[7,8] Such formulations containing microgranules coated with ethylcellulose release mesalazine continuously throughout the gastrointestinal tract under any pH conditions and regardless of the presence of food. The extended release tablets with MMX enable release when pH ≥ 7, providing slow release of mesalazine concentrations throughout the colon, with limited systemic absorption; the half-life for elimination of mesalazine and its major metabolite is 7 to 9 hours and 8 to 12 hours, respectively. The pharmaceutical forms for topical anorectal use are mesalamine enemas and suppositories.[7,9]

### Mechanism of action

The exact mechanism of action of mesalazine is still uncertain, but it possibly blocks cyclooxygenase and inhibits prostaglandin production by the colonic mucosa.

In relation to pharmacokinetic parameters, it is known to be absorbed rapidly after oral ingestion, allowing only 20% reach the terminal ileum and colon; excretion occurs rapidly through the kidneys (13 to 30%), primarily of the metabolite N-acetyl-5-aminosalicylic acid, and through the feces (72%). Mesalazine and its metabolites do not cross the blood-brain barrier, and minimum amounts cross the placenta. Mesalazine is found 43% bound to plasma proteins, and its metabolite, N-acetyl-acid-5-aminosalicylic acid, 78%. The half-life for elimination of mesalazine is about 1 hour and its metabolite, 10 hours.[10] The use of mesalazine is considered safe during pregnancy and breastfeeding.[11]

### Indications

Oral pharmaceutical forms are the first line treatment for mild to moderate UC, and forms for topical use, such as suppositories and enemas are effective in active proctitis and distal UC with a response rate of 75 to 90%.[12]

### Posology

The usual dose to treat the active disease can range from 800 to 2400 mg/day up to a maximum of 4000 to 4800 mg/day, divided or in a single daily dose. The usual dosage for extended-release tablets (MMX) is 2400 to 4800 mg/day in a single dose, preferably at the same time.

### Administration

The coated tablets should be swallowed whole with the aid of liquid. The granules (sachets) should be placed under the tongue and swallowed with liquids; the granules cannot be mixed with liquids to facilitate administration.

Among the preparations for topical anorectal use, enemas, at the usual dosage of 4 g must be used at bedtime with a retention time of at least 8 hours; while suppositories, at the usual dose of 500 to 1000 mg should be used 2 or 3 times/day, with a retention time of at least 3 hours. The response to topical treatment occurs in 3 to 21 days and, upon the remission of symptoms, lower doses are established for maintenance.

### Drug interactions

Mesalazine can change or enhance the effects of many other drugs. The action of oral hypoglycemic agents, such as sulfonylureas, may have its hypoglycemic effects intensified in the presence of mesalazine. The effects of oral coumarin anticoagulants are altered when administered with mesalazine, increasing the risk of bleeding. There is a reduction in the diuretic action of furosemide and spironolactone, and the tuberculostatic action of rifampicin. Concomitant administration with mercaptopurine or azathioprine may increase the risk of leucopenia and nephrotoxicity.

Some special recommendations are made relating to the use of lactulose, which may lower the pH of the colon and interfere with the absorption of the tablets and the use of antacids, which can alter the bioavailability of extended release formulations by interfering with the pH for dissolution of mesalazine.[10]

### Adverse effects

These are relatively infrequent (10%) and unimportant. The most common effects include headache, dyspepsia, nausea, abdominal pain and rash.[8,12]

## Sulfasalazine

Sulfasalazine (SSZ) is a combination of 5-ASA with the sulfapyridine molecule through an azo bond. The azo bond prevents absorption in the stomach and small intestine. When ingested, SSZ is broken down by the colon through bacterial action on sulfapyridine (almost completely absorbed) and 5-ASA, which is the active portion of the drug in the large intestine.[9,12]

### Formulations

SSZ is found in tablet form with gastro-resistant coating, or with gastrointestinal intolerance or intolerance to the metabolites of sulfasalazine.

Mechanism of action

The mechanism of action is not clear; it may reduce inflammation through the removal of free radicals, by inhibiting the production of prostaglandins and leukotrienes and/or decreasing neutrophil chemotaxis and generation of superoxides.[9]

In relation to the pharmacokinetic parameters, approximately 20 to 30% of the orally administered drug is absorbed in the small intestine. Most of it is captured by the liver and excreted in the urine and the remainder reaches the colon where it is cleaved by intestinal bacteria into sulfapyridine and 5-ASA.[12] Sulfamate pyridine, which is highly lipid soluble, is absorbed rapidly from the colon and has bioavailability of 60%, suffering extensive hepatic metabolism through acetylation, hydroxylation and conjugation reactions with glucuronic acid; it is then excreted in the urine and to a lesser extent, in feces.[12] 5-ASA is poorly absorbed in the colon and excreted extensively in the feces.[7]

Indications

SSZ is indicated for treatment of mild to moderate UC and CD.

Posology/administration

The usual dose is 3 to 6 g/day in 4 divided doses, with food. To avoid potential adverse effects, the dosage should be increased gradually from a starting dose of 500 mg, 2 times/day.[12]

Drug interactions

SSZ inhibits the transport and absorption of folate. This effect can lead to a folate deficiency and consequent megaloblastic anemia. Supplementation of 1 mg/day of folate is therefore recommended in patients using sulfasalazine.[7]

Among other drug interactions, sulfasalazine can strengthen the effect of hypoglycemic agents and anticoagulants. The plasma levels of digoxin may be reduced in the presence of sulfasalazine. Cyclosporine may have its effects reduced through the interference of sulfasalazine on cytochrome P450, responsible for the metabolism of cyclosporine.[10]

Adverse effects

These occur in 10 to 45% of patients with UC and are mainly related to the sulfa fraction, responsible for 15 to 30% of the adverse effects, which may be related to the posology, hypersensitivity or idiosyncratic effects, or the effects anticipated with the use of the drug. During pregnancy it should be used with care, given that it interferes with the metabolism of folic acid.[11]

When related to posology (dose), the commonly observed effects are headaches, nausea, vomiting, pancreatitis, dyspepsia, diarrhea, anemia (hemolytic or megaloblastic) and fatigue, which can be minimized by reducing the dose or administering it together with food.

When the adverse effects are related to hypersensitivity or idiosyncrasy with the use of sulfasalazine, discontinue use of the drug immediately. The reported effects include severe skin rash, hepatitis, pancreatitis, pneumonitis, agranulocytosis and aplastic anemia.

Effects commonly expected with use of the drug include abdominal pain, fever, hemolytic anemia, interstitial nephritis, arthralgia, redness/rash, alopecia, diarrhea and oligospermia (reversible and rare).[8,12] Around 8% of patients taking sulfasalazine are intolerant to it and the use of olsalazine or mesalazine are indicated as an alternative treatment.

## Antibiotics

Antibiotics are indicated for the control of inflammatory activity in moderate to fulminant CD, as adjuvants to other medicines or treatment of specific complications of the disease or for prophylaxis of recurrences postoperatively. They are indicated when there is suspicion of infectious complications, such as abscesses or infections, and treatment of fistulas. Reports of toxicity with prolonged use, such as peripheral neuropathy with the use of metronidazole, as well as inducing antimicrobial resistance potential are reasons for contraindication of the use of isolated antimicrobials for treatment of CD. The use of antibiotics for the treatment of UC does not have a defined role; there are reports of an increased risk of developing pseudomembranous colitis associated with antibiotics. Metronidazole and ciprofloxacin stand out as the most frequently used drugs.

**Metronidazole**

Metronidazole is classified as an antibiotic belonging to the class of nitroimidazoles which presents a spectrum of activity covering anaerobic microorganisms exclusively.

Formulations

It is found in the form of a coated tablet, oral suspension and injectable solution.

Mechanism of action

The antibiotic action occurs after its diffusion in the body, where it ends up interacting with the DNA and causing a loss of the helical structure, resulting in inhibition of protein synthesis and cell death of susceptible microorganisms.[13]

Pharmacokinetic parameters

After oral administration, the drug is completely absorbed with oral bioavailability of 100%, reaching peak plasma concentrations within 1 to 3 hours, with a half-life of about 7 hours. Metronidazole undergoes hepatic metabolism, generating two metabolites: an acid with bactericidal activity of 3% and another which is alcoholic with 30% activity when compared to metronidazole. Less than 20% binds to plasma proteins and is distributed to the saliva, bile, seminal fluid, liver, lung and vaginal secretions, as well as crossing the blood brain barrier. Most of it is excreted through urine (60 to 80%) and, to a lesser extent, feces (6 to 15%).[9,10]

Indications

Metronidazole is indicated for treatment of CD, particularly in patients with perianal and colonic disease or fistula that are non-responsive to other previous treatment (sulfasalazine, corticosteroids), in order to prolong the time until endoscopic and clinical recurrence. In the case of perianal fistula, the treatment can be performed with metronidazole or ciprofloxacin.

Posology

Generally, metronidazole is used for a period of 3 months at the dose of 10 to 20 mg/kg/day, or 250 to 500 mg per dose, 2, or 3 times/day orally.

### Administration

The preferred route of administration is oral for CD, with or without the presence of food, with the aid of liquid.[10]

### Drug interactions

It has drug interactions with oral coumarin anticoagulants such as warfarin, where use should be monitored, as the anticoagulant effect may be boosted. Use with lithium may increase its plasma levels, leading to toxic effects (weakness, excessive thirst, tremors, and mental confusion). Cyclosporine and carbamazepine may also have increased plasma levels and their use should be monitored.[9,10]

### Precautions

Patients treated with the antibiotic should not consume alcoholic beverages up to 3 days after stopping using the drug. Patients should also be monitored periodically for signs of neuropathy. If identified, they should immediately discontinue use of the medication.

Neuropathy is associated with both oral and intravenous use at high doses ($\geq$ 1.5 g/day), and is associated with prolonged use of the antibiotic (> 30 days). The probable mechanism is the inhibition of neuronal protein synthesis, resulting in peripheral axonal degeneration.[10]

Use is contraindicated in the first trimester of pregnancy, since the drug crosses the placental barrier and is excreted in breast milk.[10,11]

### Adverse effects

Metronidazole causes few effects at therapeutic doses. There are reports of mild gastrointestinal symptoms, and a bitter metallic taste in the mouth, as well as symptoms related to the central nervous system (CNS) such as dizziness, headache and sensitive neuropathies.

### Ciprofloxacin

Ciprofloxacin belongs to the fluoroquinolone class.

### Formulations

It is found in formulations for parenteral and oral use. For use in the treatment of CD, ciprofloxacin is found in the form of a capsule or coated tablet with immediate release for oral use.

### Mechanism of action

It relies on the inhibition of DNA gyrase enzyme activity or topoisomerase II, an essential enzyme for bacterial survival. By inhibiting it, the DNA molecule occupies a large space inside the bacterium, and its free ends cause uncontrolled synthesis of messenger RNA and proteins, resulting in bacterial death. It acts on enterobacteria, staphylococci, haemophilus, neisseria and *P. aeruginosa*.

In relation to pharmacokinetic parameters, it is known that oral absorption is rapid (50 to 85%) and occurs in 1 to 2 hours with oral bioavailability of 60 to 80%. It is distributed widely through the tissues with concentrations in the prostate, feces, bile, lung, neutrophils and macrophages exceeding serum concentrations, while in the saliva, bones and cerebrospinal fluid the concentrations are lower. Hepatic metabolism forms active metabolites; the half-life of elimination varies from 3 to 5 hours and about 20 to 40% of the drug is bound to plasma proteins. Excretion occurs in both the urine (30 to 50%) and feces (15 to 45%).[10] The drug crosses the blood brain barrier in varying amounts.

### Indications

Like metronidazole, it is indicated for the treatment of active CD.

### Posology

Dose of 500 mg every 12 hours for 6 to 12 weeks.[7,8]

### Administration

The presence of food does not significantly affect absorption. However, with dairy products, a 1 hour interval is suggested in administration.

## Drug interactions

There is an interaction with warfarin and oral hypoglycemic agents, with changes in their plasma levels. Antacids with calcium, magnesium and aluminum greatly reduce the absorption of ciprofloxacin. Concomitant use with simvastatin should be monitored because of the risk of myopathy or rhabdomyolysis.[7,10]

## Precautions

It should be used with caution in patients with severe kidney disease, in the use of tricyclic antidepressants, antipsychotics and/or antiarrhythmics due to the risk of prolonging the AT interval.

## Adverse effects

These are rare and the most common are skin rash, diarrhea, nausea, vomiting, headache, increase in transaminases and abdominal pain. Monitoring the development of signs of peripheral neuropathy caused by fluoroquinolones, both orally and intravenously, is required, discontinuing use immediately in case they occur.[14]

## Corticosteroids

Corticosteroids, steroid hormones, are divided into mineralocorticoids and glucocorticoids, according to the sodium retention ratio and effects on the metabolism of carbohydrates and duration of action. Glucocorticoids have important anti-inflammatory and immunosuppressive actions. They are produced and secreted by the adrenal cortex and their circulating concentration is regulated by adjusting the hypothalamic-pituitary-adrenal axis, where release is conditioned by stress (positive feedback), and inhibition (negative feedback) is performed in response to the circulating hormones.[15]

They are recommended for treating patients with active CD or UC. The mechanism of action, in general, is based on the suppression of inflammatory processes in their early stages, inhibiting the capacity to recruit monocytes and neutrophils to the inflammatory site. They also reduce the dilation of the microvasculature and inherent permeability, reducing edema and migration of leukocytes to the inflammatory process. In late-onset manifestations, they interfere with the activation of fibroblasts, vascular proliferation and collagen deposition, as well as influencing the immune response of T lymphocytes, reducing the release of mediators, inhibiting the activation of surface molecules and cytokines by reducing the activity of nuclear factor kappa B (NF-kβ).[15,16] They also present properties of inhibiting the synthesis of prostaglandins and leukotrienes by inhibiting the release of arachidonic acid from phospholipids.

In CD, they are not effective in maintaining remission or the treatment of fistulas, but can be used during pregnancy in the control of active disease.[11] They are not indicated as maintenance therapy, due to the severe and irreversible side effects with prolonged treatment. These effects include electrolyte disorders, osteoporosis, aseptic necrosis, myopathy, peptic ulcers, cataracts, endocrine disorders, infections and psychiatric disorders.[8]

The adverse effects that can occur with the use of corticosteroids for short periods are mood changes, insomnia, dyspepsia, weight gain, edema, elevated glucose levels, acne and swelling in the face (moon face).

Corticosteroids can be administered orally (prednisone, prednisolone and budesonide), topically through the rectum (prednisone, budesonide, methylprednisolone and hydrocortisone) and intravenously (methylprednisolone and hydrocortisone). Rectal topical drugs are indicated for the treatment of disease related to the rectum and descending colon, and intravenous use is indicated for the treatment of severe or serious illness.

The dosage should be appropriate to the clinical situation of each individual. However, after the therapeutic response, there should be a gradual reduction of the dose of 10 mg/week until 0.5 mg/kg/day, and then 5 mg/week until complete withdrawal. Patients using corticosteroids for long periods should receive supplemental calcium and vitamin D, and an ophthalmologic evaluation is recommended.[8] Among corticosteroids, prednisone and methylprednisolone can be highlighted.

**Prednisone**

Formulations

It is found in the form of an immediate-release tablet.

Pharmacokinetic parameters

It presents a peak plasma concentration within 1.3 to 2 hours after oral administration and bioavailability of 92%, with about 70% bound to plasma proteins with a distribution volume of 0.4 to 1 L/kg. It undergoes hepatic metabolism, where it is completely metabolized to form the active metabolite prednisolone, which has a concentration 4 to 6 times greater than prednisone. The prednisolone then also undergoes hepatic metabolism, with sulfates and conjugated glucoronide released in order to be excreted via urine. The elimination half-life of prednisone is 2 to 3 hours.[10]

### Indication
It is indicated for the treatment of moderate to severe CD and severe UC.[11,12]

### Posology
The initial dose may vary from 1 to 2 mg/kg/day or 40 to 60 mg/day, 1 or 2 times/day orally.[8,11,12] After clinical improvement, it should be reduced by 5 to 10 mg/week to 20 mg/day, then reduced to 2.5 to 5 mg/week until complete withdrawal.[11]

### Drug interactions
The majority are related to interference with cytochrome P450 metabolism. Some drugs that interact with prednisone are antifungal medication and antiretroviral protease inhibitors, which may increase the concentrations of prednisone and, thereby, promote an increase in adverse effects (e.g. Cushing's syndrome).

The use of fluoroquinolones should be monitored for signs of tendon rupture, particularly in the elderly. Warfarin can have its anticoagulant effects altered by co-administration with prednisone.

### Precautions
Prednisone is the most recommended corticosteroid for pregnant women.[11] However, it should be used with caution in patients with a history of hepatic problems, osteoporosis, those using oral anticoagulants or with hypersensitivity to the drug.

### Budesonide
Budesonide is a synthetic corticosteroid with potent glucocorticoid action. Due to the extensive hepatic metabolism, it is associated with a lower rate of

adverse effects associated with corticosteroids, including a lower reduction in bone mineral density.[8]

## Formulations

It is found in the form of an enema for rectal topical use (dispersible tablet) and extended release ileal capsules for oral use. The extended release capsules contain gastro-resistant granules for ileal release, which are practically insoluble in gastric juice. They present the property of extended release in order to adjust the release of budesonide in the ileum and the ascending colon.

## Pharmacokinetic parameters

Budesonide has a bioavailability of about 10 to 20% after oral administration, 85 to 90% bound to plasma proteins and the half-life is 2 to 3.6 hours. The volume of distribution is approximately 2 to 3 L/kg and the maximum plasma concentration is reached in 1.5 hours (rectal) and 3.5 hours (oral). It undergoes extensive hepatic metabolism, forming two metabolites with low glucocorticoid action: 6-beta-hydroxybudesonide and 16-alpha-hydroxyprednisolone with majority excreted by the kidneys (60%) and the remainder through the feces (15.1 to 19.6%) and bile.[10]

## Indications

It is indicated as an anti-inflammatory treatment for mild to moderate CD involving the ileum or colon. It is not recommended in maintenance treatment of the disease and in the treatment of UC (topical).[15]

## Posology

Extended release preparations induce remission of mild to moderate CD in 50 to 70% of cases, at the initial dose of 9 mg for 8 to 16 weeks. After the initial treatment, the dosages are reduced at the rate of 3 mg, using 6 mg per day for 3 months.[8,10]

Likewise, for inducing remission of symptoms of mild to moderate UC, the dose is 9 mg/day for 8 weeks.[10] The appropriate release of the amount of the drug occurs in the specific inflamed portion of the intestine, minimizing the systemic side effects of the extensive first pass metabolism that inactivates the derivatives.[12]

For patients that have developed steroid dependence, such as in the case of prednisone, and to prevent recurrence of active disease after surgery, the recommended dosage is 6 mg, once/day. Enemas are indicated for the topical treatment of distal UC, at a dose of 2 mg (suspension 0.02 mg/mL) once/day in the evening for 4 weeks.

### Administration

The extended release capsules should be administered whole with water, once/day in the morning. In case of problems with swallowing the capsules, they can be opened and their contents mixed in apple sauce without being ground or crushed for immediate use.[10] For enemas, after preparation of the dispersible tablet in the diluent solution, use is immediate. The use of enemas during pregnancy and lactation is not recommended.

The precautions for the use of budesonide follow the same recommendations for the use of other corticosteroids, such as monitoring patients with a history of diabetes, glaucoma, liver problems, hypertension, osteoporosis and others. For enemas, these include lactose present in the formulation and should be avoided in patients intolerant to lactose.

### Adverse effects

The commonly related effects are headache, nausea, vomiting, skin rashes, muscular pain, agitation and diarrhea.

### Hydrocortisone

Hydrocortisone may be used in cases of ulcerative colitis associated with aminosalicylates or used in isolated form.

### Formulations

It is found for intravenous use and ointment/foam for topical rectal use. The foam is not available in Brazil. It is also available in preparations in the form of suppositories and rectal suspension (enema).

### Pharmacokinetic parameters

Following administration, it is rapidly absorbed, with 90% of the drug binding to plasma proteins. It presents extensive hepatic metabolism and renal excretion.

Indications

It is indicated in the treatment of mild to moderate UC.

Posology/administration

The use of hydrocortisone is combined with aminosalicylates at a dose of 100 mg with 1 hour of retention, or 80 mg in the form of topical rectal foam, once or twice/day, both for 2 or 3 weeks and, in severe cases, for 2 to 3 months.[10,12] In severe UC, the same dosage, 100 mg twice/day, may be administered in the form of an enema for emergency treatments or tenesmus; or parenterally at a dose of 300 mg/day, divided every 8 hours, or continuous infusion with response between 7 and 10 days.[8,11]

Drug interactions

Patients using oral anticoagulants and potassium-sparing diuretics, which may potentiate hypokalemia, should be monitored.

Precautions

Fasting blood glucose and serum sodium should be monitored. When treatment with hydrocortisone is prolonged for more than 72 hours, there is a risk of hypernatremia. In this case, it is recommended to replace the hydrocortisone with methylprednisolone, which produces little or no sodium retention.

Adverse effects

These are similar to those related to other corticosteroids.

**Methylprednisolone**

This presents a potent anti-inflammatory action, with advantages even at lower doses. Another advantage is presenting a significant difference in the anti-inflammatory and mineralocorticoid activity, with less water and sodium retention.[11]

Formulations

It is found in the form of succinate for the preparation for intravenous, intramuscular or intrarectal (enema) administration, and acetate for intramuscular and intrarectal administration.

Pharmacokinetic parameters

Methylprednisolone is well absorbed and has a peak concentration within 31 minutes, with a half-life of 2 to 3 hours and volume of distribution of 1.5 L/kg. It presents extensive hepatic metabolism and renal excretion. The clearance in obese patients is lower.[10] It crosses the placental barrier.

Indications

It is indicated for the treatment of moderate to severe CD and UC.

Posology/administration

In both CD and UC, the usual dose ranges from 10 to 40 mg/dose via the intravenous or intramuscular route, based on the clinical response. If higher doses are required for severe cases, 30 mg/kg intravenously in infusion is recommended from 30 minutes, every 4 or 6 hours, for 48 to 72 hours.[7,10] In severe UC, in the event of improvement with parenteral use, oral prednisone should be started or, otherwise, if no significant improvement occurs, surgery or treatment with cyclosporine or tumor necrosis factor anti-(anti-TNF) should be considered. For the treatment of UC via the rectal route, methylprednisolone can be administered as a retention enema with a dose of 40 to 120 mg, for 2 weeks or more.

Drug interactions

Similarly to prednisone, methylprednisolone can interact with antifungal medication and antiretroviral protease inhibitors, increasing plasma concentrations of methylprednisolone and also altering the effects of warfarin.

It may also increase plasma levels of cyclosporin and tacrolimus and increase the effects of hypokalemia with potassium-sparing diuretics. Antacids, anticonvulsants and rifampicin can decrease serum concentrations of methylprednisolone.[10]

Adverse effects

The most common adverse effects reported are hypertension, hypernatremia, gastrointestinal disorders and muscle weakness.

## Immunomodulators (immunosuppressants)

Immunomodulators are drugs that reduce the underlying inflammation of CD and UC. They are associated with the development of high blood pressure, hyperglycemia, liver disease and nephropathy. Another relevant factors is that immunosuppressive drugs interact with other drugs, and may alter the expected therapeutic effect.

Among the immunomodulators suitable for the treatment of IBD, we can highlight zathioprine, mercaptopurine, methotrexate (MTX), cyclosporine, and tacrolimus.

Azathioprine, mercaptopurine and methotrexate are examples of antimetabolites derived from thiopurine drugs, which act by inhibiting the activity of lymphocytes T and B and natural killer cells. Cyclosporine and tacrolimus, initially studied as antimicrobial treatment, were subsequently found to have immunosuppressant properties, by inhibiting the synthesis of interleukin 2 in the G0 phase of the cell cycle.[16]

The precautions with the use of these drugs include weekly hematologic monitoring for the first few weeks, then every 2 weeks for 8 weeks, and in the maintenance phase every 1 to 3 months. Monitoring of plasma serum levels of the drug should be conducted, especially regarding interactions with drugs and food, as well as monitoring of hepatic and renal function and electrolytes.[8]

### Azathioprine

Azathioprine (AZA) is an imidazolyl derivative of 6-mercaptopurine (or mercaptopurine) and acts as an immunosuppressant antimetabolite. It is converted in a non-enzymatic manner into 6-mercaptopurine.

Formulations

It is found in the form of tablets. In relation to pharmacokinetic parameters, azathioprine is well absorbed orally, with oral bioavailability of 47.4% and rectal bioavailability of 1.3 to 5.3%. Around 30% of azathioprine and mercaptopurine bind to plasma proteins; the time to reach maximum concentration occurs within 1 to 2 hours. It undergoes extensive hepatic metabolism by oxidation and methylation reactions, forming the active metabolites 6-mercaptopurine and 6-thioguanine nucleoside. The excretion has a half-life of elimination of

5 hours.[10] It crosses the placental barrier and is excreted in breast milk and is not recommended for use by pregnant women.

### Indications
It is indicated for the treatment of CD in the maintenance of remission or reduction in steroid use, and in the treatment of ulcerative colitis.

### Posology/administration
For DC, the dose is 2 to 3 mg/kg/day orally, and for UC, 1.5 to 2.5 mg/kg/day orally. In the case of gastrointestinal symptoms it is recommended to administer the tablets with food or shortly after, or in divided doses.[10,13]

### Adverse effects
The most common are nausea and vomiting. Some products may contain lactose in the composition and use in patients intolerant to lactose should be conducted with caution.

## Mercaptopurine
### Formulations
It is found in the form of tablets.

### Mechanism of action
Mercaptopurine (6-MP) is an antimetabolite analog of purines, cycle-specific to the S phase of cell division, which acts by inhibiting DNA synthesis, and, to a lesser extent, RNA.

### Pharmacokinetic parameters
Around 50% of the drug is absorbed variably by the oral route, with bioavailability ranging between 5 and 37%, due to pre-metabolism of mercaptopurine in the intestine and in the liver by xanthine oxidase. It achieves peak plasma concentrations within 1 to 2 hours after oral administration. In relation to metabolism following oral administration, mercaptopurine undergoes extensive first pass metabolism in two main ways: one involves methylation of the sulfhydryl group and subsequent oxidation of the methylated derivatives thereof,

and the other is oxidation by xanthine oxidase forming inactive metabolites (6-methylmercaptopurine and 6-thiouric acid) and, possibly, active metabolite (6-thioguanine).[4,7] Around 19% binds to plasma proteins with a volume of distribution of 0.9 L/kg. Excretion is renal (46%) during the first 24 hours, with half-lives of elimination of 21 to 90 minutes.[10]

## Posology

The treatment of active CD, maintenance of remission or decrease in use of corticosteroids is from 1 to 1.5 mg/kg/day orally, with the dosage adjusted according to monitoring tests. For the treatment of UC, the initial dose is 50 mg/day, which can be adjusted higher or lower according to clinical response and tolerance. For the remission maintenance phase, the recommended dose is 1.5 mg/kg/day.[10,13]

## Administration

It should be administered one hour before or two hours after food/milk, given that there is a 30 to 50% reduction in the absorption of the drug in the presence of food and dairy products.

## Drug interactions

There is an interaction with allopurinol, since this inhibits xanthine oxidase, increasing the immunosuppressive and toxic effects of mercaptopurine. Patients on allopurinol for treating gout or hyperuricemia should be monitored, and mercaptopurine doses reduced by 25 to 33% in relation to the usual dosage.[7,10]

Co-administration with mesalazine, olsalazine or sulfasalazine can trigger bone marrow suppression by inhibiting the enzyme thiopurine-methyltransferase (TPMT), an enzyme involved in the metabolism process by aminosalicylates. Therefore, concomitant use should be monitored, considering a reduction of the mercaptopurine dose.

Use associated with warfarin can reduce the anticoagulant activity and should be monitored, adjusting the dose of anticoagulant. With MTX at low doses (20 mg/m$^2$), the area under the curve (AUC) of mercaptopurine can increase around 31%, but this increase does not significantly affect the treatment if the patient is monitored.

Precautions

Hepatotoxicity and hyperuricemia are effects to be monitored with the use of the drug. Hepatotoxicity may occur at any dose; however, it is more frequent when exceeding a daily dose of 2.5 mg/kg. The formulation may contain lactose in the composition and use in patients intolerant to lactose should be conducted with caution.

Adverse effects

The most common effects reported with the use of mercaptopurine and which should be monitored are leukopenia, thrombocytopenia, hepatotoxicity and hyperuricemia.[7,10]

**Methotrexate**

Formulations

It is found in injectable form for intramuscular and subcutaneous administration.

Mechanism of action

MTX is an antimetabolite that interferes with DNA synthesis, cellular repair and replication by inhibiting the enzyme dihydrofolate reductase which is responsible for the reduction of dihydrofolic acid into folinic acid (active intracellular metabolite).[10] At low doses, it displays anti-inflammatory activity, including inhibition of TNF expression in monocytes and macrophages.[8]

The mechanism of action in CD is uncertain, with some possibilities. One is increasing the extracellular concentration of adenosine with high anti-inflammatory activity. Another probable mechanism includes the inhibition of vital methylation reactions for cellular activities and replication and cell apoptosis activated by T cells.

Pharmacokinetic parameters

Bioavailability is 76 to 100%, when administered intramuscularly; about 50% bins to plasma proteins. It undergoes hepatic metabolism, forming active metabolites, polyglutamates and 7-hydroxymethotrexate; most of excretion (48 to 100%) is renal, with small amounts in feces and bile.[10]

### Indications

It is indicated for the treatment of mild to moderate CD and treatment of fistulas.

### Posology/administration

The recommended dosage for treatment for inducing remission or reduction in steroid use is 25 mg, once/week, intramuscularly or subcutaneously. For maintenance of remission, a 15 mg dose, once/week, intramuscularly, is indicated.[13] In intolerant patients or those resistant to azathioprine or mercaptopurine, MTX should begin at a dose of 25 mg/week via the intramuscular route.[11]

### Drug interactions

MTX can reduce the immunological response of vaccines. Serious allergic reactions in case of concomitant use with live virus vaccines may also occur. The effects of oral anticoagulants, such as warfarin, may be increased with concomitant use of MTX, increasing the risk of bleeding.

### Precautions

Use is contraindicated during pregnancy, due to its teratogenic effects. As a folate antagonist, supplementation with folic acid at a dose of 1 to 2 mg/day is recommended for the use of MTX for the purpose of preventing nausea, vomiting, abdominal pain and other effects.[7] Hematological and liver monitoring should be performed every 1 to 3 months.[8]

### Adverse effects

The most common in patients with CD are skin rash, abdominal pain, diarrhea, headache, nausea, vomiting, increase in transaminases and photosensitivity.[7,8,10]

Serious secondary effects are rare and include leukopenia, hypersensitivity interstitial pneumonitis and liver fibrosis. In any situation the use of the drug should be discontinued.[7,12]

## Cyclosporin

Cyclosporin is a drug with immunosuppressive properties, originally isolated from the fungus *Tolypocladium inflatum*.

## Formulations

It is found in the forms of a gelatin capsule (soft), gelatin capsule in microemulsion, oral solution and solution for parenteral use (intravenous).

The microemulsion formulations enables less variability in the pharmacological kinetics of the drug with a more consistent absorption profile, allowing less influence of the presence of food.

## Mechanism of action

Cyclosporin acts by inhibiting the production and release of interleukin 2, reversibly reducing the activity of T-lymphocytes.[15]

## Pharmacokinetic parameters

The oral absorption of cyclosporine is erratic and variable, occurring in the duodenum and jejunum. The modified pharmaceutical form (microemulsion) is less dependent on food and bile acids, having a 30% higher absorption compared to the unmodified form. Around 90% of cyclosporin binds to plasma proteins, mainly lipoproteins. It is widely distributed by blood cells, kidney, liver, pancreas, and synovial fluid and has a volume of distribution between 3.9 and 4.5 L/kg. It undergoes extensive hepatic metabolism, producing at least 25 metabolites. Excretion takes place through urine and feces (6%), with a half-life of 19 hours for the unmodified form (gelatin capsules), and 8.4 hours for the modified form (capsule in microemulsion).[10,16]

## Indication/posology

Cyclosporin is indicated for the treatment of severe or fulminant UC at a dose of 2 to 4 mg/kg/day as a continuous infusion, passing onto oral administration as soon as possible, at a dose of 2.3 3 mg/kg every 12 hours.[13]

In severe to fulminant and fistulizing CD, the recommended dosage is 2 to 4 mg/kg/day as continuous infusion for 1 to 2 weeks, passing onto oral administration as soon as possible at 6 to 8 mg/kg/day (maximum 10 mg/kg/day) for 4 to 6 months.[10,17]

Administration

Cyclosporin must always be administered in the same way in relation to times and as to the presence of food in order to avoid plasma fluctuations of the drug. The gelatin capsules should be administered whole with water.[10]

Drug interactions

Both the plasma concentration of cyclosporin and the associated drug may be changed. Among the examples of significant interactions, we can mention concomitant use of infliximab, which may decrease plasma concentrations of cyclosporin, requiring dosage adjustment.

MTX may increase in serum exposure in the presence of cyclosporin, which may lead to levels of toxicity.

Concomitant use with warfarin can reduce anticoagulant activity. Glucocorticoids, metronidazole and clarithromycin are examples of drugs that may increase the plasma concentrations of cyclosporin, leading to the onset of adverse effects and requiring dosage adjustments.[10]

Precautions

The cyclosporin dose should be adjusted according to serum, which then must be regularly monitored. Its use requires close monitoring of blood pressure, blood count and creatinine.[13]

Cholesterol, uric acid, potassium and magnesium should also be monitored. Cyclosporin should be reduced by 25 to 50% if there is any relevant change in any of these values in relation to pretreatment levels; if this reduction is not effective or the change is serious, cyclosporin should be discontinued.[11]

Adverse effects

The problems derived from use of cyclosporin include possible serious side effects that are relatively frequent, affecting 50% of patients. Among the most cited adverse effects are nephrotoxicity, hypertension, hepatotoxicity, diarrhea, abdominal pain, tremors, numbness and edema.[13,15]

## Tacrolimus

Tacrolimus has immunomodulatory properties similar to cyclosporin. It is 100 times more potent than cyclosporin and is not dependent on bile or the integrity of the mucosa for absorption. These properties enable good oral absorption, despite low absorption in the intestine in CD.

### Formulations

It is found in the form of immediate release oral capsules and solution for parenteral use.

### Mechanism of action

Tacrolimus inhibits the activation of T lymphocytes, possibly by binding to the intracellular protein FKBP-12.[16]

### Pharmacokinetic parameters

Absorption is incomplete and variable from the gastrointestinal tract, and is affected by the presence of food, reaching between 5 and 67%, with oral bioavailability between 17 and 31%. The peak concentration following oral administration occurs between 30 minutes and 6 hours. Plasma protein binding is approximately 99%, distributed widely by erythrocytes, breast milk, lung, kidney, pancreas, liver, placenta, heart and spleen.

It undergoes extensive hepatic metabolism by cytochrome P450, originating 8 possible metabolites, with the main ones being 13-desmethyl tacrolimus and 31-desmethyl tacrolimus (active). Excretion is conducted 92.6% in feces with less than 1% in the urine and bile. The half-life of elimination is variable, between 17 and 31 hours.[10,16]

### Indications

Tacrolimus is indicated in fistulizing CD, in combination with other immunosuppressants.

### Posology

A dose of between 0.2 and 0.27 mg/kg/day, every 12 hours has been reported for the treatment of fistulas in CD.[17]

Administration

Tacrolimus should be administered while fasting since the presence of food interferes by decreasing the rate and extent of absorption by approximately 27%.[10]

Drug interactions

As tacrolimus is primarily metabolized by the CYP3A enzyme system, drugs that inhibit these enzymes may decrease the metabolism or increase bioavailability of tacrolimus resulting in increased plasma concentrations. Thus, some clinically relevant interactions have been cited, such as the concomitant use of tacrolimus with clarithromycin, where there is risk of QT prolongation.

Combined use with infliximab or adalimumab may result in reduction of plasma concentrations of tacrolimus, with a possible reduction of therapeutic efficacy. In combination with metronidazole or methylprednisolone (or glucocorticoid), there may be an increase in serum concentrations of tacrolimus, triggering significant adverse effects (nephrotoxicity, hyperkalemia, hyperglycemia), which is reversible with dose adjustment. Tocilizumab can also cause changes in serum concentrations of tacrolimus.[10]

Adverse effects

The most common effects observed during use are hypertension, edema, localized pain, arthralgia, headache, insomnia, fever, rash/pruritis, amended electrolyte feed, diarrhea, nausea, vomiting, dyspepsia, trembling, weakness, and increased creatinine. These effects are reversible when reducing the dose.

## Biological therapy

Biological therapy is often reserved for more serious patients with CD and UC, or those who have not responded to other treatments. Patients who respond to biological therapies show significant improvement in clinical symptoms, improvement in quality of life, reduced disability, fatigue and depression, and fewer surgeries and hospitalizations.

The intestinal mucosa is in equilibrium with the inflammatory mechanisms regulated by the production of proinflammatory cytokines, such as TNF-alpha and others. However, in IBD, this mechanism is unbalanced. Biological drugs act on specific cytokines by blocking the inflammatory activity.

Infliximab, adalimumab and certolizumab pegol are prominent treatments for IBD.

### Infliximab

Infliximab is a chimeric murine monoclonal antibody targeting the proinflammatory cytokine TNF-alpha.

### Formulations

The drug is found in the form of lyophilized powder for preparation of intravenous solution.

### Mechanism of action

Infliximab binds to TNF-alpha bound to active T cells or other immune cells, promoting apoptosis or destruction of these cells through antibody-dependent cellular toxicity.[18]

### Pharmacokinetic parameters

In CD patients, infliximab starts acting within approximately 1 to 2 weeks and half-life of elimination in adults from 7 to 12 days.[13,17] Distribution occurs predominantly through the vascular compartment, with no excretory route reported.[10,17]

### Indications

It is indicated for the treatment of moderate to severe CD, including fistulizing disease, and moderate to severe UC in adults, with no response to treatment with corticosteroids and/or immunosuppressants, or contraindications/intolerance to such treatments.[10,17] The same criteria apply to the treatment of children between 6 and 17 years old with CD.

### Posology

In combination with classical immunosuppressive drugs, the dose of infliximab for the indications stated is 5 mg/kg as a single dose through slow intravenous infusion (at least 2 hours) in weeks 0, 2 and 6, then a maintenance dose of 5 mg/kg every 8 weeks. The dose may be increased up to 10 mg/day in patients that responded to treatment but lost the response. If there is no response by the 14th week, suspension of treatment should be considered.[13]

Drug interactions

It is known that in inflammatory states, the formation of cytochrome P450 enzymes is suppressed by high levels of cytokines such as TNF-alpha. When administering infliximab, a TNF-alpha antagonist, the formation of cytochrome P450 enzymes may be restored. Thus, possible drug interactions should be monitored, especially drugs metabolized by cytochrome P450 and infliximab due to the risk of loss of therapeutic efficacy caused by plasma variations of these drugs or the need for dose adjustment during therapy with the biological agent. Tacrolimus, warfarin, cyclosporine, sirolimus, phenytoin and pimozide can be cited among the relevant drugs.[10,17]

Precautions

Use is contraindicated in patients with congestive heart failure (NYHA class III-IV), serious infections (tuberculosis, sepsis and opportunistic infections), liver disease, and a history of hematological abnormalities.[16] There is a recommendation that the candidates for therapy with infliximab should be tested for tuberculosis, and patients who test positive should be treated prophylactically with isoniazid.[12] It is also worth considering the empirical use of antifungal agents in patients at risk for opportunistic fungal infections.[13] The concomitant use of infliximab with abatacept, TB vaccine (BCG), certolizumab pegol, natalizumab, vaccines (inactivated or living organisms) and etanercept should be avoided due to the risk of infection.[11]

Adverse effects

Despite the efficiency of infliximab in IBD and it being considered a safe therapy in the long term, some serious side effects can occur, such as acute reactions to the infusion (headache, rash, nausea, diarrhea, arthralgia, cough, dyspnea, flushing, chills, dizziness), manifestations of serious infections, drug induced lupus, delayed hypersensitivity reactions, demyelination, potential increased risk of incidence of non-Hodgkin's lymphoma. However, the cause has not yet been established, as well as in cases of heart failure and death.[12]

The use of medication prior to infusion such as antihistamines, paracetamol and corticosteroids can prevent unwanted reactions. The infusion should be slow, for at least 2 hours or more in the case of patients with hypersensitivity.[13]

**Adalimumab**
Adalimumab is a recombinant human monoclonal antibody of immunoglobulin (IgG1) humanized against TNF-alpha.

Formulations
It is found in the form of a solution for injection for subcutaneous use.

Mechanism of action
It binds to TNF-alpha, blocking its interaction with the cell receptors p55 and p75 (TNF surface receptors) present on the cell surface. Furthermore, it induces lysis of cells that express TNF-alpha.[18]

Pharmacokinetic parameters
After administration, the peak concentration occurs within 5 days, with bioavailability of 64%. As for the volume of distribution (4.7 to 6 L), adalimumab is distributed similarly through the vascular and extravascular fluid.[17] Excretion is decreased in patients over 40 years of age, and the elimination half-life is 14 days, with no excretory route reported.[16] Secretion in breast milk is unknown and the use is not recommended during lactation.

Indications
Adalimumab is indicated for the treatment of moderate to severe CD and UC, in patients with loss of response to or intolerance of conventional therapy, including corticosteroids, azathioprine or mercaptopurine, or loss of response to or intolerance of infliximab.[13,16]

Posology/administration
The administration is subcutaneous, starting with a dose of 160 mg (week 0, and may be administered as four injections on the same day or two injections on 2 consecutive days), followed by a dose of 80 mg (week 2). Subsequently,

the maintenance treatment is 40 mg (week 4). For patients that do not respond to treatment with adalimumab by week 4, the dose of 40 mg can be continued through to week 12. Aminosalicylates, corticosteroids and immunosuppressants may be continued if necessary. In relation to administration, the site of application should be rotated, giving preference to the abdomen and anterior surface of the thigh.[16]

## Precautions

Use with caution in patients with chronic or repeated infections and with left ventricular dysfunction. There are reports of reactivation of tuberculosis with adalimumab, as well as harming defense against malignancies and opportunistic infections. Concomitant use with vaccines with live organisms, abciximab, certolizumab pegol, natalizumab and tacrolimus (topical) should be avoided, as this can cause thrombocytopenia or decreased therapeutic effect. Use with caution in patients with chronic or repeated infections and with left ventricular dysfunction.[13]

## Drug interactions

There is potential interaction with decreased plasma concentrations of tacrolimus, sirolimus, cyclosporine, pimozide.[10]

## Adverse effects

The most common reactions are at the injection site and infections. There are also reports of headache, rash, upper respiratory tract infections, flu-like symptoms, hypertension, nausea, increased alkaline phosphatase, hematuria, arrhythmia, confusion, fever, cellulitis, pancytopenia, dehydration and asthma.

## Certolizumab pegol

### Formulations
It is found in syringes filled with solution for subcutaneous use only.

### Mechanism of action
Certolizumab pegol is constituted by a Fab fragment of a humanized recombinant antibody against TNF-alpha bound to polyethylene glycol (PEG), selectively neutralizing TNF-alpha.

Pharmacokinetic parameters

After administration, the peak concentration is between 54 and 171 hours, with bioavailability between 76 and 88% (average of 80%) with an elimination half-life of 14 days.[16] Excretion of certolizumab is renal (17 mL/h).[10] Secretion in breast milk is unknown and its use is not recommended during lactation.[3]

Indications

It is indicated for the treatment of moderate to severe CD, with inadequate response to conventional therapy, either as a treatment for induction or maintenance, regardless of previous therapies, associated with a good safety profile.

Posology/administration

This drug is administered subcutaneously at a dose 400 mg at weeks 0, 2 and 4. The maintenance of treatment is performed with a dose of 400 mg subcutaneously every 4 weeks. The site of application should be rotated, giving preference to the abdomen and anterior surface of the thigh. Do not administer to sites where the skin is reddened or has hardened tissue.[13]

Drug interactions

There are no reports of significant drug interactions; MTX, corticosteroids, nonsteroidal anti-inflammatory drugs (NSAIDs), analgesics, aminosalicylates or anti-infectives did not change and did not have their pharmacokinetics altered by possible interaction with certolizumab pegol.

Precautions

Patients receiving certolizumab have an increased risk of serious infections, especially if they are receiving other immunosuppressive agents such as methotrexate or corticosteroids. There are reports of reactivation of tuberculosis with the use of certolizumab pegol. Consider the risk/benefit for use in patients with chronic infections. Lymphoma and other malignancies may also occur in patients with hepatitis B using certolizumab.[13] Use with caution in patients with heart failure or blood disorders.

## Adverse effects

The most common are hypersensitivity reactions, nasopharyngitis, urinary tract infection, arthralgia, dyspnoea, rash, headache, nausea, hypertension, fever, fatigue and erythema at the injection site.[10,17]

## REFERENCES

1.  Adams SM, Bornemann PH. Ulcerative colitis. AM Fam Physician 2013; 87(10):699--705.

2.  Cheifetz AS. Management of active Crohn disease. JAMA 2013; 309(20):2150-8.

3.  Ford AC, Moayyedi P, Kirsner SBHJB. Ulcerative colitis, clinical review. BMJ 2013; 346:1-9.

4.  Katz S. My treatment approach to the management of ulcerative colitis. Mayo Clin Proc 2013; 88(8):841-53.

5.  Mehta SJ, Silver AR, Lindsay JO. Review article: strategies for the management of chronic unremitting ulcerative colitis. Aliment Pharmacol Ther 2013; 38:77-97.

6.  Ordás I, Eckmann L, Talamini M, Baumgart D, Sandborn W. Ulcerative colitis. Lancet 2012; 380:1606-19.

7.  Up to Date. Inflammatory bowel disease. Available at: www.uptodate.com/contents/search. Accessed in: 01/2014.

8.  Kenneth R, McQuaid MD. Gastrointestinal disorders. In: Papadakis M, McPhee SJ, Rabow MW. Current medical diagnosis & treatment. 53.ed. Maidenhead: McGraw-Hill Medical, 2014.

9.  Rang HP, Dale MM, Ritter JM, Flower RJ, Henderson G. Trato gastrointestinal. In: Rang HP, Dale MM. Farmacologia. 7.ed. Rio de Janeiro: Elsevier, 2011; p.360-71.

10. DRUGDEX® System. Greenwood Village (USA): Truven Health Analytics. Available at: www.micromedexsolutions.com.

11. Brasil. Ministério da Saúde. Doença de Crohn/Retocolite ulcerativa. Protocolos clínicos e diretrizes terapêuticas. 2010; 2:125-46.

12. Hardman JG, Limbird LE. Goodman & Gilman. As bases farmacológicas da terapêutica. 11.ed. Maidenhead: McGraw Hill, 2006.

13. Dana WJ, Fuller MA, Goldman M et al. Drug information handbook – a comprehensive resource for all clinicians and healthcare professionals. 22.ed. Hudson: Lexicomp, 2013-2014.

**14.** US Food and Drug Administration (FDA). Drug Safety Communication. FDA requires label changes to warn of risk for possibly permanent nerve damage from antibacterial fluoroquinolone drugs taken by mouth or by injection. 2013. Available at: www.fda.gov/Drugs/DrugSafety/ucm365050.htm.

**15.** Fuchs DF, Wannacher L. Farmacologia clínica: fundamentos da terapêutica racional. 4.ed. Rio de Janeiro: Guanabara Koogan, 2010.

**16.** Silveiro S, Guimarães JF, Soares AA. Corticosteroides sistêmicos. In: Barros E. Medicamentos na prática clínica. Porto Alegre: Artmed, 2010. p.304-19.

**17.** American Society of Health-System Phamacists. AHFS Drug Information 2013. Bethesda; 2013.

**18.** Ordás I, Mould DR, Feagan BG, Sandborn WJ. Anti-TNF monoclonal antibodies in inflammatory bowel disease: pharmacokinetics-based dosing paradigms. Clin Pharmacol Ther 2012; 91(4):635-46.

# SURGICAL TREATMENT IN ULCERATIVE COLITIS

MAGALY GEMIO TEIXEIRA

## INTRODUCTION

The probability of surgical indication in patients with ulcerative colitis (UC) is directly proportional to time of follow-up and extent of disease. Approximately one third of patients with UC will undergo surgery over the course of the disease.

## INDICATIONS FOR SURGICAL TREATMENT

Surgical treatment may be elective or emergency. The main indications for elective procedure are presented below.

### Clinical intractability

This is the most frequent indication. It is characterized by:

- lack of response to properly conducted medical treatment;
- proper answer, but at the expense of significant side effects;
- corticosteroid dependence;
- drug intolerance;

- need for frequent hospitalization and blood transfusions;
- progressive loss of quality of life.

The presence of extra-intestinal manifestations (e.g., pyoderma gangrenosum) can corroborate the indication for surgical treatment. Other extraintestinal manifestations, such as sclerosing cholangitis and ankylosing spondylitis, have independent course and do not change with treatment of colonic disease.

### Suspected or confirmed cancer

The colitis is considered a premalignant condition. The presence of cancer becomes more frequent after 8 to 10 years of disease in cases of universal involvement.

The ability to predict the presence and location of cancer in UC is of fundamental importance to indicate and perform surgery. The study of 348 surgical specimens with preoperative dysplasia showed the presence of cancer in 51 patients (15%), and dysplasia in only 172 (49%). In the cases of high-grade dysplasia, cancer was found in 29% of patients, compared with 3% in the cases of low-grade dysplasia (3 fold increase). Dysplasia-associated lesion or mass (DALM) was associated with cancer in 25% of patients, compared with 8% in patients with flat lesions (P < 0.001). In low-grade dysplasia, there was no difference when associated with flat lesions or mass.

In conclusion, the risk of cancer for patients with high-grade dysplasia or elevated lesion (DALM) is substantial and surgical treatment should be considered, since this is the only option that can eliminate the risk of cancer.[1]

### Growth retardation

This indication should be done while the patient is still able to recover growth.

UC can manifest initially as acute and severe symptoms requiring hospitalization and even emergency surgery. However, it should be noted that even when patients have a confirmed diagnosis of inflammatory bowel disease (IBD), other acute situations not necessarily related to the disease (e.g., intestinal infections) also occur. Also bear in mind that these patients have extraintestinal manifestations that can lead to acute symptoms (e.g., renal colic and acute cholelithiasis). The drugs used in treatment can also be responsible for complications, such as peptic ulcer perforation and pancreatitis. Obstructive acute abdomen may be caused by the adhesions resulting from surgical procedures performed in the past or the presence of tumor. Patients with IBD are at increased risk for developing acute mesenteric ischemia.[2]

Indications for emergency procedures are listed below.

## Massive bleeding

This situation is uncommon. Other causes of bleeding, such as bleeding ulcer in patients treated with corticosteroids, should be excluded.

## Obstruction

Usually caused by neoplasia. However, obstruction may be high. This situation occurs when there is a perforation of the colon, blocked by small bowel loops.

## Toxic megacolon

Signs and symptoms include increased number of bowel movements, which become liquid and accompanied by blood, fever over 38.6°C (101.5°F), tachycardia above 100 bpm, colicky abdominal pain, tenesmus, paleness, leukocytosis, dehydration or shock. Abdominal examination can show distension, pain on palpation (localized or diffuse) with abdominal noises or not, depending on the developmental stage at which the patient is examined. Dilatation of the upper transverse colon greater than 7 cm occurs due to inflammation and destruction of the colonic musculature, and/or myenteric and submucosal nerve plexuses.

Transmural ulceration can determine colon perforation in the absence of colonic dilatation, which may or may not be blocked by the omentum or other adjacent structures, or leak into the peritoneal cavity, causing peritonitis.

Endotoxemia or shock is the result of toxins released into the systemic circulation and results in hypoperfusion, tissue ischemia, multiple organ failure, hypotension, and hyperdynamic state. The most common laboratory findings are anemia, leukocytosis > 10,500/mm³, hypoalbuminemia < 3 g/dL, hypogammaglobulinemia, hyponatremia, hypokalemia, hypochloremia, and hypoprothrombinemia.

A plain abdominal radiography can characterize the thickening of haustra, which tend to disappear as the expansion of the colon is accentuated. Dilation usually begins at the splenic flexure and extends toward the proximal transverse colon, either reaching the cecum or not. If perforation occurs, it can be diagnosed by the presence of free air or the outline of the inner and outer walls of the colon, indicating blocked perforation.

In case of suspected toxic megacolon, barium enema and colonoscopy are contraindicated because they can cause perforation or, if already present, unblock the lesion with consequent leakage of barium, causing peritonitis with very serious consequences.

Computed tomography is useful in detecting abdominal complications, thus contributing to the management of these patients. Continuous monitoring of the patient is mandatory and should include clinical, laboratory and radiological examinations every 12 hours.

In recent years, a higher incidence of UC has been observed in patients above 60 years of age. The prognosis of patients operated on urgently is poor, with mortality rates of up to 26.7% compared to 0.88% undergoing elective surgery. The most common causes of death were related to infection of the respiratory tract and sepsis. For this reason, clinicians and surgeons should work together to optimize the time of surgical indication in this group of patients.[3]

## PREOPERATIVE PREPARATION

Correction of metabolic, electrolyte and nutritional disorders, as well as blood transfusion should be performed, if necessary.

In broad-spectrum antibiotic therapy, patients with IBD are immunosuppressed by their own illness or the use of drugs, and therefore are more susceptible to infectious complications.

Nonsteroidal anti-inflammatory drugs (NSAIDs) should be avoided due to risk of exacerbating IBD.

When the patient is already treated with corticosteroids, the dose should be increased during the acute phase, both in order to treat the flare and because of the risk of adrenal insufficiency, if the patient has been using this medication for a long time.

Prophylactic measures should be taken for gastric protection, such as $H_2$, blockers, and deep vein thrombosis, using pneumatic compression boots or low-dose heparin subcutaneously.

## SURGICAL TREATMENT

Surgery can be performed by laparotomy or videolaparoscopy, video-assisted or by using a hand portal device.[4,5] The choice of access route will depend on the surgeon and his or her familiarity with the technology involved, in addition to the clinical condition of the patient.

### Subtotal colectomy and ileorectal anastomosis (Figure 28.1)

### Indications

- Rectum preserved: in UC, the rectum the rarely preserved; even in the few cases in which this occurs, there is a possibility that, with time, disease relapses requiring subsequent resection in up to 50% of patients after 10 years;
- non-acceptance of ileostomy by the patient.

### Contraindications

- Rectal cancer: in this situation, amputation of the rectum is required;
- anal incontinence: this situation can be aggravated after performing colectomy.

### Advantages

- Simpler technique; does not require specialists to be performed;
- the fact that the rectum is not dissected prevents nerve damage that could lead to erectile and bladder dysfunction.

### Disadvantages

- Risk of cancer in the remaining rectum;
- the rectum remains inflamed, which implies long-term follow-up and maintenance of clinical treatment.

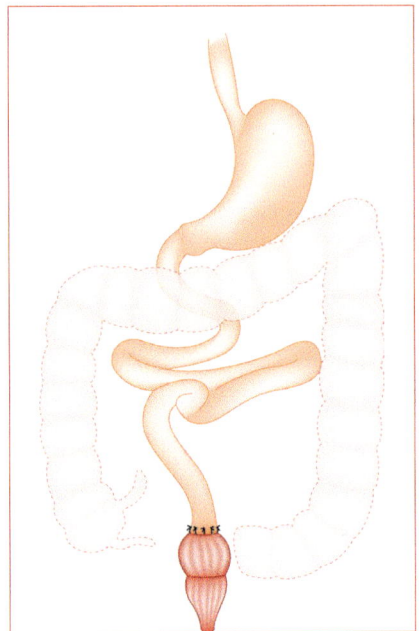

**Figure 28.1** Schematic figure for total colectomy with ileorectal anastomosis.

## Total rectocolectomy and ileoanal anastomosis with ileoanal reservoir (IAR)

This is the procedure of choice. The possibilities for reservoir construction are many, but the most widely used is J-pouch reservoir (Figure 28.2).

### Indications
- Universal UC unresponsive to clinical treatment;
- absence of anal incontinence;
- absence of rectal tumor.

### Contraindications
- Anal incontinence;
- rectal cancer;
- perianal disease manifested as fissures, ulcers, and fistulas, with suspected perianal CD;

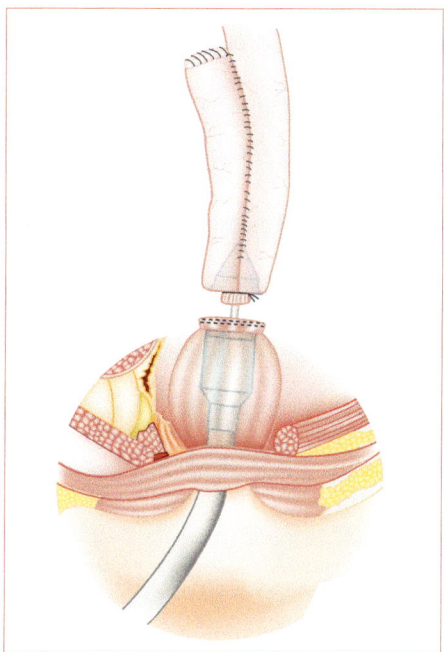

**Figure 28.2** Schematic figure for ileal J-pouch.

- patient's inability to adapt to the new situation. On average, patients undergoing this surgical technique have 5 to 7 bowel movements/day and at least one at night. The sensation of desire to evacuate changes. Difficult belching of gas.

## Advantages
- Eliminates colonic disease;
- prevents the emergence of cancer;
- maintains anal evacuation with continence.

## Disadvantages
- This is a more complex technique and must be performed by specialists familiarized with the procedure;
- the reservoir can be a site of complications during follow-up. The most frequent complication is pouchitis, which can determine the loss of the reservoir and need for definite ostomy.

The fact that the ileal pouch comprises 30 to 40 cm of the terminal ileum may cause these patients to develop vitamin B12 and iron deficiency, malabsorption of bile acids, and other problems. Therefore, they should be monitored on the long run.[6]

### Total proctocolectomy and terminal ileostomy
This procedure was practically abandoned after the introduction of the ileal pouch, but can be used in special situations.

## Indications
- Tumors in the middle and distal third of the rectum;
- failure of other techniques.

## Contraindication
- The only contraindication is the patient refusing to undergo terminal ileostomy.

## Advantages
- Completely eliminates the risk of colon cancer;
- there is no recurrence of colonic disease.

## Disadvantage
- Permanent ileostomy is the main disadvantage of the procedure, since it is not well accepted by most patients, especially young people.

### INFLUENCE OF BIOLOGICAL THERAPY IN THE SURGICAL TREATMENT OF ULCERATIVE COLITIS
The introduction of biological medication in the treatment of IBD has raised some questions, discussed below.

### Would preoperative use increase the risk of postoperative complications?
Most authors consider that there is increase in complications.[7,8] However, some have demonstrated that previous use of infliximab increases the incidence of pelvic sepsis.[9] For this reason, they suggest that surgery is

performed in two stages, i.e., colectomy with closure of the rectal stump and ileostomy, followed by proctectomy and pouch construction with or without ileostomy.

### Would it change the surgical conduct?

If using biological therapy improved the conditions of the rectum, it would be possible to indicate colectomy followed by ileorectal anastomosis instead of rectocolectomy with ileal pouch. This technique would mainly benefit young women who want a family, since there may be changes in fertility associated with ileal pouch.

### In cases of severe pouchitis, could it spare the pouch?

Barreiro-de Acosta et al.[10] studied 33 patients with pouchitis who were treated with infliximab and achieved complete response at 26 and 52 weeks, 33 and 27%, respectively, and partial response in 33 and 18%. Thirteen patients were withdrawn from the study for loss of efficacy or adverse effects, demonstrating the possibility of some results in the short and medium term.

Adalimumab may also be an alternative for patients with chronic pouchitis refractory to treatment with infliximab.[11]

### EMERGENCY CONDUCT

In emergency situations, the best approach is subtotal colectomy, closure of the rectal stump and terminal ileostomy. After the patient's recovery, reconstruction of intestinal transit using one of the techniques mentioned above should be planned. The rectum should be left complete in order to facilitate the next measure. However, in the event of major bleeding, the rectum should be removed, and closure performed close to the anal canal.

Systematic review of the literature from 1975 to 2007, including 29 studies with a total of 2,714 patients, of which 1,257 were urgently operated, revealed hospital mortality at 8 and 5.2% within 30 days after surgery, and morbidity at 50.8%. Most complications were infectious or thromboembolic. Indication for treatment of megacolon between 1975 and 1984 reached 71.1% and fell to 21.6% between 1995 and 2005. Mortality fell from 10 to 1.8% in the same periods, respectively.[12]

Although the study has shown a significant drop in the incidence of cases of toxic megacolon, representing improved diagnosis and treatment of IBD, morbidity and mortality are still high, pointing out the severity of this complication. The results of surgical treatment in IBD emergencies, in recent decades, have improved considerably, but are still associated with high levels of morbidity and mortality.

## REFERENCES

1. Kiran RP, Ahmed Ali U, Nisar PJ, Khoury W, Gu J, Shen B et al. Risk and location of cancer in patients with preoperative colitis-associated dysplasia undergoing proctocolectomy. Ann Surg 2013 Apr 10. [Epub ahead of print].

2. Ha C, Magowan S, Accort NA, Chen J, Stone CD. Risk of arterial thrombotic events in inflammatory bowel disease. Am J Gastroenterol 2009; 104:1445-51.

3. Ikeuchi H, Uchino M, Matsuoka H, Bando T, Hirata A, Takesue Y et al. Prognosis following emergency surgery for ulcerative colitis in elderly patients. Surg Today 2014; 44(1):39-43.

4. Holubar SD, Larson DW, Dozois EJ, Pattana-Arun J, Pemberton JH, Ana RR. Minimally invasive subtotal colectomy and ileal pouch-anal anastomosis for fulminant ulcerative colitis: a reasonable approach. Dis Colon Rectum 2009; 52:187-92.

5. Watanabe K, Funayama Y, Fukushima K, Shibata C, Takahashi K, Sasaki I. Hand-assisted laparoscopic vs. open subtotal colectomy for severe ulcerative colitis. Dis Colon Rectum 2009; 52:640-5.

6. Buckman SA, Heise CP. Nutrition considerations surrounding restorative proctocolectomy. Nutr Clin Pract 2010; 25:250-6.

7. Uchino M, Ikeuchi H, Matsuoka H, Bando T, Ichiki K, Nakajima K et al. Infliximab administration prior to surgery does not increase surgical site infections in patients with ulcerative colitis. Int J Colorectal Dis 2013; 28(9):1295-306.

8. Billioud V, Ford AC, Tedesco ED, Colombel JF, Roblin X, Peyrin-Biroulet L. Preoperative use of anti-TNF therapy and postoperative complications in inflammatory bowel diseases: a meta-analysis. J Crohns Colitis 2013; 7(11):853-67.

9. Eshuis EJ, Al Saady RL, Stokkers PC, Ponsioen CY, Tanis PJ, Bemelman WA. Previous infliximab therapy and postoperative complications after proctocolectomy with ileum pouch anal anastomosis. J Crohns Colitis 2013; 7:142-9.

10. Barreiro-de Acosta M, García-Bosch O, Souto R, Mañosa M, Miranda J, García-Sanchez V et al. Grupo joven GETECCU. Efficacy of infliximab rescue therapy in patients with chronic refractory pouchitis: a multicenter study. Inflamm Bowel Dis 2012; 18:812-7.

11. Barreiro-de Acosta M, García-Bosch O, Gordillo J, Mañosa M, Menchén L, Souto R et al. Grupo Joven GETECCU. Efficacy of adalimumab rescue therapy in patients with chronic refractory pouchitis previously treated with infliximab: a case series. Eur J Gastroenterol Hepatol 2012; 24:756-8.

12. Teeuwen PH, Stommel MW, Bremers AJ, van der Wilt GJ, de Jong DJ, Bleichrodt RP. Colectomy in patients with acute colitis: a systematic review. J Gastrointest Surg 2009; 13:676-86.

# SURGICAL TREATMENT IN CROHN'S DISEASE

CARLOS WALTER SOBRADO
WILTON SCHMIDT CARDOZO

## INTRODUCTION

Crohn's disease (CD) is a chronic inflammatory bowel disease (IBD) that can compromise the entire digestive tract, with associated systemic manifestations and which, although described in 1932, has not yet had its pathogenesis fully clarified, or its cure. It progresses in flares, with periods of calm and exacerbation of symptoms, leading to a significant decrease in quality of life.[1-3] Despite its unknown etiology, the distribution and behavior of the disease can be well characterized. Treatments, both clinical and surgical, as well as response to treatment mainly depend on the distribution of the disease in the digestive system and its behavior: inflammatory, stricturing or fistulizing.

It is known that the clinical and surgical treatments evolve with frequent relapses and 80% of patients require some type of surgery at some point of disease progression.[4] Therefore, surgical treatment of CD aims to treat complications, severe as well as chronic, with consequent relief of symptoms and improvement of quality of life.

By virtue of the high rate of recurrence and the need for repeated intestinal resections, which can lead to short bowel syndrome, it is important that the surgeon is aware that surgery should be as economic as possible.

## SURGICAL INDICATIONS

The indications for CD surgery can be didactically divided into elective and emergency treatment (Table 29.1).

| Table 29.1 Surgical indications | |
| --- | --- |
| **Elective** | **Emergency** |
| Clinical intractability | Acute intestinal obstruction |
| Delayed growth | Intestinal perforation |
| Fistulizing disease (abdominal and perineal) | Massive hemorrhage |
| Palpable abdominal mass | Abdominal abscess |
| Malignization | Acute colitis/toxic megacolon |
| | Acute ileitis |

### Elective surgery

### Clinical intractability

Clinical intractability is the most common indication for surgery in patients with CD and is difficult to define as it depends on many factors such as location of the disease, response to the correct medication and appropriate doses, treatment time, regular administration, adverse effects, nutritional status, patient age, delayed growth, cortico-dependence, extraintestinal manifestations, among others.

The persistence of symptoms or their worsening after 3 to 6 months of intensive treatment under specialist medical care is sufficient to consider the possibility of surgical treatment.[5]

Another indication of elective surgery, which didactically may be included in this group, is delayed growth, despite adequate therapy. CD may occur in prepubescence and adolescence (before 18 years of age) in approximately 18 to 25% of cases, with frequent growth deficits in this age range.[6,7] Despite the need for more aggressive therapy in the pediatric age range (enteral nutrition, corticosteroids, immunomodulator and biological therapy), the height/weight growth deficit is frequent. Thus, surgical treatment is indicated, and should be performed before the onset of puberty.[8-10] Extra-intestinal manifestations (EIM) occur in 30 to 65% of those with CD[11-13] and some correlate with the activity of the intestinal disease, such as arthralgias and migratory peripheral arthritis, erythema nodosum, pyoderma gangrenosum, oral lesions and episcleritis/uveitis. In such cases, the intensive treatment of intestinal disease can reverse the clinical manifestations of EIM and bowel resection surgery may be indicated among conservative measures. For EIM which progresses independently from intestinal disease activity (primary sclerosing cholangitis, spondyloarthropathy, urologic complications), intestinal resection should not be indicated for control.[14]

### Fistulizing disease (abdominal and perineal)

Abdominal disease

This consists of external and internal fistulas, which shall be discussed ahead.

External fistulas

The appearance of a fistula on the abdominal wall may be spontaneous, or more often secondary to intraperitoneal complication from surgical procedures. In general, the internal fistula orifice is situated in the anastomosis area or near to it, and the outer orifice on the wound or surgical scar.

Symptoms are characterized by the outflow of air or enteric material, which may be minimal or tolerable by the patient or in large quantities, causing serious dermatitis, indicating the degree of disease activity, and requiring a more aggressive approach with resection of the affected segment.

Spontaneous enterocutaneous fistula is generally associated with abdominal abscess and phlegmon, resulting from extensive ileocecal disease, and is best treated with laparotomy and bowel resection. It is most often located in the anterior abdominal wall and can also be found in the lumbar region, groin, popliteal fossa and base of the thigh (Figure 29.1).

Enterocutaneous fistulas are very symptomatic; in cases of high output, they cause metabolic, electrolyte and body image changes. Until recently, they were treated surgically as soon as they were diagnosed. With the advent of biological therapy, which has been shown to be efficient for the control

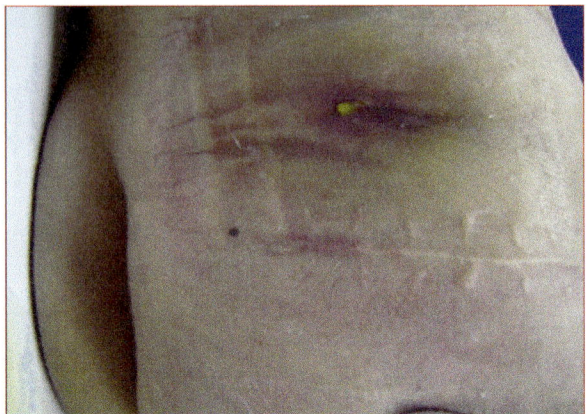

**Figure 29.1** Fistulous orifice near the right hip joint in a patient with CD.

and closure of fistulas in 30 to 50% of patients, conservative management has been chosen initially in selected cases. Biological therapy is recommended (infliximab, adalimumab or certolizumab pegol) for approximately 6 months before an assessment can be made. If the response is not favorable (closure of the fistula and clinical improvement), surgical treatment is indicated.

Internal fistulas

Internal fistulas are the most common, usually oligosymptomatic and hidden, and the diagnosis is made by radiological examination, during surgery, or through pathological examination. In a series of 59 patients with internal fistulas at St. Mark's Hospital, the diagnosis was made via radiological study in 54% of cases, and during surgery in 27%.[15]

The most common location is between the sigmoid colon and ileum (59%) and, in most of these cases, only the terminal ileum is diseased.[16] Enteroenteric and enterocolic fistulas should be treated with surgery, due to the higher risk of malignancy in the segment excluded from the intestinal transit, especially in the case of long segments and long duration of disease. Greenstein and Sachar found an average time of 18 years from evolution to the appearance of cancer.[17] The fistula may be single or multiple and affect the small intestine, colon, bladder, vagina, stomach, and duodenum.

Enterovesical fistulas are uncommon, with an incidence of 1.7% at the Cleveland Clinic and 2.4% at the Birmingham Hospital.[18,19] They usually affect the bladder ceiling, but may also affect the base of the bladder. They are more frequent in males, given that the uterus interposed before the bladder hinders formation.

The main symptoms are pneumaturia and fecaluria, as well as presentation of dysuria, pollakiuria and choluria, with frequent recurrent urinary infections. The diagnosis is clinically suspected and confirmed by contrast studies (intestinal transit, cystography or CT scan with contrast) or via endoscopic examination (cystoscopy). Treatment is always surgical, due to the risk of loss of renal function caused by repeated urinary tract infections. In most cases, the fistular orifice in the bladder is not identified, and the closure is carried out with separated absorbable sutures and an omental pedicle flap positioned, as well as maintaining a urinary catheter for 7 to 10 days. Resection of the ileal segment affected by the disease is mandatory, with primary anastomosis conducted, preferably side-to-side, mechanical or manual.

Entero/colon or rectovaginal fistulas are rare and extremely uncomfortable. Few symptoms may be shown, producing symptoms (release of gas and feces through the vagina) only when feces are liquid although the majority of the time, daily outflow of feces through the vaginal introitus is observed. Treatment of enterocutaneous or colovaginal fistula is always surgical, with resection of the affected bowel segment and closure of the vaginal orifice. In selected cases, an omental pedicle segment may be interposed or transposition of the gracilis muscle.

Gastro, duodenoenteric or colic fistulas are rare but very symptomatic. These fistulas usually occur with diarrhea, epigastric pain, nausea, fecaloid vomiting, and weight loss associated with nutritional disorders. They are most commonly associated with ileitis/Crohn's ileocolitis. According to Greenstein et al., these fistulas were found in 0.8% of patients with ileocolitis, and 0.6% of Crohn's colitis patients (more common in the transverse colon).[20] They require early surgical treatment and, as the disease is usually located in the ileum or colon, treatment should be focused on resection of the bowel segment affected and where the fistula is located. With regard to the stomach and the duodenum, the edges of the fistula's orifice are revived, with posterior suture.

In 107 (17%) of 661 operations performed in patients with DC in the service of colon, rectum and anus surgery at the Hospital das Clinicas, University of São Paulo Medical School (HC-FMUSP) in the period from 1984 to 2004, the primary (main) or secondary indication was the presence of fistula (Table 29.2).[21]

**Table 29.2** Distribution of patients with simple fistulas submitted to surgical treatment

| Location | Small intestine | Colon | Rectum |
|---|---|---|---|
| Enteric | 9 | 2 | 0 |
| Cutaneous | 52 | 6 | 1 |
| Vaginal | 1 | 0 | 12 |
| Bladder | 11 | 1 | 0 |

Note: 12 patients had complex fistulas and are not included in this table.

### Malignization

Patients with long-lasting CD (> 8 to 10 years) are at increased risk of developing cancer in the digestive tract, which generally occurs in younger age groups. Few studies have addressed this issue in depth, but it seems that the incidence of colorectal cancer (CRC) in CD is similar to that observed in patients with ulcerative colitis (UC).[22,23] Crohn's colitis carriers with over 8 years from the onset of symptoms should be submitted to CRC screening with colonoscopy every 1 to 2 years, depending on the extent and activity of the disease. The colonoscopy should be performed when the disease is in remission in order to avoid diagnostic errors, more specifically, confusion in the interpretation of dysplastic lesions. Biopsies are recommended in the 4 quadrants of the colon and rectum at 10 cm intervals, as well as the collection of material in areas of stricture, chronic fistulas, lesions or mass - dysplasia associated lesions or masses (DALM). In contrast to sporadic adenomas, in which dysplastic changes generally occur in polyps in DALM, the dysplasias occur in flat lesions or masses and are commonly associated with invasive carcinoma, which is infiltrative and has a mucinous component.[24] Yamazaki et al. found 6.8% neoplasia in 132 patients with Crohn's colitis, which stresses the importance of endoscopic study with serial biopsies in this population, especially in areas with strictures.[25] Polyps, as well as flat lesions must be resected and/or biopsied to exclude dysplasia and carcinoma. The presence of adenocarcinoma, high grade dysplasia associated with DALM or low grade dysplasia in various segments of the colon (multifocal) is an absolute indication for total colectomy.[26] On the other hand, CD patients with long evolution, with lesions located in the upper gastrointestinal tract, ileum and ileocecal region should undergo biopsy of suspicious lesions via endoscopy or during surgery. Reports of adenocarcinoma in areas of previous stenoplasty with long evolution have been observed. Consequently, the areas of stricture and prior strictureplasty must always be carefully inspected and, if necessary, biopsied or resected.[27-30]

### Palpable abdominal mass

The indication of laparotomy in the presence of abdominal plastron in patients with CD is controversial, and for every 25 surgeons surveyed, only 6 performed this procedure systematically.[31] The presence of a persistent, painful and fixed abdominal mass is usually more frequent in the right side of the abdomen, and may be due to extensive inflammation in the terminal ileum or internal fistulas, increased mesenteric ganglia, mesenteric panniculitis, intestinal adhesions or abscesses. Investigation with imaging methods such as ultrasound (US) and computed tomography (CT) is of fundamental importance for defining the best approach to be taken. In cases of abscesses, collections may be located above and next to the abdominal wall, between loops, in the pelvis or retroperitoneal (psoas muscle). In selected cases, these collections can be drained with the help of US or CT, always with antibiotic coverage, and the material being sent for analysis (bacterioscopy, culture and antibiogram), and surgery, when indicated, can be performed under better clinical conditions. In cases of plastron without the presence of collections, clinical treatment may be elected, with laparotomy and bowel resection indicated in cases symptomatic of failure.

### Emergency

The main emergency surgical indications are: intestinal obstruction, perforation with peritonitis, bleeding, abdominal abscess, acute colitis, toxic megacolon and acute ileitis.

## Intestinal obstruction

The presence of bowel obstruction symptoms is frequent in patients with CD, usually associated with deep ulcers and edema of the intestinal wall, reacting to the inflammatory process, which generally improves with clinical treatment. It is very important to remember that in addition to inflammatory edema, other causes may be responsible for intestinal obstruction, such as fibrosis, adhesions resulting from the transmural nature of the disease, fistulas and extrinsic compressions (abscesses). Partial or total obstruction of the small intestine is the most common indication for emergency surgery in those with CD.[32] In diseases with long-term progression, intestinal wall thickening associated with fibrosis adhesions and the involvement of adipose tissue may be responsible for concentric strictures which are common in the jejuno-ileum segment in the colon. These fibrotic strictures cause major bowel distension, chronic obstructive symptoms with dilation upstream, and bacterial overgrowth, with high risk of perforation and peritonitis, which is better managed with laparotomy.

The presence of fibrotic strictures increases with the duration of the disease, with lesions more frequent in the upper gastrointestinal tract, that is, proximal to the terminal ileum.[33,34] Acute attacks of obstruction should always be evaluated using imaging methods (abdominal x-ray, US and CT), aiming at a precise diagnosis of the site and the cause of the stricture, in addition to the exclusion of cancer. It is noteworthy that 7% of the causes of stricture in the large intestine in long-term DC carriers are cancer.[25] Long or multiple strictures tend to improve slightly with clinical treatment and generally require surgical treatment for the relief of symptoms. These patients, with chronic and intermittent symptoms of bowel obstruction (pain, nausea, bloating, etc.) are often malnourished because they avoid eating as a result of colic attacks.[35]

The preoperative preparation is very important in such cases, with correction of anemia, nutritional and electrolyte disorders, and the use of broad spectrum antibiotics (Gram-negative and anaerobic) for 7 to 10 days. Corticosteroid therapy may be required during and after surgery in patients who used steroids in the 4 to 6 months prior to the operation.

In summary, one can say that patients with symptomatic strictures in the small or large intestine that did not improve with intensive medical treatment should be referred to surgery. In cases of bowel obstruction associated with abscess, antibiotics must first be introduced, then perform percutaneous drainage, and surgery – if necessary – should be conducted at a subsequent time. CD patients with long-term progression (> 8 years) and strictures in the colon, even if asymptomatic, and that cannot be adequately evaluated via biopsy and/or cytology should also undergo surgery.

### Perforation

Free perforation in CD is a rare complication, with surgery indicated in 1 to 6% of the cases, which can occur both in Crohn's colitis and in lesions in the small intestine, which are the most common.[17,36-38] The perforation site in the intestinal loop is near to the narrowed area, and when this occurs in the free peritoneum it does not generally cause diagnostic doubts. However, in some patients taking corticosteroids at high doses or with associated immunosuppression symptoms, performing an abdominal CT is useful for the diagnosis of perforation, which may reveal the presence of air and free fluid in the peritoneal cavity.

For perforations of the small intestine, surgery must be performed for resection of the perforated segment. Simple suturing of the lesion should be avoided due to the high morbidity and mortality, and may be performed in selected cases, always with protective upstream bypass.[17] Greenstein et al. observed mortality indexes of 4% in cases of resection and proximal ileostomy and 41% for simply suturing of the lesion. In cases of colorectal perforation, the treatment should be primary bowel resection without primary anastomosis, that is, realization of Hartmann's procedure or mucous fistulas. Primary anastomosis with upstream derivative colostomy can be an alternative for selected cases.

### Hemorrhage

Massive hemorrhage, which leads to the need for emergency surgery in patients with CD, is a rare complication reported in 1 to 2% of cases. It is important to exclude other common causes of bleeding in the gastrointestinal tract such as erosive gastritis, gastric ulcer, vascular ectasia, diverticular disease, ulcerative colitis, hemorrhoidal disease and coagulopathy. Cirocco et al. identified four cases (0.6%) of massive hemorrhage in 631 hospital admissions for DC. These cases are more common in males (1.5 to 2 times), with an average age of 35 years, disease progression over 4.5 to 5 years, being the small intestine, or more precisely the terminal ileum, the most frequent site (66%).[39] In contrast, in the experience of Belaiche et al. the colon was the most frequent site, and deep ulcers were the most common cause (95%) with the need of surgical treatment in 20.5% of cases.[40]

In bleeding due to CD, the presence of a deep ulcer in the bowel wall with submucosal vessels erosion is generally observed. In hemodynamically stable patients with minor bleeding, laboratory tests (blood count, coagulation, etc.) are advised, as well as passage of a nasogastric tube with analysis of aspirated liquid, endoscopic studies (upper endoscopy, enteroscopy and colonoscopy) and capsule endoscopy, which are usually sufficient measures to clarify the site of bleeding, helping to plan the definitive therapy. When colonic lesions are suspected, colonoscopy is usually indicated since it enables identification of the site of bleeding and also may have therapeutic activity (sclerotherapy, clips). If bleeding is voluminous and colonoscopy does not identify the site, digital angiography or scintigraphy should be performed. Remzi et al. described a technique of digital angiography with methylene blue injection in the pre-operative period, with the objective of identifying the bleeding site precisely. Patients with profuse gastrointestinal bleeding that cannot be treated with endoscopy and/or interventional radiology (vasopressin infusion or embolization) who have frequent recurrences should undergo surgery.[40,41] In the presence of hemodynamic changes that are difficult to control (need for more than 6 red blood cell units/24 hours), patients may be submitted to laparotomy with intraoperative endoscopic study, aiming to identify the site of bleeding, and subsequent resection of the diseased segment.[42,43]

### Abdominal abscess

The presence of an intra-abdominal abscess associated with CD is usually due to extensive lesions located in the ileal or ileocecal region, caused by micro or macro perforation of the bowel. This septic complication should be suspected in patients with abdominal pain, failure to eliminate of gas and stools, fever and presence of a painful abdominal mass, usually located in the right iliac fossa or right flank. A CT scan of the abdomen and pelvis is the examination of choice, which, in addition to identifying the collection (location, volume, and presence of septation) can also be therapeutic, helping to perform percutaneous drainage with collection of material for bacterioscopy, culture and antibiogram, always under antibiotic coverage. Percutaneous drainage prevents emergency surgery in 90% of cases.[44]

The presence of retroperitoneal abscess or abscess in the psoas muscle is the exception for drainage and requires surgical treatment. For persistent infectious symptoms associated with clinical deterioration, surgical drainage with bowel resection should be performed immediately; otherwise, septic shock and death may occur.

In 382 patients with CD, Yamaguchi et al. found abscesses in 9.9% (35 patients), after 10 years of follow-up, and 25% after 20 years of disease progression. 60% of the 35 patients with intra-abdominal abscess had previous operations. The locations of suppurative collections were: abdominal (40%), intraperitoneal (29%), retroperitoneal or iliopsoas (26%) and the subphrenic space (6%). Almost all abscesses occurred close to the anastomosis, and 65.7% of them were located in the right side of the abdomen. Conservative treatment (antibiotics and drainage guided by CT) was efficient in 20% (7/35) of cases and surgical drainage was required in 80% (28/35).[45] When analyzing 51 patients with pelvic abdominal abscesses, Garcia et al. compared the treatment outcomes of two groups undergoing percutaneous drainage *versus* surgical drainage. After an average follow-up period of 3.8 years, 12% of those undergoing drainage with bowel resection surgery relapsed and, in the percutaneous drainage group, abscess recurrence occurred in 56%. In the conservative treatment group, relapse occurred earlier, within the first 3 months.[46]

### Acute colitis, fulminant or toxic megacolon

CD patients suffering from severe acute or fulminant colitis have the same risk as patients with UC of developing toxic megacolon with or without perforation. Severe acute colitis is a clinical condition characterized by more than 6 liquid evacuations with blood per day, fever (> 37.5 °C), tachycardia (> 90 bpm), anemia (Hb <75% of normal) and ESR (> 30 mm) in the first hour. Fulminating colitis has an increased severity and also higher risk of perforation than severe colitis and is characterized by more than 10 liquid evacuations per day with blood, fever (> 37.5 °C), tachycardia (> 90 bpm) and ESR (> 30 mm) in the first hour. In addition to the criteria mentioned above, toxic megacolon presents pain, bloating and dilated colon, with transverse colon having an expansion of over 6 cm in radiological imaging of the abdomen. Acute severe or fulminant colitis patients are initially treated with intravenous corticosteroids, cyclosporine or biological therapy. In the absence of a response, with clinical deterioration or suspected perforation, performing surgery with subtotal colectomy and terminal ileostomy is indicated. Anticholinergic and antidiarrheal drugs or narcotics should be avoided in these situations due to the risk of clinical deterioration. Total or subtotal colectomy with terminal ileostomy and closure of the rectal stump (Hartmann's procedure) or mucous fistula is the best approach. In this way, the entire inflamed colon is removed while avoiding dissection of the pelvis, with reconstruction of the transit conducted subsequently, after 10 to 12 weeks.

### Acute ileitis

The clinical profile of acute ileitis is very similar to that of acute appendicitis, characterized by pain in the right iliac fossa, fever, nausea and failure to eliminate gas and feces. Imaging methods such as US and preferably CT are necessary to clarify the location and extent of the inflammatory process, as well as the presence of associated collections. In many situations, the diagnosis is only confirmed during laparotomy or laparoscopy, when viewing the normal cecal appendix and inflamed terminal ileum with thickened walls, which may also include the presence of mesenteritis and local adhesions. If ileal inflammation is extensive, with local substricturing and upstream dilation, ileocolectomy should be chosen. In selected cases, the ileum can be preserved and intensive treatment for the underlying disease initiated.

In the case of exclusive involvement of CD in the appendix, appendectomy is conducted alone, without the need for bowel resection, an option that is generally not accompanied by increased morbidity.

In cases of non-granulomatous appendicitis without the presence of CD in the terminal ileum, only an appendectomy is conducted.

## GENERAL CONSIDERATIONS AND SURGICAL TECHNIQUES

CD patients requiring surgical treatment should, whenever possible, undergo staging of the extent of the involvement of the digestive tract and its severity (presence of fistulas, abscesses, strictures), which is rarely feasible in emergency situations. However, some measures are essential before the patient is taken to the operating room, such as correction of anemia, the electrolyte disorders of hypoalbuminemia, administration of antibiotics and prophylaxis of thromboembolic events. The patient should also be aware of the possibility of undergoing an ostomy, and the stoma site should be marked preoperatively. The patient should be physically and psychologically prepared for the surgery. As these patients may require multiple surgeries, as well as performing other derivations, it should be done, whenever possible, with a median laparotomy incision, preserving the side walls of the abdomen.

### Surgical techniques

Approximately half of patients with CD have involvement of the ileocecal region. In 30 to 40%, the lesions are confined to the small intestine, and in 20 to 25% they are limited to the colon, which explains the variability of the clinical symptoms and surgical options (Table 29.3).[47]

| Table 29.3 Patterns of involvement in CD[47] | |
|---|---|
| **Patterns of involvement** | **Location of involvement** |
| Ileocecal region | 40 to 50% |
| Small intestine | 30 to 40% |
| Large intestine | 20 to 25% |

The choice of surgical techniques in patients with CD depends on the affected region, whether it is the small intestine or colon, and can be divided into groups of procedures with bowel resection and procedures without intestinal resection, such as stenoplasty, internal derivations and ostomies.

### Small intestine Crohn's disease

The resection techniques in small intestine CD should be limited to resection of the affected segment macroscopically, given it has been well documented that wide resections, in addition to not being more effective, may culminate in short bowel syndrome.[48,49] Fazio et al., in a randomized prospective study, found no difference with respect to recurrence rates when comparing two groups of patients undergoing intestinal resection: one with margins of 2 cm and the other with 12 cm.[50]

A wide mobilization of the ascending and transverse colon is important in lesions located in the ileocecal region, so that anastomosis can be performed without tension, while maintaining good vascularization. Wherever possible, restoration of the intestinal transit should be performed through wide anastomosis and on healthy tissue, which can be done end-to-end or side-to-side, manually or mechanically. Preference is given to mechanical side-to-side anastomosis using a 75 or 100 mm linear stapler. Although there are controversies about the best configuration of the anastomosis, some studies have shown that side-to-side is associated with a lower risk of recurrence.[51]

In selected cases, when the wall of the loop is very thick, it is performed manually, since most of the staplers were not designed to operate under these conditions. Although many studies do not show differences in complication rates between the manual or mechanical techniques,[52,53] three have shown that anastomosis with a stapler is safer and has lower morbidity.[51,54,55]

Removal of the mesentery with the draining lymph node chains is not recommended, because in addition to increasing morbidity, it does not prevent recurrences. An important technical precaution is the closure of the mesenteric gap in order to prevent internal hernia and adhesions. Derivative ostomy for protection of the anastomosis is usually not necessary and only performed in cases of major contamination of the peritoneal cavity, incomplete drainage of collections, long surgeries associated with voluminous bleeding and need for transfusions, hypoalbuminemia (< 2 to 2.5 g/dL) and prolonged use of corticosteroids at high doses. Intestinal resection can be performed via laparotomy or laparoscopic video access, since both have similar results regarding the rate of dehiscence, abscesses, blood loss and infection of the wound when performed to treat uncomplicated ileocecal disease. With laparoscopy, the surgical time is greater, as well as the cost, but pain after surgery is less pronounced, the cosmetic appearance is better and return to daily activities is sooner.[56]

For patients with multiple strictures in the small intestine, strictureplasty is indicated, with the technique used depending upon their degree and extent. Strictureplasty, was first described in India by Katariya et al. in 1977 to be used in patients with narrowing resulting from intestinal tuberculosis and later adapted to CD patients.[57]

With this technique, it is possible to enlarge the area with stenosis, without intestinal resection and immediate relief of the obstructive symptoms.[58-60] It is worth noting that patients undergoing strictureplasty do not present higher rates of complication or reoperation than those with segmental resection.[61] Before indicating plastic surgery for strictures in patients with long-lasting CD (greater than 8-10 years), neoplasia must be ruled out by means of biopsies and frozen section examination, especially if the patient is using biological therapy associated with an immunosuppressant for many years – the same conduct should be adopted in chronic penetrating disease.

The main contraindications for strictureplasty are: presence of purulent coarse or fecal contamination, abscesses, doubtful segment vitality, area with internal or external fistula, stricture very close to the anastomosis area, multiple narrowing areas in short segment of intestine, strictures larger than 30 cm and hypoalbuminemia (<2 g/dL). Strictureplasty should be performed in areas with inactivity or low disease activity and is most precisely indicated in cases of short strictures associated with obstructive conditions, especially if the patient has previous resections, since a new enterectomy can produce absorption deficits and malnutrition. Recently, new techniques have been described for cases of long and multiple strictures, such as the Finney, Jaboulay and Michelassi techniques.

The Heineke-Mikulicz technique is suitable for cases of short strictures, those less than 10 cm in length, and its performance is similar to that of pyloroplasty. A longitudinal incision is made over the narrowed area of the antimesenteric border, which extends over 1 to 2 cm in normal tissue, for subsequent closure in the transverse direction with slow absorption sutures, in one or two suture lines (Figure 29.2).

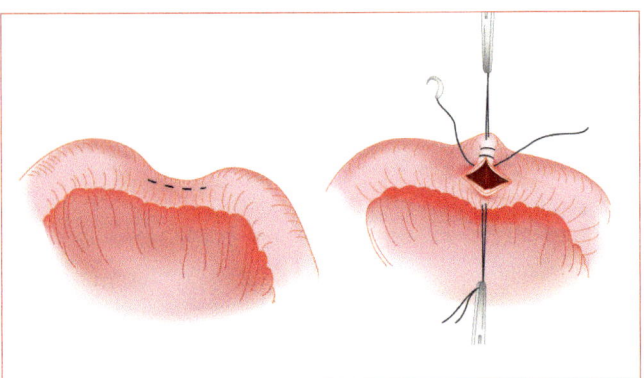

**Figure 29.2** Strictureplasty using the Heineke-Mikulicz technique.

A metal clip can be placed near the mesentery at the site of the strictureplasty to facilitate identification of the site of bleeding in cases of postoperative intestinal bleeding. In such cases, mesenteric angiography may still precisely locate the bleeding site and act therapeutically through the infusion of vasopressin without the need of opening the strictureplasty in cases of reoperation.[62]

The Finney technique is indicated in strictures measuring 10 to 20 cm in length and consists of a longitudinal incision over the entire affected extension that extends by 1 to 2 cm in normal tissue after the loop is approximated and folded over itself. The posterior wall is then sutured initially, followed by the anterior wall. This procedure may be undertaken using manual or mechanical suture (Figure 29.3).

The Jaboulay technique is similar to the Finney technique. However, in this case, a segment of the loop is excluded from the transit; therefore, it has been used less often (Figure 29.4).

The Michelassi technique is indicated in very long strictures with much thickened walls, and consists of a total section of the intestine at the midpoint of the narrowed area. Subsequently, the loops are placed on top of one another and a longitudinal enterotomy is performed along the whole length of the stenosis. Next, a side-to-side isoperistaltic entero-enterostomy is performed (Figures 29.5 and 29.6).

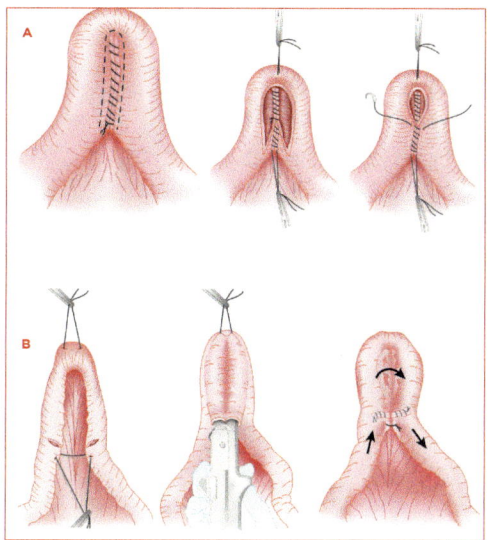

**Figure 29.3** Strictureplasty using the Finney technique. A: manual anastomosis. B: mechanical anastomosis.

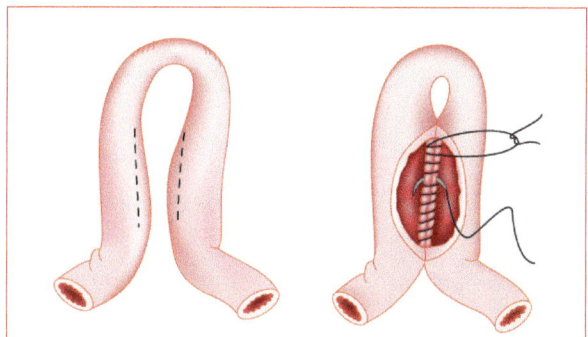

**Figure 29.4** Strictureplasty using the Jaboulay technique (manual anastomosis).

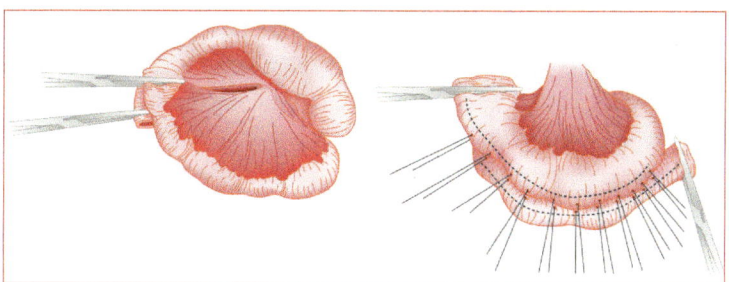

**Figure 29.5** Strictureplasty using the Michelassi technique.

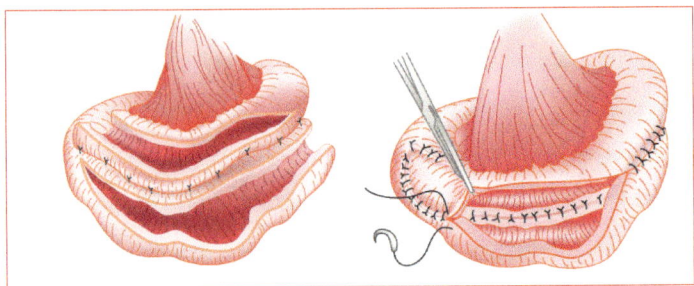

**Figure 29.6** Strictureplasty using the Michelassi technique.

In a meta-analysis in which 15 studies were reviewed with 506 patients who underwent 1,825 strictureplasties, it was observed that the Heineke-Mikulicz technique was used in 83%, the Finney in 15% and the others in 2%. Complications occurred in 66 patients (13%), the main ones being abscess, fistulas, bleeding and infection of the wound. It was found that patients undergoing the Finney technique had lower reoperation rates than the Heineke-Mikulicz technique, and that marked weight loss, as well as active disease were negative predictive factors.[58] A meta-analysis study conducted by Yamamoto et al. described a complication rate ranging from 1 to 14%. This study reviewed 3,250 strictureplasties in 1,112 patients, where the techniques used most were: Heineke-Mikulicz (81%), Finney (10%), Michelassi (5% – see Figure 29.6) and others (4%). More than 90% of the enteroplasty procedures were performed on strictures located in the jejunum and ileum. Recurrence of stenosis was observed in 3% and septic complications in 4% of patients.[63] Experiments with side-to-side isoperistaltic enteroplasty techniques in 6 centers involving 184 patients were reviewed by Michelassi et al. and more than 90% of the procedures were performed on the jejunum or ileum; complication rates ranged from 5.7 to 20.8%. Of the 184 patients with an average follow-up of 35 months, 48 (26%) underwent reoperation.[64]

### Large intestine Crohn's disease

Exclusive involvement of the colon in CD occurs in approximately 20 to 25% of cases and the most commonly used techniques are segmental colectomy, total colectomy with ileorectal anastomosis and total proctocolectomy with definitive ileostomy. The choice of technique to be employed depends upon the degree and extent of damage, as well as the segment involved.

Segmental colectomy with primary anastomosis is the procedure indicated for cases of lesion in a single colonic segment, with low complication rate and high recurrence rate reaching 62%, especially in cases of left colectomy.[65,66] In cases of left-sided colitis, left hemicolectomy or total colectomy may be used with comparable operative complication rates, though segmental resection has earlier recurrences.[67] Despite being controversial, with the use of biological therapy before and after surgery, more economic resections also in the colon have been recommended. In the case of involvement of two or more colic segments, total colectomy should be the conduct of choice.[67] In severe proctitis refractory to conservative measures, with strictures or fistulas and associated with anorectal/perinea involvement, proctocolectomy or proctectomy with ostomy are indicated. In special situations with extensive colon and rectum involvement without perineal damage and without involving the small intestine, total proctocolectomy with production of an ileal J-pouch has been advocated by some authors, despite higher complication rates and worse functional outcomes, as well as pouch removal rate.[68-73] While there are some reports favoring the realization of the ileal pouch in CD, it is still the exception, indicated in selected cases (teenagers/young people with pancolitis, without affecting the small intestine and perineum), and those who refuse ostomy and are aware of the risks. Biological therapy, and the use of immunosuppressants, should be part of the postoperative treatment in order to prevent recurrence.[74] Total proctocolectomy with permanent terminal ileostomy can also be performed in the described situation and in cases of extensive perineal pancolitis associated with disease and/or severe fecal incontinence.

Laparoscopic video access can be used in cases of elective intestinal resection, but with greater surgical time and cost. Recovery is faster and cosmetic results are better.[75-77]

## GASTRODUODENAL CROHN'S DISEASE

Exclusive involvement of the gastroduodenal region is very rare (less than 1%), and 96% of cases are associated with lesions of the intestinal tract.[78] The main complication is sub-strictures/strictures due to deep ulceration, and surgery is necessary in cases refractory to conservative measures (proton pump inhibitor, prokinetics, biological and/or an immunosuppressant therapy).

Gastroenterostomy or duodenojejunal anastomosis (bypass) and strictureplasty are the most used surgical options, with preference given to the Heineke-Mikulicz technique, which should be performed whenever possible as it keeps the pylorus in transit, which in addition to being more physiological also helps to control diarrhea, especially in patients who have previous bowel resection.

Gastroenterostomy was widely used in the past but its functional results are not good, as it is often accompanied by nausea and vomiting.

Yamamoto et al. conducted a study of 10 patients with gastroduodenal CD and obstructive symptoms, submitting them to strictureplasty and, in 4 of these, the operation included the pylorus. Good results were observed in 8 (80%), and 2 underwent reoperation, 1 underwent gastrojejunostomy and another to duodenojejunal anastomosis.[78]

Fistulas involving the duodenum, ileum and colon must be treated with surgery, with separate structures involved and orifices closed primarily after revival of the borders, with good results in most cases.[79]

### Anorectal/perineal Crohn's disease

Perineal manifestations in CD are common and can occur in 20 to 80% of cases, usually associated with lesions in the colon and less often with proximal lesions.[80,81] Perianal involvement in CD may precede, be simultaneous, or appear after gastrointestinal involvement. When it appears as the first manifestation, the diagnosis of anal lesions may not raise suspicions about the presence of IBD. Drug treatment (antibiotics, biological therapy, hyperbaric oxygen therapy, etc.) should be indicated, and surgical treatment, when necessary, should be conservative, avoiding manipulation of the anal sphincters due to the high risk of progression with fecal incontinence. The main anal lesions associated with CD are: perianal skin tags, fissures, hemorrhoids, abscesses, anal and anovaginal fistulas, strictures and carcinomas.

Careful evaluation of the type of lesion, its extent and severity should be performed whenever possible, before the surgery, using endoanal US, MRI and proctologic examination under anesthesia.[82,83] The objectives of surgical treatment in patients with perianal disease are: drainage of collections with or without placement of a seton, avoiding sphincterotomy in the prevention of anal incontinence, treatment of the underlying disease and improved quality of life.

If asymptomatic, perianal skin tags, fissures and hemorrhoids should not be resected. Inflammatory skin tags, even if symptomatic, should be managed using conservative measures initially, which consists of sanitation with water, local heat and topical corticosteroids. Surgery should only be indicated in cases of severe pain, difficult hygiene and infection.

Fissures may be single or multiple. In most cases they are painless, and in 50% of cases they heal spontaneously upon control of the underlying disease and topical treatment with corticosteroids. Refractory cases should be initially treated with curettage and local infiltration of corticosteroids at the base and the edges of the wound. If unsuccessful, new application of corticosteroids may be performed and, as a last resort, fissurectomy without sectioning the sphincter is indicated.

Anal or anorectal ulceration usually manifests with pain, edema and abundant secretion, and should be treated like fissures. Some deep ulcers may develop infections and sepsis of the perirectal tissues, leading to hardening and destruction of the sphincter and impaired continence, requiring more aggressive approaches with proctectomy and creation of an ostomy.

Hemorrhoidal disease is usually not associated with CD activity and is often an incidental association and, therefore, should be treated using standard clinical methods.

Anorectal stricture usually results from the healing of ulcers, fissures and anal fistulas, and is often characterized by difficulty in expelling gases and feces. Short strictures (less than 2 cm) and distal strictures are managed by anal dilation, scar tissue section and corticosteroid infiltration, with good response. More extensive strictures may require proctectomy.[84]

Anorectal abscesses are common occurrence in patients with CD, in 23 to 62% of patients. The treatment is simple drainage of the pus, without the need of performing a fistulotomy in the acute phase. Drains should be avoided and are indicated only in cases of deep abscesses located in the ischioanal fossa or supralevator. The use of broad-spectrum antibiotics (ciprofloxacin and metronidazole) for 7 to 10 days is advised. The recurrence of the abscess may be seen in 13 to 21% of patients within the first year of follow-up.[85] The incidence of perianal fistula is common in CD, with rates ranging from 9 to 56% of cases (Table 29.4).[86]

**Table 29.4** Incidence of anal fistulas in CD patients[86]

| Author | Incidence (%) |
| --- | --- |
| Kodner and Fry | 14 to 38 |
| Markowitz et al. | 11 |
| Lockhart-Mumery | 10 |
| Wolff et al. | 14 |
| Bernard et al. | 24 |
| Van Dogen and Lubbers | 28 |
| Radcliffe et al. | 9 |
| Goebbell | 30 |
| Kangas | 11 |
| Palder et al. | 14 |
| Nordgren et al. | 20 |
| Halme and Sainio | 31 |
| Sugita et al. | 56 |
| Platell et al. | 31 |

Low intersphincteric fistulas associated with CD heal spontaneously with conservative treatment in 20 to 30% of cases.[87,88] When there are persistent fistulas with abundant symptoms, fistulotomy with curettage may be performed, a procedure that should be performed preferably during periods of inactivity of the disease. In women with anterior fistulas extra care should be taken due to the risk of anal incontinence, especially if multiparous.

In complex fistulas that may involve the anal sphincter in varying degrees, treatment should be initially clinical and aggressive in order to reduce the infective process, so that the most conservative interventions can then be made without the risk of progressing to fecal incontinence. Fistulectomy, as well as fistulotomy should be avoided because of the risk of incontinence, with the most appropriate treatment being mucosal flap advancement associated with curettage of the bed. Healing rates with this procedure vary from 50 to 80%.[89,90] In case of multiple fistulous orifices or lack of viable tissue to perform the flap advancement, a non-cutting seton may be used, which aims to keep the fistula path open for continuous drainage of secretions, until complete healing. The seton may be placed on multiple paths. The isolated use of a collagen plug has been recommended by some with mixed results, with the best success rates observed when associated with other techniques.[91,92]

Rectovaginal or anovaginal fistulas occur in 3.5 to 23% of women with CD and, whenever possible, should be treated with surgery. In the case of absence of proctitis, the recommended technique is mucosal flap advancement. In our personal experience, we performed the flap advancement technique associated with collagen plug placement in 5 patients with complete closure of fistulas in 4 (80%), with an average follow-up of 16 months (10-38) (Figures 29.7 and 29.8 ). None of the patients had inflammatory activity (CRP <5) at the time of surgery or active disease in the rectoscopy, and they were all maintained with anti-TNF therapy.

Proctectomy is reserved for cases of deep ulceration associated with severe proctitis, severe injury to the sphincters and cases of failure using other techniques.

Currently, aggressive drug therapy (biologic and/or immunosuppressive therapy) is recommended after surgical treatment in order to prevent recurrences.[93-95]

Surgery for CD is known not to be curative, with recurrence being the rule. Symptomatic relapse occurs in 40 to 80% of cases, and endoscopic relapse figures reach 90% after 5 years of monitoring.[96] Various studies have identified some risk factors for early recurrence following surgery in patients with CD, such as smoking, previous operations (including appendectomy), penetrating abdominal and perineal disease, previous extensive resections and myenteric plexitis in the surgical specimen.[97,98] In this group of patients, prophylaxis for recurrence has been shown to be effective in controlled studies and meta-analyzes.[99,100] With relation to other factors, such as sex, age upon appearance of the disease, surgical margins, colon disease, type of operation and presence of granulomas, the data are conflicting.[98] Prophylaxis should be started 2 weeks after surgery and maintained for at least 2 years. The drugs of choice for chemoprophylaxis are immunosuppressants (azathioprine/6-mercaptopurine) or anti-TNF (infliximab, adalimumab or certolizumab pegol).[100]

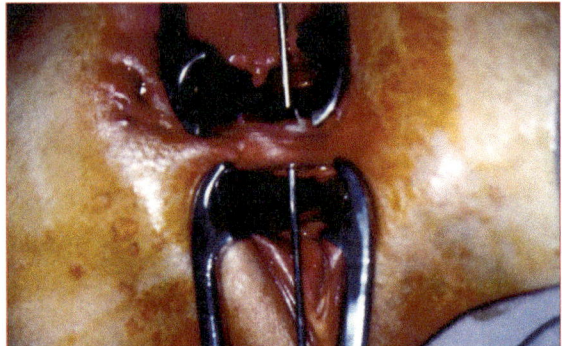

**Figure 29.7** Patient with Crohn's disease and rectovaginal fistula.

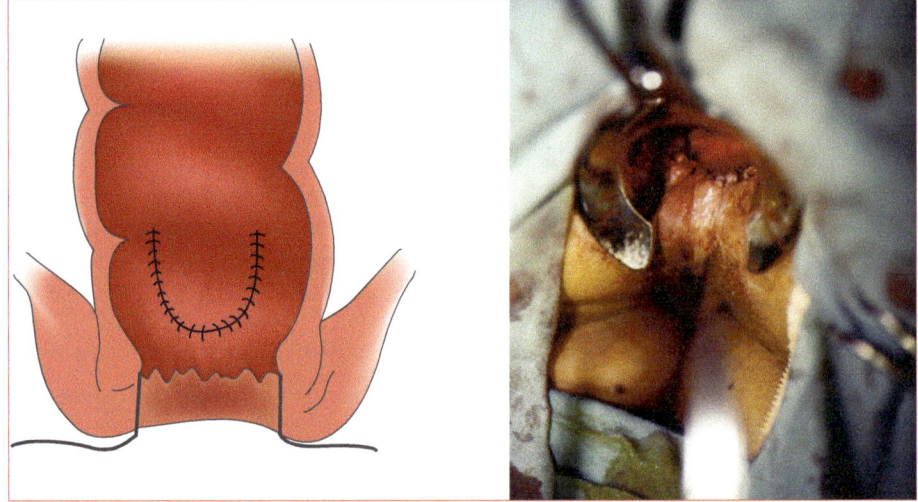

**Figure 29.8** Final appearance of the rectovaginal fistula correction using the flap advancement technique.

## REFERENCES

1. Crohn BB, Ginzburg L, Oppenheimer GD. Regional ileitis: a pathologic and clinical entity. JAMA 1932; 99:1323-9.

2. Greenstein AJ, Janowitz H, Sachar DB. The extra-intestinal complications of Crohn's disease and ulcerative colitis: a study of 700 patients. Medicine 1976; 55:401-12.

**3.** Shivananda S, Peña AS, Nap M, Weterman IT, Mayberry JF, Ruitenberg EJ et al. Epidemiology of Crohn's disease in region Leiden, the Netherlands. A population Study from 1979 to 1983. Gastroenterology 1987; 93:66-74.

**4.** Binder V, Both H, Hansen PK, Hendriksen C, Kreiner S, Torp-Pedersen K. Incidence and prevalence of ulcerative colitis and Crohn's disease in the country of Copenhagen 1962 to 1978. Gastroenterology 1982; 83:563-8.

**5.** Hultén L. Surgical treatment of Crohn's disease of small bowel or ileocecum. World J Surg 1988; 12:180-5.

**6.** Motil KJ, Grand RJ, Davis-Kraft L, Ferlic LL, Smith EO. Growth failure in children with inflammatory bowel disease: a prospective study. Gastroenterology 1993; 105:681-91.

**7.** Savage MO, Beattie RM, Camacho-Hubner C, Walker-Smith JA, Sanderson IR. Growth in Crohn's disease. Acta Paediatr Suppl 1999; 88:89-92.

**8.** Besnard M, Jaby O, Mougenot JF, Ferkdadji L, Debrun A, Faure C et al. Postoperative outcome of Crohn's disease in 30 children. Gut 1998; 43:634-8.

**9.** Sentongo TA, Stettler N, Christian A, Han PD, Stallings VA, Baldassano RN et al. Growth after intestinal desection for Crohn's disease in children, adolescents, and young adults. Inflamm Bowel Dis 2000; 6:265-9.

**10.** Dokucu AI, Sarnacki S, Michel JL, Jan D, Goulet O, Ricour C et al. Indications and results of surgery in patients with Crohn's disease with onset under 10 years of age: a series of 18 patients. Eur J Pediatr Surg 2002; 12:180-5.

**11.** Teixeira MG, Brunetti C, Gonçalves SR et al. Manifestações extra-intestinais da doença de Crohn. Rev Bras Coloproct 1988; 8:7-10.

**12.** Van Bodegraven AA, Pena AS. Treatment of extraintestinal manifestations in inflammatory bowel disease. Curr Treat Options Gastroenterol 2003; 6:201-12.

**13.** Juillerat P, Mottet C, Froehlich F, Felley C, Vader J-P, Burnand B et al. Extraintestinal manifestations of Crohn's disease. Digestion 2005; 71:31-6.

**14.** Caprilli R, Gassull MA, Escher JC, Moser G, Munkholm P, Forbes A et al. European Crohn's and Colitis Organization. European evidence based consensus on the diagnosis and management of Crohn's disease: special situations. Gut 2006; 55(Suppl.1):i36-58.

**15.** Fazio VW, Wilk P, Turnbull Jr. RB, Jagelman DG. The dilemma of Crohn's disease: Ileosigmoid fistula complicating Crohn's disease. Dis Colon Rectum 1977; 20:381-6.

**16.** Glass RE, Ritchie JK, Lennard-Jones JE, Hawley PR, Todd IP. Internal fistulas in Crohn's disease. Dis Colon Rectum 1985; 28:557-61.

17. Greenstein AJ, Sachar DB. Cancer in Crohn's disease. In: Allan RN (ed.). Inflammatory bowel disease. New York: Churchill Livingstone, 1983.

18. Heyen F, Ambrose NS, Allan RN et al. Enterovesical fistulas in Crohn's disease. Ann R Coll Surg Engl 1989; 71:101-4.

19. Mcnamara MJ, Fazio VW, Lavery IC, Weakley FL, Farmer RG. Surgical treatment of enterovesical fistulas in Crohn's disease. Dis Colon Rectum 1990; 33:271-6.

20. Greenstein AJ, Sachar DB, Mann D, Lachman P, Heimann T, Aufses Jr AH. Spontaneous free perforation and perforated abscess in 30 patients with Crohn's disease. Ann Surg 1987; 205:72-6.

21. Teixeira MG. Tratamento das fístulas intestinais por doença de Crohn. In: Habr-Gama A, Rodrigues JG, Cecconello I, Zilberstein B, Machado MCC, Saad WA (eds.). Atualização em cirurgia do aparelho digestivo e coloproctologia. São Paulo: Frontis Editorial, 2005. p.485-8.

22. Winawer S, Fletcher R, Rex D et al. For the U.S Multisociety Task Force on colorectal cancer. Colorectal cancer screening and surveillance: clinical guidelines and rationale-update based on new evidence. Gastroenterology 2003; 124:544-60.

23. Maykel JA, Hagerman G, Mellgren AF, Li SY, Alavi K, Baxter NN et al. Crohn's colitis: the incidence of dysplasia and adenocarcinoma in surgical patients. Dis Colon Rectum 2006; 49:950-7.

24. Blackstone MO, Riddell RH, Rogers BHG, Levin B. Dysplasia-associated lesions or mass (DALM) detected by colonoscopy in long-standing ulcerative colitis: an indication for colectomy. Gastroenterology 1981; 80(2):366-74.

25. Yamazaki Y, Ribeiro MB, Sachar DB, Aufses Jr AH, Greenstein AJ. Malignant colorectal strictures in Crohn's disease. Am J Gastroenterol 1991; 86:882-5.

26. Itzkowitz SH, Present DH. Crohn's and Colitis Foundation of America Colon Cancer in IBD Study Group. Consensus conference: colorectal cancer screening and surveillance in inflammatory bowel disease. Inflamm Bowel Dis 2005; 11:314-21.

27. Marchetti F, Fazio VW, Ozuner G. Adenocarcinoma arising from a strictureplasty site in Crohn's disease: report of a case. Dis Colon Rectum 1996; 39:1315-21.

28. Yamamoto T, Bain IM, Allan RN, Keighley MR. An audit of strictureplasty for small-bowel Crohn's disease. Dis Colon Rectum 1999; 42:797-803.

29. Jaskowiak NT, Michelassi F. Adenocarcinoma at a strictureplasty site in Crohn's disease: report of a case. Dis Colon Rectum 2001; 44:284-7.

**30.** Partridge SK, Hodin RA. Small bowel adenocarcinoma at a strictureplasty site in a patient with Crohn's disease: report of a case. Dis Colon Rectum 2004; 47:778-81.

**31.** Kisner JB. Current medical and surgical opinions on important therapeutic issues in inflammatory bowel disease. A special 1979 survey. Am J Surg 1980; 140:391-5.

**32.** Alexander-Williams J. Overview of surgical management and directions of future research. In: Allan RN (ed.). Inflammatory bowel disease. New York: Churchill Livingstone, 1983.

**33.** Louis E, Collard A, Oger AF, Degroote E, El Yafi FAN, Belaiche J. Behaviour of Crohn's disease according to the Vienna classification: changing pattern over the course of the disease. Gut 2001; 49:777-82.

**34.** Freeman HJ. Natural history and clinical behavior of Crohn's disease extending beyond two decades. J Clin Gastroenterol 2003; 37:216-9.

**35.** Alos R, Hinojosa J. Timing of surgery in Crohn's disease: a key issue in the management. World J Gastroenterol 2008; 14:5532-9.

**36.** Bundred NJ, Dixon JM, Lumsden AB, Gilmour HM, Davies GC. Free perforation in Crohn's colitis. A ten year review. Dis Colon Rectum 1985; 28:35-7.

**37.** Freeman HJ. Spontaneous free perforation of the small intestine in Crohn's disease. Can J Gastroenterol 2002; 16:23-7.

**38.** Greenstein AJ, Mann D, Sachar DB, Aufses Jr AH. Free perforation in Crohn's disease: I. A survey of 99 cases. Am J Gastroenterol 1985; 80:682-9.

**39.** Cirocco WC, Reilly JC, Rusin LC. Life-threatening hemorrhage and exsanguination from Crohn's disease. Report of four cases. Dis Colon Rectum 1995; 38:85-95.

**40.** Belaiche J, Louis E, D'Haens G, Cabooter M, Naegels S, De Vos M et al. Acute lower gastrointestinal bleeding in Crohn's disease: Characteristics of a unique series of 34 patients. Belgian IBD Research Group. Am J Gastroenterol 1999; 94:2177-81.

**41.** Kostka R, Lukas M. Massive, life-threatening in Crohn's disease. Acta Chir Belg 2005; 105:168-74.

**42.** Robert JR, Sachar DB, Greenstein AJ. Severe gastrointestinal hemorrhage in Crohn's disease. Ann Surg 1991; 213:207-11.

**43.** Driver CP, Anderson DN, Keenan RA. Massive intestinal bleeding in association with Crohn's disease. J R Coll Surg Edinb 1996; 41:152-4.

**44.** Doemeny JM, Burke DR, Meranze SG. Percutaneous drainage of abscess in patients with Crohn's disease. Gastrointest Radiol 1988; 13:237-41.

**45.** Yamaguchi A, Matsui T, Skurai T, Ueki T, Nakabayashi S, Yao T et al. The clinical characteristics and outcome of intra-abdominal abscess in Crohn's disease. J Gastroenterol 2004; 39(5):441-8.

**46.** Garcia JC, Persky SE, Bonis PA, Topazian M. Abscesses in Crohn's disease: outcome of medical versus surgical treatment. J Clin Gastroenterol 2001; 32:409-12.

**47.** Judge TA, Lichtenstein GR. Inflammatory bowel disease. In: Friedman SL (ed.). Current diagnosis & treatment in gastroenterology. New York: McGraw Hill, 2003.

**48.** Bechi P, Tonelli L. Results in the radical surgical treatment of Crohn's disease. Int Surg 1982; 67:325-8.

**49.** William P, Homan D, Dineen P. Comparision of the results of resection, bypass and bypass with exclusion for ileocecal Crohn's disease. Ann Surg 1978; 187:530-5.

**50.** Fazio VW, Marchetti F, Church M, Goldblum JR, Lavery C, Hull TL et al. Effect of resection margins on the recurrence of Crohn's disease in the small bowel. A randomized controlled trial. Ann Surg 1996; 224:563-71.

**51.** Hashemi M, Novell JR, Lewis AA. Side-to-side stapled anastomosis may delay recurrence in Crohn's disease. Dis Colon Rectum 1998; 41:1293-96.

**52.** Munoz-Juarez M, Yamamoto T, Wolff BG, Keighley MR. Wide-lumen stapled anastomosis vs. conventional end-to-end anastomosis in the treatment of Crohn's disease. Dis Colon Rectum 2001; 44:20-5.

**53.** Tersigni R, Alessandroni L, Barreca M, Piovanello P, Prantera C. Does stapled functional end-to-end anastomosis affect recurrence of Crohn's disease after ileocolonic resection? Hepatogastroenterology 2003; 50:1422-5.

**54.** Yamamoto T, Bain IM, Mylonakis E, Allan RN, Keighley MRB. Stapled functional end-to--end anastomosis versus sutured end-to-end anastomosis after ileocolonic resection in Crohn's disease. Scand J Gastroenterol 1999; 34:708-13.

**55.** Resegoti A, Astegiano M, Farina EC, Ciccone G, Avagnina G, Giustetto A et al. Side--to-side stapled anastomosis strongly reduces anastomotic leak rates in Crohn's disease surgery. Dis Colon Rectum 2005; 48:464-8.

**56.** Tilney HS, Constantinides VA, Heriot AG, Nicolaou M, Athanasiou T, Ziprin P et al. Comparision of laparoscopic and open ileocecal resection for Crohn's disease: a meta-analysis. Surg Endosc 2006; 20:1036-44.

**57.** Katariya RN, Sood S, Rao PG, Rao PL. Strictureplasty for tubercular strictures of the gastro-intestinal tract. Br J Surg 1977; 64:496-8.

**58.** Tichansky D, Cagir B, Yoo E, Marcus SM, Fry RD. Strictureplasty for Crohn's disease: meta-analysis. Dis Colon Rectum 2000; 43:911-9.

**59.** Dietz DW, Laureti S, Strong AS, Hull TL, Church J, Remzi FH et al. Safety and long-term efficacy of strictureplasty in 314 patients with obstructing small bowel Crohn's disease. J Am Coll Surg 2001; 192:330-7.

**60.** Roy P, Kumar D. Strictureplasty. Br J Surg 2004; 91:1428-37.

**61.** Ozuner G, Fazio VW, Lavery I, Milsom JW, Strong SA. Reoperative rates for Crohn's disease following strictureplasty. Dis Colon Rectum 1996; 39:1199-203.

**62.** Ozuner G, Fazio VW. Management of gastrointestinal hemorrhage after strictureplasty for Crohn's disease. Dis Colon Rectum 1995; 38:297-300.

**63.** Yamamoto T, Fazio VW, Tekkis PP. Safety and efficacy of strictureplasty for Crohn's disease: a systematic review and meta-analysis. Dis Colon Rectum 2007; 40:1968-86.

**64.** Michelassi F, Taschieri A, Tonell F, Sasaki I, Poggioli G, Fazio V et al. An international multicenter, prospective, observacional study of the side-to-side isoperistaltic strictureplasty in Crohn's disease. Dis Colon Rectum 2007; 50:277-84.

**65.** Himal HS, Belliveau P. Prognosis after surgical treatment for granulomatous enteritis and colitis. Am J Surg 1981; 142:347-9.

**66.** Mayberry JF, Rhodes J. Epidemiological aspects of Crohn's disease: a review of the literature. Gut 1984; 25:886-99.

**67.** Tekkis PP, Purkayastha S, Lanitis S, Athanasiou T, Heriot AG, Orchard TR et al. A comparison of segmental vs. subtotal/total colectomy for colonic Crohn's disease: a meta-analysis. Colorectal Dis 2006; 8:82-90.

**68.** Regimbeau JM, Panis Y, Pocard M, Bouhnik Y, Lavergne-Slove A, Rufat P et al. Long--term results of ileal pouch-anal anastomosis for colorectal Crohn's disease. Dis Colon Rectum 2001; 44:769-78.

**69.** Hartley JE, Fazio VW, Remzi FH, Lavery IC, Church JM, Strong SA et al. Analysis of the outcome of ileal pouch-anal anastomosis in patients with Crohn's disease. Dis Colon Rectum 2004; 47:1808-15.

**70.** Sagar PM, Dozois RR, Wolff BG. Long-term results of ileal pouch-anal anastomosis in patients with Crohn's disease. Dis Colon Rectum 1996; 39:893-8.

**71.** Braveman JM, Schoetz DJ, Marcello PW, Roberts PL, Coller JA, Murray JJ et al. The fate of the ileal pouch in patients developing Crohn's disease. Dis Colon Rectum 2004; 47:1613-9.

**72.** Brown CJ, Maclean AR, Cohen Z, Macrae HM, O'Connor BI, McLeod RS. Crohn's disease and indeterminate colitis and the ileal pouch-anal anastomosis: outcome and patterns of failure. Dis Colon Rectum 2005; 48:1542-9.

**73.** Tekkis PP, Heriot AG, Smith O, Smith JJ, Windsor AC, Nicholls RJ. Long-term outcomes of restorative proctocolectomy for Crohn's disease and indeterminate colitis. Colorectal Dis 2005; 7:218-23.

**74.** Reese GE, Lovegrove RE, Tilney HS, Yamamoto T, Heriot AG, Fazio VW et al. The effect of Crohn's disease on outcomes after restorative proctocolectomy. Dis Colon Rectum 2007; 50(2):239-50.

**75.** Milson JW, Hammerhofer KA, Bohm B, Marcello P, Elson P, Fazio VW. Prospective, randomized trial comparing laparoscopic vs. conventional surgery for refractory ileocolic Crohn's disease. Dis Colon Dis 2001; 44:1-8.

**76.** Msika S, Ianelli, Deroide G, Jouët P, Soulé JC, Kianmanesh R et al. Can laparoscopic reduce hospital stay in the treatment of Crohn's disease? Dis Colon Rectum 2001; 44:1661-6.

**77.** Tabet J, Hong D, Kim CW, Wong J, Goodacre R, Anvari M. Laparoscopic versus open bowel resection for Crohn's disease. Can J Gastroenterol 2001; 15:237-42.

**78.** Yamamoto T, Allan RN, Keighley MRB. An audit of gastroduodenal Crohn's disease: clinic/pathologic features and management. Scand J Gastroenterol 1999; 34:1019-24.

**79.** Jacobson IM, Schapiro RH, Warshaw AL. Gastric and duodenal fistulas in Crohn's disease. Gastroenterology 1985; 89:1347-52.

**80.** Rankin GB, Watts HD, Melnyk CS, Kelley Jr ML. National Cooperative Crohn´s Disease Study: extraintestinal manifestations and perianal complications. Gastroenterology 1979; 77:914-20.

**81.** Singh B, McC Mortensen NJ, Jewell DP et al. Perianal Crohn's disease. Br J Surg 2004; 91:801-14.

**82.** Schwartz DA, Wiersema MJ, Dudiak KM, Fletcher JG, Clain JE, Tremaine WJ et al. A comparison of endoscopic ultrasound, magnetic resonance imaging, and exam under anesthesia for evaluation of Crohn's perianal fistulas. Gastroenterology 2001; 121:1064-72.

**83.** West RL, Dwakasing S, Felt-Bersma RJ, Schouten WR, Hop WC, Hussain SM et al. Hydrogen peroxide-enhanced three-dimensional endoanal ultrasonography and endoanal magnetic resonance imaging in evaluation perianal fistulas: agreement and patient preference. Eur J Gastroenterol Hepatol 2004; 16:1319-24.

**84.** Harper PH, Fazio VW, Lavery I, Jagelman DG, Weakley FL, Farmer RG et al. The long-term outcome in Crohn's disease. Dis Colon Rectum 1987; 30:174-9.

**85.** Platell C, Mackay J, Collopy B, Fink R, Ryan P, Woods R. Anal pathology in patients with Crohn's disease. ANZ J Surg 1996; 66(1):5-9.

**86.** Nivatvongs S, Gordon PH. Crohn's disease. In: Principles and practice of surgery for the colon, rectum and anus. Ed. Informa Healthcare 2007; 27:819-99.

**87.** Buchmann P, Keighley MRB, Allan RN, Thompson H, Alexander-Williams J. Natural history of perianal Crohn's disease: ten years follow-up. Am J Surg 1980; 140(5):642-4.

**88.** Allan A, Keighley MRB. Management of perianal Crohn's disease. World J Surg 1988; 198-202.

**89.** Makowiec F, Jehle EC, Becker HD, Starlinger M. Clinical course after transanal advancement flap repair of perianal fistula in patients with Crohn's disease. Br J Surg 1995; 82:603-6.

**90.** Joo JS, Weiss EG, Nogueras JJ, Wexner SD. Endorectal advancement flap in perianal Crohn's disease. Am Surg 1998; 64:147-50.

**91.** Sandborn WJ, Fazio VW, Feagan BG, Hannauer SB. American Gastroenterological Association Clinical Practice Committee. AGA technical review on perianal Crohn's disease. Gastroenterology 2003; 125:1508-30.

**92.** Schwandner O, Stadler F, Dietl O, Wirsching RP, Fuerst A. Initial experience on efficacy in closure of cryptoglandular and Crohn's transsphincteric fistulas by the use of the anal fistula plug. Int J Colorectal Dis 2008; 23(3):319-24.

**93.** Regueiro M, Mardini H. Treatment of perianal fistulizing Crohn's disease with infliximab alone or as an adjunct for exam under anesthesia with seton placement. Inflamm Bowel Dis 2003; 9:98-103.

**94.** van der Hagen SJ, Baeten CG, Soeters PB, Russel MG, Beets-Tan RG, van Gemert WG. Anti-TNF-alpha infliximab used as induction treatment in case of active proctitis in a multistep strategy followed by definitive surgery of complex anal fistulas in Crohn's disease: a preliminary report. Dis Colon Rectum 2005; 48:758-67.

**95.** Colombel JF, Schawartz DA, Sandborn WJ, Kamm MA, D'Haens G, Rutgeerts P et al. Adalimumab for the treatment of fistulas in patients with Crohn's disease. Gut 2009; 58:940-8.

**96.** Moss AC. Prevention of postoperative recurrence of Crohn's disease: what does the evidence support? Inflamm Bowel Dis 2013; 19:856-9.

**97.** Welsch T, Hinz U, Loffler T, Muth G, Herfarth C, Schmidt J et al. Early re-laparotomy for post-operative complications is a significant risk factor for recurrence after ileocaecal resection for Crohn's disease. Int J Colorectal Dis 2007; 22:1043-9.

**98.** Vaughn BP, Moss AC. Prevention of post-operative recurrence of Crohn's disease. World J Gastroenterol 2014; 20(5):1147-54.

**99.** Caprilli R, Taddei G, Viscido A. In favour of prophylatic teratment for post-operative recurrence in Crohn's disease. Ital J Gastroenterol Hepatol 1998; 30:219-25.

**100.** Van Assche G, Dignass A, Reinisch W, van der Woude CJ, Sturm A, De Vos M et al. The second European evidence-based Consensus on the diagnosis and management of Crohn's disease: special situations. J Crohn's Colitis 2010; 4:63-101.

## BIBLIOGRAPHY

**1.** Bufo AJ, Feldman S, Daniels GA, Lieberman RC. Stapled strictureplasty for Crohn's disease: a new technique. Dis Colon Rectum 1995; 38:664-7.

**2.** Cristaldi M, Sanpietro GM, Danelli PG, Bollani S, Bianchi Porro G, Taschieri AM. Long-term results and multivariate analysis of prognostic factor in 138 consecutive patients operated for Crohn's disease using bowel-sparing techniques. Am J Surg 2000; 179:266-70.

**3.** Ikeuchi H, Kusunoki M, Yamamura T. Long-term results of stapled and hand-sewn anastomoses in patients with Crohn's disease. Dig Surg 2000; 17:493-6.

**4.** Lee ECG, Papioannou N. Minimal surgery for chronic obstruction in patients with extensive or universal Crohn's disease. Ann R Coll Surg Engl 1982; 64:229-33.

**5.** Michelassi F, Hust RD, Mellis M, Rubin M, Cohen R, Gasparitis A et al. Side-to-side isoperistaltic strictureplasty in extensive Crohn's disease. Ann Surg 2000; 232:401-8.

**6.** Michelassi F. Side-to-side isoperistaltic strictureplasty fos multiple Crohn's strictures. Dis Colon Rectum 1996; 39:345-9.

**7.** Scarpa M, Angriman I, Barollo M, Polese L, Ruffolo C, Bertin M et al. Role stapled and hand-sewn anastomoses in recurrence of Crohn's disease. Hepatogastroenterology 2004; 51:1053-7.

**8.** Stebbing JF, Jewell DP, Kettlewell MG, Mortensen NJ. Recurrence and reoperation after strictureplasty for obstructive Crohn's disease: long-term results. Br J Surg 1995; 82:1471-4.

**9.** Whelan PJ, Saibil FG, Harrison AW. New options in the surgical management of Crohn's disease. Can J Surg 1987; 30:133-6.

**10.** Yamamoto T, Allan RN, Keighley MRB. Strategy for surgical management of ileocolonic anastomotic recurrence in Crohn's disease. World Surg 1999; 23:1055-61.

# TREATMENT OF PERIANAL FISTULIZING CROHN'S DISEASE

CARLOS WALTER SOBRADO
WILTON SCHMIDT CARDOZO

## INTRODUCTION

Crohn's disease (CD) is a chronic inflammatory condition that can affect the entire digestive tract; it is characterized by periods of exacerbation and remission, and its etiopathogenesis, even today, is not fully elucidated.

CD is characterized by chronic focal or segmental, transmural, persistent and progressive inflammation, which can progress despite the absence of symptoms, culminating with intestinal lesions and complications such as strictures, abscesses and fistulas.

Progressive transmural inflammation leads to the formation of abscesses and fistulas, which can affect intestinal segments only (internal fistulas), or intestinal segments and other organs (e.g., vagina, bladder), intestinal wall, skin (enterocutaneous, colocutaneous and peristomal fistulas [external fistulas]) or the perianal area.

In 1932, the original description by Crohn, Ginzburg and Oppenheimer involved 14 patients with terminal ileitis who underwent surgical treatment, and did not report the presence of perianal features.[1] Six years later, Penner and Crohn reported the presence of perianal fistula associated with regional ileitis, and from that moment on it became clear that the presence of perianal lesions was a frequent complication and should always be investigated in these patients, which was confirmed by many publications thereafter.[2-4]

The appearance of a perianal fistula is, undoubtedly, one of the most serious manifestations of CD, a factor that aggravates symptoms due to secretion and local discomfort, with apparent worsening of quality of life, being an indicator of poor prognosis and increasing treatment costs.[5]

In addition to fistulas, other perianal manifestations, such as plicomas (hemorrhoidal skin tags), fissures, hemorrhoids, ulcers, strictures, tumors and abscesses may arise during the course of CD, affecting 25 to 60% of patients.[6] Regarding the presence of perianal fistulas in CD, the cumulative risk ranges from 20 to 35% in 10 years; two population studies have shown incidence of 23% (Stockolm County Study) and 38% (Olmsted County Study).[7.8]

These perianal complications are most commonly associated with lesions in the colon and rectum and, less frequently, with proximal lesions. Furthermore, they can precede, be simultaneous or appear before any gastrointestinal symptoms. In 54 to 68% of cases, perianal manifestation occurs after the diagnosis of bowel lesions and, in 20 to 36% of cases, it precedes the disease in the gastrointestinal tract.[9-11] Hellers et al. found an incidence of anal fistula of 12% in patients with ileal lesions; 42% in patients with lesions in the colon and spared rectum; and 92% when the rectum was affected.[7]

Perianal CD can be classified as fistulizing and non-fistulizing; the latter can be divided into 5 groups: infectious (abscesses), ulcerations (fissures, ulcers), stricturing, inflammatory (skin tags, hemorrhoids) and neoplastic.

The presence of invasive cancer in areas with chronic long-lasting inflammatory bowel disease in patients with CD using immunosuppressive drugs for a long time must always be considered. Cancer occurrence in patients with perianal CD is around 0.7%, with incidence of adenocarcinoma and squamous cell carcinoma at equal frequencies.[12-14]

Anal fistula in CD can be caused by infection of the anal glands (cryptoglandular) or as a result of transmural nature of the inflammatory process leading to development of deep ulcers in the anal canal and rectum, and fistulas that arise as a result of increase in intraluminal pressure and fecal trauma.

Infectious factors, namely luminal and skin bacteria and altered gut flora, have also been blamed for the development and maintenance of fistulas in CD.[15]

## CLASSIFICATION

Numerous classifications have been proposed for anal fistulas. The most used is that proposed by Parks in 1976, based on the relationship between tract and anal sphincter muscles.[16] The original Parks classification subdivides the fistulas into four categories, with subsequent inclusion of a submucosal subtype[17] (Table 30.1 and Figure 30.1).

More recently, the American Gastroenterological Association (AGA) proposed a classification that has been widely used in daily clinical practice, which subdivides the fistulas into two categories: simple and complex.[18]

Simple fistulas have an external opening, an internal opening and a single tract; they are not associated with abscesses, strictures, neoplasia and/or irradiated tissue. In simple fistulas, the tract affects less than 1/3 of the external anal sphincter (EAS). They include the following varieties: submucosal, intersphincteric and low trans-sphincteric (< 1/3 of the EAS), and do not affect other organs (bladder, prostate, and vagina).
Generally, simple fistulas respond better to surgical and drug therapies, and do not progress with fecal incontinence.[19-21]
Complex fistulas, in turn, present high tracts and affect more than 1/3 of EAS, including high trans-sphincteric, suprasphincteric and extrasphincteric fistulas. They often have multiple tracts, several internal and external openings, and can be associated with abscesses, strictures in the anal canal and rectum, or rectovaginal fistulas, being more difficult to control and showing higher rate of complications.

Because of the association with other perianal disorders (fissures, strictures, ulcers), other classifications have been proposed, such as The Cardiff Classification System, which did not gain acceptance and has not been used in daily clinical practice.[22]

Another classification used in some studies to evaluate perianal CD is the Perianal Disease Activity Index (PDAI), which is based on five criteria: anal secretions, pain, type of disease, sexual activity, and degree of local

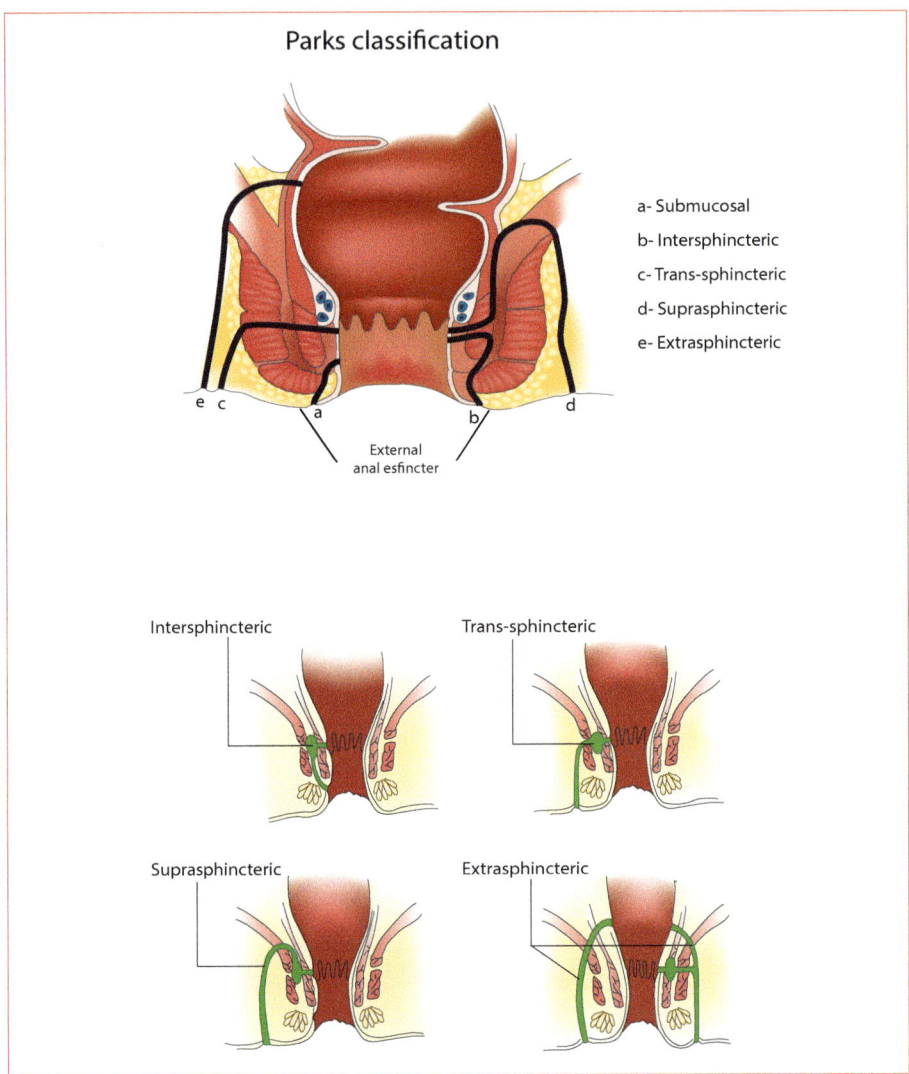

**Figure 30.1** Modified Parks classification.

| Table 30.1 Classification of anal fistulas as proposed by Parks et al. (modified) | |
|---|---|
| Submucosal fistula | Originates in the dentate line (anal crypt) has submucosal tract and does not affect the anal sphincter |
| Intersphincteric fistula | Originates in the dentate line (anal crypt) and the tract involves the IAS only |
| Trans-sphincteric fistula | Originates in the dentate line (anal crypt) and the tract involves the IAS and EAS. They can be subdivided into low, when the tract includes less than 1/3 of EAS, or high |
| Suprasphincteric fistula | Originates in the dentate line (anal crypt) and the tract extends cranially along the intersphincteric space, passes above the anorectal ring and descends through the ischiorectal space, draining out to the perineal skin |
| Extrasphincteric fistula | Originates in the skin or rectum; the tract crosses the ischiorectal fossa, draining out to the perineal skin |

IAS: internal anal sphincter; EAS: external anal sphincter.

induration. Each criterion is graded with points, from 0 (no symptoms) to 4 (intense/severe), but also needs validation.[23] The advantage of the PDAI is that in addition to clinical data, it also examines aspects related to quality of life.

## DIAGNOSIS

The main symptoms are anal pain, purulent or seropurulenta discharge, anal itching, burning and fever. Another frequent complaint is worsening of sexual activity.

On proctological examination, the external opening of the fistula is observed with pus drainage, perianal dermatitis, hyperemia, abscesses, inflammatory skin tags, fissures, ulcers, strictures, and more. The presence of anal fissures in atypical location (lateral or anterior), ulcers positioned above the dentate line, complex fistulas in patients with intestinal lesions and/or extraintestinal manifestations adds to the diagnostic suspicion. Often, proctological examination is only performed using anesthesia, on account of pain.

It is very important to make a careful assessment of fistulas and their relationship to the anal sphincter muscles, the degree of inflammatory activity and involvement of the rectum and anal canal, as well as the differential diagnosis of other diseases such as hidradenitis suppurativa, syphilis, neoplasms and lymphogranuloma venereum. Examination under anesthesia is the gold standard to confirm the presence of CD through biopsies, rule out other related disorders (cancer), drain abscesses/sepsis and properly assess the site's anatomy.

Endoscopic studies are important to evaluate the presence of inflammation and the degree of inflammatory activity in the colon and rectum, and to investigate areas of dysplasia and perform biopsies.

Fistulography has not been used due to pain and also the possibility of infection spreading locally.

The most frequently used imaging methods are endoanal ultrasound (US), computed tomography (CT) and magnetic resonance imaging (MRI), with the aim of staging CD and evaluating the degree of anorectal/perineal involvement, which facilitates surgical planning.

Endoanal US has been used to assess the presence of abscesses or deep collections, fistulas and internal opening (hard to diagnose on proctological examination); its accuracy varies from 56 to 100% when performed with hydrogen peroxide.[24-26] It is an operator-dependent method but costs less than MRI, can be used in the outpatient setting, and, combined with the examination under anesthesia, has 100% accuracy.[27] Its limitations include the presence of strictures in the anal canal, local pain and difficulty in assessing proximal lesions. Endoanal US has also been used to monitor therapeutic response, assessing perineal healing and inflammatory activity in fistulous tracts. Some studies show high rates of disagreement between clinical manifestations (symptoms) and the inflammatory activity inside the fistula, a factor responsible for complications and early recurrence.[28,29] Ardizzone et al. studied the use of endoanal US in 30 patients with CD and perianal fistula treated with anti-TNF for 10 weeks. They found that those with persistent active disease in the tracts, regardless of symptoms, had higher rates of recurrence and complications.[30]

Schwartz et al. performed a triple blind prospective study comparing US, MRI and examination under narcosis in patients with perianal CD. They noted that any method used alone had an accuracy rate higher than 91%; however, the combination of two methods led to 100%.[27]

Meta-analysis study comparing the accuracy of endoanal US and MRI in the assessment of anal fistula concluded that both have comparable sensitivity, but MRI has greater specificity.[31]

In a study with 71 patients with recurrent fistulas that used MRI to guide the surgical planning, reduced recurrence occurred in 75%. In cases where there was disagreement between surgical and MRI findings, the recurrence rate was 52%, and MRI identified the site of recurrence in all.[32]

Another study using MRI preoperatively to assist in the treatment of nonspecific and specific fistulas allowed the surgeon to modify the surgical plan in 21% of the first, and when associated with CD, the benefit was 40%.[33]

In addition to assisting in the evaluation of affected structures in CD patients with fistulas, MRI has also been used to assess and monitor the response to combination therapy (seton and biological agents) in these patients as it helps to detect disease activity in asymptomatic patients, minimizing recurrence and guiding the withdrawal of the seton.[34-36] It is known that the persistence of active inflammation not only increases the risk of perineal sepsis and abscess formation, but is also a cause of recurrent fistulizing disease, with the possibility of progression to serious complications such as anal stricture or fecal incontinence. Hence the importance of these imaging methods in detecting unapparent disease, guiding therapy and minimizing recurrence.

### Guidelines from the The Second European Evidence-based Consensus on the Diagnosis and Management of Crohn's Disease: Special Situations (ECCO)

The ECCO consensus recommended that MRI should be the examination used to evaluate fistulizing CD as it is noninvasive and has high accuracy, and it should not be indicated routinely for simple fistulas (strength of recommendation [SR] B, level of evidence [LE] B).

Endoanal US requires experience, it is operator-dependent, but can be equivalent to MRI if performed in accordance with the full protocol, using hydrogen peroxide in patients without rectal stricture (SR B, LE 2b).

Fistulography is not recommended (SR C, LE C). Examination under anesthesia is considered the gold standard only with experienced proctologists as it allows both the diagnosis and treatment of fistulizing disease (drainage and seton). This test requires an informed consent (IC) from the patient (SR D, LE 5).[37]

## TREATMENT

Treatment of perianal fistulas in CD has as a principle the initial treatment of perineal sepsis associated with control of the inflammatory process. This combined goal has shown the highest success rates. Keep in mind that surgical treatment will be employed in at least 80 to 90% of patients over the years, since it is a chronic and recurring disease, which has no cure.

Some placebo-controlled studies have shown that only 10 to 15% of patients will improve without any specific treatment (placebo group).[38,39]

### Assessment of mucosal healing

Regardless of the therapy employed, some patients may show improvement with decreased seropurulent secretion and local discomfort, which must obviously be distinguished from "cure", i.e. complete closure of fistulas (complete remission). Most trials with biological therapy use clinical response (defined as >50% of draining fistula for any two consecutive visits over a minimum of 4 weeks), and not clinical remission (complete healing of fistulas, which is defined as lack of secretion on digital compression of the perineal area for any two consecutive visits over a minimum of 4 weeks) as outcome.[38,40-42] Therefore, the most widely used tool to access treatment outcome is the Fistula Drainage Assessment (FDA), which classifies fistulas as open (no response or partial response) or closed, not taking into account local pain, among other factors.[38] A more detailed assessment of the deep healing of perianal fistulas in CD patients with clinical data and imaging methods (MRI), i.e. deep remission, should be the target for monitoring and evaluation. Closing and re-epithelialization of external openings in clinical studies, despite being evidence of clinical improvement, do not represent complete healing and "cure" of fistulas.

### Pharmacological treatment

#### Corticosteroids and salicylic derivatives

Corticosteroids and 5-ASA derivatives are not effective in controlling fistulizing disease, although they may be used in the treatment of luminal disease.[43]

## Antibiotics

Antibiotics (combined ciprofloxacin and metronidazole) are the first line treatment for perianal sepsis (abscesses and fistulas), typically in association with surgical drainage. Perineal disease usually improves after 2 to 4 weeks, and, in selected cases, can extend the antibiotic therapy for 3 to 4 months.[44,45] Solomon et al., using ciprofloxacin and metronidazole in 14 patients with fistulizing CD, observed clinical response in 62%.[45] A double-blind, placebo-controlled RCT comparing ciprofloxacin, metronidazole and placebo in 25 patients with fistulizing CD, treated for 10 weeks, found the following remission and clinical response: ciprofloxacin group (30% and 40%); metronidazole group (0% to 14%) and placebo (12.5% and 12.5%). No differences were observed in CDAI, PDAI and IBDQol scores during the study period.[46] Antibiotics have also been indicated as a "bridge" therapy, in combination with immunosuppressants (azathioprine – AZA), with good results.[47] The combination of ciprofloxacin with infliximab (IFX) *versus* IFX alone was tested in a double-blind, placebo-controlled trial in patients with perianal fistulas. No improvement was observed in the healing rate with the addition of antibiotic to IFX, as assessed by proctological examination and endoanal US.[48] In conclusion, antibiotics improve the symptoms resulting from perineal sepsis; however, used alone, they present low rates of healing, early recurrence and adverse effects. Adverse effects are more common with metronidazole and include metallic taste, glossitis, nausea, flatulence and peripheral neuropathy.

## Immunosuppressants

Regarding immunosuppressants (AZA, 2 to 2.5 mg/kg/day, and 6-mercaptopurine [6-MP], 1 to 1.5 mg/kg/day), until now, there is no RCT with perianal fistula healing considered as the primary endpoint. The effectiveness of immunosuppressants in fistulizing perianal CD has been suggested by meta-analysis of controlled trials in which the healing of fistulas is assessed as a secondary endpoint. In this meta-analysis of five controlled studies, the healing rate was 54% in the immunosuppression group and 21%

in the placebo group.[49] Camus et al. reported that 38% of patients with fistulas had sustained remission after 10 years of follow-up with AZA treatment and, in this group, they also observed decreased need for surgical treatment. Present et al., using 6-MP in fistulizing CD, found 31% of healing.[42] It can be concluded that low rate of healing, long time required for therapeutic response (14 to 16 weeks), side effects and high recurrence rate are factors that limit the use of immunosuppressants in perianal fistulizing CD.

### Methotrexate

Methotrexate is a folate antagonist, has immunomodulatory action and interferes with DNA synthesis. It has been indicated for patients intolerant to AZA/6-MP at a dose of 15 to 25 mg, intramuscularly every week. Adverse events occur in 10 to 25% of cases and include diarrhea, nausea, stomatitis, hair loss, hypersensitivity pneumonia, leukopenia, elevated transaminases and liver cirrhosis. Retrospective and uncontrolled studies with small case series have shown partially effective response in the control of perianal CD.[50,51] Schroder et al. studied 12 consecutive patients using IFX (5 mg/kg/dose) as induction therapy and as a "bridge" for intramuscular methotrexate (20 mg/week), and reported 33% rate of complete response, with closure of fistulas.[51] The lack of controlled studies in fistulizing CD, the absence of long-term response and the adverse effects of methotrexate do not support its indication as first-line treatment.

### Tacrolimus

Tacrolimus is an immunomodulatory agent, more precisely a macrolide antibiotic, used in liver transplants to prevent rejection. Randomized placebo-controlled study in fistulizing CD patients treated with tacrolimus (0.2 mg/kg/day) for 10 weeks showed good initial clinical response. However, complete closure of the fistula (remission) with secretion drainage cessation occurred in only 10%.[52] A study in patients with fistulas and unresponsive to IFX using tacrolimus at a dose of 0.05 mg/kg every 12 hours found complete response in 40% after follow-up of 6 to 24 months.[53] In this study, Gonzales Lama et al. called attention to the need for therapy for a longer time to achieve complete healing. Sandborn et al., in a placebo-controlled study using tacrolimus orally for 10 weeks in 42 patients with CD and perianal fistulas, found more evident clinical improvement (clinical response) in the tacrolimus group (45% *versus* 9%), but no improvement in healing rates (clinical remission) of anal lesions.[39] The topical use of tacrolimus was tested in 19 patients with perianal disease, classified as fistulizing or ulcerated lesions. The benefit was evident in ulcers but not in the fistulizing group, and adverse effects were not reported in this formulation.[54] In general, the main adverse effects include headache, tremor, paresthesias, insomnia and renal failure. It is therefore important to monitor the serum levels of the drug and renal function. It can be concluded that, based on the results of clinical trials, tacrolimus is reserved for patients unresponsive to biological therapy.

### Cyclosporine

To date, there are no randomized, double-blind, placebo-controlled studies having the closing of fistulizing CD as their primary endpoint. Some open studies with case series showed good initial response, but high rates of relapse and serious and severe adverse effects have limited its use.[55,56] Many patients show good initial response to intravenous medication, but exhibit early relapse when treatment is changed to oral or when the drug is withdrawn.

### Biological therapy

The introduction of biological therapy in 1998 for the treatment of CD was undoubtedly a turning point in the management of this disease and represents a major breakthrough in controlling fistulizing disease. The first placebo-controlled study was conducted by Present et al. in 1999, using IFX in 94 patients with CD and perianal fistulas. These 94 patients were treated with IFX at a dose of 5 mg/kg at weeks 0, 2 and 6. Clinical response (50% reduction of draining fistula for a minimum of two consecutive visits over an interval of 4 weeks) was observed in 68%, and remission with complete closure of fistulas in 55% in the IFX group and 13% in the placebo group, after 14 weeks of follow-up.[38]

In the ACCENT II study, 282 patients received IFX (5 mg/kg) at weeks 0, 2 and 6, and the responders were randomized at week 14 to receive placebo or IFX every 8 weeks until week 54. The clinical response rate at week 54 was 46% (IFX group) and 23% (placebo group), and complete closure of the fistula (remission) was achieved by 36% (IFX) and 19% (placebo).[57] Regarding rectovaginal fistulas, the initial response rate was 64%.

In 2005, Lichtenstein et al. assessed the impact that maintenance therapy had on these patients (ACCENT II) regarding the need for hospitalization and surgical treatment, and quality of life. They reported a reduction of over 50% in hospital admissions and surgeries, and, in larger operations (abdominal), the reduction exceeded 80%.[58] A similar result was found by Ng et al., who, in addition to confirming a reduction in hospitalizations and the number of surgeries with anti-TNF therapy, found improvement in the quality of life of patients with perianal fistulas.[59] A question that still needs to be properly answered is the duration of therapy with anti-TNF in patients with penetrating disease, since it is known that with the cessation of the drug, the risk of recurrence is higher in fistulizing perianal disease than in luminal inflammatory disease. Domenech et al. followed for 12 months patients with luminal and fistulizing perianal disease with complete response after IFX therapy at induction and maintenance doses. They reported that relapse was observed in 66% of patients with perianal fistulizing CD, and only 17% of the luminal disease group.[60]

Observational study conducted in Hungary that included 148 patients with CD and perianal fistulas treated with IFX found that, after 12 months of follow-up, 49% were in clinical remission with complete closure of fistulous openings.[61]

With respect to IFX, some studies revealed that, when combined with an immunosuppressant, response to the drug is more efficient in perianal CD. However, the risk *versus* benefit ratio should always be considered, especially in young individuals.[62,63] Response in perianal disease is better than that observed in enteric/colocutaneous fistulas and/or internal fistulas.[64] IFX has also been used in the fistula's opening and tract after careful curettage, in cases where the disease is limited to the anal region and in those with contraindications to systemic use, or as a booster in intravenous therapy. In 2005, Poggioli et al. treated 15 patients with local infusions of IFX (3-12 applications) with complete healing of fistulas occurring in 10 out of 15 (66.7%).[65] Asteria et al. observed clinical response in 6 (54.5%) of 11 patients treated locally with IFX, and 4 (4/11 = 36.3%) remained with healed fistulas (remission) after a mean follow-up of 10 months.[66]

With regard to adalimumab (ADA), its effectiveness was evaluated by a randomized study (Crohn's Trial of the Fully Human Antibody Adalimumab for Remission Maintenance – CHARM) with 113 patients with fistulizing CD; clinical response and remission were analyzed as secondary outcome variables. Healing of the fistulas was observed in 39% of patients at week 56 (placebo = 13%).[67] In this study, early healing (week 26) occurred in 30% and was a positive predictor of sustained response (week 56). After 24 months of follow-up, 60% (23/37) of these patients remained with their fistulas closed.[68] In the ADHERE study, an open extension that included all patients who had successfully completed (responders) the CHARM study, 117 patients with moderate to severe CD (CDAI between 220 and 450) presenting fistulizing disease with drainage in the study baseline were evaluated. The authors found that 75% of fistulas remained closed in the long term (2 years), and ADA was well tolerated with a good safety profile.[69] In addition to the use of ADA as the drug of choice, other studies were performed including patients intolerant or unresponsive to IFX. In an open study of 22 patients unresponsive or intolerant to IFX, induction with ADA was performed (160/80 mg) and remission with healed fistulas was achieved in 23% after 4 weeks.[70] Tonelli et al. used local infusions with ADA in 12 patients with perianal fistulas (3 anovaginal, 7 trans-sphincteric, and 2 complex fistulas), of which 9 were female patients. The average number of infusions per patient was 7 (4 to 16), and the mean follow-up period was 17 months (5 to 30). Complete fistula healing was observed in 75%, and 3 (25%) showed improvement with decreased drainage volume. It was concluded that high concentrations of ADA in the fistula's openings and tract promotes healing.[71]

Certolizumab pegol studies (Pegylated Antibody Fragment Evaluation in Crohn's Disease Safety and Efficacy – NEED 2) to evaluate efficacy in the treatment of fistulizing CD have been performed, with healing of fistulas as a secondary outcome. In the induction period, 73% of the certolizumab pegol group maintained at least 50% of fistulas closed at week 26, *versus* 39% in the placebo group. In the maintenance period (week 26), 67% continued with healed fistulas, *versus* 31% in the placebo group. Despite the numerical advantage, there was no statistical difference.[72] Subsequent open-label extension studies using subcutaneous certolizumab pegol reported clinical response (decreased secretion, with closure of more than 50% at two consecutive visits over an interval of 3 weeks) in 54% of patients at week 26.[73]

Although there is still no study comparing the effectiveness of different anti-TNFs in healing of abdominal or perineal fistulizing disease, a meta-analysis published in 2008 with three different anti-TNF drugs showed good response in the control of fistulas compared to placebo.[74]

### Hyperbaric oxygen therapy

Hyperbaric oxygen (HBO) therapy using $O_2$ at 100% in chambers with a pressure of two atmospheres (ATA) in 90- to 120-minute sessions have been used in severe perianal CD patients with encouraging results. On average, 20 to 40 sessions of hyperbaric oxygen therapy are required depending on the disease progression. These studies are small series of uncontrolled cases that do not allow definitive conclusions about the actual effectiveness.[75,76] Noyer and Brandt conducted a retrospective study and evaluated the response to HBO therapy of 22 patients with active and refractory perianal CD. Clinical response was seen in 73% of patients.[77] Its mechanism of action is multifactorial and is characterized by improved oxygenation of inflamed tissue, assisting phagocytosis, a bactericidal effect, and the reduction of proinflammatory cytokines (IL-1, IL-6 and TNF), while also stimulating healing. Due to a lack of randomized controlled studies, it has not been indicated as a routine therapy.

### Surgical treatment

Despite the good initial results observed with the use of biologic therapy in the treatment of fistulizing CD, the latest data do not show very clearly a decrease in the number of surgeries for perianal fistulas. The mere presence of penetrating disease is a predictor of worse outcome, and the vast majority of authors recommend a more aggressive approach (*top-down*). On the other hand, some studies have shown that biological therapy alone may be associated with an increase in new or recurrent abscesses, when it is not preceded by surgery with drainage and seton placement.[20,43,57] This is probably due to the closure of the external opening of the fistula without proper healing of the fistula tract, with consequent accumulation of secretion and development of the abscess.[21,35,38,57] Repeated abscesses and complex fistulas, as well as numerous drainages performed in emergency situations, lead to destruction of the anal sphincters, with consequent anal incontinence and/or strictures.

Importantly, surgical treatment of perianal disease in patients with CD should always be conservative, avoiding extensive tissue resections and large sphincterotomies, limiting the conduct to the treatment of perineal sepsis with placement of loose setons (non-cutting) (Figures 30.2 and 30.3).

Drainage should be carried out immediately in the presence of abscess. In cases of simple anal fistulas (intersphincteric and low), if asymptomatic, nothing should be done; but if they are symptomatic, a non-cutting seton should be placed. Fistulotomy or fistulectomy should be avoided due to the risk of incontinence, but can be performed in selected cases. In complex fistulas, the procedure of choice is placement of seton, which also treats the intestinal lesion (small intestine, colon, or rectum), if any.

The surgeon should make a thorough assessment under anesthesia, trying to identify superficial and deep collections, inspecting fistulas, performing adequate drainage and placing loose setons in fistulous tracts, always with antibiotic coverage. Early seton removal should be avoided due to the risk of early recurrence of perianal sepsis. After drainage of sepsis and control of local infection, the underlying disease therapy (with immunosuppressant or anti-TNF) should be started early, trying to control the chronic inflammatory process. According to the ECCO consensus, maintenance therapy should be continued for at least 12 months.[37]

The best results are obtained with combined therapy: surgery to control infection and biological agents to control inflammation.

Regueiro and Mardini, from the University of Pittsburg, in a study with 32 patients with fistulizing perianal CD treated with surgery and IFX *versus* IFX alone, found better results with respect to healing (100% *versus* 82.6%), lower recurrence (44% *versus* 79%) and longer time to recurrence (13.5 months *versus* 3.6 months) in the combined therapy group.[20]

A retrospective study conducted at the University of Calgary (Canada) evaluated 29 consecutive patients with fistulizing perianal disease (21 perianal, rectovaginal 8 and 4 combined). The authors concluded that surgery with placement of loose and comfortable setons, combined with infusion of IFX and maintenance with immunosuppressants, resulted in complete healing in

67%, and partial healing in 19%. They also found worse outcome in cases of rectovaginal fistula.[21] Schwartz et al. conducted a retrospective study of case series using combined therapy (surgery + biological agent) and found early response (14 weeks) with fistula closure in 86%, which persisted in 76% after a mean follow-up of 68 weeks.[78]

El-Gazzaz et al. divided 218 patients with CD and perianal fistulas into two groups, according to therapy: group A (surgical) including 117 (53.7%) patients, and group B (surgery + biological agent) including 101 patients, with a mean follow-up 3.2 years. The authors obtained better response with combination therapy compared with surgical treatment alone (71.3% *versus* 35.9%).[79] They also noted the presence of granulomas in 11.5% (25/218) of these patients and, after performing a separate analysis of these cases, they found better response to biological therapy in this group, characterized by greater healing rate in fistula tracts and less need for ostomy or proctectomy.

Although some authors report the same efficacy of surgery alone compared with combination therapy, these studies are mostly retrospective, series of cases, with patients treated with IFX only as induction therapy.[80-82]

Gaertner et al., in a retrospective study of 226 consecutive patients with fistulizing CD (surgery alone = 147 and surgery + IFX = 79), found healing rates similar to those found with biological therapy with or without surgery. Complete healing of fistulas occurred in approximately 60% of patients.[81]

Bouguen et al., in a bicenter study of 156 patients treated for long periods with seton, IFX and immunosuppressants, noted that the cumulative chance of closure of the perianal fistulizing disease was 73% after 5 years, and 88% in 10 years. They concluded that combination therapy, placement of seton for less than 34 weeks and prolonged biological therapy were the factors associated with better results.[83] Sciaudone et al. analyzed 35 consecutive patients with complex fistulas and CD, which were prospectively divided into three groups: IFX, surgery and combined therapy. Best results were obtained with the combined therapy: healing ratio (63.6% *versus* 69.9% and 78.5%), and longer time to recurrence (2.6 months *versus* 3.6 months *versus* 10.1 months).[84]

It can be concluded that with regard to treatment of perianal fistulizing CD, combination therapy is the gold standard. An initial approach with surgery is important to confirm the case as CD, to rule out the presence of cancer in long-lasting disease (> 8 years), to drain the anorectal sepsis, to inspect the fistulas, and to evaluate the anal canal and anal sphincters, so that inflammation can be addressed with drug therapy (immunosuppressant, anti-TNF). The presence of active proctitis associated with perianal or rectovaginal fistulizing disease is undoubtedly a negative predictor of response to biological therapy.[21,35]

In refractory cases with persistent tract and absence of proctitis, other surgical options may be indicated, such as rectal advancement flap, fillers (fibrin sealant and collagen plug), and more, with contrasting results.[85]

Colostomy or proctectomy is reserved for cases of severe proctitis associated with stenosis, destruction of the sphincter with fecal incontinence, recurrent and wide rectovaginal fistula, and in cases of failure with other therapies.[86]

## CONCLUSIONS

Perianal fistulas occur in 20 to 35% of CD patients, especially in cases of colonic involvement.

The presence of penetrating disease is a factor of poor prognosis, with more aggressive phenotype.

Proper early diagnosis of perianal CD (type of fistula, tract, collections) and staging (presence of concomitant intestinal disease) are critical for therapeutic planning, minimizing iatrogenic complications.

The management of perianal CD requires a multidisciplinary approach, using imaging methods (endoanal US, MRI and CT), surgery (drainage and seton) and appropriate drug therapy.

Drug therapy is performed with antibiotics (ciprofloxacin, metronidazole), immunosuppressants (AZA and 6-MP) and anti-TNF therapy.

In cases of failure, other medical (oral and topical tacrolimus, oral and intravenous methotrexate or cyclosporine, hyperbaric oxygen therapy) or surgical (flap advancement, fibrin glue, collagen plug, ostomy and proctectomy) options can be used in order to improve the quality of life.

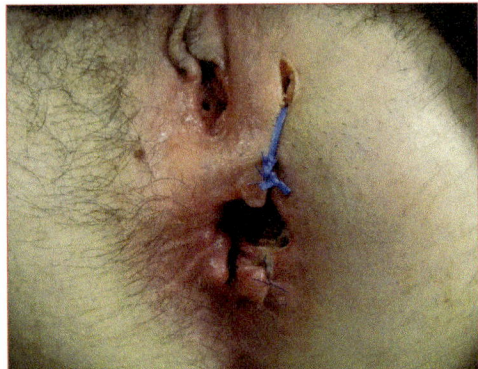

**Figure 30.2** Examination under anesthesia with drainage of perineal sepsis and placement of a non-cutting seton (loose).

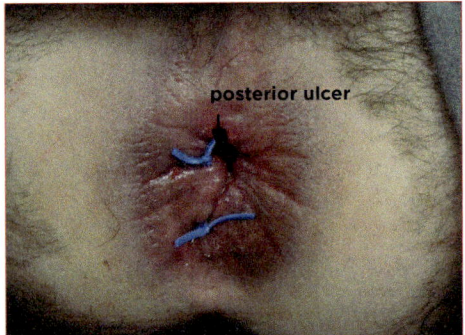

**Figure 30.3** Patient with perianal complex disease, with multiple fistulas and ulcer on the posterior wall.

## REFERENCES

1. Crohn BB, Ginzburg L, Oppenheimer GD. Regional ileitis: a pathological and clinical entity. JAMA 1932; 99:1323-9.

2. Penner A, Crohn BB. Perianal fistulae as a complication of regional ileitis. Ann Surg 1938; 108:867-73.

3. Fielding JF. Perianal lesions in Crohn's disease. J R Coll Surg Edinb 1972; 17:32-7.

4. Lockhart-Mummery HE. Symposium-Crohn's disease: anal lesions. Dis Colon Rectum 1975; 18:200-3.

5. Jaisson-Hot I, Flourié, Descos L, Colin C. Management for severe Crohn's disease: a lifetime cost-utility analysis. Int J Technol Assess Health Care 2004; 20:274-9.

6. Platell C, Mackay, Collopy B, Fink R, Ryan P, Woods R. Anal pathology in patients with Crohn's disease. Aust N Z J Surg 1996; 66(1):5-9.

7. Hellers G, Bergstrand O, Ewerth S, Hostrom B. Occurrence and outcome after primary treatment of anal fistulae in Crohn's disease. Gut 1980; 21:525-7.

8. Schwartz D, Loftus EV Jr, Tremaine W, Panaccione R, Harmsen WS, Zinsmeister AR et al. The natural history of fistulising Crohn's disease in Olmsted County, Minnesota. Gastroenterology 2002; 122:875-80.

9. Sangwan YP, Schoetz Jr DJ, Murray JJ, Roberts PL, Coller JA. Perianal Crohn's disease. Results of local surgical treatment. Dis Colon Rectum 1996; 39:529-35.

10. Williams DR, Coller JA, Corman ML et al. Anal complications in Crohn's disease. Dis Colon Rectum 1981; 24:22-4.

11. Keighley MR, Allan RN. Current status and influence of operations on perianal Crohn's disease. Int J Colorectal Dis 1986; 1:104-7.

12. Sjodahl RI, Myrelid P, Soderholm JD. Anal and rectal cancer in Crohn's disease. Colorectal Dis 2003; 5:490-5.

13. Laurent S, Barbeaux A, Detroz B, Detry O, Louis E, Belaiche J et al. Development of adenocarcinoma in chronic fistula in Crohn's disease. Acta Gastroenterol Belg 2005; 68:98-100.

14. Singh B, Mc Mortensen NJ, Jewell DP et al. Perianal Crohn's disease. Br J Surg 2004; 91:801-14.

15. West RL, Van der Woude CJ, Endtz HP, Hansen BE, Ouwedijk M, Boelens HA et al. Perianal fistulas in Crohn's disease are predominantly colonized by skin flora: implications for antibiotics treatment? Dig Dis Sci 2005; 50:1260-3.

16. Parks AG, Gordon PH, Hardcastle JD. A classification of fistula-in-ano. Br J Surg 1976; 63:1-12.

17. Rizzo JA, Naig AL, Johnson EK. Anorectal abscess and fistula-in-ano: Evidence-Based management. Surg Clin N Am 2010; 90:45-68.

18. American Gastroenterological Association. American Gastroenterological Association technical review on perianal Crohn's disease. Gastroenterology 2003; 125:1508-30.

19. Bell SJ, Williams AB, Wiesel P, Wilkinson K, Cohen RC, Kamm MA. The clinical course of fistulating Crohn's disease. Aliment Pharmacol Ther 2003; 17:1145-51.

20. Regueiro M, Mardini H. Treatment of perianal fistulizing Crohn's disease with infliximab alone or as an adjunct to exam under anesthesia with seton placement. Inflamm Bowel Dis 2003; 9:98-103.

21. Topstad DR, Panaccione R, Heine JA, Johnson DR, MacLean AR, Buie WD. Combined seton placement, infliximab infusion and maintenance immunessuppresives improve healing rate in fistulizing anorectal Crohn's disease: a single center experience. Dis Colon Rectum 2003; 46:577-83.

22. Hughes LE. Clinical classification of perianal Crohn's disease. Dis Colon Rectum 1992; 35:928-32.

23. Irvine EJ. Usual therapy improves perianal Crohn's disease as measured by a new disease activity index. J Clin Gastroenterol 1995; 20:27-32.

24. Orsoni P, Barthet M, Portier F, Panuel M, Desjeux A, Grimaud JC. Prospective comparison of endosonography, magnetic resonance imaging and surgical findings in anorectal fistula and abscess complicating Crohn's disease. Br J Surg 1999; 86:360-4.

25. Stewart LK, McGee J, Wilson SR. Transperineal and transvaginal sonography of perianal inflammatory disease. Am J Roentgenol 2001; 177:627-32.

26. Sloots CE, Felt-Bersma RJ, Poen AC, Cuesta MA, Meuwissen SG. Assessment and classification of fistula-in-ano in patients with Crohn's disease by hydrogen peroxide enhanced transanal ultrasound. Int J Colorectal Dis 2001; 16:292-7.

27. Schwartz DA, Wiersema MJ, Dudiak KM, Fletcher JG, Clain JE, Tremaine WJ et al. A comparison of endoscopic ultrasound, magnetic resonance imaging and exam under anesthesia for evaluation of Crohn´s perianal fistulas. Gastroenterology 2001; 121:1064-72.

28. Rasul I, Wilson S, Cohen Z et al. Infliximab therapy for Crohn´s disease fistulae: discordance between perianal ultrasound findings and clinical response. Gastroenterology 2001;120:A619-A623.

29. Van Bodegraven AA, Sloots CE, Felt-Bersma RJ, Meuwissen SG. Endosonographic evidence of persistence of Crohn's disease-associated fistulas after infliximab treatment, irrespective of clinical response. Dis Colon Rectum 2002; 45:39-45.

30. Ardizzone S, Maconi G, Colombo E, Manzionna G, Bollani S, Bianchi Porro G. Perianal fistulae following infliximab treatment: clinical and endosonographic outcome. Inflamm Bowel Dis 2004; 10:91-6.

31. Siddiqui MRS, Ashrafian H, Tozer P, Daulatzai N, Burling D, Hart A et al. A diagnostic accuracy meta-analysis of endoanal ultrasound and MRI for perianal fistula assessment. Dis Colon Rectum 2012; 55:576-85.

32. Buchanan G, Halligan S, Williams A, Cohen CR, Tarroni D, Phillips RK et al. Effect of MRI on clinical outcome of recurrent fistula-in-ano. Lancet 2002; 360:1661-2.

33. Beets-Tan RG, Beets GL, van der Hoop AG, Kessels AG, Vliegen RF, Baeten CG et al. Preoperative MRI imaging of anal fistulas: does it really help the surgeon? Radiology 2001; 218:75-84.

34. Van Assche G, Vanbeckevoort D, Bielen D, Coremans G, Aerden I, Noman M et al. Magnetic resonance imaging of the effects of infliximab on perianal fistulizing Crohn's disease. Am J Gastroenterol 2003; 98:332-9.

35. Ng SC, Plamondon S, Gupta A, Burling D, Swatton A, Vaizey CJ et al. Prospective evaluation of anti-tumor necrosis factor therapy guided by magnetic resonance imaging for Crohn's perineal fistulas. Am J Gastroenterol 2009; 104:2973-86.

36. O'Malley RB, Al-Hawary MM, Kaza RK, Wasnik AP, Liu PS, Hussain HK. Rectal imaging: Part 2, perianal fistula evaluation on pelvic MRI-What the radiologist needs to know. AJR 2012; 199:W43-W53.

37. Van Assche G, Dignass A, Reinisch W, van der Woude CJ, Sturm A, De Vos M et al. The second European evidence-based consensus on the diagnosis and management of Crohn's disease: special situations. J Crohn's Colitis 2010; 4:63-1001.

38. Present DH, Rutgeerts P, Targan S, Hanauer SB, Mayer L, van Hogezand RA et al. Infliximab for the treatment of fistulas in patients with Crohn's disease. N Engl J Med 1999; 340(18):1398-405.

39. Sandborn WJ, Present DH, Isaacs KL, Wolf DC, Greenberg E, Hanauer SB et al. Tacrolimus for the treatment of fistulas in patients with Crohn's disease: a randomized placebo-controlled trial. Gastroenterology 2003; 125:380-8.

40. Sandborn WJ, Fazio VW, Feagan BG, Hanauer SB. AGA technical review on perianal Crohn's disease. Gastroenterology 2003; 125:1508-30.

41. Bernstein LH, Frank MS, Brandt LJ, Boley SJ. Healing of perianal Crohn's disease with metronidazole. Gastroenterology 1980; 79:599-605.

42. Present DH, Korelitz BI, Wisch N, Glass JL, Sachar DB, Pasternack BS. Treatment of Crohn's disease with 6-mercaptopurine. A long-term, randomized, double-blind study. N Engl J Med 1980; 302:981-7.

43. Tozer PJ, Burling D, Gupta A, Phillips RK, Hart AL. Review article: medical, surgical and radiological management of perianal Crohn's fistulas. Aliment Pharmacol Ther 2011; 33:5-22.

**44.** Jacobovits J, Schuster MM. Metronidazole therapy for Crohn's disease and associated fistulae. Am J Gastroenterol 1984; 79:533-40.

**45.** Solomon MJ, McLeod RS, O'Connor BJ et al. Combination ciprofloxacin and metronidazole in severe perineal Crohn's disease. Can J Gastroenterol 1993; 7:571-3.

**46.** Thia KT, Mahadevan U, Feagan BG, Wong C, Cockeram A, Bitton A et al. Ciprofloxacin or metronidazole for the treatment of perianal fistulas in patient with Crohn's disease: a randomized, double-blind, placebo-controlled pilot study. Inflamm Bowel Dis 2009; 15:17-24.

**47.** Dejaco C, Harrer M, Waldhoer T, Miehsler W, Vogelsang H, Reinisch W. Antibiotics and azathioprine for the treatment of perianal fistulas in Crohn's disease. Aliment Phamacol Ther 2003; 18:1113-20.

**48.** West RL, Van der Woude CJ, Hansen BE, Felt-Bersma RJ, van Tilburg AJ, Drapers JA et al. Clinical and endosonographic effects of ciprofloxacin treatment of perianal fistulae in Crohn's disease with infliximab: a double-blind, placebo-controlled study. Aliment Pharmacol Ther 2004; 20:1329-36.

**49.** Pearson DC, May GR, Fick GH, Sutherland LR. Azathioprine and 6-mercaptopurine in Crohn's disease. A meta-analysis. Ann Intern Med 1995; 122:132-42.

**50.** Mahadevan U, Marion JF, Present DH. Fistula response to methotrexate in Crohn's disease: a case series. Aliment Pharmacol Ther 2003; 18(10):1003-8.

**51.** Schröder O, Blumenstein I, Schulte-Bockholt A, Stein J. Combining infliximab and methotrexate in fistulizing Crohn's disease resistant or intolerante to azathioprine. Aliment Pharmacol Ther 2004; 19(3):295-301.

**52.** Sandborn WJ, Feagan BG, Hanauer SB, Lochs H, Löfberg R, Modigliani R et al. A review of activity indices and efficacy endpoints for clinical trials of medical therapy in adults with Crohn's disease. Gastroenterology 2002; 122:512-30.

**53.** Gonzalez Lama Y, Abreu LE, Vera MI, Pastrana M, Tabernero S, Revilla J et al. Long-term oral tacrolimus in refractary to infliximab fistulizing Crohn's disease: a pilot study. Inflamm Bowel Dis 2005; 11:8-15.

**54.** Hart AL, Plamondon S, Kamm MA. Topical tacrolimus in the treatment of perianal Crohn's disease: exploratory randomized controlled trial. Inflamm Bowel Dis 2007; 13:245-53.

**55.** Lichtiger S. Cyclosporine therapy in inflammatory bowel disease: open-label experience. Mt Sinai J Med 1990; 57:315-9.

**56.** Hanauer SB, Smith MB. Rapid closure of Crohn's disease fistulas with continuous intravenous cyclosporine A. Am J Gastroenterol 1993; 88:646-9.

**57.** Sands BE, Anderson FH, Bernstein CN, Chey WY, Feagan BG, Fedorak RN et al. Infliximab maintenance therapy for fistulizing Crohn's disease. N Engl J Med 2004; 350:876-85.

**58.** Lichtenstein GR, Yan S, Bala M, Blank M, Sands BE. Infliximab maintenance treatment reduces hospitalizations, surgeries, and procedures in fistulizing Crohn's disease. Gastroenterology 2005; 128:862-9.

**59.** Ng SC, Plamondon S, Gupta A, Burling D, Kamm MA. Prospective assessment of the effect on quality of life of anti-tumor necrosis factor therapy for perianal Crohn's fistulas. Aliment Pharmacol Ther 2009; 30:757-66.

**60.** Domenech E, Hinojosa J, Nos P, Garcia-Planella E, Cabré E, Bernal I et al. Clinical evolution of luminal and perianal Crohn's disease after inducing remission with infliximab: how long should be patients be treated? Aliment Pharmacol Ther 2005; 22:1107-13.

**61.** Miheller P, Lakatos PL, Horvath G, Molnár T, Szamosi T, Czeglédi Z et al. Efficacy and safety of infliximab induction therapy in Crohn's disease in central Europe: a Hungarian nationwide observational study. BMC Gastroenterol 2009; 9:66-72.

**62.** Lichtenstein GR, Diamond RH, Wagner CL, Fasanmade AA, Olson AD, Marano CW et al. Clinical trial: benefits and risks of immunomodulators and maintenance infliximab for IBD-subgroup analyses across four randomized trials. Aliment Pharmacol Ther 2009; 30:210-26.

**63.** Sokol H, Seksik P, Carrat F, Nion-Larmurier I, Vienne A, Beaugerie L et al. Usefulness of co-treatment with immunomodulators in patients with inflammatory bowel disease treated with scheduled infliximab maintenance therapy. Gut 2010; 59:1363-8.

**64.** Parsi MA, Lashner BA, Achkar JP, Connor JT, Brzezinski A. Type of fistula determines response to infliximab in patients with fistulous Crohn's disease. Am J Gastroenterol 2004; 99:445-9.

**65.** Poggioli G, Laureti S, Pierangeli F, Bazzi P, Coscia M, Gentilini L et al. Local injection of infliximab for the treatment of perianal Crohn's disease. Dis Colon Rectum 2005; 48:768-74.

**66.** Asteria CR, Ficari F, Bagnoli S, Milla M, Tonelli F. Treatment of perianal fistulas in Crohn's disease by local injection of antibody to TNF-alpha accounts for a favorable clinical response in selected cases: a pilot study. Scand J Gastroenterol 2006; 41:1064-72.

**67.** Colombel JF, Sandborn WJ, Rutgeerts P, Kamm MA, Yu AP, Wu EQ et al. Adalimumab for the maintenance of clinical response and remission in patients with Crohn's disease: the CHARM trial. Gastroenterology 2007; 132(1):52-65.

**68.** Panacione R, Colombel JF, Sandborn WJ, Rutgeerts P, D'Haens GR, Robinson AM et al. Adalimumab sustains clinical remission and overall clinical benefit after 2 years of therapy for Crohn's disease. Aliment Pharmacol Ther 2010; 31(12):1296-309.

**69.** Colombel JF, Schwartz DA, Sandborn WJ, Kamm MA, D'Haens G, Rutgeerts P et al. Adalimumab for the treatment of fistulas in patients with Crohn's disease. Gut 2009; 58:940-8.

**70.** Hinojosa J, Gomollon F, Garcia S, Bastida G, Cabriada JL, Saro C et al. Efficacy and safety of short-term adalimumab treatment in patients with active Crohn's disease who lost response or showed intolerance to infliximab: a prospective open-label, multicentre trial. Aliment Pharmacol Ther 2007; 25:409-18.

**71.** Tonelli F, Giudice F, Asteria CR. Effectiveness and safety of local adalimumab injection in patients with fistulizing perianal Crohn's disease: a pilot study. Dis Colon Rectum 2012; 55:870-5.

**72.** Schreiber S, Khaliq-Kareemi M, Lawrance IC, Thomsen OØ, Hanauer SB, McColm J et al. Maintenance therapy with certolizumabe pegol for Crohn's disease. N Engl J Med 2007;357:239-50.

**73.** Schreiber S, Lawrance IC, Thomsen OØ, Hanauer SB, Bloomfield R, Sandborn WJ. Randomised clinical trial: certolizumab pegol for fistulas in Crohn's disease-sugroup results from a placebo-controlled study. Aliment Pharmacol Ther 2011; 33:185-93.

**74.** Peyrin-Biroulet L, Deltenre P, de Suray N, Branche J, Sandborn WJ, Colombel JF. Efficacy and safety of tumor necrosis factor antagonists in Crohn's disease: meta-analysis of placebo-controlled trials. Clin Gastroenterol Hepatol 2008; 6:644-53.

**75.** Colombel JF, Mathieu D, Bouault JM, Lesage X, Zavadil P, Quandalle P et al. Hyperbaric oxygenation in severe perineal Crohn's disease. Dis Colon Rectum 1995; 38:609-14.

**76.** Lavy A, Weisz G, Adir Y, Ramon Y, Melamed Y, Eidelman S. Hyperbaric oxygen for perianal Crohn's disease. J Clin Gastroentrol 1994; 19:202-5.

**77.** Noyer CM, Brandt LJ. Hyperbaric oxygen therapy for perineal Crohn's disease. Am J Gastroenterol 1999; 94:318-21.

**78.** Schwartz DA, White CM, Wise PE, Herline AJ. Use of endoscopic ultrasound to guide combination medical and surgical therapy for patients with perianal fistulas. Inflamm Bowel Dis 2005; 11:727-32.

**79.** El-Gazzaz G, Hull T, Church JM. Biological immunomodulators improve the healing rate in surgically treated perianal Crohn's fistulas. Colorectal Dis 2012; 14:1217-23.

**80.** Hyder SA, Travis SP, Jewell DP, McC Mortensen NJ, George BD. Fistulating anal Crohn's disease: results of combined surgical and infliximab treatment. Dis Colon Rectum 2006; 49:1837-41.

**81.** Gaertner WB, Decanini A, Mellgren A, Lowry AC, Goldberg SM, Madoff RD et al. Does infliximab infusion impact results of operative treatment for Crohn's perianal fistulas? Dis Colon Rectum 2007; 50:1754-60.

**82.** Uchino M, Ikeuchi H, Bando I, Matsuoka H, Takesue Y, Takahashi Y et al. Long term efficacy of infliximab maintenance therapy for perianal Crohn's disease. World J Gastroenterol 2011; 17:1174-9.

**83.** Bouguen G, Siproudhis L, Gizard E, Wallenhorst T, Billioud V, Bretagne JF et al. Long-term outcome of perianal fistulizing Crohn's disease treated with infliximab. Clin Gastroenterol Hepatol 2013; 11:975-81.

**84.** Sciaudone G, Stazio CD, Limongelli P, Guadagni I, Pellino G, Riegler G et al. Treatment of complex perianal fistulas in Crohn disease: infliximab, surgery or combined approach. Can J Surg 2010; 53:299-304.

**85.** Gaertner WB, Madoff RD, Spencer MP, Mellgren A, Goldberg SM, Lowry AC. Results of combined medical and surgical treatment of rectovaginal fistula in Crohn's disease. Colorectal Dis 2011; 13(6):678-83.

**86.** Siemanowski B, Regueiro M. Management of perianal fistula in Crohn's disease. Inflamm Bowel Dis 2008; 14:S266-S268.

# NUTRITION IN INFLAMMATORY BOWEL DISEASE

MARIA DE LOURDES TEIXEIRA DA SILVA
MARIA IZABEL LAMOUNIER DE VASCONCELOS

## INTRODUCTION

Malnutrition is frequent in inflammatory bowel disease (IBD) and stems from a systemic inflammatory response, inadequate oral intake, reduction of absorption and deficiencies in vitamins and minerals (Table 31.1). The prevalence of malnutrition increases in hospitalized patients due to greater severity of the disease and comorbidities, and varies from 44 to 48%.[1] Interest in the nutritional status of patients with Crohn's disease (CD) is justified to the measure in which nutritional losses, when accompanying their evolution, can bring disastrous consequences for the patient. Malnutrition has a negative effect on clinical evolution and post-operative complication rates and mortality.

Nutritional therapy is considered a supportive treatment in IBD when aimed at correcting malnutrition, restoring macro and micronutrient deficiencies and reversing their metabolic and pathological consequences.[2] This goal can be achieved in all patients with IBD.

It is considered a primary treatment when it can be replaced by conventional pharmacological treatment without loss of efficacy, as shown in studies of pediatric patients with CD.[3]

| Table 31.1 Multifactorial causes of protein-calorie malnutrition in DC |
|---|
| Systemic inflammatory response |
| Increase of inflammatory cytokines (TNF-alpha, IL-1, IL-6) |
| Increase in adipokines (leptin, adipokines, resistins) |
| Decreased oral intake |
| Anorexia |
| Change in taste |
| Abdominal pain or abdominal discomfort, diarrhea, nausea, vomiting |
| Iatrogenic dietary restriction |
| Inadequate dietary supply |
| Malabsorption |
| Changes to the mucosa |
| Decreased absorptive surface: surgical resection or progression of extensive disease |
| Acceleration of intestinal transit |
| Bacterial overgrowth |
| Bile salt deficiency |
| Excessive losses through the intestinal mucosa |
| Protein-losing enteropathy |
| Bleeding |
| Loss of intestinal fluid in the case of fistulas |
| Increased caloric needs |
| Hypercatabolic state: worsening crisis, fever, sepsis, fistula |
| Period of growth in children |
| Iatrogenic causes |
| Surgical complications |
| Drugs: corticosteroids, sulfasalazine, cholestyramine, 5-aminosalicylate, metronidazole |

TNF-alpha: tumor necrosis factor alpha; IL-1: interleukin-1; IL-6: interleukin 6.

## PREVALENCE OF MALNUTRITION

Weight loss in the recent past is assumed as a significant index of malnutrition when it equals or exceeds 10% in the last 6 months. A loss of more than 10% of the normal bodyweight has been observed in 70% patients with CD. Malnutrition in IBD occurs in 23.7 to 82% of cases.[4] The systemic inflammatory response, alongside low oral intake, is the main cause of the increased rate of malnutrition.[5]

Anorexia is secondary to an increase in pro-inflammatory cytokines (tumor necrosis factor [TNF alpha], interleukin-1 [IL-1], interleukin-6 [IL-6]) and adipokines (leptin, adiponectin and resistins). It is associated also with malabsorption of nutrients during periods of activity or remission of the disease.[5]

The resting energy expenditure (REE) may vary depending on inflammatory activity, extent of disease or nutritional status. Changes in the metabolism of substrates, with reduction in the oxidation of carbohydrates and increased oxidation of lipids, are similar to the alterations that occur during fasting and are not specific of the disease, being reversible with proper nutritional support. The low body mass index (BMI) and weight reduction reflect the losses in nutritional status, and also the difficultly to control the disease.[6] The ingestion of 25 to 30 kcal/kg/day is generally adequate to meet energy and nutrient requirements.

The impact of CD on nutritional status is relevant at all ages, especially children, as growth implies a greater metabolic energy demand. Children and adolescents present a reduction in growth speed in 15 to 50% of cases, especially in CD.

The reduction of oral intake of food, the main cause of malnutrition, results from fear of triggering abdominal pain or diarrhea, or via dietary restrictions, prescribed or otherwise, in the presence of acute crisis. In CD, there can be bowel obstruction and even a compromised upper gastrointestinal tract which can contribute to lower food intake. The classical recommendation of diet low in fiber, low in sugar or lactose-free can contribute, iatrogenically, to the decreased intake of food, as it has lower palatability and social acceptance, especially among children. The concept of therapeutic fasting for bowel rest, use of corticosteroids and intravenous fluids may reduce food intake by up to 44% of the expected theoretical value.[7] Alongside the insufficient caloric protein intake, malnutrition can result from increased losses or malabsorption due to the extent of the inflamed area, prior surgical resections, fistulas, bacterial overgrowth, sub-stenosis or resection of the ileocecal valve, which cause a functional or anatomic reduction of intestinal absorptive surface and/or acceleration of the intestinal transit.

The pharmacological agents used in the treatment of IBD can also contribute negatively to the loss of nutrients or nutritional alterations with chronic use (Table 31.1). In CD, treatment with corticosteroids increases food intake (protein and calories) but it does not result in a positive nitrogen balance.

The evaluation of a meta-analysis of seven observational studies including 1,532 patients with IBD showed increased risk of general and infectious postoperative complications associated with the use of corticosteroids in the preoperative period. The complications further increased in patients receiving more than 40 mg of oral corticosteroids. However, the use of immunomodulators (azathioprine, cyclosporine and infliximab) in the preoperative period did not increase the risk of postoperative complications, as confirmed in a recent meta-analysis.[8]

Mortality from CD is associated with protein–calorie malnutrition alongside hydro-electrolyte imbalance and factors associated with the severity of disease. Malnutrition is a poor prognostic factor, with direct and indirect contribution to the results of hospitalized IBD patients, including increased mortality, longer hospital stays, greater need for parenteral nutrition, increased incidence of digestive fistula and a need for intestinal resection for obstruction.[1] Preoperative malnutrition increases surgical complications, including anastomotic dehiscence.

## NUTRITIONAL DEFICIENCIES

Deficiencies of vitamins and micronutrients are common, especially in the acute phase of IBD or after extensive surgery (Table 31.2). Special attention should be paid to deficiency of calcium and vitamin D, due to the high prevalence of osteopenia.[9]

**Table 31.1** Nutritional changes caused by drugs used to treat IBD

| Actions | Nutritional risk |
|---|---|
| **Corticosteroid** | |
| Inhibits lymphocyte activation | > Risk of infection |
| Inhibits IL-2 release | < Wound healing |
| Anti-inflammatory properties | Catabolism |
| Stabilization of the lysosomal membrane | Hyperglycemia |
| | Arterial hypertension |
| | Hyperlipemia |
| | Sodium retention |
| | Hydro-electrolytic disorders |
| | Hyperphagia |
| | Increased calciuria |
| | Peptic ulcer |
| **Sulfasalazine** | |
| Competes with the absorption of folic acid | Folic acid deficiency |
| Anti-inflammatory properties | Gastric discomfort |
| **Azathioprine** | |
| Inhibits purine synthesis | Nausea and vomiting |
| Blocks the proliferation of lymphocytes | Sore throat |
| Anti-inflammatory properties | Change in taste |
| | Macrocytic anemia |
| **Cholestyramine** | |
| Malabsorption of fats | Deficiency of vitamins A, D, E, K |
| Malabsorption of fat-soluble vitamins | Hydro-electrolytic disorders |
| Malabsorption of calcium | |

In Table 31.3, the main deficiencies that occur in IBD along with the mechanisms involved are stated.

Alterations of specific micronutrients may occur in IBD with repercussions on lipid peroxidation, protein degradation and cell death, thereby modifying the inflammatory mechanisms and the evolution of the disease.[10] On the other hand, early recognition of protein-calorie malnutrition allows the administration of nutritional therapy in the oral, enteral and parenteral forms, seeking not only to recover the nutritional status but also exert adjuvant action in CD treatment.

Zinc deficiency in IBD is frequent, due to inadequate oral intake, decreased absorption, increased needs and losses. The low serum zinc and hypoalbuminemia is associated with the activity of inflammatory disease and impairment of the small intestine.[11] It is worth noting, however, that in acute colon disease, this association does not exist, but rather with a deficiency of vitamin A, as zinc deficiency can cause a deleterious effect on the hepatic synthesis of the retinoid-binding protein. Antioxidant and cellular immunity activities stand out among zinc's functions. It is assumed that the generation of free radicals may have importance in the pathophysiology of IBD. As such, replacement of not only zinc, but also copper, selenium and antioxidant vitamins, which are recognized sweepers of free radicals of oxygen, can prevent or improve mucosal damage in IBD.

When there has been a resection of the last 60 cm of the distal ileum or extensive inflammation occurs in this segment, or bacterial overgrowth, there is a loss of absorption of vitamin B12 and bile salt. Bile salt not absorbed suffers deconjugation in the colon, with consequent increase in the secretion of water, electrolytes and worsening of diarrhea. The reduction of bile salt, due to the absence of resorption in the enterohepatic cycle causes malabsorption of fats and steatorrhea in up to 40% of patients with CD.

Deficiencies of vitamin B12, folic acid and iron are responsible for nutritional anemia, so common in those with IBD.

## Anemia

Anemia is a frequent complication in IBD, with a variation of 10 to 73% of cases. It is often considered as inevitable, or is sometimes not emphasized or

**Table 31.2** Nutritional deficiencies in CD compared to the UC

| Deficiency | CD | UC |
|---|---|---|
| Weight loss | 65 to 75 | 18 to 62 |
| Impaired growth | 50 | 15 |
| Delayed puberty | 30 | 20 |
| Hypoalbuminemia | 25 to 80 | 25 to 50 |
| Loss of intestinal protein | 75 | Yes |
| Negative nitrogen balance | 69 | Yes |
| Anemia | 60 to 80 | 66 |
|   Iron | 25 to 50 | 81 |
|   Folic acid | 56 to 62 | 30 to 41 |
|   Vitamin B12 | 48 | 5 |
| Calcium | 13 | Yes |
| Magnesium | 14 to 33 | Yes |
| Potassium | 5 to 29 | Yes |
| Vitamin A | 11 to 50 | 93 |
| Vitamin D | 23 to 75 | 35 |
| Vitamin E | 0 | 40 |
| Vitamin C | 12 | - |
| Vitamin K | Yes | - |
| Zinc | 50 | Yes |
| Selenium | 35 to 40 | Yes |
| Metabolic bone disease | 30 to 50 | - |

Source: adapted from Heizer, 1998 and Han et al. 1999.

**Table 31.3** Main nutritional deficiencies in IBD patients along with the respective mechanisms

| | |
|---|---|
| Folic acid | Folic acid deficiency is common in the UC (up to 60%) than in CD (up to 40%), largely due to increased everyday use of sulfasalazine in the UC. Sulfasalazine interferes with intestinal absorption of folic acid, causing macrocytic anemia. Methotrexate can also cause folic acid deficiency and consequent mucositis, especially oral. In these two situations, the patient must replace folic acid at a dose of 1 to 5 mg/day orally |
| Calcium | The use of corticosteroids interferes with calcium absorption, which should be remembered especially in the treatment of children and adolescents, due to interference in development and growth. Calcium malabsorption is also described in patients with CD. Calcium deficiency is related to the development of osteopenia and osteoporosis and risk of fractures |
| Copper | Patients with severe diarrhea, fistulas and stomas are at risk of developing copper deficiency |
| Iron | In addition to reduced intake, chronic or acute iron loss through diarrhea or rectal bleeding causes iron deficiency in patients with IBD (iron, microcytic and hypochromic anemia). This tends to be more common in UC (up to 70%) than in CD (up to 40%). Iron deficiency causes a strong impact on the quality of life of patients. For iron deficiency in these patients, the suppression of erythropoietin production by proinflammatory cytokines (e.g. IL-6) and the change in iron metabolism caused by increase in proinflammatory cytokines, reactive oxygen metabolites and nitric oxide also contribute |

(continues)

**Table 31.3** Main nutritional deficiencies in IBD patients along with the respective mechanisms

| | |
|---|---|
| Homocysteine | Homocysteine levels are usually high in both children and adults with IBD, particularly in CD, further increasing the existing thrombotic risk |
| Lipoproteins | These transport fats and fat soluble vitamins in the circulation, contributing to the integrity of the cell membrane. LDL-cholesterol and lipoprotein AI and B are usually low in CD, and the levels do not correlate with disease activity |
| Magnesium and phosphorus | These decrease through a reduction of oral intake, by chronic loss (diarrhea, fistulas) and intestinal malabsorption. They contribute to osteopenia and osteoporosis. Phosphorus plays an important role in the maintenance of bowel function (e.g. absorption processes and recognition of antigens) and balance of the intestinal microbiota (balance between non-pathogenic and potentially pathogenic bacteria) |
| Niacin | Niacin deficiency causes pellagra. It can be observed in patients with CD and important malabsorption It is characterized by scaly and hyperchromic lesions on the body, particularly in areas exposed to the sun. It may be accompanied by fatigue and psychiatric symptoms (e.g. slowness, confusion) |
| Proteins | Hypoalbuminemia is relatively common in patients with IBD (up to 65% of cases). The decrease in food intake, intestinal losses and increased catabolism contribute to hypoproteinemia. The dosage of alpha-1-antitrypsin in feces is useful in the assessment of intestinal protein loss. Protein depletion is associated with higher post-surgical mortality |
| Selenium | This is reduced in patients with short bowel syndrome and those submitted to long-term enteral nutrition |

(*continues*)

**Table 31.3** Main nutritional deficiencies in IBD patients along with the respective mechanisms

| | |
|---|---|
| Vitamin A | Part of the group of fat-soluble vitamins and reduced in patients with intestinal malabsorption (steatorrhea) and those with low intake. A deficit causes dry mouth, dry eye, night blindness and increases the risk of fractures |
| Vitamin B12 | Especially in patients with CD which endangers the ileum or resection patients. The reduced intake also contributes to the deficit of vitamin B12, with consequent megaloblastic anemia. Intramuscular replacement in such cases is recommended |
| Vitamin K | Besides being essential in coagulation processes, vitamin K is also a cofactor in the carboxylation of osteocalcin, a crucial protein for incorporating calcium into bone. Vitamin K promotes carboxylation of glutamic acid residues in proteins containing glutamic acid, such as osteocalcin. It is therefore related to the development of osteoporosis |
| Antioxidant vitamins | Patients with IBD are subject to intense oxidative stress, which leads to the consumption of antioxidants elements such as ascorbic acid, alpha and beta-carotene and lycopenes. Supplementation with vitamins E and C favors the reduction of oxidative stress |
| Fat-soluble vitamins (A, D, E and K) | Especially in patients with CD and malabsorption and/or those with ileal involvement (or resection). The lack of reabsorption of bile salts in the terminal ileum causes a reduction in the pool of these salts and impaired absorption of lipids, including fat soluble vitamins. The use of cholestyramine (causing the chelation of bile acid) also impairs the absorption of these vitamins. Vitamin D is the fat soluble vitamin most often affected, which further contributes to the development of osteoporosis |
| Zinc | Up to 50% of CD patients may develop a zinc deficit. Zinc reduction contributes to a reduction in appetite. In parallel, there is a reduction in alkaline phosphatase, which is a zinc metalloenzyme |

LDL: low density lipoprotein.

diagnosed.[12] The difficulty begins with the lack of consensus on the definition of anemia, given that different hemoglobin levels are considered. However, in recent years, the treatment of anemia has been evaluated as a specific treatment in these patients, with the objective of reducing the severity of the disease and improve quality of life.

The most common symptoms are fatigue, malaise, weakness, nausea, irritability and difficulty in concentration. Anemia is caused by several factors, such as chronic blood loss, inadequate intake of nutrients, iron, folic acid and vitamin B12 absorption failure, as well as reduction of erythropoiesis, whether through the occurrence of chronic disease or increased cytokine or even the use of sulfasalazine. Iron, vitamin B12 and folic acid are the essential factors to erythrocyte production.

The main characteristics of anemia, such as sources, signs and symptoms of deficiency and treatment are described in Table 31.4.

### Iron deficiency anemia

Iron deficiency is the most common cause of anemia in IBD. On average, the daily intake is 15 mg of iron, with absorption of 1-2 mg in the proximal duodenum. The balance is obtained by the daily loss of 1-2 mg of iron from the gut, skin and sweat, in addition to the monthly loss in women of 15-20 mg due to menstruation (1 mL of blood contains 0.5 mg of iron). In a situation of iron deficiency, the absorption can reach 3-4 mg/day. However, blood loss greater than 10 mL/day should result in iron deficiency.[13] Chronic blood loss from the gastrointestinal tract, often hidden, is the main cause of this deficiency. The increased desquamation of inflamed intestinal epithelium may facilitate iron loss greater than 1 mg/day. Malabsorption of iron is less common.

Iron deficiency is treated with supplements such as ferrous sulfate, starting with high doses (300 mg/day), though these patients frequently do not tolerate oral iron. For this reason, or when the response to oral iron is not adequate, the administration of parenteral iron is often necessary.

### Anemia of chronic disease

Anemia is often associated with inflammatory, infectious or neoplastic diseases, conditions that manifest in the first few months of disease progression, and

**Table 31.4** Main characteristics of anemias

| Characteristics | Iron deficiency anemia/chronic disease anemia |
|---|---|
| Sources | Ferric or non-heme iron = grains, vegetables, fruits, egg*,**,*** |
| | Ferrous or heme iron = red meat, liver, fish, poultry - is |
| | better absorbed*,** |
| Symptoms/signs | Asymptomatic, mucocutaneous pallor, fatigue, glossitis, |
| | brittle and fragile nails |
| Treatment | Oral = 50 to 100 mg of elemental iron 2 times/day#, 4 to 6 mg/kg/day |
| | – maximum of 200 mg/day |
| | Parenteral (IV/IM) = 200 mg iron saccharate + SF 250 mL, 2 times/ |
| | week for 2 weeks, and then 1 time/week |
| Refractory anemia | Does not respond to 2 months of iron: common in chronic disease anemia |
| | Occurs in 56% of those with CD |
| | Erythropoietin 150 IU/kg subcutaneous, 2 to 3 times/week |
| | + Oral or parenteral iron for 6 to 12 weeks |
| **Characteristics** | **Megaloblastic anemia (B12 and folate deficiency)** |
| Sources | Vitamin B12 = meat, egg, fish, milk and dairy products |
| | Folate = spinach, beans, liver and peanuts |
| Symptoms/signs | Fatigue, irritability, headache, vertigo, tinnitus, palpitations, anorexia, |
| | weight loss, difficulty concentrating, atrophy of tongue papillae, hyper- |
| | pigmentation of the nails) |
| | Vitamin B12 deficiency: neuropsychiatric disorders, paresthesia of extre- |
| | mities, abnormal gait, urinary and fecal incontinence, impotence, irritabi- |
| | lity, memory loss, disorientation, change in taste, smell and sight, depres- |
| | sion, hallucination |
| Treatment | Vitamin B12 = 1000 mcg, IM, once/day for 1 week |
| | 1000 mcg, MI, once/day for 8 weeks |
| | 1.000 mcg, MI, once/month for life, in the case of bowel resection |
| | Folate = 1 to 2 mg, once/day |

#Elemental iron (%): ferrous sulfate, 20% polymaltose iron, 30%, ammonium ferric citrate, 16.5%; ferrous gluconate, 12%.

*Vitamin C and hydrochloric acid assist absorption.

**Phytates and oxalates cause impairment.

***Meat increases absorption by 4 times.

the so-called anemia of chronic disease (ACD), secondary to iron deficiency. There is a correlation between disease activity and the magnitude of the anemia, which can be explained as a consequence of the activation of a chain of cytokines, particularly TNF, IL-1 and interferon (INF).

This anemia is associated with decreased concentration of serum iron and total iron binding capacity, with normal or elevated ferritin, but normal or increased bone marrow iron. The three important pathophysiological mechanisms in the pathogenesis of ACD are: reduction in survival of erythrocytes, the inadequate response of bone marrow relating to anemia and defects in iron metabolism.[14] The reduction of erythrocyte survival is caused by extra-globular hemolysis, with early removal of circulating erythrocytes, attributed to the hyperactive state of the mononuclear phagocyte system. The inadequate bone marrow response is due to low secretion of erythropoietin (EPO) and the reduction of the bone marrow response to EPO. There is also a reduction of erythropoiesis due to lower supply of iron to the bone marrow, related to activation of macrophages and the release of inflammatory cytokines (TNF alpha, IFN-gamma, IL-1). Around 25% of the iron is mainly stored in the spleen and liver. When necessary, this iron returns to the plasma and travels to the bone marrow for the formation of new erythrocytes. In ACD, there is a disturbance in the reuse of this iron, which is kept in the form of a deposit, thanks to increased synthesis of lactoferrin, promoted by IL-1. Lactoferrin has a higher affinity to iron, but does not transfer it to erythropoietic cells and is retained by macrophages.

In general, 56% of patients with CD and 25% of those with UC who suffer from ACD do not respond to supplementation with oral or parenteral iron. In this case, EPO associated with iron should be recommended, with good results after 6 to 12 weeks of treatment.[15]

### Vitamin B12 Deficiency Anemia

Vitamin B12 deficiency is often a consequence of active disease in the terminal ileum or intestinal resection of this segment. Patients with an ileal resection greater than 60 cm invariably have a deficiency of vitamin B12; those with a resection less than or equal to 60 cm have 50% malabsorption of B12. Deficiency of vitamin B12 only manifests itself after a few years of malabsorption. For this reason, prophylactic vitamin B12 replacement therapy should be undertaken in cases of ileal disease or resection. Less common causes of B12 deficiency in CD are inflammation or gastric resection and bacterial overgrowth.

### Folic Acid Deficiency Anemia

Deficiency of folic acid, as well as in the case of vitamin B12, manifests as megaloblastic anemia and occurs due to inadequate intake, malabsorption or interference of drug treatment.

Sulfasalazine causes folate malabsorption due to competitive inhibition of the enzyme folate conjugase in the jejunum. Some studies have shown a protective effect of folate supplementation against angiodysplasia, polyps and cancer in patients with prolonged IBD. Another adverse effect of folate deficiency is increased homocysteine that is associated with thromboembolic events in IBD. The high level of homocysteine induces a hypercoagulable state, present in 26.5% of patients with IBD, compared with 3.3% in controls.[16] Folate and vitamin B12 are cofactors in the homocysteine-methionine metabolic pathway.

Azathioprine and methotrexate inhibit enzymes involved in DNA synthesis. CD patients should routinely receive 1 mg/day of folic acid.

### Energy expenditure in IBD

Studies of the energy expenditure in IBD present conflicting results. Rigaud et al.,[17] using indirect calorimetry, found increased resting energy expenditure in patients with a CD activity index above 200 and a relationship between energy expenditure and acute phase proteins (C-reactive protein and serum mucoproteins). On the other hand, studies have shown that in the absence of fever or sepsis, resting energy expenditure in patients with IBD that are not malnourished are the same as in the control group. Patients with less than 90% of their ideal weight showed a slight increase in energy expenditure.

Stokes and Hill,[18] using a combination of body scan techniques (DEXA) with neutron activation analysis and tritiated water dilution demonstrated that patients with CD did not have increased requirements (33 kcal/kg/day) compared with normal individuals.

Different observations of energy expenditure in IBD can be attributed to methodological distinctions and the severity of patients. Therefore, there is no consensus in relation to caloric protein needs, but studies show that they are variable according to the nutritional condition and the activity of the disease, reaching about 45 kcal/kg/day. The supply of 25 to 35 kcal/kg of bodyweight/ day[19] can be used as a goal for maintaining or gaining weight, respectively, maintaining the supply of 1.2 to 1.5 g of protein/kg of bodyweight/day.

Some authors recommend daily supplementation of multivitamins and minerals. Vitamin B12 should be measured regularly in patients with CD with or without resection in the distal ileum. Patients using corticosteroids should receive 1500-2000 mg of elemental calcium. In cases of multiple and extensive short bowel resections, vitamin D should be measured and supplemented (25,000 to 50,000 U/week), if below normal.

Nutritional intervention should be promptly instituted if the caloric and protein intake is insufficient to maintain the weight of adults or growth in children.

## NUTRITIONAL ASSESSMENT

The best way to conduct a nutritional assessment in a patient with inflammatory disease has not yet been defined, due to the changes in intra and extracellular fluids and metabolism at the cellular level, due to the stress response.[20]

Anthropometric techniques are used primarily for the diagnosis of nutritional status. However, they provide little information on functional and metabolic disorders. In daily clinical practice, under the term anthropometry, techniques that are simple to execute are grouped together for the assessment of body composition, such as weight, height, BMI, cutaneous folds, wrist circumference, circumference of the arm, muscular circumference of the arm and muscular area of the arm.[21] Changes to the anthropometric measurements may result from excessive or decreased body water, without changes to fat and/or muscular mass.[22,23]

Auxiliary laboratory methods in nutritional evaluation arise when early biochemical alterations, previous cell or organ damage become evident. The biochemical serum parameters are primarily proteins used in planning the stabilization program of the nutritional status in patients, determining if there is a risk of complications and monitoring the nutritional therapy. In general, the results of biochemical tests are abnormal (e.g. serum albumin, which may be low in response to treatment of hypovolemia, without necessarily representing a state of malnutrition).[24,25]

All serious forms of malnutrition result in a reduction of the patient's immunologic responses. As malnutrition increases, reductions occur in the total lymphocyte count that compromise the lymphocyte response to phytohemagglutinin, with decrease in neutrophil chemotaxis, IgG and C3 deficiencies and reduction of cutaneous reactivity to various allergens used in the tests. These changes are reversible with nutritional replacement. The immunologic tests also suffer the same possible influences of other diseases, without representing malnutrition. Considering these factors, nutritional assessment of critically ill patients should be based on the clinical information, especially a diagnosis for the cause of the hypermetabolic state and the inflammatory response to stress.[23,24,26,27]

The study of the body composition is relevant in the assessment and control of acute or chronically malnourished patients. It can be assumed that the body composition is the bodily distribution of ingested nutrients and is closely related to the biochemical, metabolic and mechanical functions of the body. Part of the body composition of the normal and healthy human population is composed of adipose tissue, which is equivalent to 10-25% of body weight for males and 18-30% for females. Theoretically, the remaining portion of the body composition is lean body mass (LBM), comprising 75-85% of the body weight.[20]

Bioelectrical impedance analysis (BIA) is a noninvasive method that is fast, sensitive, painless, relatively accurate, and used to assess body composition.[28]

## NUTRITIONAL ASSESSMENT METHODS
### Subjective global assessment

Subjective global assessment (SGA) of the nutritional status, as per Detsky et al.,[29] consists of simple but relevant questions about clinical history and a physical exam. It appears as a reliable, useful, valid, practical and economical instrument. The parameters used are:

- weight loss in the last six months prior to assessment (expressed in kilograms and as a proportional loss from usual weight) and alteration in recent weeks;

- food intake in relation to the patient's usual pattern (quantity and quality);
- presence of persistent significant gastrointestinal symptoms (nausea, vomiting, diarrhea);
- assessment of the functional capacity of the patient realizing day-to-day activities;
- loss of muscle mass and/or subcutaneous fat;
- presence of edemas (sacral region and ankles).

### Anthropometry

As already mentioned, anthropometry involves simple techniques for evaluation of body composition, such as weight, height, BMI, cutaneous folds, circumference of the arm, muscular circumference of the arm and muscular area of the arm.[20,22]

The ideal method for assessing body composition must be reliable, easy to perform, inexpensive and non-invasive. Unfortunately, this method has not yet been developed and there is no consensus on the best method to use. Anthropometry should be the most obvious choice, but for the anthropometric techniques to be reliable, there is still a long way to go, seeking to individualize the needs and peculiarities of the various population groups.[20,22]

It is evident that the criteria 'nourished' and 'malnourished' cannot be based on the appearance of the patient. If cases of extreme malnutrition were recognized in this manner, it would still be undoubtedly difficult to diagnose protein malnutrition (Kwashiorkor).[20,22]

Blackburn et al. are responsible for the popularization of a nutritional assessment method for hospitalized patients, which is based on the evaluation of the components of body composition[30] obtained through examinations and measurements of the patient's nutritional profile, which may alert the professional to a state of malnutrition capable of influencing morbidity and mortality. This is a measurement of the human body or its parts, including measurements such as weight, height, skin fold thickness, and circumferences of the limbs. The estimates of the composition of bodyweight are necessary to determine and monitor the nutritional status of patients.[31]

## Biochemical assessment

The parameters most often used are:[32,33]

- albumin: is the most abundant protein that circulates in plasma and extracellular fluids. It has long half-life (20 days) and indicates chronic malnutrition. Levels ≤ 3.5 g/dL are considered abnormal;
- transferrin: iron carrier protein has a half-life of 4 to 8 days. It can be determined directly by radioimmunoassay or indirectly, from the total iron binding capacity (TIBC) using the formula: transferrin = (0.8 × CTLF) – 43. Normal value: 200 to 400 mg/dL;
- pre-albumin or transthyretin and retinol binding protein: a half-life of 12 hours to 2 days. They appear as significant indicators of malnutrition, responding sharply when energy and protein intake are low, but with an increased cost. Reference values: pre-albumin, 17 to 42 mg/dL; and retinol binding protein, 2.6 to 7.6 mg/dL.

### Nitrogen balance (NB)

This consists of calculating the difference between the nitrogen introduced into the body or ingested (Ni) and that eliminated or excreted (Ne), as shown in Table 31.5.

| Table 31.5 Nitrogen balance calculation |
| --- |
| Molecular weight of urea = 60 |
| Molecular weight of nitrogen in urea molecule = 28 |
| NUU = urinary urea × 28/60 = urinary urea x 0.46 |
| Ne = NUU + f |
| NB = Ni - Ne |
| NB = g of ingested protein/6.25 - (urinary urea x 0.46 + f) |

NUU: nitrogen in urinary urea; Ne: endogenous nitrogen; f: unmeasured losses > 2 to 4 g; NB: nitrogen balance; Ni: nitrogen in ingested food.

Source: Waitzberg DL, 2009.

### Bioelectrical impedance (BIA)

It is based on passing a low-intensity electric (800 mA) and fixed frequency (50 kHz) current through the individual's body, determining the resistance (impedance) offered by the various tissues of the body. If, for example, 75% of the musculature (MCM) is formed of water, the hydration rate of the fat is virtually zero, this results in easier conduction of an electric current through muscle rather than the fat. Thus, with support in appropriate equations, the percentage of fat, lean mass, total body water, extracellular water, intracellular water, third space, phase angle and daily basal energetic metabolism of the person can be estimated. The BIA analysis is an assessment method of body composition of acceptable accuracy in healthy people, chronic disease and mild to moderate obesity.[28]

## NUTRITIONAL NEEDS IN IBD

Energy expenditure is not significantly high in patients with inactive CD.[34] Resting energy expenditure is increased during the active disease, although the total energy expenditure is not significantly increased. This may be related to decreased physical activity during the periods where the disease exacerbates.[18]

Protein needs are high in CD because of increased losses related to intestinal inflammation, or even the presence of fistulas. Postoperative patients can benefit from an increased protein intake. The protein recommendations for patients with CD are 1 to 1.5 g/kg,35 although there are no randomized studies that have investigated the ideal protein intake in this population.[36]

Measurements such as weight, height and cutaneous folds are associated with countless errors in technique and interpretation. The measurement equipment used, although simple and inexpensive, require frequent calibration and maintenance. Examiners need to be suitably trained to take measurements. The standard value tables have been developed from a healthy population, excluding the elderly, critically ill patients and victims of chronic diseases; therefore, they should be interpreted with caution.

Although there are other ways to assess body composition, a gold standard method to determine the nutritional status is not available. There is no single method without at least one important limitation to this assessment. Various studies show that the utilization of a single method alone is not capable of classifying the patient reliably. Because it is easy to conduct and not invasive, anthropometry should always be considered when undertaking an assessment the nutritional status of a patient.

Pioneering work by many researchers has led us to understand importance of the role of malnutrition/nutrition in patients with IBD. It is known that providing a nutritional therapy to patients with IBD may change the nutritional results. Although there are few prospective and controlled studies which conclusively demonstrate that nutritional therapy improves the morbidity and mortality in this population, the lack of data does not mean that it is ineffective. Similarly, since the nutritional assessment is not easy to conduct, may suffer various influences and is not a gold standard method, monitoring and accompanying of patients with IBD is of paramount importance.

## NUTRITIONAL THERAPY

Nutritional therapy (NT) should be recommended in order to prevent retarded growth in children with mild to moderate IBD and encourage quality of life. In a systematic review, Newby et al.,[37] assessed interventions for children with CD and delayed growth. Three randomized trials were identified, given that 2 of them compared enteral nutrition therapy (ENT) with corticosteroids for induction of remission. In both studies, the growth score was significantly increased in the group that received ENT. A carefully made surgical recommendation is also favorable to growth in children in the pre-pubescent phase with refractory CD.

The influence of NT was tested on the quality of life (QoL) of children and adolescents with CD followed with exclusive ENT in two studies. Gailhoustet et al.[38] compared adolescents with exclusive ENT *versus* adolescents with corticosteroids through questionnaires and daily monitoring. The authors emphasized the difficulties associated with the use of a nasogastric tube and the suspension of the oral diet, as well as the side effects of corticosteroids in the other group. However, the improvement related to intestinal symptoms and sensation of wellbeing seems to overcome the negative aspects related to nasogastric tube. Afzal et al.[39] showed improvement in QoL scores in 24 of 26 children with active CD treated with exclusive ENT, with 90% remission.

The main indications for NT in patients with a diagnosis of IBD are described in Table 31.2.

The main results of nutritional interventions in parenteral nutrition (PNT) or ENT, in accordance with the activity or remission of the disease, will be discussed ahead.

### Parenteral nutrition therapy (PNT)

#### PNT in the acute phase

PNT has been used in worsening outbreaks of IBD over the last 30 years as an adjuvant treatment, especially in cases resistant to corticosteroids, with good results in relation to the remission of the disease and avoidance surgical treatment.[40]

PNT is indicated in cases where it is not possible to perform ENT: intestinal obstruction, short bowel with severe malabsorption, severe hydro-electrolyte disturbances, severe intestinal dysmotility, high flow fistulas or anastomotic dehiscence. PNT may also be used upon the occurrence of intolerance to ENT or where it is impossible to maintain an adequate oral diet associated with the absence of an access route for ENT.

Patients with severe nutritional risk, characterized by a loss of more than 10% of body weight in 3 to 6 months, BMI less than 18.5 kg/m$^2$ and/or serum albumin less than 30 g/L showed increased complications in post-operative period.[41]

| Table 31.2 Main indications for NT in IBD |
| --- |
| Serious and repeated crises |
| Prevention and treatment of malnutrition |
| Preoperative preparation |
| Digestive fistulas |
| Anatomical or functional short bowel syndrome |
| Improved growth and development of children and adolescents |
| Improved quality of life |
| Specific indication for enteral nutrition in CD |
| Treatment of the acute phase |
| Preoperative nutrition |
| Treatment of remission |

Uncontrolled studies on patients with IBD have suggested that the use of PNT before surgery reduces complications and the extent of intestinal resection; however, it often leads to longer hospital stay. Niu et al.[42] showed that combined NT (ENT and PNT) was superior to isolated ENT or PNT in the perioperative period of 165 patients with CD. There was reduction in morbidity, hospital stay and cost-effectiveness in the group receiving combined NT.

In the 1970s and 1980s, oral fasting associated with PNT was advocated for the treatment of remission of acute outbreaks in CD, as an efficient alternative to avoid surgery or high doses of corticosteroids. It was believed that this measure enabled the inflamed mucosa to heal. At least two prospective studies have analyzed the effect of PNT and maintaining bowel rest. However, the only prospective, randomized and controlled study comparing PNT, ENT or oral diet did not show advantages in keeping the patient on PNT and an oral fast to ensure bowel rest.[43] The authors demonstrated that NT facilitated the remission of the diseases with or without fasting.

Later studies have questioned the damages of fasting, which promoted intestinal hypoplasia not only in the diseased areas, but throughout the intestine. The classic PNT regimes are free from glutamine, the preferred energy source of enterocytes. Interest has turned to the use of the gastrointestinal tract for nutritional recovery and maintenance of tropism, as early as possible. Frascio et al. compared the effect of PNT with a polymeric enteral diet in patients with active CD.[44] The results favored the use of the enteral diet, showing once again that not only the maintenance of fasting but also the enterotropic effect of polymeric diet is unnecessary.

It has already been shown that PNT for a short period does not promote weight gain or restores total nitrogen stocks, but prevents protein loss and improves certain physiological variables such as respiratory function and the function of the skeletal musculature. These authors demonstrated that nutritional recovery was delayed. However, the complications associated with PNT do not justify its use to induce remission of CD, whenever enteral nutrition can be used. Therefore, it can be concluded that bowel rest is not necessary to achieve clinical remission. PNT remains a useful weapon for patients that do not tolerate intestinal feeding and require NT.

### PNT in the remission phase

PNT is not recommended for maintaining remission. In the long term, PNT should be reserved for situations of chronic malnutrition associated with intestinal failure, previous bowel resections, and in such cases could improve rehabilitation and social reintegration. However, PNT at home presents complications which may compromise quality of life, irrespective of CD, such as catheter sepsis, venous thrombosis and hepatic dysfunction. IBD has an independent risk factor for the development of venous thromboembolism.[45]

### Crohn's disease and digestive fistula

Transmural inflammation in CD may be associated with the formation of enterocolic, enterovesical, rectovaginal and perianal fistulas. The latter is the most common and can affect one third of patients with CD, requiring surgical drainage whenever associated with suppuration and abscess.

In general, high flow fistulas (> 500 mL/day) involve the proximal intestine and can cause hydro-electrolyte disorder. CD is one of the main factors that negatively influence the spontaneous closure of enterocutaneous fistulas, alongside malnutrition.

PNT and bowel rest are preferential measures to obtain the healing of the fistula or surgical preparation in case of high flow fistulas, as well as in situations in which ENT is not tolerated or distal enteral access cannot be obtained.[46]

ENT can be used in cases of low flow fistulas located in the terminal ileum or colon, in the same way as fistulas in proximal location, if distal enteral access can be obtained for ENT.

A retrospective analysis of patients with fistula in the small intestine treated with total PNT revealed lower mortality rates, higher spontaneous closing rates and higher surgical closure rates in comparison to historical controls, but the results are much lower when compared with patients without CD. However, the overall closure rate of 35% in this study, achieved with total PNT, was not maintained, since half of the patients' fistulas reopened within 3 months.

The most efficient pharmacological treatment is conducted with infliximab (5 mg/kg), which led to spontaneous closure of all fistulas in 55% of CD patients, compared with 13% for controls.[47]

**The role of NT as a primary or adjunct (supplementary) measure in IBD**

The main objective of an adequate nutritional approach is to prevent patients with IBD developing into malnutrition and its consequences. Furthermore, NT has immunomodulatory and anti-inflammatory implications, and may in certain situations represent a primary measure in the therapeutic treatment of IBD, particularly CD. In patients that are refractory to conventional therapy, NT may also be an attractive adjuvant measure. Its importance in preparing patients for surgery is also well recognized. In patients already malnourished upon entry, NT is critical to the recovery of nutritional status. Most of the time, oral supplementation or an exclusive diet with polymeric formulas is possible.[48,49]

In outpatients during the acute phase of IBD, the realization of a high calorie, high-protein, low-fat and normoglycemic diet is recommended. Preference should be given to maltodextrins and linear polymers of glucose (complex sugars). Milk and dairy products should be avoided, as well as excess sucrose and fiber, due to the risk of abundant fermentation and worsening diarrhea. It is convenient to split up the diet. An anti-fermentation diet is advisable (avoiding simple sugars and lots of pasta). As the patient improves, they must proceed to return to a normal, balanced and healthy diet, without major restrictions. Except in the case of intolerance to lactose, dairy products should return to patient diets shortly after the disease enters into remission. The prescription of dietary supplements and vitamins is common in patients during maintenance.

In hospitalized patients or those in outpatient care that have an important nutritional deficit, exclusive or supplementary ENT (partial ENT) via a probe or orally is indicated as long as there are no contraindications to its use.

**Enteral nutrition therapy (ENT)**

ENT has shown itself to be effective in promoting improvement of nutritional status, and effective compared to corticosteroids in the induction of remission of active CD. This alternative to conventional pharmacological treatment, successfully used, has been considered primary therapy.

ENT as the primary treatment has a favorable impact on the inflammatory process, inducing remission, treating malnutrition and preventing side effects from the agents that modulate and suppress immunity (5-ASA, corticosteroids, azathioprine, etc.).

### ENT in the active phase

ENT or oral nutritional supplementation (ONS) is effective in the treatment of the acute phase of CD. In adults, however, the remission rate with exclusive ENT is high, regardless of the formula; corticosteroids, in turn, are more effective in inducing remission. In children with active CD, ENT is the best choice for first-line treatment and is considered as a level A recommendation.

The effectiveness of ENT as primary treatment for active CD has been evaluated since the 1980s, when authors found the same remission index in children treated with conventional therapy (corticosteroids) or an elemental diet for 4 weeks. Since then, three meta-analyses and two systematic reviews have already been conducted in adult patients with active CD, comparing the results of different formulas and between these formulas, and corticosteroids.[50-54] The rate of clinical remission after 4 to 8 weeks of treatment was close to 60% (ranging from 36 to 80%) in the case of exclusive ENT, regardless of the formula, and 80% (ranging from 70 to 90%) for corticosteroids, a statistically significant difference in favor of corticosteroids, albeit with the cost of more side effects. In both treatments, the effectiveness is much better than the spontaneous remission in 3-4 months observed in natural history studies of CD, ranging from 18 to 42%.[55] Furthermore, the average clinical remission rate of 60% for ENT in adults with active CD is equivalent to the rate obtained with budesonide and, in some studies, with prednisone.[56]

In pediatrics, corticosteroids are associated with even more important adverse effects, especially delayed growth and pubertal development, moon face and acne. ENT is effective in inducing the remission of active CD. Heuschkel et al.[57] analyzed five randomized studies on 147 children and showed that ENT was as effective as corticosteroids in inducing remission of active DC. The improvement in growth and development, without the side effects of corticosteroids, makes ENT the best choice for primary first-line treatment in children with active CD. Later, in 2007, a new meta-analysis in pediatrics, with 11 randomized and controlled studies of 394 cases, compared the effectiveness of ENT with corticosteroids. There was no difference between the groups in the induction of remission.[58]

Borrelli et al.,[59] in a pediatric study, compared exclusive treatment for 10 weeks, with a polymeric diet (oral) *versus* corticosteroids. Although clinical remission rates have been similar in both ENT and corticosteroid groups (ENT = 79% *versus* corticosteroid = 67%, non-significant), endoscopic and histological remissions were higher with the diet (74% for ENT and 33% for corticosteroids, p <0.05). Similarly, Canani et al.,[60] besides confirming this data, revealed that the growth of children was significantly higher with the exclusive use of ENT compared with the use of corticosteroids, for 8 weeks. In addition, there was no statistically significant difference in the results when using a polymeric diet (orally), oligomeric diet (based on oligopeptides, using a probe) or monomeric diet (basic amino acids, using a probe). Next, patients in remission with corticosteroids or exclusive NT (polymeric, oligomeric or monomeric) were observed using mesalazine for maintenance (50 to 75 mg/kg/day, orally) for 12 months. After this period, about 80% of patients who achieved remission with diet remained in remission *versus* only 30% of the patients that used corticosteroids for induction of remission.

Thus, in children and adolescents, the use of exclusive ENT for 4 to 8 weeks can be considered an effective primary measure, superior to corticosteroid treatment in the treatment of active CD. This, incidentally, has been one of the main options for the treatment of mild to moderate CD (and eventually severe) in children and adolescents in countries like England, Italy, Canada and Japan.[56,61] It is noteworthy that in these countries, the diet or the cost of treatment is reimbursed by public organs, which usually does not occur in other countries.

In adults, exclusive NT has been used more often in refractory cases, including biological therapy, and preparing for surgery.[48] On the other hand, supplementary ENT to the oral diet as an adjunct to medication has been widely used, with satisfactory results.[61]

In relation to the use of ENT (orally or by probe) in the maintenance treatment of CD, Yamamoto et al.,[61] in a recent revision evaluated 10 studies that addressed the use of ENT in the maintenance of CD and which met the inclusion criteria set by the authors. Some points of this work deserve consideration:

- Division of the studies chosen into two groups:
  - Category A (seven works): for studies that compared a group receiving enteral nutrition *versus* another that did not receive ENT;

- Category B (three works): when the studies investigated the impact of the dose or amount of enteral formula given per day in the evolution of patients.
- The studies were further divided into those in which remission was achieved by clinical measures (e.g. ENT or PNT, corticosteroids and possibly infliximab or various combinations of the three measures) and those in which remission was obtained by surgery (e.g. resections, estenoplastia or combination of the two).
- 6 of the 10 studies were retrospective and four were prospective, with the majority (eight works) originating from Japan.
- The number (n) of patients studied varied from 39 to 218.
- There was much variation in the definition of remission and postoperative recurrence, as well as the scheme for induction of remission and maintenance drugs, which prevented the authors from conducting a formal meta-analysis.
- All patients received ENT (polymeric, oligomeric or monomeric) as an additional measure to the oral diet as well as maintenance drugs (e.g. salicylic derivatives, azathioprine or combination of the two); in general, ENT was offered orally, in the case of polymeric diets (better palatability), and via probe in oligomeric or monomeric diets (elemental). In the case of using a probe, usually the patients were instructed to place the probe via the nasogastric route at night, when they received the diet; most of the patients were instructed to ingest 40 to 50% of their daily caloric needs, calculated for the ideal weight, in the form of enteral nutrition, and the remainder in the form of a normal diet, usually low-fat.
- In some studies, patients that adhered better to ENT during induction of remission were chosen to continue using the same therapy in the maintenance phase; those who did not adhere were allocated to the group without ENT.[61]

Despite the limitations described, the results obtained from the use of ENT to treat maintenance of CD, including the post-operative period, were very promising. In all studies that evaluated NT for maintenance of patients with remission induced medically, there were higher rates of clinical remission after one year of follow-up in patients who used NT (percentage of clinical remission with additional NT: average of 70%, variation of 48 to 94% *versus* without supplementary NT: average of 35%, variation of 21 to 50%). In one of the studies,[62] the endoscopic activity score was higher, after one year, in the group that did not receive enteral nutrition (p = 0.04). In general, the effect of supplementary maintenance NT was dose-dependent, with better efficiency when the patient ingested an amount equal to or greater than 30 kcal/kg/day of optimal weight or equal to or higher than 1200 kcal/day.[50]

In the case of patients operated on (in the post-operative period), the cumulative rate of re-operation at 5 years was significantly lower in the group receiving supplementary NT for maintenance *versus* the group without maintenance NT (with NT, n = 180, accumulated re-operation rate = 16% *versus* without maintenance NT, n = 38, accumulated re-operation rate = 38%, p = 0.016).[63] In another study,[64] which evaluated the use of supplemental NT in the post-operative period, clinical remission and endoscopic rates after one year were 95 and 70% in the group receiving supplementary NT *versus* 65 and 30% in the group without nutritional therapy for maintenance, respectively (p=0.048 for clinical remission and p = 0.027 for endoscopic remission). There was also a dose-dependent effect in the postoperative period.[61]

Yamamoto et al.[62] quantified IL-1-beta, IL-6 and TNF-alpha in the tissue (ileocolonoscopy biopsies) after 6 and 12 months of use of supplementary NT for maintenance in CD, or not. The levels of these cytokines progressively increased in the group that did not receive ENT, which did not occur in the group receiving supplementary ENT.

Finally, Yamamoto et al.[61] found that patients with involvement of the small intestine tended to respond better to ENT, a fact also observed by another group.[65]

It is noteworthy that ENT for maintenance should preferably be offered as an adjunct measure to treatment with maintenance drugs, such as immunosuppressants, azathioprine or 6-mercaptopurine. The interruption of the enteral diet is associated with high rates of relapse at 1 year, when the patient does not receive any kind of maintenance therapy (about 30-50% of relapse in children and adolescents and 60-90% in adults).[59,60,66] Unfortunately, there are no well-designed studies on the association of NT and biological therapy, nor on results regarding what happens if biological therapy is used initially followed by NT plus immunosuppressants.[67]

In the case of UC, there is little information on the use of ENT, whether in the acute phase or the remission phase. However, in practice, the guidelines recommended for CD are followed. Gonzalez-Huix et al.[68] assessed the use of ENT *versus* PNT in patients with severe acute exacerbation of UC. Both groups received corticosteroids, 1 mg/kg/day. The frequency of postoperative infection and complications related to nutritional support were significantly higher in the group receiving full PNT. Thus, ENT is considered the mode to be chosen in the case of UC requiring NT, since there are no contraindications (e.g. toxic megacolon, intestinal perforation).

The results have not been different when offering a monomeric (or elemental diet, based on amino acids), oligomeric (based on oligopeptides) or polymeric (based on full protein) diet.[60] Considering that polymeric diets are isosmolar and more palatable, which favors oral consumption, the polymeric diet is recommendable for ENT in IBD.[48,49] However, note that some patients with CD and significant lesions of the small intestine will certainly benefit more from oligomeric diets, through the greater absorption of dipeptides and tripeptides in situations of major intestinal lesion.[69] The same applies to those who do not tolerate the polymeric diet for some reason.

Enteral diets are usually free from lactose and sucrose, to decrease diarrhea. The sources of carbohydrates are generally maltodextrins and linear polymers of glucose.

What draws attention in studies of the adult population is that even in conditions where corticosteroids were superior to diets in the induction of remission, the nutritional benefit of dietary treatment should be valued. When the results are similar, the superiority of nutritional treatment becomes evident, since many of these patients are malnourished. Furthermore, the side effects of corticosteroids are also reflected in nutritional deterioration because it is a drug that increases nitrogen loss and reduces the absorption of calcium. The similarity of the results in pediatrics also clearly shows the advantage of using the dietary treatment, considering that in this group, the side effects of corticosteroids are even more disastrous. Thus, the beginning of the treatment of active CD with enteral nutrition seems to be an interesting alternative, especially in malnourished and pediatric patients. If the result is not favorable, the association with drug treatment should prevail.

### Mechanism of action of ENT

The therapeutic mechanism of ENT in active CD is not completely understood. One of the classical theories of pathogenesis of IBD points to the possibility of intraluminal antigens causing an acute crisis. These antigens act by determining an exaggerated or anomalous immune response, with release of the inflammatory mediators responsible for the clinical and histological manifestations of the disease. For this reason, total bowel rest prevailed as the recommendation in handling of acute attacks of CD for a long time. PNT was associated under these conditions to allow the preservation or improvement of nutritional status.

In the 1980s, studies evaluating the use of an elemental diet in the induction of remission of DC explained their action by several mechanisms. The fact that these diets are free from whole protein or peptides, but are based on amino acids, caused low antigenicity and lower immunological hyperactivity. The elemental diet does not require digestion and allows absorption in the upper segment of the small intestine, keeping just the distal portion at rest, generally the site of the inflammatory process. Furthermore, these diets are practically sterile. Thus, it was believed that the association of the reduction of both the bacterial load and the antigenic nature of the diet determined lower intestinal permeability. An additional advantage is the low lipid content of these diets (1 to 10%), and particularly reduced amount of linoleic acid. A precursor to arachidonic acid, linoleic acid is a substrate for the synthesis of eicosanoids with large inflammatory activity, such as leukotriene B4, thromboxane A2 and prostaglandin E2.[70] This metabolic pathway is the same as that explaining the action of some drugs used in the treatment of IBD.

However, the exact mechanism of action of the enteral diet remains unknown. Oligomeric diets were also tested, and it was believed that their results would be better than the elemental diet, considering some of the physiological benefits of these diets. The absorption of nitrogen is better from dipeptides and tripeptides than mixing isonitrogenous free amino acids. The osmolality of the elemental diet is higher than that of the oligomeric diet, which may exacerbate diarrhea due to increased osmotic load. Thus, theoretically, the peptide-based diet must be nutritionally more effective than the elemental one. However, large European studies[71,72] showed the effectiveness of the enteral diet was inferior to pharmacological treatment.

The polymeric diet with whole protein was also investigated in controlling the inflammatory activity of patients. Many studies have evaluated the effects of a polymeric diet and compared it with the elemental diet in the induction of remission. The analysis can be complicated due to the use of different diets, lack of standardization the indices of inflammatory activity in CD, as well as difference in the fat content of diets, although the results were similar in many studies. Only one prospective and controlled study compared the effectiveness of a polymeric diet with corticosteroids and showed the same capacity to induce remission.[73] However, these authors consider that the lipid source was responsible for the good results of the polymeric diet used in this study and concluded that the polymeric diet is safe, well tolerated and as effective as pharmacological treatment.

The favorable results of the polymeric diet on acute crises in CD make clear that the mechanism of action of the diets has no relationship with the proteins, as previously thought. More recently, however, the possible influence of different compositions of fat used in the formulations of ENT has been assessed.

Traditionally, elemental diets have a very low lipid content (0.6 to 1.3% of total calories), and perhaps this explains their good results. In a study by Gonzales-Huix,[73] the polymeric diet had, on the contrary, a high lipid (33% of total calories) and monounsaturated fatty acid (oleic acid) content. Monounsaturated fatty acids are considered as neutral fats since they are not precursors of prostaglandins and leukotrienes.

Giaffer et al.[74] compared the elemental diet with a high fat polymeric diet (36% of total calories), rich in linoleic acid. The patients that received the polymeric diet presented the worst results. Thus, the understanding of the role of enteral feeding in the treatment of CD becomes clearer, as well as the importance of determination of lipid formulation (saturated, monounsaturated, polyunsaturated), because it is possible to modulate the synthesis of eicosanoids, and immunomodulatory mechanisms may influence the results.

The fatty acid of the omega-3 series is metabolized into prostaglandin E3 (PGE3) and leukotriene B5, 30 times less potent than leukotriene B4. Thus, the long-term intake of fish oil, a source of omega 3, promotes increased production of leukotriene B5 and reduction of leukotriene B4, and therefore its utilization is useful, with the possibility of reduction of omega 6. Diets high in monounsaturates, in turn, do not interfere with the synthesis of eicosanoids, and are beneficial in these patients.[75]

Diets with a high lipid content (12 to 30% of total calories) are associated with less favorable results. In patients treated with an intermediate amount of fat, but with a large proportion of monounsaturated fat, the results were positive.[73,75]

Bamba et al. compared the efficacy of elemental diets with different percentages of fat in a prospective, multi-center, randomized, controlled study conducted in Japan.[76] The three groups were divided into low (1.15%), medium (6.21%) or high fat content (11.27%). Clinical remission was obtained 4 weeks later in 80, 40 and 25% in the groups with low, medium or high in fat, respectively. The authors concluded that the high lipid content, mainly consisting of polyunsaturated fatty acid (PUFA) and omega-6 long chain triglycerides (LCT) in enteral diets reduces the therapeutic effect in the treatment of active CD.

Although the exact mechanism explains the anti-inflammatory effect of ENT on the mucosa in active CD, it is possible for a reduction of protein loss, intestinal permeability, fecal excretion of leukocytes and the production of inflammatory cytokines to occur. With respect to the location of the lesion, the results are better in ileal rather than in colon disease.[77]

### ENT in the remission phase

Supplementary ENT may be effective in maintaining remission in CD. In a systematic review, Akobeng et al.[78] evaluated the effectiveness of ENT in preventing crises from worsening and maintaining remission, similar to what occurs with immunosuppressants. Only two studies were included, and therefore assessed differently. In a study by Takagi et al.,[79] patients who received half of their daily nutritional needs as EN (elemental diet) and half as normal diet had significantly lower relapses of worsening crises than the group that only received a normal diet. In a study by Verma et al.,[80] patients who received 35 to 50% of nutritional needs with the elemental or polymeric diet for maintenance of remission for 12 months showed the same effectiveness and possibility of withdrawal from corticosteroids. The authors of this review suggest that supplementary ENT can be effective in maintaining the remission of CD, but studies with larger numbers of patients are needed.

Verma et al.[81] assessed the use of oral nutritional supplements in maintaining remission in CD, compared to an unrestricted oral diet. The authors concluded that the use of an oral nutritional supplement in addition to the usual, safe diet was well tolerated and effective in maintaining the remission of CD.

Harries et al.[82] showed that the daily intake of a 600 kcal oral supplement is possible in patients with inactive CD. The oral intake with nutritional supplement at an amount exceeding this is possible only for a very short period in active CD.[82]

The induction of remission was also evaluated after ileal or ileocolonic surgical resection by Yamamoto et al.[62] The authors investigated the impact of the ENT in the clinical and endoscopic recurrence of 40 consecutive patients randomized to receive partial ENT or a free diet. The authors concluded that long-term supplementation can significantly reduce clinical and endoscopic recurrence.

There are no specific studies to evaluate the role of ENT in maintaining remission in patients with UC.

**Anti-inflammatory and immunomodulatory mechanisms of action of ENT**
Various studies have revealed that the exclusive use of ENT reduces the amount of proinflammatory cytokines (IL-1, IL-2, IL-8, interferon gamma [IFN-gamma], TNF- alpha, etc.) in the blood and/or in the intestinal mucosa and, at the same time, increases the tissue content of anti-inflammatory cytokines (e.g. TGF-beta, insulin-like growth factor 1 [IGF-1], IGF binding protein 3 [IGFBP -3], increase in the antagonist ratio for t for IL-1/interleukin-1 beta [IL-1ra/IL-1-beta]).[59,60,66,83] Furthermore, intestinal permeability, usually greatly increased in IBD allowing uncontrollable entry

of antigenic components, can become normal with exclusive ENT.[48] One study,[84] where biopsies of the inflamed mucosa of CD patients were incubated with different enteral diets and elements present in some enteral diet (control, monomeric diet, polymeric diet and whey) revealed an increase of IL-1ra/IL-1 beta in the middle, which characterizes a protective effect, in sequence from the highest to the lowest value: IL-1ra/IL-1-beta ratio = 142.8 with the polymeric diet; ratio with incubation with whey = 95.7; relationship with monomeric diet = 89.6; and in the control group = 45.7. It is therefore clear that "something" exists in enteral diets that provides an anti-inflammatory and immunomodulatory effect. Furthermore, apparently whey exhibits anti-inflammatory and protective effects.

At least four hypotheses have been suggested to explain the anti-inflammatory and immunomodulatory effects of enteral feeding in IBD. It is possible that all the mechanisms act together, culminating in the healthy effect of enteral feeding in IBD. Such hypotheses are described below.

## Improvement of nutritional condition

It has long been known that a well-nourished individual responds better from an immunological point of view. For a long time, the effect of diet on nutritional status, assessed by anthropometric measures such as weight gain, BMI measurement, recovery of cutaneous folds and circumference of the arm, among others, was considered mainly responsible for the anti-inflammatory action of enteral feedings. However, in 2004, Bannerjee et al.,[83] in a study with exclusive enteral nutrition in patients with active DC, showed that the anti-inflammatory action of the enteral diet precedes the nutritional recovery assessed by the usual methods of anthropometry. The evidence of inflammatory activity, such as the erythrocyte sedimentation rate (ESR), CRP, inflammatory activity index, IL-6 and IGF-1 were already reduced in the first week of treatment. The improvement of anthropometric data only started to happen from the second to third week of treatment. It is worth noting that it is possible, from a cellular point of view, that the energy gain and mitochondrial recovery obtained with nutritional replacement have a share in the anti-inflammatory response promoted by the enteral diet, which would make this a very plausible hypothesis. However, this mechanism has not been duly proven.

Reduction of the luminal antigen load ("bowel rest")

Today, IBD is understood as a condition with genetic susceptibility, involving several etiopathogenic factors such as increased intestinal permeability (barrier defect) and imbalance of immunoregulation in the intestinal mucosa. There are various antigenic elements that could overstimulate the immune system. Bacterial antigens (e.g. lipopolysaccharides [LPS], peptidoglycan [PGN]) have been the most studied and are the basis for the theory that patients with IBD react violently to their own bacterial antigens or, in other words, do not tolerate their own intestinal microbiota. More recently special attention has been given to food antigens which, as in the case of bacterial antigens, also make up this aggressor contingent. This group includes so-called xenobiotics, a generic name that encompasses a wide range of food antigens represented by additives (colorants, anti-caking agents etc.) and preservatives present in processed foods that may be consumed in a normal diet.

In the western world especially, there is wide exposure to a number of inorganic microparticles, used as preservatives and additives in foodstuffs with inflammatory power, commonly known as nanoparticles of titanium, aluminum and silicon oxides, among others.[48] Furthermore, certain deviations from the normal diet (e.g. rich in polyols and short-chain refined carbohydrates, excess polyunsaturated fat n-6) may help to enhance the inflammatory process and increase intestinal permeability. Thus, it is assumed that the anti-inflammatory power of enteral diets results from the low content of elements with antigenic power. However, if that was the main reason in the comparison between the monomeric, oligomeric and polymeric diets, the best therapeutic response should occur with monomeric or elemental diet, based on amino acids, which is known to be the least antigenic.[56] In fact, children with severe allergies to cow's milk protein, while not responding to therapy with hydrolyzed milk (about 5 to 10% of children with this type of allergy do not respond to exclusive treatment with hydrolyzed milk) become responsive after the introduction of an amino acid based formula. However, in patients with active CD, remission was also achieved irrespective of the formulation used – monomeric, oligomeric or polymeric.[60]

Therefore, although enteral diets are less antigenic, this does not seem to be the only or fundamental mechanism to explain the anti-inflammatory effectiveness of enteral formulas.

### Effects of ENT on the intestinal microbiota

As already mentioned, the intestinal microbiota appears to play a crucial role in the pathophysiological context of IBD. Quantitative and qualitative changes (e.g. reduction of *Lactobacillus* and *Bifidobacterium* and increase in potentially pathogenic anaerobic bacteria) of the intestinal microbiota have been described, especially in CD.[85,86] These changes have led some authors to study the impact of ENT on the intestinal microbiota. In fact, when compared with PNT, ENT promoted a reduction in the ratio between anaerobic/aerobic bacteria (0.42 *versus* 201, $p < 0.05$) and promoted reasonable growth (log CFU/g) of *Lactobacillus* spp and *Bifidobacterium* spp.[87] In the case of PNT, colonization by *Lactobacillus* spp was significant and *Bifidobacterium* spp was undetectable. The impact of TNE on the intestinal microbiota of adult patients with IBD has not been adequately studied.[53] However, in children, its use caused a favorable impact on intestinal microbiota in patients with CD.[56]

### Appropriate mixture of lipids

The fatty acids present in enteral diets may exert anti-inflammatory and immunomodulatory effects, in addition to interfering in intestinal permeability. Gassull et al.[88] note that the positive outcome of enteral diets in IBD is related, to a large extent, with the appropriate blend of fatty acids with a final anti-inflammatory result. In an excellent experimental study, Sadeghi et al.,[89] showed the different effects of lipids on the intestinal inflammatory process. Medium chain triglycerides (MCT) and fish oil (rich in omega 3 fatty acids) have an anti-inflammatory effect; oleic acid (from olive oil) has intermediate anti-inflammatory activity; linoleic acid has pro-inflammatory action, but should be part of the lipid profile of diets in low amounts, because it is an essential fatty acid. Thus, an appropriate mixture of lipids (with a predominance of lipids with anti-inflammatory action), with a final anti-inflammatory result should be present in the enteral diets used for the treatment of IBD.[88,89] Ideal sources of fatty acids, in this anti-inflammatory context, include canola (e.g. oleic acid, linolenic acid) and olive oil (rich in oleic acid and phenols, which have powerful antioxidant action).[88,89]

## SPECIFIC NUTRIENTS

It has been shown that the enteral diet may induce remission of disease in adults and children, promoting growth in the latter. However, the effectiveness of enteral feeding appears to depend on the patient's acceptance and their ability to consume them for extended periods. It has not been determined whether the success of NT is its elemental nature; this depends on the nutrient content or the pharmacological effect of the nutrients. The role of NT seems clear, but the use of exclusion diets, short-chain fatty acids, glutamine, fish oils or probiotics in CD still needs more clinical evidence about their efficacy, and are used more often on an experimental basis.

The benefit of specialized formulas (modification of the lipid content, inclusion of glutamine, omega 3 fatty acid and TGF-beta) has not yet been proven.

### Omega 3 fatty acid

Fish oil derived from n-3 PUFA inhibits leukotriene B4, is a potent pro-inflammatory eicosanoid synthesized from fatty acids, involved in the pathogenesis of IBD. Active CD presents a reduction of antioxidant capacity as well as an abnormal lipid profile, conducive situations for the role of n-3 PUFA. Supplementation with liquid formulations containing antioxidants and n-3 PUFA leads to a reduction in the proportion of arachidonic acid and increase of docosapentaenoic acid (DHA) and eicosapentaenoic acid (EPA) in plasma phospholipids and adipose tissue, suggesting this favors the anti-inflammatory profile. Studies by Lorenz et al.[70] and Lorenz-Meyer et al.[90] did not show a reduction in the remission of CD when supplemented with fish oil.

The study by Lorenz-Meyer[90] was a randomized, placebo-controlled study of 204 patients with CD, included after an acute relapse to receive fish oil, a diet low in carbohydrates or a placebo for 1 year.

In many studies, side effects such as halitosis, belching and diarrhea have been described with interference in the acceptance of the treatment.

Belluzzi et al.[91] showed that supplementation with 9 capsules of enteric dissolution of fish oil, equivalent to 2.7 g of n-3 PUFA was an effective measure for maintenance treatment of remission in patients with CD compared with a placebo. This presentation is shown to be advantageous in optimizing the absorption of EPA and DHA without side effects. The authors conducted a placebo-controlled, double-blind study for 1 year, with these dissolution enteric capsules to assess the maintenance of remission in 78 CD patients with a high risk of relapse. After 1 year, 23 patients (59%) from the group receiving omega 3 remained in remission compared with 10 patients (26%) in the placebo group.

However, these three controlled studies have limitations for various reasons, such as a low number of patients, intolerance to fish oil capsules (halitosis, bad taste) and use of corn or olive oil in the control group, which could favor the result. Corn oil can be metabolized into gamma-linolenic acid (GLA), a precursor of monoenoic eicosanoid and 3-series leukotrienes, and fish oil may act to sweep up free radicals. These biological activities may reduce chronic inflammation and interfere in the results.

A systematic review and meta-analysis of these studies concluded that enteric capsules of omega 3 fatty acid can be effective when administered for maintenance of remission in CD.[91-93] However, the studies are inconclusive and there is still a lack of data to recommend routine use. The effectiveness of lipid emulsion enriched with omega 3 fatty acid has not yet been proven in the PNT of patients with active CD. Studies have evaluated the effectiveness of omega 3 fatty acids in UC and have not found clinical relevance.[94]

### Glutamine

Glutamine is a nonessential amino acid, but may have increased needs in hypercatabolic states. It is an important nutrient for enterocytes and lymphocytes, and thus can improve intestinal integrity, reduce bacterial translocation and maintain the level of intestinal and extraintestinal immunoglobulin A.

In a clinical study, Scheppach et al.[95] showed that glutamine or glutamine dipeptides are trophic to the intestinal mucosa, with increased villus height and reduction in intestinal permeability. However, results of clinical studies with glutamine supplementation have not demonstrated that it restores intestinal permeability or induces remission.[95] Thus, the available data suggest that, as a trophic nutrient for the intestinal mucosa, glutamine may have a potential benefit in CD, but more studies to determine its actual effect are required.

Many controlled studies have assessed and compared the effectiveness of formulas based on free amino acids, peptides or whole protein in active CD. No difference in relation to the formulas has been found.[96] Formulas with protein intact and added glutamine has not shown superiority with relation to the standard formula in active CD, from a clinical and nutritional point of view.[95,97] However, in recent years, the association of glutamine with arginine in acute colitis in CD has been assessed, as nutrients capable of reducing inflammation of the mucosa. The combined effect of glutamine and arginine cause a decrease in TNF-alpha production, inflammatory cytokines and possibly by regulating the expression of NFkB and p38 MAPK in colon biopsies of patients with active CD.[98] To do so, appropriate concentrations of these amino acids at the inflamed site are required, from oral or enteral intake at high doses of these nutrients.

There is only one prospective, controlled study of parenteral glutamine used in 24 patients with active IBD, of which 19 had CD. However, there was no improvement with the addition of glutamine at 0.3 g/kg of L-alanyl-L-glutamine in PNT, with relation to intestinal permeability, plasma concentration, nutritional parameters, disease activity and hospital stay.[99]

## Probiotics

Probiotics are live microorganisms which make up some of the foods that provide benefit to the intestines, since they can replace the pathogenic bacteria through competition, which may favor active CD, given that inflammation promotes a change in the intestinal microflora.

Probiotics are not pathogenic and, in general, are of the genus *Bifidobacterium* or *Lactobacillus*, but may also include other bacteria such as non-invasive coliforms and organisms that are not bacteria, such as *Saccharomyces boulardii*.

O'Mahony et al.[100] suggest that in CD, the probiotic *L. salivarius* has equivalent efficacy to 5-ASA, a classic pharmacological treatment. Studies with small numbers of patients have shown promising results with the use of probiotics in CD. In a prospective preliminary study, randomized and controlled, conducted on small number of patients with active CD, Grupta et al.[101] indicated the efficacy of capsules of *Lactobacillus GG* administered twice/day for 6 weeks. After four weeks, there was an improvement in permeability and the function of the intestinal barrier, as well as disease activity. However, Prantera et al.[102] repeated this study on a larger number of patients with CD, and the results did not favor the use of probiotics in CD for prevention of recurrence or reduction of inflammation.

Recently, a meta-analysis of eight studies did not demonstrate the efficacy of probiotics in the maintenance of remission and prevention of clinical and endoscopic recurrence in CD.[103] Another meta-analysis showed that the use of probiotics associated with conventional treatment does not improve the remission index in mild to moderate UC, but is more effective in severe and extensive forms of the disease. It is necessary to increase case series for more definitive conclusions.[104] However, a meta-analysis of 13 studies conducted on patients with UC showed that probiotics are more effective than a placebo in maintaining remission.[105]

### Butyrate

Short chain fatty acids (SCFA), including butyrate, propionate and acetate are created in the colon as a result of bacterial fermentation of dietary fiber by bacteria in the intestinal lumen. The SCFAs are rapidly absorbed by the intestinal mucosa and are important sources of metabolic substrates for colonocytes, such as tropism for the mucosa, stimulating the absorption of water and sodium, in addition to stimulating enzymes to repair the mucosa.

The SCFAs, particularly butyrate, have anti-inflammatory effects, given that they decrease the expression of pro-inflammatory cytokines by inhibiting the activation of nuclear factor kappa-B in patients with CD.[106] However, prospective, randomized studies with a suitable case series should be performed to determine the favorable effect on CD and UC.

## TGF-B2

The transforming growth factor B2 (TGF-B2) is a polypeptide normally found in milk, able to enrich enteral feeding, due to their action in modulating intestinal immunity, as it antagonizes TNF-alpha.

An oral polymeric formula rich in TGF-B2 was evaluated as the only form of nutrition for 8 weeks in 29 children with active CD.[66] Of these, 79% achieved complete remission, with improved laboratory inflammation indicators, endoscopic healing and reduction of pro-inflammatory cytokines (interferon-gamma, interleukin-8) in the terminal ileum and colon. Enteral formulas supplemented with TGF-B2 have not been evaluated in controlled clinical studies. A reduction in mucosal inflammation, reduction of pro-inflammatory cytokines in the ileum and colon, and increased TGF-B2 m-RNA have been found. However, the clinical benefits of these modified formulations remain unproven in the absence of adequate clinical studies.[107]

## EXCLUSION DIET

The exclusion of dietary antigens from diets may favor the control of CD activity, although more controlled studies are necessary with a larger number of patients to demonstrate its real effectiveness.

Many years ago, Alun et al.[108] demonstrated in 20 consecutive patients with active CD that certain foods not tolerated by patients, when eliminated from diets and reintroduced individually, caused a recurrence of symptoms. Jones[109] evaluated 36 patients randomized to receive ENT with an elemental diet or PNT, and showed that both were efficient in the induction of remission. They were then followed by diets excluding foods considered intolerable, and the rate of recurrence after 1 year was 11%, considered an important strategy in the maintenance of CD remission. Riordan et al.[110] showed a reduction of remission with the exclusion of nutrients compared with corticosteroids (62 *versus* 79%). Two studies have shown high prevalence of lactose intolerance in patients with CD.[110,111]

Mishkin et al.[111] found lactase deficiency in 40% of patients with CD compared with 29% of controls and 13% of patients with UC. In CD limited to the terminal ileum, the chance of this deficiency is higher than in Crohn's colitis or ulcerative colitis. The prevalence of lactose malabsorption in CD

(46.9%) was similar in a German study[95] with a higher frequency in active CD (83.3%). Among the possible causes of this malabsorption, bacterial overgrowth and increased transit time are prominent. In UC, there is no evidence of lactose malabsorption.

Thus, by virtue of the quality of the studies, there is no evidence that dietary modifications to the oral diet has benefits related to maintaining remission, with reservations associated with lactose, which, in the event of doubts, should always be investigated. Various dietary changes have been investigated as adjuvant treatment, such as modifications to the dietary fiber content and simple sugars, but showed no benefit in the maintenance of remission.

## REFERENCES

1. Nguyen GC, Munsell M, Harris ML. Nationwide prevalence and prognostic significance of clinically diagnosable protein-calorie malnutrition in hospitalized inflammatory bowel disease patients. Inflamm Bowel Dis 2008; 14:1105-11.

2. van Heel DA, Fisher SA, Kirby A, Daly MJ, Rioux JD, Lewis CM. Inflammatory bowel disease susceptibility loci defined by genome scan meta-analysis of 1952 affected relative pairs. Hum Mol Genet 2004; 13(7):763-70.

3. Costas Armada P, Garcia-Mayor RV, Larranaga A, Seguin P, Perez Mendez LF. Rate of undernutrition and response to specific nutritional therapy in Crohn's disease. Nutr Hosp 2009; 24(2):161-6.

4. Valentini L, Schaper L, Buning C, Hengstermann S, Koernicke T, Tillinger W et al. Malnutrition and impaired muscle strength in patients with Crohn's disease and ulcerative colitis in remission. Nutrition 2008; 24(7-8):694-702.

5. Reimund JM, Arondel Y, Escalin G, Finck G, Baumann R, Duclos B. Immune activation and nutritional status in adult Crohn's disease patients. Dig Liver Dis 2005; 37(6):424-31.

6. Sousa Guerreiro C, Cravo M, Costa AR, Miranda A, Tavares L, Moura-Santos P et al. A comprehensive approach to evaluate nutritional status in Crohn's patients in the era of biologic therapy: a case-control study. Am J Gastroenterol 2007; 102(11):2551-6.

7. Gassul MA, Cabré E, Vilar LI. Nível de ingesta hospitalar y su papel en el desarollo de malnutricion calórico-proteica en pacientes gastroenterológicos hospitalizados. Med Clin (Barc) 1985; 85:85-90.

8.  Subramanian V, Saxena S, Kang JY, Pollok RC. Preoperative steroid use and risk of postoperative complications in patients with inflammatory bowel disease undergoing abdominal surgery. Am J Gastroenterol 2008; 103(9):2373-81.

9.  Filippi J, Al-Jaouni R, Wiroth JB, Hebuterne X, Schneider SM. Nutritional deficiencies in patients with Crohn's disease in remission. Inflamm Bowel Dis 2006; 12(3):185-91.

10. Waitzberg DL, Teixeira da Silva ML. Diagnóstico das alterações nutricionais da doença inflamatória intestinal. In: Habr-Gama A. CBC – Clínica Brasileira de Cirurgia. Doença inflamatória intestinal. São Paulo: Atheneu, 1997.

11. Fleming CR, Huizenga KA, McCall JT, Gildea J, Dennis R. Zinc nutrition in Crohn's disease. Dig Dis Sci 1981; 26:865-70.

12. Wilson A, Reyes E, Ofman J. Prevalence and outcomes of anemia in inflammatory bowel disease: a systematic review of the literature. A J Med 2004; 116(Suppl.7A):44S-9S.

13. Cronin CC, Shanahan F. Anemia in patients with chronic inflammatory bowel disease. Am J Gastroenterol 2001; 96(8):2296-8.

14. Means Jr. RT. Advances in the anemia of chronic disease. Int J Hematol 1999; 70:7-12.

15. Schreiber S, Howaldt S, Schnoor M, Nikolaus S, Bauditz J, Gasche C et al. Recombinant erythropoietin for the treatment of anemia in inflammatory bowel disease. N Engl J Med 1996; 334(10):619-23.

16. Papa A, De Stefano V, Danese S, Chiusolo P, Persichilli S, Casorelli I et al. Hyperhomocysteinaemia and prevalence of polymorphisms of folate-metabolism-related enzymes in patients with inflammatory bowel disease. Am J Gastroenterol 2001; 96:2677-82.

17. Rigaud D, Cerf M, Angel Alberto L, Sobhani I, Carduner MJ, Mignon M. Increase of resting energy expenditure during flare-ups in Crohn disease. Gastroenterol Clin Biol 1993; 17(12):932-7.

18. Stokes MA, Hill GL. Total energy expenditure in patients with Crohn's disease: measurement by the combined body scan technique. JPEN 1993; 17:3-7.

19. Graham TO, Kandil HM. Nutritional factors in inflammatory bowel disease. Gastroenterol Clin North Am 2002; 31(1):203-18.

20. Vasconcelos MIL. Avaliação de pacientes hospitalizados. In: Tirapegui J, Ribeiro SML (eds.). Avaliação nutricional: teoria e prática. Rio de Janeiro: Guanabara Koogan, 2009.

21. Silva MKS, Félix DS. Uso da antropometria na avaliação do estado nutricional. Rev Bras Nutr Clin 1998; 13:74-80.

22. Vasconcelos MIL. Avaliação nutricional antropométrica. In: Magnoni D, Cukier C (eds.). Nutrição na insuficiência cardíaca. São Paulo: Sarvier, 2002.

23. Valdés MP, Savino P, Pimiento S, Escallón J. Evaluación nutricional en pacientes con soporte metabólico y nutricional. Lectura Nutr 1997; 4:28-38.

24. Dias MCG, Horie LM, Waitzberg DL. Exame físico e antropometria. In: Waitzberg DL (ed). Nutrição oral, enteral e parenteral na prática clínica. 4.ed. São Paulo: Atheneu, 2009.

25. Vasconcelos MIL. Nutrição enteral. In: Cuppari L (ed.). Nutrição clínica no adulto. 2.ed. Barueri: Manole, 2005.

26. Riella MC. Avaliação nutricional e metabólica. In: Riella MC (ed.). Suporte nutricional parenteral e enteral. Rio de Janeiro: Guanabara Koogan, 1993.

27. Heymsfield SB, Tighe A, Wang AM. Nutritional assessment by anthropometric and biochemical methods. In: Shils ME, Olson JA, Shike M (eds.). Modern nutrition in health and disease. Philadelphia: Lea & Febiger, 1994.

28. Coppini LZ, Horie LM, Waitzberg DL. Impedância bioelétrica. In: Waitzberg DL (ed.). Nutrição oral, enteral e parenteral na prática clínica. 4.ed. São Paulo: Atheneu, 2009.

29. Detsky AS, McLaughlin JR, Baker JP, Johnston N, Whittaker S, Mendelson RA et al. What is subjective global assessment of nutritional status? JPEN J Parenteral Enteral Nutr 1987; 11:8-13.

30. Blackburn GL, Bistrian BR, Maini BS, Schlamm HT, Smith MF. Nutritional and metabolic assessment of the hospitalized patient. JPEN 1977; 1:11.

31. Bernard MA, Jacobs DO, Rombeau JL. Necessidades nutricionais. In: Suporte nutricional e metabólico de pacientes hospitalizados. Rio de Janeiro: Guanabara Koogan, 1988.

32. Burgos MGPA, Salviano FN, Belo GMS, Bion FM. Doenças inflamatórias intestinais: o que há de novo em terapia nutricional? Rev Bras Nutr Clin 2008; 23(3):184-9.

33. Bottoni A, Rodrigues RC, Bottoni A, Nogueira RJN. Exames laboratoriais. In: Waitzberg DL (ed.). Nutrição oral, enteral e parenteral na prática clínica. 4.ed. São Paulo: Atheneu, 2009.

34. Chan AT, Fleming CR, O'Fallon WM, Iluizenga KA. Estimated versus measured basal energy requirements in patients with Crohn's disease. Gastroenterology 1986; 91:75-8.

**35.** Eiden K. Nutritional considerations in inflammatory bowel disease. Pract Gastroenterol 2003; XXVII:33.

**36.** American Society for Parenteral and Enteral Nutrition. Nutrition Support Core Curriculum: a case-based approach – the adult patient. Gastrointestinal Disease 2007.

**37.** Newby EA, Sawczenko A, Thomas AG, Wilson D. Interventions for growth failure in childhood Crohn's disease. Cochrane Database Syst Rev 2005; (3):CD003873.

**38.** Gailhoustet L, Goulet O, Cachin N, Schmitz J. Study of psychological repercussions of 2 modes of treatment of adolescents with Crohn's disease. Arch Pediatr 2002; 9(2):110-6.

**39.** Afzal NA, Van Der Zaag-Loonen HJ, Arnaud-Battandier F, Davies S, Murch S, Derkx B et al. Improvement in quality of life of children with acute Crohn's disease does not parallel mucosal healing after treatment with exclusive enteral nutrition. Aliment Pharmacol Ther 2004; 20(2):167-72.

**40.** Scolapio JS. The role of total parenteral nutrition in the management of patients with acute attacks of inflammatory bowel disease. J Clin Gastroenterol 1999; 29:223-4.

**41.** Lindor KD, Fleming CR, Ilstrup DM. Preoperative nutritional status and other factors that influence surgical outcome in patients with Crohn's disease. Mayo Clin Proc 1985; 60(6):393-6.

**42.** Niu LY, Gong JF, Wei XW, Zhu WM, Li N, Li JS. Effects of perioperative combined nutritional support in Crohn disease. Zhonghua Wai Ke Za Zhi 2009; 47(4):275-8.

**43.** Greenberg GR, Fleming CR, Jeejeebhoy KN, Rosenberg IH, Sales D, Tremaine WJ. Controlled trial of bowel rest and nutritional support in the management of Crohn's disease. Gut 1988; 29(10):1309-15.

**44.** Frascio F, Giacosa A, Martines D, Sukkar SG, Naccarato RI. The bowel rest: a key factor in the management of active Crohn's disease? Rivista Italiana Di Nutrizione Parenterale Ed Enterale 1997; 15(12):90-6.

**45.** Miehsler W, Reinisch W, Valic E, Osterode W, Tillinger W, Feichtenslager T et al. Is inflammatory bowel disease an independent and disease specific risk factor for thromboembolism? Gut 2004; 53:542-8.

**46.** Michetti P, Peppercorn MA. Medical therapy of specific clinical presentations. Gastroenterol Clin North Am 1999; 28(2):353-70.

**47.** Present DH, Rutgeerts P, Targan S, Hanauer SB, Mayer L, van Hogezand RA et al. Infliximab for the treatment of fistulas in patients with Crohn's disease. N Engl J Med 1999; 340:1398-405.

**48.** Tárrago CP, Maestu AP, de La Torre AM. Tratamiento nutricional en la enfermedad inflamatoria intestinal. Nutr Hosp 2008; 23:417-27.

**49.** ASPEN. Guidelines for the use of parenteral and enteral nutrition in adult and pediatric patients. JPEN J Parenter Enteral Nutr 2002; 26(Suppl 1):1S-138S.

**50.** Fernandez-Banares F, Cabre E, Esteve-Comas M, Gassull MA. How effective is enteral nutrition in inducing clinical remission in active Crohn's disease? A meta-analysis of the randomized clinical trials. JPEN J Parenter Enteral Nutr 1995; 19(5):356-64.

**51.** Griffiths AM, Ohlsson A, Sherman PM, Sutherland LR. Meta-analysis of enteral nutrition as a primary treatment of active Crohn's disease. Gastroenterology 1995; 108(4):1056-67.

**52.** Messori A, Trallori G, D'Albasio G, Milla M, Vannozzi G, Pacini F. Defined-formula diets versus steroids in the treatment of active Crohn's disease: a meta-analysis. Scand J Gastroenterol 1996; 31(3):267-72.

**53.** Zachos M, Tondeur M, Griffiths AM. Enteral nutritional therapy for inducing remission of Crohn's disease. Cochrane Database Syst Rev 2001; 3.

**54.** Zachos M, Tondeur M, Griffiths AM. Enteral nutritional therapy for induction of remission in Crohn's disease. Cochrane Database Syst Rev 2007; 1.

**55.** El-Matary W. Enteral nutrition as a primary therapy of Crohn's disease: the pediatric perspective. Nutr Clin Pract 2009; 24:91-7.

**56.** Gassull MA. Can nutritional therapy replace pharmacologic therapy in pediatric Crohn's disease? Nat Clin Pract Gastroenterol Hepatol 2009; 6:80-1.

**57.** Heuschkel RB, Menache CC, Megerian JT, Baird AE. Enteral nutrition and corticosteroids in the treatment of acute Crohn's disease in children. J Pediatr Gastroenterol Nutr 2000; 31(1):8-15.

**58.** Dziechciarz P, Horvath A, Shamir R, Szajewska H. Meta-analysis: enteral nutrition in active Crohn's disease in children. Aliment Pharmacol Ther 2007; 26(6):795-806.

**59.** Borrelli O, Cordischi L, Cirulli M, Paganelli M, Labalestra V, Uccini S et al. Polymeric diet alone versus corticosteroids in the treatment of active pediatric Crohn's disease: a randomized controlled open-label trial. Clin Gastroenterol Hepatol 2006; 4:744-53.

**60.** Canani BR, Terrin G, Borrelli O, Romano MT, Manguso F, Coruzzo A et al. Short- and long-term therapeutic efficacy of nutritional therapy and corticosteroids in paediatric Crohn's disease. Dig Liver Dis 2006; 38:381-7.

**61.** Yamamoto T, Nakahigashi M, Umegae S, Matsumoto K. Enteral nutrition for the maintenance of remission in Crohn's disease: a systematic review. Eur J Gastroenterol Hepatol 2010; 22:1-8.

**62.** Yamamoto T, Nakahigashi M, Saniabadi AR, Iwata T, Maruyama Y, Umegae S et al. Impacts of long-term enteral nutrition on clinical and endoscopic disease activities and mucosal cytokines during remission in patients with Crohn's disease: a prospective study. Inflamm Bowel Dis 2007; 13(12):1493-501.

**63.** Ikeuchi H, Yamamura T, Nakano H, Kosaka T, Shimoyama T, Fukuda Y. Efficacy of nutrition therapy for perforating and non-perforating Crohn's disease. Hepatogastroenterology 2004; 51:1050-2.

**64.** Yamamoto T, Nakahigashi M, Umegae S, Kitagawa T, Matsumoto K. Impact of long-term enteral nutrition on clinical and endoscopic recurrence after ressection for Crohn's disease: a prospective, non-randomized, parallel, controlled study. Aliment Pharmacol Ther 2007; 25:67-72.

**65.** Afzal NA, Davies S, Paintin M, Arnaud-Battandier F, Walker-Smith JA, Murch S et al. Colonic Crohn's disease in children does not respond well to treatment with enteral nutrition if the ileum is not involved. Dig Dis Sci 2005; 50:1471-5.

**66.** Fell JM, Paintin M, Arnaud-Battandier F, Beattie RM, Hollis A, Kitching P et al. Mucosal healing and a fall in mucosal pro-inflammatory cytokine mRNA induced by a specific oral polymeric diet in paediatric Crohn's disease. Aliment Pharmacol Ther 2000; 14:281-9.

**67.** Damião AOMC. Doença inflamatória intestinal: terapia biológica. J Bras Gastroenterol 2009; 9:4-7.

**68.** González-Huix F, Fernández-Bañares F, Esteve-Comas M, Abad-Lacruz A, Cabré E, Acero D et al. Enteral versus parenteral nutrition as adjunct therapy in acute ulcerative colitis. Am J Gastroenterol 1993; 88:227-32.

**69.** Damião AOMC. Dieta oligomérica: conceito e aplicações clínicas. Rev Visão Méd Oncologia (RVMO) 2008; 6:13-6.

**70.** Lorenzs R, Weber PC, Szimnau P. Suplemmentation with n-3 fatty acids from fish oil in chronic inflammatory bowel disease a randomized placebo-controlled, double blind cross-over trial. J Intern Med 1989; 225(731):S32-7.

**71.** Malchow H, Steinhardt HJ, Lorenz-Meyer H, Strohm WD, Rasmussen S, Sommer H. Feasibility and effectiveness of a defined-formula diet regimen in treating active Crohn's disease. European Cooperative Crohn's Disease Study III. Scand J Gastroenterol 1990; 25:235-44.

72. Lochs H, Steinhardt HJ, Klaus-Wentz B, Zeitz M, Vegelsang H, Sommer H. Comparison of enteral nutrition and drug treatment in active Crohn's disease. Gastroenterology 1991; 101:881-8.

73. Gonzales-Huix F, de Leon R, Fernandez-Banares F, Esteve M, Cabre E, Acero D. Polymeric enteral diets as primary treatment of active Crohn's disease: a prospective steroid controlled trial. Gut 1993; 34:778-82.

74. Giaffer MH, North G, Holdsworth CD. Controlled trial of polymeric versus elemental diet in treatment of active Crohn's disease. Lancet 1990; 335:816-9.

75. Stein J. Chemically defined structured lipids: current status and future directions in gastrointestinal diseases. Int J Colorectal Dis 1999; 14(2):79-85.

76. Bamba T, Shimoyama T, Sasaki M, Tsujikawa T, Fukuda Y, Koganei K et al. Dietary fat attenuates the benefits of an elemental diet in active Crohn's disease: a randomized, controlled trial. Eur J Gastroenterol Hepatol 2003; 15(2):151-7.

77. Griffiths AM. Enteral nutrition in the management of Crohn's disease JPEN J Parenter Enteral Nutr 2005; 29(4 Suppl.):S108-12; discussion S112-7, S184-8.

78. Akobeng AK, Thomas AG. Enteral nutrition for maintenance of remission in Crohn's disease. Cochrane Database Syst Rev 2007.

79. Takagi S, Utsunomiya K, Kuriyama S, Yokoyama H, Takahashi S, Iwabuchi M et al. Effectiveness of an "half elemental diet" as maintenance therapy for Crohn's disease: a randomized-controlled trial. Aliment Pharmacol Ther 2006; 24(9):1333-40.

80. Verma S, Holdsworth CD, Giaffer MH. Does adjuvant nutritional support diminish steroid dependency in Crohn disease? Scand J Gastroenterol 2001; 36(4):383-8.

81. Verma S, Kirkwood B, Brown S, Giaffer MH. Oral nutritional supplementation is effective in the maintenance of remission in Crohn's disease. Dig Liver Dis 2000; 32(9):769-74.

82. Harries AD, Jones LA, Danis V, Fifield R, Heatley RV, Newcombe RG et al. Controlled trial of supplemented oral nutrition in Crohn's disease. Lancet 1983; 1(8330):887-90.

83. Bannerjee K, Camacho-Hübner C, Babinska K, Dryhurst KM, Edwards R, Savage MO et al. Anti-inflammatory and growth-stimulating effects precede nutritional restitution during enteral feeding in Crohn's disease. J Pediatr Gastroenterol Nutr 2004; 38:270-5.

84. Meister D, Bode J, Shand A, Ghosh S. Anti-inflammatory effects of enteral diet components on Crohn's disease-affected tissues in vitro. Dig Liver Dis 2002; 34:430-8.

85. Swidsinski A, Ladhoff A, Pernthaler A, Swidsinski S, Loening-Baucke V, Ortner M et al. Mucosal flora in inflammatory bowel disease. Gastroenterology 2002; 122:44-54.

86. Kleessen B, Kroesen AJ, Buhr HJ, Blaut M. Mucosal and invading bacteria in patients with inflammatory bowel disease compared with controls. Scand J Gastroenterol 2002; 37:1034-41.

87. Schneider SM, Le Gall P, Girard-Pipau F, Piche T, Pompei A, Nano JL et al. Total artificial nutrition is associated with major changes in the fecal flora. Eur J Nutr 2000; 39:248-55.

88. Gassull MA, Fernández-Bañares F, Cabré E, Papo M, Giaffer MH, Sánchez-Lombraña JL et al. Fat composition may be a clue to explain the primary therapeutic effect of enteral nutrition in Crohn's disease: results of a double blind randomized multicentre European trial. Gut 2002; 51:164-8.

89. Sadeghi S, Wallace FA, Calder PC. Dietary lipids modify the cytokine response to bacterial lipopolysaccharide in mice. Immunology 1999; 96:404-10.

90. Lorenz-Meyer H, Bauer P, Nicolay C, Schulz B, Purrmann J, Fleig WE et al. Omega-3 fatty acids and low carbohydrate diet for maintenance of remission in Crohn's disease. A randomised controlled multi-centre trial. Study Group Members (German Crohn's Disease Study Group). Scand J Gastroenterol 1996; 31:778-85.

91. Belluzzi A, Brignola C, Campieri M, Pera A, Boschi S, Miglioli M. Effect of an enteric-coated fish-oil preparation on relapses in Crohn's disease. N Engl J Med 1996; 334(24):1557-60.

92. MacLean CH, Mojica WA, Newberry SJ, Pencharz J, Garland RH, Tu W et al. Systematic review of the effects of n-3 fatty acids in inflammatory bowel disease. Am J Clin Nutr 2005; 82(3):611-9.

93. Turner D, Zlotkin SH, Shah PS, Griffiths AM. Omega 3 fatty acids (fish oil) for maintenance of remission in Crohn's disease. Cochrane Database Syst Rev 2007.

94. Turner D, Zlotkin SH, Shah PS, Griffiths AM. Omega 3 fatty acids (fish oil) for maintenance of remission in Crohn's disease. Cochrane Database Syst Rev 2009.

95. Scheppach W, Loges C, Bartram P, Christl SU, Richter F, Dusel G et al. Effect of free glutamine and alanyl-glutamine dipeptide on mucosal proliferation of the human ileum and colon. Gastroenterology 1994; 107:429-34.

**96.** Akobeng AK, Miller V, Stanton J, Elbadri AM, Thomas AG. Double-blind randomized controlled trial of glutamine-enriched polymeric diet in the treatment of active Crohn's disease. J Paediatr Gastroenterol Nutr 2000; 30:78-84.

**97.** Verma S, Brown S, Kirkwood B, Giaffer MH. Polymeric versus elemental diet as primary treatment in active Crohn's disease: a randomized, double-blind trial. Am J Gastroenterol 2000; 95(3):735-9.

**98.** Den Hond E, Hiele M, Peeters M, Ghoos Y, Rutgeerts P. Effect of long-term oral glutamine supplements on small intestinal permeability in patients with Crohn's disease. JPEN J Parenter Enteral Nutr 1999; 23(1):7-11.

**99.** Lecleire S, Hassan A, Marion-Letellier R, Antonietti A, Savoye G, Feysot CB et al. Combined glutamine and arginine decrease proinflammatory cytokine production by biopsies from Crohn's patients in association with changes in nuclear factor-kB and p38 mitogen-activated protein kinase pathways. J Nutr 2008; 138:2481-6.

**100.** O'Mahony L, McCarthy J, Feeney M. Immunologic response to a novel probiotic organism in patients with active Crohn's disease. Gastroenterol 2000; 116:A4763.

**101.** Grupta P. Is lactobacillus GG helpful in children with Crohn's disease? Results of a preliminary open-label study. J Pediatr Gastroenterol Nutr 2000; 31(4):453-7.

**102.** Prantera C, Scribano ML, Falasco G, Andreoli A, Luzi C. Ineffectiveness of probiotics in preventing recurrence after curative resection for Crohn's disease: a randomised controlled trial with Lactobacillus GG. Gut 2002; 51(3):405-9.

**103.** Rahimi R, Nikfar S, Rahimi F, Elahi B, Derakhshani S, Vafaie M et al. A meta-analysis on the efficacy of probiotics for maintenance of remission and prevention of clinical and endoscopic relapse in Crohn's disease. Dig Dis Sci 2008; 53(9):2524-31.

**104.** Mallon P, McKay D, Kirk S, Gardiner K. Probiotics for induction of remission in ulcerative colitis. Cochrane Database Syst Rev 2007.

**105.** Sang L-X, Chang B, Zhang W-L, Wu X-M, Li X-H, Jiang M. Remission induction and maintenance effect of probiotics on ulcerative colitis: a meta-analysis. World J Gastroenterol 2010; 16(15):1908-15.

**106.** Segain JP, Raingeard de la Bletierre D, Bourreille A, Leray V, Gervois N, Rosales C et al. Butyrate inhibits inflammatory responses through NF-kB inhibition: implications for Crohn's disease. Gut 2000; 47:397-403.

**107.** Beattie RM, Schiffrin EJ, Donnet-Hughes A, Huggett AC, Domizio P, MacDonald TT et al. Polymeric nutrition as the primary therapy in children with small bowel Crohn's disease. Aliment Pharmacol Ther 1994; 8(6):609-15.

**108.** Alun JV, Dickinson RJ, Workman E, Wilson AJ, Freeman AH, Hunter JU. Crohn's disease: maintenance of remission by diet. Lancet 1985; ii:177-80.

**109.** Jones VA. Comparison of total parenteral nutrition and elemental diet in induction of remission of Crohn's disease: long-term maintenance of remission by personalized food exclusion diets. Dig Dis Sci 1987; 32(Suppl. 12):S100-S7.

**110.** Riordan AM, Hunter JO, Cowan RE. Treatment of active Crohn's disease by exclusion diet: East Anglia multi-centre controlled trial. Lancet 1993; 342:1131-4.

**111.** Mishkin B, Yalovsky M, Mishkin S. Increased prevalence of lactose malabsorption in Crohn's disease patients at low risk for lactose malabsorption based on ethnic origin. Am J Gastroenterol 1997; 92(7):1148-53.

## BIBLIOGRAPHY

**1.** Detsky AS, McLaughlin JR, Baker JP, Johnston N, Whittaker S, Mendelson RA et al. What is subjective global assessment of nutritional status? JPEN J Parenter Enteral Nutr 1987; 11(1):8-13.

**2.** Dichi I, Burini RC. Desnutrição proteico-energética na doença inflamatória intestinal. Rev Bras Nutr Clin 1996; 11(1):8-15.

**3.** Flora APL, Dichi I. Aspectos atuais na terapia nutricional da doença inflamatória intestinal. Rev Bras Nutr Clin 2006; 21(2):131-7.

**4.** Griffiths AM. Doença intestinal inflamatória. In: Shils ME, Olson JA, Shike M, Ross AC (eds.). Tratado de nutrição moderna na saúde e na doença. 9.ed. Barueri: Manole, 2003.

**5.** Junior PEP, Habr-Gama A, Teixeira MG, Ferrini MT, Rodrigues JJG. Moléstia inflamatória intestinal. In: Waitzberg DL. Nutrição oral, enteral e parenteral na prática clínica. 3.ed. São Paulo: Atheneu, 2001.

**6.** Kondrup J, Rasmussen HH, Hamberg O, Stanga Z. Nutritional risk screening (NRS 2002): a new method based on an analysis of controlled clinical trials. Clin Nutr 2003; 22(3).

**7.** Polk DB, Hattner JAT, Kerner JA. Improved growth and disease activity after intermittent administration of a defined formula diet in children with Crohn's disease. JPEN 1992; 16:499-504.

**8.** Preece M. Growth retardation among children and adolescents with inflammatory bowel disease. In: Davidson M (ed.). National Foundation for ileitis and colitis. New York, 1983.

9. Ritchie JK, Wadsworth J, Lennard-Jones JE, Rogers E. Controlled multicentre therapeutic trial of an unrefined carbohydrate, fiber rich diet in Crohn's disease. Br Med J (Clin Res Ed) 1987; 295(6597):517-20.

10. von Tirpitz C, Kohn C, Steinkamp M, Geerling I, Maier V, Moller P et al. Lactose intolerance in active Crohn's disease: clinical value of duodenal lactase analysis. J Clin Gastroenterol 2002; 34(1):49-53.

# INFLAMMATORY BOWEL DISEASE DURING PREGNANCY

GENOILE OLIVEIRA SANTANA
BRUNO CÉSAR DA SILVA

## INTRODUCTION

Inflammatory bowel disease (IBD) usually affects individuals during the reproductive stage of their lives. Thus, it is common for young women with Crohn's disease (CD) or ulcerative colitis (UC) to demonstrate a desire to become pregnant.

For lack of information and clarifications, many of them are discouraged to pursue this dream. On the other hand, when they decide to have a baby, they end up being improperly instructed to reduce or even stop taking prescribed medications.

When it comes to pregnancy, it is known that there are doubts and fear in women with IBD. The risk that the drugs may offer and the consequences that the disease can cause the fetus are present matters in this scenario.[1]

## HEREDITY

In association with environmental factors, hereditary characteristics increase the risk of emergence of IBD. Children of people with IBD are 2-3 times more likely to develop these diseases than the general population. If one parent is a carrier, the risk of the offspring being affected by IBD is from 8 to 11%, and from 20 to 35% in cases where both the father and the mother are affected.[2] In another study, the authors found that the risk of the child developing UC when a parent is a carrier is 1.6%, and 5.2% of developing CD when one parent has the disease. As a rule, hereditary factors seem to be more involved with the development of CD.[3]

## FERTILITY IN PATIENTS WITH IBD

Generally, men and women with IBD do not show significant differences in fertility (ability to conceive) compared with the general population, except for some subgroups of patients with IBD, such as those composed of individuals undergoing certain surgical procedures or those who have depressive disorders. During the period with active disease, fertility seems slightly compromised, although studies have conflicting results as to this information.

The rate of female infertility in the general population is estimated at 5 to 14%. Studies show that this number goes up to 48% in women who underwent colectomy with pouch reconstruction. The use of drugs does not seem to exert influence on fertility of women; however, sulfasalazine and methotrexate can cause a reversible infertility in men.[1]

## EFFECTS OF IBD ON PREGNANCY

The fear that women with IBD have to bear children with serious problems appears to be greater than the actual risk. It is believed that, in this context, the chance of complications in childbirth is related to disease activity. Several studies point to a higher incidence of premature births, cesarean and low birth weight in the subgroup of women suffering from CD or UC. It is noteworthy that most of these premature births occur after the 35th week of pregnancy and usually present favorable a outcome for both the mother and the baby. Control of disease activity in the months before conception and during pregnancy is important to minimize the risk of complications.[2]

While most published data does not associate IBD with increased risk of congenital abnormalities, some studies indicate a slight increase in congenital malformations in patients with UC, but not with CD. An Israeli retrospective study published recently found that women with IBD who were using drug therapy during pregnancy had higher rate of congenital abnormalities, primarily related to the limbs and central nervous system. Because of the small sample size, however, it was not possible to conclude whether the association would be with the use of medications or disease activity.[4] It is believed that this risk, if present, is low and therefore has not been subject to medical recommendations to prevent reproduction in this context.[1]

In pregnant women with IBD, increased occurrence of other gestational diseases such as eclampsia and gestational diabetes has not been observed.[3]

## EFFECTS OF PREGNANCY ON IBD

In general, the disease behavior does not change. There is no concrete data associating IBD with an increased number of relapses during pregnancy, or showing that pregnancy influences the degree of disease activity. However, in two studies, it was observed that, within 3 years of birth, there was a lower rate of disease relapse. The hypothesis is that, similarly to rheumatoid arthritis, pregnant women with IBD undergo a process of immunosuppression that affects the pathogenesis of immune-mediated diseases, reducing their intensity.[5,6]

## CHOICE ON TYPE OF DELIVERY

In general, the decision between cesarean and vaginal delivery should be based purely on obstetric reasons. Some surgeons prefer elective cesarean section to prevent damage to the anal sphincter. Vaginal delivery and episiotomy can lead to the development or worsening of perianal CD. Attention should be paid to two specific situations: active perianal disease and presence of ileoanal pouch. In both cases, a c-section is preferable.[7]

## IBD THERAPY IN PREGNANCY

As mentioned, many patients, families and physicians mistakenly believe that pregnant women with IBD should avoid the use of medications. However, most drugs used to treat IBD are considered safe during pregnancy. Studies on the subject have shown that the existence of disease activity brings more risk to the fetus than treatment.[8]

There are several classes of drugs used to achieve remission, or during the maintenance phase. Surveys were conducted over the past few years to assess the risk that each of them offers to the fetus. In order to guide the management of IBD safely, each of the classes used for treatment will be analyzed below. In addition to information published in the medical literature, it is important to know that the US Food and Drug Administration (FDA) is responsible for the classification of drugs according to the risk they offer during pregnancy (Table 32.1). This classification has been helping doctors of all specialties to make safe decisions when initiating or discontinuing therapy during pregnancy.

### Aminosalicylates

Sulfasalazine and mesalazine are drugs classically used in the therapy of IBD. They are classified as category B by the FDA.

Initial data suggested that sulfasalazine could lead to cardiovascular, genitourinary and neurological congenital anomalies. However, subsequent studies did not show teratogenic effect. As sulfasalazine can lead to folic acid deficiency, neural tube malformations can occur, especially in the first trimester of pregnancy. This problem can be prevented with the administration of 2 mg/day of folic acid.[9]

There are infrequent reports of development of nephrotoxicity in the newborn when mesalazine is used at doses greater than 3 g/day.[8]

| Table 32.1 Classification of drugs used in pregnancy according to the FDA | |
|---|---|
| **Category** | **Definition** |
| A | No risk in controlled human studies. Animal studies without evidence for teratogenicity and studies in pregnant women showed no risk to the fetus |
| B | No evidence of risk in humans; or animal studies are negative, but without adequate studies in human; or animal findings demonstrate risk, but not in humans |
| C | Risk not ruled out. Lack of human studies; and animal studies are either positive for fetal risk; or studies are still insufficient. However, the benefits may justify the potential risks |
| D | Positive evidence of risk. Research and post-marketing data show risk to the fetus. However, the potential benefits may be more important that the risks |
| X | Contraindicated in pregnancy. Studies in animals or humans have demonstrated fetal abnormalities |

## Corticosteroids

Corticosteroids are included in category C of the FDA. They can cross the placental barrier and achieve variable serum concentrations in the fetus, depending on the type of corticosteroid used. There are studies that indicate increased risk of congenital anomalies, especially when this class of drugs is used in the first trimester of pregnancy. Among the problems observed, the higher incidence of cleft palate closure abnormalities should be emphasized. However, the data showed conflict with other research that did not corroborate these findings.[7,10]

There is no statistically significant increase in the occurrence of preterm births, babies with low birth weight or miscarriage.

Due to these controversies, the use of corticosteroids, especially in the first trimester of pregnancy should be discussed with the family, analyzing the risks and benefits.[3]

## Antibiotics

Metronidazole (category B) and ciprofloxacin (category C) as well as amoxicillin/clavulanate are the antibiotics used more frequently in patients with IBD.

There are reports that metronidazole given in the first trimester of pregnancy increases the chance of abnormality in the cleft palate.[11] Other studies, however, do not support this finding. The use of metronidazole is recommended during pregnancy; however, it should not be used in the first trimester.[12]

Ciprofloxacin is associated with the onset of arthropathy in children. Nevertheless, some studies indicate that the use during pregnancy does not seem to have important consequences for the baby. Even so, since there are alternative therapies with proven safety, it is suggested to avoid the use of fluoroquinolones in pregnant women with IBD.[3]

The combination amoxicillin/clavulanate is safe throughout pregnancy and should be remembered as an option at the time of prescribing antibiotics.[7]

## Immunomodulators

Among the immunomodulators, those that stand out as the most frequently used in the treatment of IBD are methotrexate, azathioprine and its metabolite, 6-mercaptopurine. The main doubts and questions about safety during pregnancy and breastfeeding refer to this class of drugs.

Methotrexate is classified by the FDA as a class X, and therefore contraindicated in pregnancy. This drug acts as an antagonist of folic acid, and its use during the period of organogenesis (first 8 weeks after conception) is associated with multiple congenital abnormalities, collectively referred to as methotrexate embryopathy. Moreover, exposure in the 2nd or 3rd trimester can lead to toxicity and fetal death. It is therefore absolutely contraindicated in pregnant women. Women of childbearing age and treated with methotrexate must receive information on the drug and additionally use one or preferably two reliable contraceptive methods. As methotrexate persists in the tissues for long periods, it is suggested that patients wait for about 6 months after discontinuation of the drug to become pregnant.[13] While most evidence condemn the use of methotrexate in pregnant women, the same cannot be said for men. Based on case reports and a recent study, exposure of men to methotrexate in the period in which conception occurs does not seem to be related to the appearance of congenital abnormalities.[14]

Azathioprine and its metabolite 6-mercaptopurine are analogs of purine (thiopurines) and interfere with the synthesis of DNA and RNA. The effect of these drugs is more evident in T lymphocytes and other cells that exhibit rapid cell division. The effectiveness is due to the cytotoxic, immunosuppressive and anti-inflammatory effect. Animal studies have shown that this class of drugs causes many teratogenic effects, including abnormalities in cleft palate and skeletal, urogenital and central nervous system malformations. However, studies in humans usually do not indicate an increased risk of such abnormalities when thiopurines are used by pregnant women.[9] Despite that, the FDA classifies these drugs as class D, on account of anecdotal studies that linked the use of these medications with increased occurrence of miscarriages.[10]

Goldstein et al. demonstrated safety for the occurrence of congenital abnormalities in babies to women who used thiopurines. However, they point out that there was a greater association with prematurity and low birth weight.[15]

In people with IBD, cohort studies did not reproduce the results obtained with animals. In one, 155 women with IBD who received 6-mercaptopurine at the time of conception or during the first trimester of pregnancy were compared with a similar group of pregnant women with IBD who were not exposed to the drug. The results showed no significant difference between the two groups.[16] In another recent study, Coelho et al. monitored 215 pregnancies from 204 women. One group was treated with thiopurines, the other was composed of women using different medications, and a third group consisted of women who did not use any drugs. At the end of the study, there were no statistically significant differences among the three groups regarding congenital abnormalities, miscarriage, prematurity and low birth weight.[17] Casanova et al., in a recent cohort study showed that the rate of pregnancies with poor outcomes and the rate of neonatal complications were lower in the group treated with thiopurines, compared with the control group. Multivariate analysis showed that the use of thiopurines was a favorable outcome predictor for pregnancy.[18]

Thus, although the FDA categorizes azathioprine and 6-mercaptopurine as class D medications, many authors consider them safe for use during pregnancy. There is no need to stop treatment with thiopurines during pregnancy in case of women who are already using the medication, if they are in remission.[8]

Cyclosporine is classically used in UC, especially in cases of severe colitis refractory to corticosteroids, when avoiding emergency surgery is desired. Use during pregnancy should be done within this clinical context, considering that performing total proctocolectomy would be more aggressive and harmful to the mother and the fetus. When used in pregnancy, cyclosporine does not seem to have teratogenic effects on the fetus. However, there is a greater tendency to complications for pregnant women.[3,12]

### Biological therapy and pregnancy

The first study in pregnant women to analyze the outcome of intentional use of infliximab for induction and maintenance of remission in CD was carried out in 2005. Ten women were included and at the end of pregnancy there were no congenital abnormalities, delayed intrauterine growth or low birth weight. Later, few reports pointed to a likely increase in the risk of congenital abnormalities. However, the correlation is weak and in most studies the drug has proved safe. According to the FDA classification, infliximab is considered a class B.[19]

In the latest consensus issued by the ECCO for CD in 2010, there is no contraindication to the use of infliximab during pregnancy. Infliximab crosses the placenta, especially in the third month of pregnancy. In newborns, the drug is found in considerable concentration; however, it is metabolized in the first 6 months of life and practically not detected at the end of the first year. Few data are available to assess the potential for damage of the drug to the immune system of a child, given that, during the first months of life, the baby's system is still developing. Currently, the interruption of infliximab is recommended between weeks 30 to 32, taking into account the risks and benefits of therapy in each situation.[3,10]

Adalimumab has a behavior similar to that of infliximab as it crosses the placental barrier, especially in the third trimester of pregnancy. Animal studies have shown no teratogenic effects or risks in the outcome of pregnancy. The most relevant studies conducted with pregnant women until then have not shown risks to the pregnancy or to the fetus. Similar to infliximab, it is recommended that treatment interruption is carried out in the period between week 30 and week 32 of gestation, with individualization of cases according to the history and clinical presentation of each patient.[10]

The first data on the use of certolizumab pegol during pregnancy, as well as research in animals, show that the drug crosses the placental barrier in much lower levels than infliximab, with low serum concentrations being found in babies and umbilical cord blood. It is known that, unlike infliximab and adalimumab, the molecular structure of certolizumab does not cross the placental barrier by active transport performed by specific receptors. Thus, the concentration of certolizumab in the umbilical cord is less than that found with the other two biological agents. It is possible that the presence of certolizumab in minimal serum concentrations in newborns results from alternative transport mechanisms that can occur throughout pregnancy. Because of the low placental transfer of this drug, current evidence does not support the cessation of certolizumab in the 3rd trimester of pregnancy. Moreover, considering this particular feature, researchers suggest that certolizumab would be a good option when a biological therapy is required in pregnant women.[20]

In babies born to mothers with IBD treated with anti-TNF during pregnancy, vaccination with live attenuated microorganisms should be avoided in the first 12 months of life.[3] There is a report of a child who developed disseminated tuberculosis after BCG vaccination in the third month of life. The mother presented CD and was treated with infliximab during pregnancy.[21]

Overall, the evidence presented so far places biological therapy in a favorable position in the treatment of IBD during pregnancy, especially regarding the lack of association with congenital abnormalities and complications of pregnancy. There is still no definitive evidence as to the interference of biological therapy in the child's development. However, it does not appear to cause long-term consequences. Thus, biological therapy, whenever required, is recommended for use during pregnancy, according to the guidelines for drug cessation around the third trimester of gestation, as shown above. One should always bear in mind that exacerbation of IBD during pregnancy can lead to negative outcomes to pregnancy and the fetus. The medical team must be able to discuss with the family the benefits and risks of biological therapy during pregnancy.

### Other drugs

Thalidomide has been used in the treatment of CD, by virtue of its anti-TNF effect. There are obvious concerns about this treatment, due to the possibility of causing serious teratogenicity. Thalidomide is absolutely contraindicated during pregnancy (category X). Women who use this medication should use two effective combined methods of contraception.[9]

Tacrolimus has been used in special situations for the treatment of UC and CD refractory to conventional therapy. The drug is category C according to the FDA classification. Information related to use during pregnancy were mainly obtained from studies with transplanted women. Current evidence have not indicated an increased risk of congenital anomalies, although a higher incidence of prematurity, neonatal nephrotoxicity and hyperkalemia can be seen.[10,22]

### BREASTFEEDING AND IBD

Similar to other diseases, the gastroenterologist is often asked by patients if breastfeeding should be avoided by mothers with IBD. Fear that breastfeeding can harm the baby must be demystified. Only in certain situations, mothers with IBD should be advised against breastfeeding.

Human milk contains many substances that can influence growth and development, as well as the function of the gastrointestinal tract of the baby. Furthermore, there are notable differences in the composition of the intestinal flora of children who were breastfed compared with those that have not.[23]

Although there is controversy in the literature, there is a belief that breastfeeding protects children from developing IBD in the future. This effect is attributed to the immunomodulatory capacity of breast milk. Although it also confers protection against UC, breastfeeding seems to be more effective against the appearance of CD.[24]

There is concern that the drugs used by the mother will harm the baby being breastfed (Table 32.2). However, several studies have shown that the use of much of the medication in the treatment of IBD is safe.

Sulfasalazine is minimally excreted in breast milk and therefore is considered a safe drug during that stage. Although there are reports of reversible diarrhea in the baby as an adverse effect, mesalazine can be used during breastfeeding. One should, however, be aware of this possible adverse effect.[9,25]

Using corticosteroids while breastfeeding is allowed. Even with the lack of concrete evidence, it is advisable, particularly when using prednisone at doses above 40 mg/day, to wait at least 4 hours after using the medication to breastfeed the child.[25]

Thiopurines are generally not detected in breast milk and, when the measurement is possible, they occur at nanomolar concentrations. However, metabolites were not detected in the few newborns studied. Therefore, maintaining treatment with thiopurines during breastfeeding is acceptable.[10]

Contradicting earlier works, Israeli researchers published studies that detected infliximab and adalimumab in the milk of mothers with IBD. However, the drug concentration was much lower than in the serum.[26,27] Still, according to current recommendations, breastfeeding should not be interrupted while using infliximab and adalimumab. These medications are detected in breast milk; however, it is believed that they suffer action of proteolytic enzymes in the digestive system of the baby. Studies so far showed no deleterious effects on the development of children breastfed by women treated with infliximab and adalimumab.[12]

Regarding the use of antibiotics, it is known that metronidazole can be toxic to newborns breastfed by mothers treated with this drug. Data on the use of ciprofloxacin are limited, but the drug seems to be safe. The combination amoxicillin/clavulanate is safer during lactation and may be a better choice of antibiotic therapy.[7]

According to current guidelines, mothers being treated with tacrolimus, thalidomide, cyclosporine or methotrexate should not breastfeed.[3,12,28]

**Table 32.2** Medications used to treat IBD

| Medication | FDA | Recommendation during pregnancy | Recommendation during breastfeeding |
|---|---|---|---|
| Adalimumab | B | Limited data in humans. Low risk. Crosses the placenta | Low concentrations in milk. Probably compatible |
| Azathioprine and mercaptopurine | D | Published data suggest low risks. Recent studies indicate safe use in pregnancy | Low secretion in breast milk. Probably compatible. Breastfeed four hours after using the medication |
| Certolizumab | B | Limited data in humans. Low risk. Less placental transfer compared to other anti-TNF | There is no evidence of secretion in breast milk. Probably compatible |
| Ciprofloxacin | C | Not recommended | Limited data in humans: probably compatible |
| Prednisone | C | Low risk. Possibility of defects of the cleft palate, especially in the 1st trimester | Compatible. Preferably, breastfeed about 4 hours after the medication |
| Cyclosporine | C | Current data has not shown an increase in congenital defects. It can cause neonatal hypertension, nephrotoxicity and hepatotoxicity | Contraindicated |

*(continues)*

**Table 32.2** Medications used to treat IBD

| Medication | FDA | Recommendation during pregnancy | Recommendation during breastfeeding |
|---|---|---|---|
| Infliximab | B | Limited data in humans. Low risk. Crosses the placenta and is found in the baby after birth | Low concentrations in milk. Probably compatible |
| Mesalazine | B | Low risk | Limited data in humans: risk of reversible diarrhea in children |
| Methotrexate | X | Contraindicated: teratogenic | Contraindicated |
| Metronidazole | B | Should be avoided in the 1st trimester of pregnancy | Limited data in humans: potentially toxic |
| Sulfasalazine | B | Low risk. Folate supplementation is necessary (2 mg/day) | Compatible |
| Tacrolimus | C | Current data has not shown an increase in congenital defects | Contraindicated |
| Thalidomide | X | Contraindicated: teratogenic | Lack of data in humans: potentially toxic |

## REFERENCES

1. Mountifield R, Bampton P, Prosser R, Muller K, Andrews JM. Fear and fertility in inflammatory bowel disease: a mismatch of perception and reality affects family planning decisions. Inflamm Bowel Dis 2009; 15:720-5.

2. Habal FM, Kapila V. Inflammatory bowel disease and pregnancy: evidence, uncertainty and patient decision-making. Can J Gastroenterol 2009; 23:49-53.

3. Mahadevan U. Pregnancy and inflammatory bowel disease. Med Clin North Am 2010; 94:53-73.

4. Dotan I, Alper A, Rachmilewitz D, Israeli E, Odes S, Chermesh I et al. Maternal inflammatory bowel disease has short and long-term effects on the health of their offspring: a multicenter study in Israel. J Crohns Colitis 2013; 7:542-50.

5. Castiglione F, Pignata S, Morace F, Sarubbi A, Baratta MA, D'Agostino L et al. Effect of pregnancy on the clinical course of a cohort of women with inflammatory bowel disease. Ital J Gastroenterol 1996; 28:199-204.

6.  Kane SV, Acquah LA. Placental transport of immunoglobulins: a clinical review for gastroenterologists who prescribe therapeutic monoclonal antibodies to women during conception and pregnancy. Am J Gastroenterol 2009; 104:228-33.

7.  Habal FM, Ravindran NC. Management of inflammatory bowel disease in the pregnant patient. World J Gastroenterol 2008; 14:1326-32.

8.  Correia LM, Bonilha DQ, Ramos JD, Ambrogini O, Miszputen SJ. Treatment of inflammatory bowel disease and pregnancy: a review of the literature. Arq Gastroenterol 2010; 47:197-201.

9.  Moffatt DC, Bernstein CN. Drug therapy for inflammatory bowel disease in pregnancy and the puerperium. Best Pract Res Clin Gastroenterol 2007; 21:835-47.

10.  Van Assche G, Dignass A, Reinisch W, van der Woude CJ, Sturm A, De Vos M et al. The second European evidence-based Consensus on the diagnosis and management of Crohn's disease: special situations. J Crohns Colitis 2010; 4:63-101.

11.  Czeizel AE, Rockenbauer M. A population based case-control teratologic study of oral metronidazole treatment during pregnancy. Br J Obstet Gynaecol 1998; 105:322-7.

12.  Ng SW, Mahadevan U. Management of inflammatory bowel disease in pregnancy. Expert Rev Clin Immunol 2013; 9:161-73; quiz 174.

13.  Silva BC, Santana GO. Methotrexate suspension before pregnancy. J Crohns Colitis 2013; 7:e151.

14.  Beghin D, Cournot M, Vauzelle C, Elefant E. Paternal exposure to methotrexate and pregnancy outcomes. J Rheumatol 2011; 38:628-32.

15.  Goldstein LH, Dolinsky G, Greenberg R, Schaefer C, Cohen-Kerem R, Diav-Citrin O et al. Pregnancy outcome of women exposed to azathioprine during pregnancy. Birth Defects Res A Clin Mol Teratol 2007; 79:696-701.

16.  Francella A, Dyan A, Bodian C, Rubin P, Chapman M, Present DH. The safety of 6-mercaptopurine for childbearing patients with inflammatory bowel disease: a retrospective cohort study. Gastroenterology 2003; 124:9-17.

17.  Coelho J, Beaugerie L, Colombel JF, Hébuterne X, Lerebours E, Lémann M et al. Pregnancy outcome in patients with inflammatory bowel disease treated with thiopurines: cohort from the CESAME Study. Gut 2011; 60:198-203.

18.  Casanova MJ, Chaparro M, Domènech E, Barreiro-de Acosta M, Bermejo F, Iglesias E et al. Safety of thiopurines and anti-TNF-$\alpha$ drugs during pregnancy in patients with inflammatory bowel disease. Am J Gastroenterol 2013; 108:433-40.

**19.** O'Donnell S, O'Morain C. Review article: use of antitumour necrosis factor therapy in inflammatory bowel disease during pregnancy and conception. Aliment Pharmacol Ther 2008; 27:885-94.

**20.** Gisbert JP. Safety of immunomodulators and biologics for the treatment of inflammatory bowel disease during pregnancy and breast-feeding. Inflamm Bowel Dis 2010; 16:881-95.

**21.** Cheent K, Nolan J, Shariq S, Kiho L, Pal A, Arnold J. Case report: fatal case of disseminated BCG infection in an infant born to a mother taking infliximab for Crohn's disease. J Crohns Colitis 2010; 4:603-5.

**22.** Kainz A, Harabacz I, Cowlrick IS, Gadgil SD, Hagiwara D. Review of the course and outcome of 100 pregnancies in 84 women treated with tacrolimo. Transplantation 2000; 70:1718-21.

**23.** Mikhailov TA, Furner SE. Breastfeeding and genetic factors in the etiology of inflammatory bowel disease in children. World J Gastroenterol 2009; 15:270-9.

**24.** Klement E, Cohen RV, Boxman J, Joseph A, Reif S. Breastfeeding and risk of inflammatory bowel disease: a systematic review with meta-analysis. Am J Clin Nutr 2004; 80:1342-52.

**25.** Farrukh A, Mayberry JF. Breastfeeding and inflammatory bowel disease. Inflamm Bowel Dis 2008; 14 Suppl 2:S39-40.

**26.** Ben-Horin S, Yavzori M, Katz L, Picard O, Fudim E, Chowers Y et al. Adalimumab level in breast milk of a nursing mother. Clin Gastroenterol Hepatol 2010; 8:475-6.

**27.** Ben-Horin S, Yavzori M, Kopylov U, Picard O, Fudim E, Eliakim R et al. Detection of infliximab in breast milk of nursing mothers with inflammatory bowel disease. J Crohns Colitis 2011; 5:555-8.

**28.** Vermeire S, Carbonnel F, Coulie PG, Geenen V, Hazes JMW, Masson PL et al. Management of inflammatory bowel disease in pregnancy. J Crohns Colitis 2012; 6:811-23.

# INFLAMMATORY BOWEL DISEASE IN CHILDREN

VERA LUCIA SDEPANIAN

## INTRODUCTION

Inflammatory bowel disease (IBD) – i.e. Crohn's disease (CD) and ulcerative colitis (UC) – has been increasingly prevalent in the pediatric population. Prevalence studies in this age range are scarce, with estimates that 20 to 30% of patients with IBD begin to show symptoms under the age of 18 years.[1]

Increased incidence of IBD in children, particularly CD, has been demonstrated in Canada, the United States, Spain, France, Northern Europe, and in Eastern Europe countries, such as Czech Republic, Croatia and Hungary.[2]

It is important to mention the limitations of studies of incidence of IBD in pediatric patients. In general, increased incidence of these diseases is observed; however, the heterogeneity of data collection techniques hinders comparison among studies.[2]

An epidemiological study[2] performed a systematic review that evaluated the incidence of IBD, CD and UC, starting in childhood. This study included 28 trials (20.1%), a total of 139 works including statistical analyses to evaluate the trends of incidence of these diseases, over time. Of the 28 studies included, nine allowed assessment of temporal trends of IBDs. Twenty-five articles evaluated the temporal trend of CD, specifically, and 20, of UC. Of the 9 articles on IBD, 7 (78%) reported temporal trend of increasing incidence of IBD, while 2 studies did not indicate any change in incidence. None of the studies reported decreased incidence (Figure 33.1).

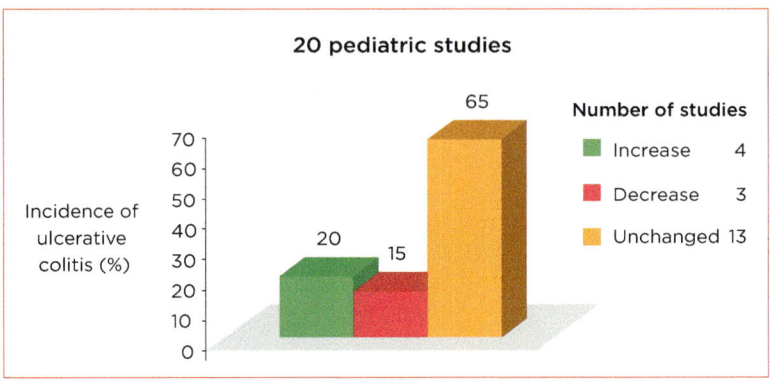

**Figure 33.1** Temporal trends of incidence of ulcerative colitis.

Source: Adapted from Benchimol et al., 2011.[2]

Regarding CD,[2] out of 25 studies analyzing temporal trends of incidence of the disease, 60% (15/25) reported significant increase, 4% (1/25) significant decrease and 36% (9/25) showed unchanged incidence (Figure 33.2).

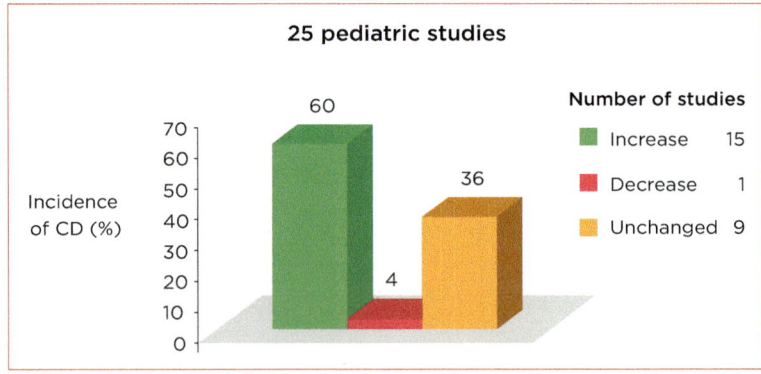

**Figure 33.2** Temporal trends of incidence of CD.

Source: Adapted from Benchimol et al., 2011.[2]

The temporal trend of UC, in turn, was different than that of CD:[2] most studies (65%; 13/20) revealed unchanged incidence, 20% (4/20) showed increased incidence, and 15% (3/20) showed reduced incidence (Figure 33.3).

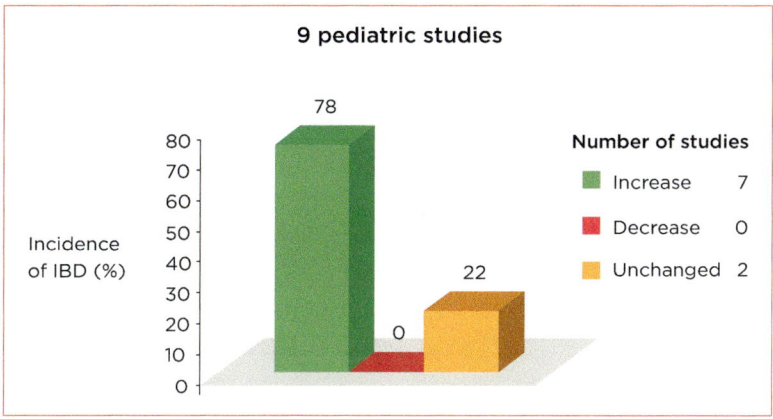

**Figure 33.3** Temporal trends of incidence of inflammatory bowel diseases.
Source: Adapted from Benchimol et al., 2011[2]

Thus, while the incidence of UC in multiple periods did not change in different countries, the incidence of CD, as well as of the IBD, increased.

### CLINICAL PICTURE

The delay in establishing the diagnosis of IBD in children is frequent and it is even harder to recognize these diseases in young children. The most common symptoms of both CD and UC are diarrhea and abdominal pain.

Children with IBD often have weight loss or poor weight gain, growth retardation, and delayed pubertal development.

Growth deficit is a factor exclusive to the pediatric age group; it occurs in 10 to 40% of patients at diagnosis and, in some patients, can be the first symptom of CD in the absence of diarrhea or abdominal pain.[3] The occurrence of growth deficit in UC is less common compared to CD. Thus, it is important to get previous weight and height data for the construction of a growth curve when evaluating the diagnostic possibility of IBD in children or adolescents. The main factors causing growth deficit are: action of proinflammatory cytokines that directly interfere with IGF-1; reduction of food intake due to anorexia mediated by cytokines and the fear of further worsening of gastrointestinal symptoms after a meal; decreased absorption in the affected small intestine; and treatment with corticosteroids.

IBD can cause delayed onset of pubertal development, and delayed or arrested sexual maturation. Secondary amenorrhea, which is a complication caused by malnutrition, can also occur in these patients. Therefore, CD should be considered as a possible diagnosis in patients with growth deficit and/or delayed pubertal development.

CD is more severe in children compared to the disease with onset in adulthood, meaning that CD is associated with more severe phenotype in this age group.[4]

When IBD occurs at a very early age in childhood, infants less than 1 year old, with severe colitis (Crohn's-like colitis), an association with deficiency of interleukin-10 is observed.[5]

The clinical manifestation of CD depends on the affected segment – small intestine, colon, predominance of extra-intestinal manifestations – and there may be a combination of these three patterns.

The different forms of presentation of CD are also noteworthy:

- non-fistulizing, non-stricturing: when the disease is characterized by inflammation without fistulas or strictures;
- stricturing: in cases of repeated narrowing of the intestine, anus and rectum documented by radiologic, endoscopic or surgical-pathological examinations, with pre-stenotic dilation or signs or symptoms of obstruction, without penetrating disease;
- penetrating: in cases of intra-abdominal or perianal fistulas.

In the case of UC, the symptoms are graded as mild, moderate or severe:

- mild: insidious onset, such as diarrhea, rectal bleeding and abdominal pain without systemic signs. In such cases, the inflammation is usually located in the distal colon;

- moderate: bloody diarrhea, colicky pain and abdominal tenderness, combined with systemic signs such as anorexia, weight loss, intermittent fever and mild anemia;
- severe: more than 6 bowel movements per day, with blood, abdominal tenderness, fever, weight loss, anemia, leukocytosis and hypoalbuminemia. These patients can show signs of toxic megacolon.

With regard to extra-intestinal manifestations, the most important in the pediatric age group are growth deficit and delayed pubertal development. Arthralgia is a common manifestation, and arthritis is present in about 25% of children with IBD. Decrease in bone mineral density can also occur in pediatric patients with IBD and can be seen in 25% of these patients (Z score <2).[6] Other extra-intestinal manifestations include recurrent aphthous oral lesions, skin lesions (such as erythema nodosum and pyoderma gangrenous), eye lesions (episcleritis and uveitis) and liver disease (primary sclerosing cholangitis – most often associated with UC than CD).

The latest proposal for IBD categorization is the Paris Classification, which is a modified Montreal Classification for pediatric patients both with CD and UC, and which can also be used by adults (Tables 33.1 and 33.2).[7]

The severity of IBD can be monitored based on disease activity indices, such as the Pediatric Ulcerative Colitis Activity Index (PUCAI)[8] for UC and the Pediatric Crohn's Disease Activity Index (PCDAI)[9] for CD.

## DIAGNOSIS

The diagnosis of IBD should be based on a combination of clinical history, physical examination, laboratory tests, endoscopy, imaging methods of the small intestine and ileocolonoscopy with histology. Multiple biopsies (2 or more per segment) should be obtained from all visible segments of the digestive tract, even in the absence of macroscopic lesions.[10]

Excluding enteric infection is critical.[10] Investigation is done by stool culture to exclude *Salmonella*, *Shigella*, *Yersinia*, *Campylobacter*, as well as toxins produced by *Clostridium difficile*. Test for *Giardia lamblia* is recommended in populations at high risk or living in endemic areas. The identification of pathogens does not necessarily exclude the diagnosis of IBD,

| Table 33.1 Paris Classification of CD | |
|---|---|
| Age at diagnosis | A1a: 0 to < 10 years |
| | A1b: 10 to < 17 years |
| | A2: 17 to 40 years |
| | A3: > 40 years |
| Location | L1: distal third of the ileum ± limited to the cecum |
| | L2: colonic |
| | L3: ileocolonic |
| | L4a: upper GI involvement proximal to the angle of Treitz |
| | L4b: upper GI involvement distal to the angle of Treitz and proximal to the distal third of the ileum |
| Behavior | B1: non-stricturing, non-penetrating |
| | B2: stricturing |
| | B3: penetrating |
| | B2B3: penetrating and stricturing, at the same time or at different times |
| Growth | G0: no evidence of growth deficit |
| | G1: evidence of growth deficit |

| Table 33.2 Paris Classification of UC | |
|---|---|
| Extension | E1: ulcerative proctitis |
| | E2: left-sided ulcerative colitis (distal to splenic flexure) |
| | E3: extensive (distal to hepatic flexure) |
| | E4: pancolitis (proximal to splenic flexure) |
| Severity | S0: not severe |
| | S1: severe (PUCAI ≥ 65)[7] |

since the first episode or relapse of IBD can be triggered by a documented enteric infection.[10]

Laboratory blood tests indicated for diagnosis are: complete blood count, at least two inflammatory markers such as CRP and ESR, albumin, transaminases and gamma-GT. Fecal calprotectin testing is better than any blood inflammatory markers to detect intestinal inflammation.

Have in mind that normal blood tests do not exclude the diagnosis of IBD, and reduced serum levels of albumin indicate protein-losing enteropathy, usually reflecting disease activity and severity, in addition to nutritional status.

Imaging studies of the small intestine are recommended for diagnosis, especially in patients with suspected CD, when the ileum cannot be intubated, and in patients apparently affected by UC with atypical presentation.[10]

## TREATMENT

The treatment of patients with IBD depends on disease severity and location. It consists of induction and maintenance of remission, aimed at mucosal healing.

There is no consensus on what the most appropriate treatment would be: conventional therapy, also known as "step-up"; or a "top-down" therapy, which uses, at first, more aggressive medication such as biological therapy. The consequences of the use of corticosteroids in pediatric patients – especially growth deficits and pubertal delay, in addition to other adverse effects including esthetic changes in both the children and adolescents – encourage strategies to prevent abuse of this type of medication.

Although the vast majority of pediatric gastroenterologists use conventional therapy to induce remission, some centers, especially Europeans, recommend exclusive enteral nutrition (polymeric diet with similar efficacy to semi-elemental and elemental diets), with cessation of the usual diet and administration of enteral formula orally or by nasogastric tube for about 6 to 8 weeks as monotherapy of choice, which would be capable of inducing the remission of CD, promoting growth and reducing the need for corticosteroids.[11,12] In children and adolescents, the results of exclusive enteral therapy in active CD are more positive and similar to those obtained with corticosteroids. It is worth mentioning that, although the polymeric feeds are iso-osmolar and more palatable, which facilitates ingestion by mouth and therefore are the most recommendable, some patients with CD and significant damage of the small intestine will certainly benefit the most from oligomeric diets due to greater absorption of di- and tripeptides. Enteral feeding can be performed orally or with a tube positioned in the stomach. In cases of duodenal-ileal or duodenocolic fistulas, the tube should be positioned in the jejunum. The mechanisms proposed for the anti-inflammatory and immunomodulatory action of enteral feeding include an impact on the intestinal microbiota, the ideal mixture of lipids, and reduced supply of antigens.[13]

Corticosteroids (prednisone, prednisolone, hydrocortisone) are effective to induce remission in mild-to-severe CD at any location, as well as in the treatment of extra-intestinal manifestations. The dose of oral prednisone indicated to induce remission is 1 to 2 mg/kg/day, not exceeding 40 mg/day.[14] Intravenous hydrocortisone can be used in severe cases. There is no evidence that continuing this treatment for more than four weeks could influence remission. Corticosteroid withdrawal should be gradual until complete cessation, quickly in 4 weeks or slowly in 12 weeks, and this does not seem to influence the extent of remission. It is important to know that corticosteroids are not effective in maintaining remission.

Thus, in active CD, both exclusive enteral nutrition and corticosteroids may be used, although exclusive enteral therapy entails a lesser proportion of side effects and allows positive impact on growth speed.[15]

Aminosalicylates, as well as antibiotics, appear to be less effective in inducing remission in patients with moderate to severe CD affecting the ileum and/or colon.

Early introduction of azathioprine (AZA) or 6-mercaptopurine (6-MP) is indicated for moderate to severe CD,[16] and these immunosuppressive drugs are also effective in maintaining remission.[17] The doses of AZA and 6-MP are 1.5 to 2.5 mg/kg/day and 0.75 to 1.5 mg/kg/day, respectively.[18] Because of the side effects of bone marrow suppression, especially leukopenia, a gradual increase of the dose is recommended, starting with 50 mg of AZA and 25 mg of 6-MP and increasing the dose by 25 mg every 1 or 2 weeks until a target dose; over that period, the occurrence of leukopenia should be monitored. Methotrexate, also an immunosuppressant, is indicated for patients with CD who need immunosuppressant therapy and were intolerant to AZA or 6-MP.

Cyclosporine, another imussupressor, often has side effects and may be useful in fistulizing CD resistant to treatment at a dose of 2 mg/kg/day, intravenously.

Biological therapy with anti-TNF monoclonal antibody has proven useful both to induce and maintain remission of CD.[19,20] This therapy is indicated for patients with perianal disease and fistulizing CD. The potential benefits of anti-TNF therapy in children with CD include: stopping/reducing corticosteroid treatment; postponing surgery; favoring growth; promoting mucosal healing and closure of fistulas.

Biological therapy available in Brazil includes infliximab, which is a chimeric IgG1 monoclonal antibody (75% human and 25% murine), and adalimumab, a human IgG1 monoclonal antibody (100% human). The use of adalimumab in pediatric patients has not been approved. Infliximab is administered by slow intravenous infusion (2 to 3 hours) at a dose of 5 mg/kg/day on day zero, week 2 and week 6, followed by a maintenance regimen every eight weeks. A Brazilian study in children and adolescents with CD and UC demonstrated that infliximab was effective in decreasing the IBD activity indices, with favorable documented response in terms of clinical manifestations, and reduced levels of corticosteroids by week 22.[21] If there is a loss of response, the interval between applications can be reduced to 6 or up to 4 weeks, if necessary, or the dose can be increased to 10 mg/kg/day.

In the case of adalimumab, the IMAgINE 1 trial[22] assessed safety and efficacy, and concluded that this immunobiological agent is capable of inducing and maintaining clinical remission in children with CD. For children weighing 40 kg or more, the suggested doses were 160 mg at week zero, 80 mg at week 2 and 40 mg from week 4 onwards. For pediatric patients weighing less than 40 kg, the suggested doses were 80 mg, 40 mg and 20 mg on weeks zero, 2 and 4 onwards, respectively.[22]

Note that, in pediatric patients, infliximab is approved for CD and UC by surveillance agencies in the US (Food and Drug Administration – FDA), Europe (European Medicines Agency – EMA) and Brazil (National Health Surveillance Agency – Anvisa). Adalimumab, in turn, has been approved by the FDA and the EMA for CD, but not by Anvisa.

Before any anti-TNF therapy, the occurrence of previous infections should be investigated, with evaluation and updating of immunizations, if necessary.[23]

There is no consensus on what infections should be screened before starting anti-TNF therapy, but there is agreement among rheumatologists and gastroenterologists that all patients should be screened for tuberculosis, hepatitis B and hepatitis C. Gastroenterologists also agree that HIV infection should be investigated. Prior to anti-TNF, investigation of TB is mandatory:[24] detailed medical history in search of occurrence of positive epidemiology for this disease; detailed physical examination; tuberculin skin test (TST) or interferon-gamma release assay must be performed for diagnosis of latent tuberculosis, as well as chest X-ray. If TST is below 5 mm or interferon-gama release assay is negative, and chest X-ray is normal, treatment with anti-TNF can be introduced. If TST is greater than 5 mm or interferon-gama release assay is positive, and chest X-ray is normal, chemoprophylaxis with isoniazid for 6 months is indicated. In this case, the anti-TNF can be started 1 month later, carefully avoiding exposure to the tuberculosis agent during treatment with anti-TNF. However, if TST is greater than 5 mm or interferon-gama release assay is positive and chest X-ray shows abnormality, treatment for tuberculosis should be done before starting anti-TNF therapy.

As for immunizations,[23] all patients starting anti-TNF therapy should receive simultaneously 23-valent pneumococcal polysaccharide vaccine and influenza vaccine with inactive virus. Those not immune to hepatitis B virus should be vaccinated against this virus which can be reactivated with anti-TNF therapy. HPV vaccination should be done in female patients with CD who will receive anti-TNF therapy. Thus vaccines allowed before and during anti-TNF therapy are: DPT vaccine, recombinant hepatitis B, hepatitis A, inactivated influenza virus, pneumococcal polysaccharide and HPV. Live vaccines are contraindicated for patients during anti-TNF treatment, including: intranasal influenza, MMR (measles, mumps and rubella), oral polio vaccine, chickenpox, BCG, yellow fever and anthrax. These live vaccines should be given 3 weeks prior to initiating therapy with immunosuppressant or anti-TNF. When a live virus vaccine is required during treatment with an immunosuppressant or anti-TNF, these drugs should be discontinued for at least 3 months, and only then, vaccination can be performed.

The benefits of anti-TNF therapy in children with CD include: stopping/reducing corticosteroid treatment; postponing surgery; favoring growth; promoting mucosal healing and closure of fistulas. The adverse effects of anti-TNF therapy are: reaction during the infusion (e.g., chest pain, throat tightness, difficulty breathing, increased or decreased blood pressure, fever and tremor); opportunistic infection; lymphoma; and demyelinating disease. The risk of opportunistic infection with anti-TNF therapy increases when using two or more drugs to treat CD. There are reports of a rare and severe lymphoproliferative disease (hepatoslenic T cell lymphoma – HSTCL) in patients with CD receiving concomitant treatment with an immunomodulator and infliximab, mostly in males under the age of 35 years.[25] Since 2006, after the first reported cases of HSTCL, monotherapy with anti-TNF became the treatment of choice for children. However, after the results of the SONIC[26] study in an adult population, which demonstrated a greater proportion (44%) of mucosal healing with combined therapy (infliximab and azathioprine) compared to monotherapy (30%), there is debate on what the most appropriate therapy for pediatric patients would be, monotherapy or combination therapy.

The indications for surgical treatment in CD include: strictures, fistulizing disease, abscess drainage, in addition to the indications for emergency surgery in UC, such as toxic megacolon, intestinal perforation and massive bleeding, noting that surgery is considered a palliative measure in the case of CD, because there can be recurrence elsewhere in the digestive tract, from mouth to anus.[18]

Since most of the children with UC have pancolitis, treatment will depend on the severity of disease.[27] An effective therapy in patients with mild UC consists of oral aminosalicylates. Sustained-release microgranules of mesalazine containing ethylcellulose to protect the drug against gastric degradation have fewer side effects and allow the active substance to be released in a continuous and prolonged manner from the proximal small intestine to the most distal parts of the large intestine. Dosage is 50 to 80 mg/kg/day.

Patients with moderate to severe UC should receive a corticosteroid, prednisone or prednisolone, at a dose and time period similar to those described in the CD, in addition to mesalazine. Those with severe UC, frequent bloody diarrhea (more than 6 times/day) and severe systemic signs should be hospitalized for treatment with intravenous corticosteroids, broad spectrum antibiotic therapy, and monitoring for the occurrence of intestinal perforation and toxic megacolon. About 50% of patients with severe UC fail to respond to corticosteroids. In this case, colectomy or potent immunomodulatory agents, such as cyclosporin or tacrolimus, should be the choice.

Patients with disease limited to the rectum (proctitis) or left colon can benefit from topical drugs such as mesalazine as enema or suppository, or corticosteroid enema.

For UC maintenance therapy, immunomodulatory agents, such as AZA or 6-MP, are useful. Biological therapy with anti-TNF monoclonal antibody is a therapeutic possibility in severe UC.

Surgery (colectomy) should be considered in UC when there is massive bleeding, active disease that does not respond to medical treatment, perforation and toxic megacolon. Since disease activity in UC is limited to the colon, colectomy is curative.

## REFERENCES

1. Heyman MB, Kirschner BS, Gold BD, Ferry G, Baldassano R, Cohen SA et al. Children with early-onset inflammatory bowel disease (IBD): analysis of a pediatric IBD consortium registry. J Pediatr 2005; 146(1):35-40.

2. Benchimol EI, Fortinsky KJ, Gozdyra P, Van den Heuvel M, Van Limbergen J, Griffiths AM. Epidemiology of pediatric inflammatory bowel disease: a systematic review of international trends. Inflamm Bowel Dis 2011; 17(1):423-39.

3. IBD Working Group of the European Society for Paediatric Gastroenterology, Hepatology and Nutrition. Inflammatory bowel disease in children and adolescents: recommendations for diagnosis – the Porto criteria. J Pediatr Gastroenterol Nutr 2005; 41(1):1-7.

4. Pigneur B, Seksik P, Viola S, Viala J, Beaugerie L, Girardet JP et al. Natural history of Crohn's disease: comparison between childhood- and adult-onset disease. Inflamm Bowel Dis 2010; 16(6):953-61.

5. Glocker EO, Frede N, Perro M, Sebire N, Elawad M, Shah N et al. Infant colitis – it's in the genes. Lancet 2010; 376(9748):1272.

6.  Lopes LH, Sdepanian VL, Szejnfeld VL, de Morais MB, Fagundes-Neto U. Risk factors for low bone mineral density in children and adolescents with inflammatory bowel disease. Dig Dis Sci 2008; 53(10):2746-53.

7.  Levine A, Griffiths A, Markowitz J, Wilson DC, Turner D, Russell RK et al. Pediatric modification of the Montreal classification for inflammatory bowel disease: The Paris classification. Inflamm Bowel Dis 2010 [Epub ahead of print] PubMed PMID:21061387.

8.  Turner D, Otley AR, Mack D, Hyams J, de Bruijne J, Uusoue K et al. Development, validation, and evaluation of a pediatric ulcerative colitis activity index: a prospective multicenter study. Gastroenterology 2007; 133(2):423-32.

9.  Hyams JS, Ferry GD, Mandel FS, Gryboski JD, Kibort PM, Kirschner BS et al. Development and validation of a pediatric Crohn's disease activity index. J Pediatr Gastroenterol Nutr 1991; 12(4):439-47.

10. Levine A, Koletzko S, Turner D, Escher JC, Cucchiara S, de Ridder L et al. The ESPGHAN revised Porto criteria for the diagnosis of inflammatory bowel disease in children and adolescents. J Pediatr Gastroenterol Nutr 2013. [Epub ahead of print] PubMed PMID: 24231644.

11. Zachos M, Tondeur M, Griffiths AM. Enteral nutritional therapy for induction of remission in Crohn's disease. Cochrane Database Syst Rev 2007; (1):CD000542.

12. Buchanan E, Gaunt WW, Cardigan T, Garrick V, McGrogan P, Russell RK. The use of exclusive enteral nutrition for induction of remission in children with Crohn's disease demonstrates that disease phenotype does not influence clinical remission. Aliment Pharmacol Ther 2009; 30(5):501-7.

13. Sauer CG, Kugathasan S. Pediatric inflammatory bowel disease: highlighting pediatric differences in IBD. Med Clin North Am 2010; 94(1):35-52.

14. Benchimol EI, Seow CH, Steinhart AH, Griffiths AM. Traditional corticosteroids for induction of remission in Crohn's disease. Cochrane Database Syst Rev 2008; 2:CD006792.

15. Van Assche G, Dignass A, Reinisch W, van der Woude CJ, Sturm A, De Vos M et al. The second European evidence-based Consensus on the diagnosis and management of Crohn's disease: special situations. J Crohn's Colitis 2010; 4:63-101.

16. D'Haens G, Baert F, van Assche G, Caenepeel P, Vergauwe P, Tuynman H et al. Early combined immunosuppression or conventional management in patients with newly diagnosed Crohn's disease: an open randomised trial. Lancet 2008; 371:660-7.

17. Prefontaine E, Sutherland LR, Macdonald JK, Cepoiu M. Azathioprine or 6-mercaptopurine for maintenance of remission in Crohn's disease. Cochrane Database Syst Rev 2009; 1:CD000067.

18. Dignass A, Van Assche G, Lindsay JO, Lémann M, Söderholm J, Colombel JF et al. The second European evidence-based Consensus on the diagnosis and management of Crohn's disease: current management. J Crohn's Colitis 2010; 4:28-62.

19. Hyams J, Crandall W, Kugathasan S, Griffiths A, Olson A, Johanns J et al; REACH Study Group. Induction and maintenance infliximab therapy for the treatment of moderate-to-severe Crohn's disease in children. Gastroenterology 2007; 132:863-73.

20. Behm BW, Bickston SJ. Tumor necrosis factor-alpha antibody for maintenance of remission in Crohn's disease. Cochrane Database Syst Rev 2008; (1):CD006893.

21. Tiemi J, Komati S, Sdepanian VL. Effectiveness of infliximab in Brazilian children and adolescents with Crohn disease and ulcerative colitis according to clinical manifestations, activity indices of inflammatory bowel disease, and corticosteroid use. J Pediatr Gastroenterol Nutr 2010; 50(6):628-33.

22. Hyams JS, Griffiths A, Markowitz J, Baldassano RN, Faubion WA Jr, Colletti RB et al. Safety and efficacy of adalimumab for moderate to severe Crohn's disease in children. Gastroenterology 2012; 143(2):365-74.e2.

23. Melmed GY. Vaccination strategies for patients with inflammatory bowel disease on immunomodulators and biologics. Inflamm Bowel Dis 2009; 15(9):1410-6.

24. Winthrop KL. Risk and prevention of tuberculosis and other serious opportunistic infections associated with the inhibition of tumor necrosis factor. Nat Clin Pract Rheumatol 2006; 2(11):602-10.

25. Kotlyar DS, Osterman MT, Diamond RH, Porter D, Blonski WC, Wasik M et al. A systematic review of factors that contribute to hepatosplenic T-cell lymphoma in patients with inflammatory bowel disease. Clin Gastroenterol Hepatol 2011; 9(1):36-41.e1.

26. Colombel JF, Sandborn WJ, Reinisch W, Mantzaris GJ, Kornbluth A, Rachmilewitz D et al.; SONIC Study Group. Infliximab, azathioprine, or combination therapy for Crohn's disease. N Engl J Med 2010; 362(15):1383-95.

27. Turner D, Travis SP, Griffiths AM, Ruemmele FM, Levine A, Benchimol EI et al. Consensus for managing acute severe ulcerative colitis in children: a systematic review and joint statement from ECCO, ESPGHAN, and the Porto IBD Working Group of ESPGHAN. Am J Gastroenterol 2011. [Epub ahead of print] PubMed PMID:21224839.

# COLORECTAL CANCER IN INFLAMMATORY BOWEL DISEASE

CARLOS WALTER SOBRADO

GUILHERME CUTAIT DE CASTRO COTTI

## INTRODUCTION

Patients with long-term inflammatory bowel disease (IBD) have an increased risk of developing cancer of the colon. In 1925, Crohn and Rosenberg called attention to the association between inflammatory regional enteritis and colorectal cancer.[1] Warren and Sommers, in 1949, warned about the possibility of patients with ulcerative colitis (UC) developing colorectal cancer (CRC) and described these histological changes as precancerous epithelial hyperplasia, found in 5% of the patients in their series.[2] In 1967, Morson and Pang, using rectal biopsy for screening, established that dysplasia was a precursor to the development of CRC, and the appearance of mucosal dysplasia would be a marker of cancer in any colonic segment.[3] In 1983, The Inflammatory Bowel Group Disease/Dysplasia Morphology Study Group (IBD/DMSG) developed a classification for dysplasias and their correlation with clinical aspects.[4]

Historically, the association between CRC and idiopathic UC was the one more widely known, but today, the clear relationship between CRC and Crohn's disease (CD) is also recognized. It is estimated that tumors of the colon and rectum in patients with inflammatory diseases correspond to 1 to

2% of all cases of CRC,[5,6] the latter being responsible for approximately 10 to 15% of deaths in patients with IBD.[7] Still, the presence of IBD is one of the most important factors in personal risk for the development of CCR.[8] The actual prevalence and incidence of CRC in patients with IBD is not known. Most of the studies available have selection biases or questionable methodology. For example, studies conducted in tertiary centers tend to overestimate the frequency of CRC in patients with IBD, since, in general, they treat more severe cases and with more extensive disease. Similarly, it is believed that population studies – which usually include patients with limited disease and patients who have undergone previous surgical treatment for IBD – may underestimate the actual values of the association.

Comparing the cases of CRC that occur in IBD with those of sporadic CRC, there are numerous distinctions. In IBD, CRC usually occurs in patients younger than the general population.[6,9] In addition, CRC often appears in IBD as flat lesions or non-polypoid dysplasia.[10] In IBD, CRC also has a higher proportion of mucinous tumors and signet-ring cells, which are multifocal and have worst prognosis. Since colitis in general is extensive – and longstanding chronic inflammation is directly related to the emergence of dysplasia – the occurrence of synchronous tumors is also more common in this population.[10]

In Figures 34.1 and 34.2, the molecular changes present in sporadic CRC can be viewed, as well as changes present in the CRC associated with chronic colitis.

### Ulcerative colitis

A meta-analysis published in 2001, comprising 116 studies including 54.478 patients with UC found that the prevalence of CRC in these patients is 3.7%, and 5.4% in those with pancolitis. In this study, the annual incidence of CRC ranged from 0.2% in the first decade, to 0.7 and 1.2% in the second and the third decades, respectively. The assessment of CRC risk in patients with UC, according to the time of disease progression, showed incidence of 1.6% after 10 years, 8.3% after 20 years and 18.4% after 30 years.[9]

A population study from Sweden reported 4 to 6 times increased risk of cancer compared to the general population.[11]

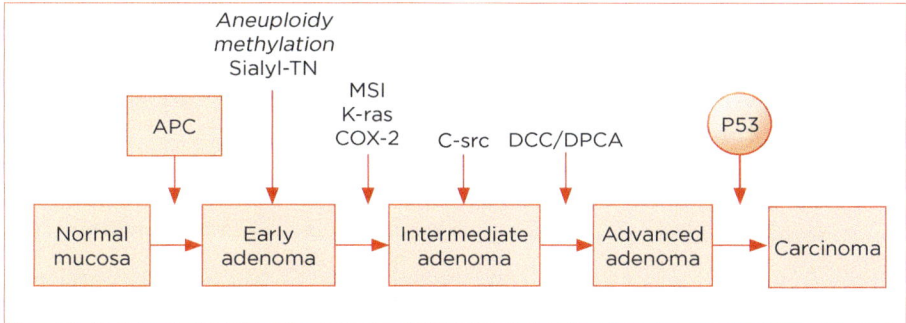

**Figure 34.1** Molecular changes seen in sporadic CRC.

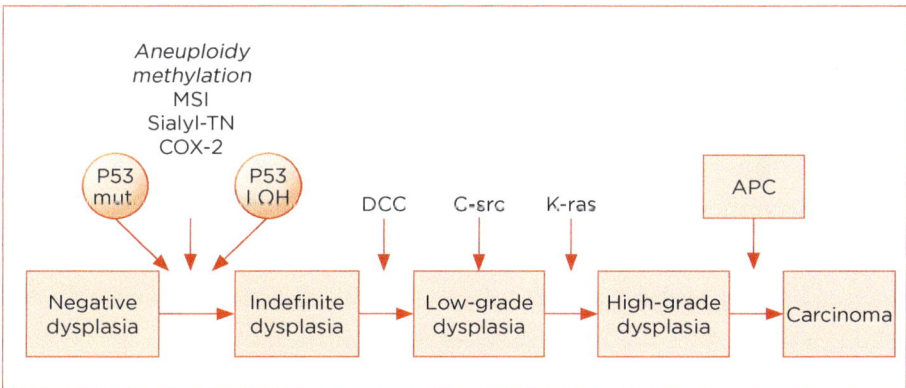

**Figura 34.2** Molecular changes seen in chronic colitis-associated CRC.

## Crohn's Disease

Data on the relationship between CRC and CD are very scarce and difficult to interpret. Methodological problems include many challenges such as the heterogeneous nature of the disease, patients without colic manifestation, and previous partial colorectal resection. Currently, it is believed that the prevalence of CRC in CD patients is lower than in UC. However, in patients with Crohn's colitis, particularly those of greater intensity and extent, data suggest that the risk of CRC is similar to that observed in UC.[12] Population studies estimate increased risk of CRC from 2 to 3 times compared to the general population.[13] Birmingham University reports about patients with

Crohn's pancolitis indicated that the risk of CRC was 18.2 (95%CI: 7.8-35.8) compared to the general population.[14] These results are very similar to those found in patients with UC of equal extent. A recent meta-analysis of 12 studies showed that the relative risk of CRC in CD is 2.5 (95%CI: 1.3-4.7) and, for patients with Crohn's colitis, 4.5 (95%CI: 1.3-14.9).[15]

Some studies have shown that cancer occurs in CD in areas with intense inflammatory process. Of 132 patients with CD presenting strictures treated at the Mount Sinai Hospital, 6.8% developed cancer, while in patients without strictures, cancer occurred in 0.7% only.[16]

The appearance of neoplasms in areas with internal or external fistulas has also been described. Although evidence is not as strong as in UC, the recommendation is that dysplasia and cancer be screened in CD patients in the following situations: with extensive colitis, over 8 years of disease progression, onset of disease at an early age and family history of CRC.

## RISK FACTORS FOR CRC IN IBD

Table 34.1 shows the main factors associated between CRC and IBD. Undoubtedly, the main risk factor associated with increased incidence of CRC in this population is the duration of colitis.[10] The occurrence of CRC is rarely seen when the duration of colitis is less than 7 or 8 years. A meta-analysis published by Eaden et al. considering a number of studies even before the advent of screening estimated a cumulative incidence of CRC in UC at 2%, with 10 years of disease progression, 8.3% with 20 years and 18.4% after 30 years.[9] Some authors believe that, because of colonoscopic surveillance, early medical treatment (chemoprevention) and surgical resection of the colon, these numbers are currently lower. For instance, Rutter et al. demonstrated, by means of a surveillance/screening program in patients with UC, incidence of 2.5% after 20 years of disease, 7.6% after 30 years and 10.8% after 40 years of follow-up.[17]

**Table 34.1** Factors associated with colorectal cancer in IBD

| Increase the risk of CRC in IBD | Reduce the risk of CRC in IBD |
|---|---|
| Long lasting colitis | Prophylactic proctocolectomy |
| Extensive colorectal involvement | Regular medical monitoring |
| Family history of CRC | Colonoscopy screening |
| Primary sclerosing cholangitis | Chemoprevention |
| Intensity of histological inflammation | |
| History of dysplasia | |
| Backwash ileitis (possibly) | |
| Young age at diagnosis of IBD (some studies) | |

CRC: colorectal cancer.

Source: Adapted from Xie et al., 2008.10

Similarly, the greater the length of the colon and rectum affected by inflammation, the greater the risk of developing CRC.[9] Thus, cases of UC with proctitis have lower incidence of CRC compared to UC that affects the left side of the colon, and this group has CRC much less often than patients with pancolitis. Moreover, currently, there are studies showing that the degree of microscopic inflammation, even in the absence of endoscopic or radiological changes, appears to increase the risk of CRC.[18] In this sense, studies begin to emerge suggesting that backwash ileitis may represent a factor associated with an increased incidence of CRC – even compared to patients with pancolitis without backwash ileitis.[19]

Patients with IBD and family history of CRC have twice the risk of developing it.[20] Another factor now known to increase the risk of CRC in patients with IBD is primary sclerosing cholangitis (PSC). Broome et al. observed that after 20 years of progression of UC, patients with PSC had an incidence of CRC of 31 *vs.* 5% in patients without PSC.[21]

### PATHOGENESIS

Although some studies have shown that CRC in CD is similar to that in UC, comparing the extent and duration of the disease, it is unclear whether the molecular changes involved in colon carcinogenesis are alike.[15] In addition to environmental factors, hereditary factors are clearly present, since, as already mentioned, patients with IBD who have family members with history of CRC are 2 times more likely to develop cancer than those with no family history.[22] In IBD patients, the risk of CRC is directly related to disease activity, duration and extent of colic inflammation. Some factors associated with inflammation, such as oxidative stress and the presence of free radicals, contribute to accelerate the molecular changes in the tissue, which does not occur in sporadic CRC.[23] Therefore, if chronic inflammation predisposes to colic carcinogenesis, reversing inflammation with consequent mucosal healing should obviously decrease the risk of CRC. In patients with IBD, especially UC, the most commonly used anti-inflammatory drugs are 5-ASA derivatives (mesalazine), corticosteroids, and purine immunomodulators (azathioprine/6MP). Many studies have shown that 5-ASA derivatives, as well as steroids, decrease the risk of CRC;[24,25] purine analogs (AZA/6-MP), in turn, do not seem to have such preventive effect.[26] Other factors, in addition to genetic predisposition and chronic inflammation, appear to play a role in the genesis of CRC, such as bacterial flora, TLR4 receptors, diet, folate deficiency and nutritional factors.[27,28]

Similarly to sporadic CRC, which follows the adenoma-carcinoma sequence, the inflammation-dysplasia-cancer[8] sequence is described in patients with IBD, with an estimate of 90% of the cases of CRC occurring in these patients.[25] In sporadic CRC, dysplasia usually occurs in elevated lesions called polyps, which are present in one or two areas of the colonic mucosa. In CRC associated with IBD, dysplasia is present as polypoid lesions or, more frequently, as flat lesions; they are usually multifocal and often difficult to detect by colonoscopy. However, it has been shown that CRC can develop in this group of patients without previous dysplasia or without the progression from low-grade to high-grade dysplasia.[29]

In fact, although the complete sequence of colorectal carcinogenesis mechanisms in patients with IBD is not yet fully elucidated, chronic inflammation is seen as the main promoter event.[6]

## DYSPLASIA

Dysplasia is defined by neoplastic changes in the epithelium with no invasion of the *lamina propria*.[4,10] From a macroscopic point of view, dysplasia during active colitis may occur as flat lesions, not visible to the naked eye (diagnosed only by biopsy), or as slightly elevated patches, or even as multiple or isolated elevated lesions. The elevated lesions observed in areas of active colitis are often referred to as dysplasia-associated lesions or masses (DALM). DALMs are subdivided into two groups: adenoma-like and non-adenoma-like.[30]

From a histological point of view, biopsies performed in patients with IBD to investigate dysplasia can have three results:

- negative for dysplasia;
- indefinite for dysplasia;
- positive for dysplasia (either low-grade or high-grade).[4]

Since the correlation among pathologists is quite variable, it is customary to recommend that the diagnosis of dysplasia is confirmed by another pathologist specializing in gastrointestinal diseases.[10,31]

## DIAGNOSIS AND WARNING SIGNS

Physicians involved in the treatment of patients with IBD should always be aware of the possibility of CRC. Risk factors, as mentioned in the previous item, must always be identified. Likewise, worsening of the general condition and abrupt changes in clinical status, including wasting syndrome, may represent the installation and progression of a CRC. Finally, strictures that remain in segments of the colon are another large group suspected for malignancy.

## PRINCIPLES OF CRC TREATMENT WITH CURATIVE INTENT

Surgery is the treatment of choice with curative intent in CRC patients, regardless of the presence of an associated IBD. When colorectal cancer associated with IBD is identified, tumor treatment becomes the main focus of attention for the patient.

Thanks to the combination of surgery with other therapeutic methods such as chemotherapy and/or radiotherapy for primary and metastatic tumors, it is possible to cure or significantly increase the survival of patients with advanced disease. In other words, patients with metastases are no longer regarded as incurable.

Resection of colorectal tumors with curative intent must meet criteria that minimize the risk of locoregional recurrence, taking into account the patterns of spread of disease:

- continuity;
- contiguity;
- lymphatic;
- hematogenous;
- emergence along suture lines or on the bloody surface of the intestine.

It is essential to remember that wherever possible – in cases of patients with primary lesions and distant metastases – surgery of the primary lesion should be conducted under the principles of oncological radicalism, given that many patients can be "rescued" for treatment with curative intent through treatment combined with chemotherapy, associated or not with radiotherapy.

The most important technical aspects related to CRC surgery are: obtaining margins (proximal, distal and circumferential), locoregional lymphadenectomy, and monoblock resection of adjacent structures, if necessary.

### Proximal and distal margins

For resection of localized tumors of the colon, it is believed that a proximal margin of 5 cm and a distal margin of 5 cm are both suitable and sufficient to prevent the occurrence of anastomotic recurrence. However, given that the realization of lymphadenectomy associated with resection of the primary lesion determines the ligature of the vessels in their respective origins, often the full extent of colonic segment resected in colectomy due to CRC reaches 30 cm or more. This fact stems from the need to use well vascularized segments to perform intestinal anastomoses with the lowest risk of dehiscence. Moreover, in tumors located in the right colon, the length of resected ileum does not seem to affect the rate of local recurrence and, thus, resection of the ileum should be as small as possible to avoid the occurrence of malabsorption syndromes.

For tumors located in the rectum, the fact that the distal margin affects the possibility of preserving the sphincter is important. That distance must always be analyzed in a recent specimen. Also, the distal margin should be measured from the surgical piece section containing the tumor and should not consider the stapling ring in the cases of mechanical colorectal anastomosis. In upper rectal tumors, the distal margin sought is 5 cm. For tumors of the extra-peritoneal rectum, it is estimated that a 2-cm margin of mucosa sectioned from the lower edge of the tumor is enough to prevent intramural spread. Nevertheless, in tumors of the distal rectum (< 5 cm from the anal verge), a distal margin of 1 cm is believed to be sufficient, especially when pre-surgical chemoradiotherapy is employed, because numerous studies attest to the fact that the spread of intramural tumor cells rarely occurs at a distance greater than 1 cm.

## Length of resection

The segment to be resected is defined based on the location of the tumor and its respective lymphatic drainage (Table 34.2). This item features one of the main peculiarities in the surgical treatment of patients with CRC and IBD. For patients with UC also diagnosed with CRC and who did not undergo any previous colorectal resection, total proctocolectomy is the surgery usually performed (maintaining the oncologic principles listed to manage the segment affected by the tumor), since the rest of the colon is greatly compromised by the inflammatory process. Sphincter preservation with ileal J-pouch reconstruction is routinely performed, with the exception of cases where the tumor occurs in the distal rectum with impairment of sphincter muscles or the anal canal. As for CD patients, in general, the extent of surgery performed is exactly the same recommended for patients who do not have IBD.

**Table 34.2** Tumor location and recommended surgery

| Tumor location | Surgery |
| --- | --- |
| Right colon | Right colectomy |
| Transverse colon | Extended right or left colectomy or transversectomy |
| Descending colon | Left colectomy |
| Sigmoid colon | Rectosigmoidectomy |
| Superior rectum (= rectosigmoid) | Rectosigmoidectomy |
| Rectum, middle third (5 to 11 cm) | Rectosigmoidectomy with low colorectal anastomosis or coloanal anastomosis<br>Manual or mechanical anastomosis<br>Cutait pull-through or Simonsen's technique surgery |
| Rectum, lower third (< 5 cm) | Abdominoperineal resection of the rectum<br>Rectosigmoidectomy with coloanal anastomosis |

### Total mesorectal excision and radial margin

The mesorectum is defined as the perirectal fat tissue, and includes lymphatic, vascular and nerve structures. The mesorectum may be affected by direct tumor invasion, lymph node and/or perineural involvement, or isolated tumor deposits. Resection of the mesorectum is an important prognostic factor, especially for patients with extraperitoneal rectal cancer. It is recognized that most local recurrences in patients with rectal cancer stems from inadequate radial margin, and not from insufficient distal margin. The practice of total mesorectal excision is associated with lower rates of local recurrence, especially by promoting appropriate clearing of the radial margin. The examination of surgical specimens shows that the involvement of the mesorectum rarely occurs at a distance greater than 4 cm beyond the lower limit of the tumor. So far, the distal margin recommended for mesorectum is 5 cm, which is adequate for upper rectal tumors, whereas tumors of the extraperitoneal rectum should always be treated with total mesorectal excision.

### Lymphadenectomy

Radical lymphadenectomy is a mandatory step in the surgical treatment with curative intent for colorectal adenocarcinoma. Because lymphatic drainage follows the colon's nutrient vessels, the extent of lymphadenectomy ends up being determined by the location in which these vessels are sectioned. Radical lymphadenectomy should be promoted as a block with the colorectal segment to be resected and is adequately performed when the surgeon performs the ligation of the nutrient vessels of the colon close to their origins (ligation of the inferior mesenteric artery close to the aorta and the branches of the superior mesenteric artery, next to their base). When the tumor is equidistant in relation to two major vessels (e.g., adenocarcinoma located in the hepatic angle, located exactly between the ileocolic trunk and the middle colic artery), both vessels need to be sectioned near their respective origins.

Lymph nodes suspected for malignant involvement, located away from the surgical resection field, should, wherever possible, be excised or biopsied for pathologic confirmation. Since lymphadenectomy has therapeutic and prognostic implications, it is believed that at least 12 lymph nodes should be evaluated by the pathologist so that the lymph node status can be known reliably.

## Monoblock resections

It is estimated that in 15% of cases colorectal tumors have become adhered to adjacent structures. In this situation, resection of the primary colorectal tumor should always be performed as a monoblock. Conversely, if the lesion is removed at the adherence site, resection is considered incomplete without oncological radicality. While there is the possibility of tumor adherences to adjacent structures, these are inflammatory in nature; the surgeon should never try to "guess" whether these are tumoral or inflammatory adherences, to avoid compromising the oncological radicality of the procedure. Studies have shown that these adherences result from direct tumor invasion between 40 and 84% of cases, depending on the series.[32,33] However, from a clinical point of view, it is impossible to predict reliably whether adherence to an adjacent organ represents direct tumor invasion or just inflammation – which always demands monoblock resection with neighboring structures adhered to the tumor.

The most important goal of monoblock resection is to achieve complete resection, with negative microscopic margins. In spite of being an adverse prognostic factor, tumors with invasion of adjacent organs may have survival similar to tumors without invasion of neighboring structures, provided that they are completely resected as a monoblock (R0 resection).

## CRC screening in patients with IBD

Histological evidence of the presence of dysplasia (intraepithelial neoplasia) in the colonic mucosa are the main markers (gold standard) to determine the risk of CRC in a patient with IBD; colonoscopy remains the main method diagnosis for dysplasia. Some authors estimate that 33 biopsies are required to achieve 90% of reliability in the detection of dysplasia in patients with UC.[34] These studies in patients with UC may not be fully translated into CD due to the more focal and limited nature of inflammation, and the presence of areas of segmental stricture and previous bowel resections. However, there are many difficulties and limitations in the interpretation of dysplasia, variations among pathologists (4 to 10% margin of error), difficulty in distinguishing areas with intense inflammatory activity, and the fact that CRC can occur in areas without previous dysplasia. Generally, there is good agreement among pathologists to exclude dysplasia from the specimens and in the diagnosis of high-grade dysplasia; however, in cases of low-grade dysplasia or indefinite dysplasia, the disagreement is high.[35] Colonoscopy with use of magnification and chromoendoscopy with serial biopsies, especially in flat areas, has been used in order to increase sensitivity and specificity.

When using these new endoscopic techniques (e.g., chromoendoscopy), biopsy is guided and there is a clear increase in the diagnosis of dysplasia. In a study of 100 patients with extensive and chronic UC, dysplasia was found in 9/157 guided biopsies, and in 0/2,904 random biopsies.[36] Another study obtained similar results, also using chromoendoscopy, and the analysis and investigation of neoplasia (dysplasia and CRC) was performed by focal endomicroscopy.[37]

Other techniques including flow cytometry, immunohistochemistry and investigation of tumor markers (e.g., expression of P53) have been used to elucidate the presence of dysplasia.

Screening, in this scenario, presupposes that the identification of associated dysplasia – and appropriate treatment at that time – would be able to reduce the incidence of CRC in the population of patients with IBD. Although many medical societies have surveillance schemes in an attempt to prevent CRC in patients with IBD, a 2006 Cochrane review on the strategies available for CRC and/or dysplasia detection in this group of patients failed to show that periodic colonoscopies are able to extend the survival of patients with extensive colitis. The very authors conclude that the tumors of patients undergoing endoscopic screening were diagnosed at earlier stages and possibly with better prognosis. However, the review of the studies cannot clearly exclude a potential lead time bias among the patients included, which is why they believe there is indirect evidence of the benefit of periodic colonoscopies in reducing mortality from CRC in patients with IBD and the cost-effectiveness of this approach. There is a tendency that patients with higher risk of developing neoplasia (dysplasia or carcinoma), with pancolitis, family history of CRC and PSC, for example, undergo careful screening. After 20 years of disease duration, there is a considerable increased risk of neoplasia; by then, screening should be intensified. On the other hand, including patients with distal proctitis in screening programs is very controversial, since the risk of CRC is similar to that of the general population.

## Specific screening situations

The diagnosis of hepatobiliary diseases is done by liver function serology tests and investigation of autoimmune disease, in addition to nuclear magnetic resonance cholangiography, which is the method of choice to identify the presence of PSC. Changes in liver function tests are common in patients with IBD and are usually associated with hepatobiliary diseases, more common in UC patients than in those with CD.

Ultrasonography with ductal changes suggests the diagnosis, but cholangiography shows multifocal strictures and dilatation of the intra- and extra-hepatic bile ducts, which practically confirms the diagnosis. Confirmation is performed by biopsy, in which the histology shows the following changes: concentric fibrosis in bile ducts, "onion skin" collagenous tissue or nonspecific changes with primary biliary cirrhosis. In rare cases, endoscopic retrograde cholangiopancreatography (ERCP) may be required for diagnosis, but its real value is therapeutic in the management of strictures of the main bile duct.[38]

The presence of PSC associated with IBD substantially increases the risk of CRC and cholangiocarcinoma.

Current guidelines recommend that patients with PSC who do not know whether they have IBD undergo colonoscopy to determine the status of the colon. For patients with IBD and PSC, CRC screening should begin at diagnosis, and annual colonoscopy is recommended. There are also groups that defend prophylactic colectomy in this group of patients requiring liver transplantation.[38]

After 7 to 8 years of progression of IBD, it is recommended that the patient undergoes a baseline colonoscopy to identify the extent and degree of activity of colitis, and possible associated neoplasia and/or dysplasia. The colon must be examined completely, with serial biopsies (4 fragments each 10 cm), particularly in areas of flat mucosa. Elevated lesions, if present, need to be resected whenever feasible.

When high-grade dysplasia is found in flat mucosa and confirmed by other gastrointestinal pathologist, the risk of CRC is very high and, therefore, total proctocolectomy is indicated. On the other hand, in the case of low-grade dysplasia in flat mucosa, treatment is controversial. Among 46 patients with low-grade dysplasia monitored at the St. Mark's Hospital, 19.4% had CRC and 39.1% progressed to high-grade dysplasia or adenocarcinoma.[17] At the Mount Sinai Hospital, advanced neoplasia was observed in 25.3% of patients with low-grade dysplasia undergoing colectomy. After 5 years of follow-up, 53% progressed to CRC.[29] A meta-analysis of 20 studies, including 508 cases of low-grade dysplasia in flat mucosa or DALM revealed incidence of CRC at 14/1,000 patient-years duration, and incidence of any advanced lesion at 30/1,000 patient-years duration. This study showed that when low-grade dysplasia is found through screening, there are 9 times more chances to develop cancer and 12 times more chances to develop elevated lesions.[39] Colectomy is recommended for low-grade dysplasia present in more than one location, found in another test after six months, or which progresses to high-grade dysplasia. In the case of indefinite dysplasia, confirmed by two pathologists, colonoscopies at short intervals (3 to 6 months) must be performed. Often indefinite dysplasia occurs in cases of intense inflammation, which is confused with areas of regenerative hyperplasia. In these circumstances, systemic and topical anti-inflammatory therapy should be intensified before performing the next endoscopic study.

Cases of polypoid lesions confined to areas without intense inflammatory activity – sporadic adenomas – should be resected endoscopically, as in the general population. If the lesion cannot be resected completely, or if dysplasia is identified in another area, colectomy is the treatment of choice.[40] In cases of low-grade dysplasia found in elevated lesion (polypoid), complete endoscopic resection is recommended, with subsequent intensive monitoring. Figure 34.3 shows one of the possible screening schemes for dysplasia and CRC in patients with IBD.

## PREVENTION AND CHEMOPREVENTION

Regarding the factors that reduce the risk of CRC in IBD, the most important are surgical removal of the colon and rectum and monitoring for life. Prophylactic proctocolectomy after 7 years of development of IBD is able to prevent almost all cases of CRC in this population. However, since many of these patients do not require extensive surgical procedure with relevant implications for their quality of life, it is recognized today that prevention and early diagnosis programs are also highly effective in this population and include a set of measures characterized by: regular medical visits, periodic colonoscopies, and chemoprophylaxis (use of medications to reduce/control the inflammatory process). The main purpose of these screening measures is early diagnosis and appropriate treatment of dysplasia.

Chemoprevention is understood as the use of drugs or dietary supplements which reduces the risk of CRC in patients with IBD. It is critically important, especially when there are still cases of CRC in patients undergoing appropriate screening. The main medications studied in this scenario are 5-aminosalicylic acid (5-ASA), nonsteroidal anti-inflammatory drugs (NSAIDs), immunomodulators, ursodeoxycholic acid, folate, calcium and statins.

5-ASA derivatives constitute the group of medications studied the most for CRC chemoprevention in patients with IBD, and have proved effective in several cohort and case-control studies. In one of the most well-controlled studies, Eaden et al. demonstrated that patients with UC in regular use of 5-ASA (2 g/day) for 5-10 years had a 75% reduction in the risk of CRC.[41] Rubin et al. reported 72% reduction in the risk of dysplasia and CRC in UC patients treated with doses greater than 1.2 g of 5-ASA per day.[42] A meta-analysis that combined nine case-control and cohort studies showed that mesalazine reduces the risk of CRC by 54%. In the analysis of dysplasia and cancer as a combined outcome, the reduction was close to 51%.[43]

As already mentioned, patients with UC and PSC exhibit very high risk of developing CRC and cholangiocarcinoma. While this population represents a very uncommon group of patients, which hinders the performance of well-controlled studies, there seems to be benefit to using ursodeoxycholic acid in the prevention of dysplasia and CRC. The use of this acid at a dose of 20 mg/kg/day can improve the levels of liver enzymes and prevent histological progression of the disease.[44,45] In addition to the advantages mentioned above, ursodeoxycholic acid can reduce the risk of dysplasia and CRC.[46]

The association with corticosteroids has been recommended, however the results are conflicting. Another immunosuppressive agent, tacrolimus, has been shown to be effective in the normalization of liver enzymes, but did not improve the histological changes.[47]

Folate supplementation in patients with IBD was also evaluated for CRC chemoprevention. There is evidence that low folate levels are associated with increased risk for CRC. Considering that patients with IBD often have low levels of folate, due to malnutrition and loss caused by reduced intestinal absorptive capacity, folate supplementation (1-2 mg/day) for at least 6 months has good theoretical support and seems to correct certain genetic abnormalities and molecular changes.[48]

Risk reduction, possibly associated with NSAIDs, immunomodulators, calcium and statins, appears to be small and not reproducible in many studies, and thus there is no current scientific evidence that supports their use in clinical practice. Corticosteroids, in turn, are associated with CRC risk reduction in patients with IBD. However, their long-term side effects contraindicate their use as chemoprevention in this scenario.

The conclusion is that many efforts and research have been done with the purpose of identifying new molecular markers in the blood and feces, which can help to understand the pathogenesis of CRC in IBD, as well as early identification of dysplasia in these patients.

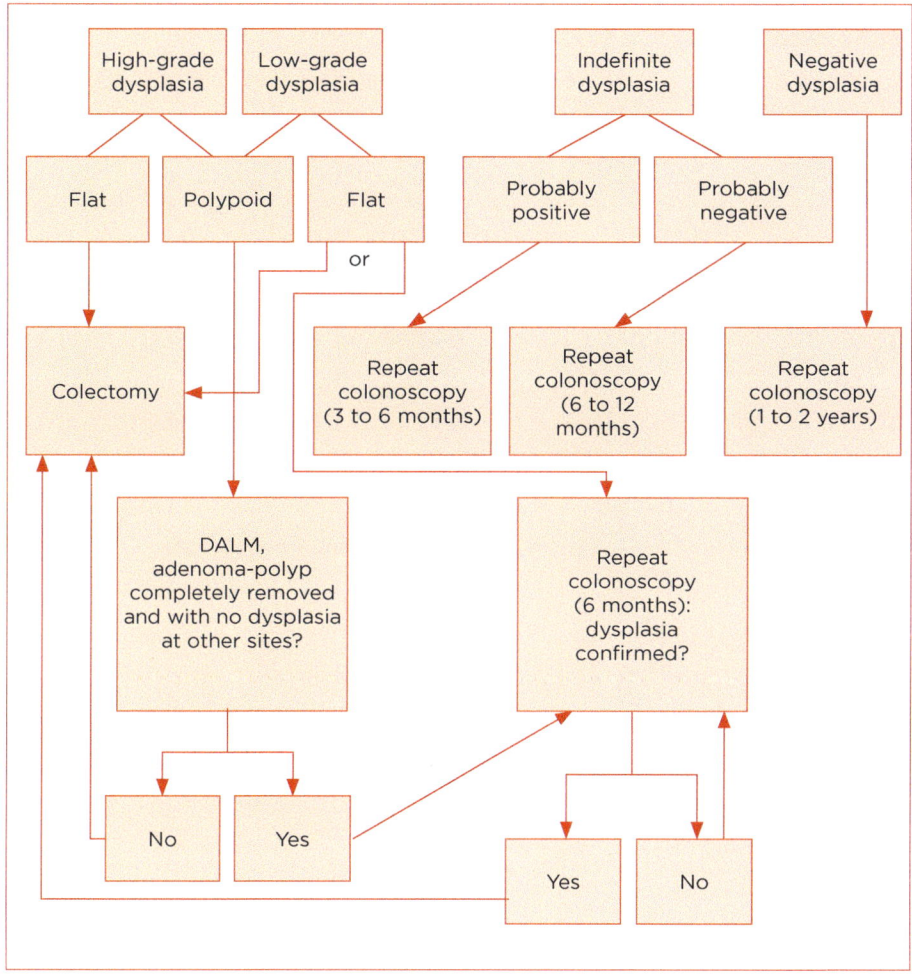

**Figure 34.3** Dysplasia and CRC screening in IBD patients.

## REFERENCES

1.  Crohn B, Rosenberg H. The sigmoidoscopic picture of chronic ulcerative colitis (non-specific). Am J Med Sci 1925; 170:220-8.

2.  Warren S, Sommers SC. Pathogenesis of ulcerative colitis. Am J Pathol 1949; 25:657-79.

3.  Morson BC, Pang LS. Rectal biopsy as an aid to cancer control in ulcerative colitis. Gut 1967; 8:423-34.

4.  Riddell RH, Goldman H, Ransohof DF, Appelman HD, Fenoglio CM, Haggitt RC et al. Dysplasia in inflammatory bowel disease: Standardized classification with provisional clinical applications. Human Pathol 1983; 14:931-66.

5.  Choi PM, Zelig MP. Similarity of colorectal cancer in Crohn's disease and ulcerative colitis: implications for carcinogenesis and prevention. Gut 1994; 35(7):950-4.

6.  Lakatos PL, Lakatos L. Risk for colorectal cancer in ulcerative colitis: changes, causes and management strategies. World J Gastroenterol 2008; 14(25):3937-47.

7.  Munkholm P. Review article: the incidence and prevalence of colorectal cancer in inflammatory bowel disease. Aliment Pharmacol Ther 2003; 18(Suppl. 2):1-5.

8.  Zisman TL, Rubin DT. Colorectal cancer and dysplasia in inflammatory bowel disease. World J Gastroenterol 2008; 14(17):2662-9.

9.  Eaden JA, Abrams KR, Mayberry JF. The risk of colorectal cancer in ulcerative colitis: a meta-analysis. Gut 2001; 48(4):526-35.

10. Xie J, Itzkowitz SH. Cancer in inflammatory bowel disease. World J Gastroenterol 2008; 14(3):378-89.

11. Ekbom A, Helmick C, Zack M, Adami HO. Ulcerative colitis and colorectal cancer. A population-based study. N Eng J Med 1990; 323(18):1228-33.

12. Sachar DB. Cancer in Crohn's disease: dispelling the myths. Gut 1994; 35(11):1507-8.

13. Bernstein CN, Blanchard JF, Kliewer E, Wajda A. Cancer risk in patients with inflammatory bowel disease: a population-based study. Cancer 2001; 91(4):854-62.

14. Gillen CD, Walmsley RS, Prior P, Andrews HA, Allan RN. Ulcerative colitis and Crohn's disease: a comparison of the colorectal cancer risk in extensive colitis. Gut 1994; 35(11):1590-92.

15. Canavan C, Abrams Kb, Mayberry J. Meta-analysis: colorectal and small bowel cancer risk in patients with Crohn's disease. Aliment Pharmacol Ther 2006; 23(8):1097-104.

**16.** Yamazaki Y, Ribeiro MB, Sachar DB, Aufses AH Jr, Greenstein AJ. Malignant colorectal strictures in Crohn's disease. Am J Gastroenterol 1991; 86(7):882-5.

**17.** Rutter MD, Saunders BP, Wilkinson KH, Rumbles S, Schofield G, Kamm MA et al. Thirty-year analysis of a colonoscopic surveillance program for neoplasia in ulcerative colitis. Gastroenterology 2006; 130(4):1030-8.

**18.** Mathy C, Schneider K, Chen YY, Varma M, Terdiman JP, Mahadevan U. Gross versus microscopic pancolitis and the occurrence of neoplasia in ulcerative colitis. Inflamm Bowel Dis 2003; 9(6):351-5.

**19.** Heuschen UA, Hinz U, Allemeyer EH, Stern J, Lucas M, Autschbach F et al. Backwash ileitis is strongly associated with colorectal carcinoma in ulcerative colitis. Gastroenterology 2001; 120(4):841-7.

**20.** Askling J, Dickman PW, Karlén P, Broström O, Lapidus A, Löfberg R et al. Family history as a risk factor for colorectal cancer in inflammatory bowel disease. Gastroenterology 2001; 120(6):1356-62.

**21.** Broomé U, Löfberg R, Veress B, Eriksson LS. Primary sclerosing cholangitis and ulcerative colitis: evidence for increased neoplastic potential. Hepatology 1995; 22(5):1404-8.

**22.** Nuako KW, Ahlquist DA, Mahoney DW, Schaid DJ, Siems DM, Lindor NM. Familial predisposition for colorectal cancer in chronic ulcerative colitis: a case-control study. Gastroenterology 1998; 115:1079-83.

**23.** Itzkowitz SH, Yio X. Inflammation and cancer. Colorectal cancer in inflammatory bowel disease: a role of inflammation. Am J Physiol Gastrointest Liver Physiol 2004; 287(1):7-17.

**24.** Rubin DT, Cruz-Correa MR, Gasche C, Jass JR, Lichtenstein GR, Montgomery EA et al. 5-ASA in colorectal cancer prevention meeting group. Colorectal cancer prevention in inflammatory bowel disease and the role of 5-aminosalicylic acid: a clinic review and update. Inflamm Bowel Dis 2008; 14:265-274.

**25.** Itzkowitz SH, Harpaz N. Diagnosis and management of dysplasia in patients with inflammatory bowel diseases. Gastroenterology 2004; 126(6):1634-48.

**26.** Matula S, Croog V, Itzkowitz S, Harpaz N, Bodian C, Hossain S et al. Chemoprevention of colorectal neoplasia in ulcerative colitis: the effect of 6-mercaptopurine. Clin Gastroenterol Hepatol 2005; 3:1015-21.

**27.** Fukata M, Chen A, Vamadevan AS, Cohen J, Breglio K, Krishnareddy S et al. Toll-like receptor-4 promotes the development of colitis-associated colorectal tumors. Gastroenterology 2007; 133:1869-81.

**28.** Kim YI, Shirwadkar S, Choi SW, Puchyr M, Wang Y, Mason JB. Effects of dietary folate on DNA strand breaks within mutation-prone exons of the p53 gene in rat colon. Gastroenterology 2000; 119:151-61.

**29.** Ullman T, Croog V, Harpaz N, Sachar D, Itzkowitz S. Progression of flat low-grade dysplasia to advanced neoplasia in patients with ulcerative colitis. Gastroenterology 2003; 125(5):1311-9.

**30.** Odze RD. Pathology of dysplasia and cancer in inflammatory bowel disease. Gastroenterol Clin North Am 2006; 35(3):533-52.

**31.** Cendan JC, Behrns KE. Associated neoplastic disease in inflammatory bowel disease. Surg Clin North Am 2007; 87(3):659-72.

**32.** Eisenberg SB, Kraybill WG, Lopez MJ. Long-term results of surgical resection of locally advanced colorectal carcinoma. Surgery 1990; 108:779-86.

**33.** Izbicki JR, Hosch SB, Knoefel WT, Passlick B, Bloechle C, Broelsch CE. Extended resections are beneficial for patients with locally advanced colorectal cancer. Dis Colon Rectum 1995; 38(12):1251-6.

**34.** Rozen P, Baratz M, Fefer F, Gilat T. Low incidence of significant dysplasia in a successful endoscopic surveillance program of patients with ulcerative colitis. Gastroenterology 1995; 108:1361-70.

**35.** Melville DM, Jass JR, Morson BC, Pollock DJ, Richman PI, Shepherd NA et al. Observer study of the grading of dysplasia in ulcerative colitis: comparison with clinical outcome. Hum Pathol 1989; 20(10):1008-14.

**36.** Rutter MD, Saunders BP, Schofield G, Forbes A, Price AB, Talbot IC. Pancolonic indigo carmine dye spraying for the detection of dysplasia in ulcerative colitis. Gut 2004; 53(2):256-60.

**37.** Kiesslich R, Goetz M, Lammersdorf K, Schneider C, Burg J, Stolte M et al. Chromoscopy-guided endomicroscopy increases the diagnostic yield of intra-epithelial neoplasia in ulcerative colitis. Gastroenterology 2007; 132(3):874-82.

**38.** Cullen SN, Chapman RW. The medical management of primary sclerosing cholangitis. Semin Liver Dis 2006; 26:52-61.

**39.** Thomas T, Abrams KA, Robinson RJ, Mayberry JF. Meta-analysis: cancer risk of low-grade dysplasia in chronic ulcerative colitis. Aliment Pharmacol Ther 2007; 25:657-68.

**40.** Rubin PH, Friedman S, Harpaz N, Goldstein E, Weiser J, Schiller J et al. Colonoscopic polypectomy in chronic colitis: conservative management after endoscopic resection of dysplastic polyps. Gastroenterology 1999; 117:1295-300.

**41.** Eaden JA, Abrams K, Ekbom A, Jackson E, Mayberry J. Colorectal cancer prevention in ulcerative colitis: a case-control study. Aliment Pharmacol Ther 2000; 14(2):145-53.

**42.** Rubin DT, LoSavio A, Yadron N, Huo D, Hanauer SB. Aminosalicylate therapy in the prevention of dysplasia and colorectal cancer in ulcerative colitis. Clin Gastroenterol Hepatol 2006; 4(11):1346-50.

**43.** Velayos FS, Terdiman JP, Walsh JM. Effect of 5-aminosalicylate use on colorectal cancer and dysplasia risk: a systematic review and meta-analysis of observational studies. Am J Gastroenterol 2005; 100(6):1345-53.

**44.** Lindor KD. The Mayo Primary Sclerosing Cholangitis-Ursodeoxycholic Acid Study Group. Ursodiol for primary sclerosing cholangitis. N Eng J Med 1997; 336:691-5.

**45.** Mitchell SA, Bansi DS, Hunt N, Von Bergmann K, Fleming KA, Chapman RW. A preliminary trial of high-dose ursodeoxycholic acid in primary sclerosing cholangitis. Gastroenterology 2001; 121:900-7.

**46.** Sjoqvist U, Tribukait B, Ost A, Einarsson C, Oxelmark L, Löfberg R. Ursodeoxycholic acid treatment in IBD-patients with colorectal dysplasia and/or DNA-aneuploidy: a prospective double-blind, randomized controlled pilot study. Anticancer Res 2004; 24:3121-7.

**47.** Van Thiel DH, Carroll P, Abu-Elmagd K, Rodriguez-Rilo H, Irish W, McMichael J et al. Tacrolimus, a treatment for primary sclerosing cholangitis: results of an open label preliminary trial. Am J Gastroenterol 1995; 90:455-9.

**48.** Cravo ML, Albuquerque CM, Salazar de Souza L, Glória LM, Chaves P, Dias Pereira A et al. Microsatellite instability in non-neoplastic mucosa of the patients with ulcerative colitis: effect of folate supplementation. Am J Gastroenterol 1998; 93:2060-4.

# IMMUNIZATION IN INFLAMMATORY BOWEL DISEASE

CRISTINA FLORES

## INTRODUCTION

The treatment of inflammatory bowel disease (IBD) often requires immunosuppressive drugs. It is estimated that 80% of patients with IBD will need to use corticosteroids at some point along their disease, 50% will require immunosuppressive drugs (6-mercaptopurine, azathioprine, methotrexate, cyclosporine and tacrolimus), and 20% will use biologic agents (infliximab, adalimumab or certolizumab pegol).[1] Furthermore, the trend towards treatment with combination therapy and the need for continued therapy to maintain remission demands more attention on prevention of infections and appropriate use of vaccines from physicians treating such patients. Concerns about the risk of infections with these medications are increasing and include pneumococcal sepsis, disseminated herpes zoster infections, severe cases of varicella, acute exacerbation of hepatitis B, and more.[2,3]

## GENERAL PRINCIPLES OF VACCINATION

Immunization is prevention of diseases and lesions caused by microorganisms through the induction of immune mechanisms.

### Passive immunity

This refers to the transfer of antibodies or effector cells ready to act against a particular microorganism. It may be natural, acquired through transplacental transfer or breastfeeding, or artificial, like human gamma globulins obtained from plasma with high titers of antibodies against a particular agent (e.g., hepatitis B, rabies, tetanus, measles and rubella).

### Active immunity

It consists in administering live attenuated or killed microorganisms, proteins, DNA or modified toxins that induce a specific immune response longer than passive immunization, but which can be shorter than the immunity acquired from a disease state.

### Types of vaccines

- Killed viruses or bacteria;
- live attenuated viruses or bacteria;
- proteins or sugars extracted from bacteria or viruses, or synthesized in a laboratory;
- toxoids used for active immunization (bacterial toxins modified to become non-toxic).

### Vaccines allowed during immunosuppression

Inactivated polio vaccine, triple bacterial against diphtheria/tetanus/pertussis, triple acellular against diphtheria/tetanus/pertussis, diphtheria/tetanus adult-type (double adult), diphtheria/tetanus infant-type (double infant), tetanus toxoid, hepatitis A, hepatitis B, *Haemophilus influenzae* type B, influenza, pneumococcal, pneumococcal conjugate vaccine, rabies.

### Vaccines contraindicated during immunosuppression

Bacillus Calmette-Guerin (BCG), measles, measles/mumps/rubella (MMR) measles/rubella (viral double), rubella, chickenpox, oral polio, yellow fever.

Tables 35.1 and 35.2 show the Brazilian basic vaccination schedule, according to the Ministry of Health.

## DEFINITION OF IMMUNOCOMPROMISED PATIENTS

- Treatment with corticosteroids (> 20 mg of prednisone or equivalent) for more than 2 weeks; or within 3 months of cessation;
- active treatment, or within three months of cessation, with azathioprine or 6-MP, methotrexate or anti-TNF agents;
- significant protein-calorie malnutrition.

The level of immunosuppression depends on the intensity, duration and type of treatment the patient is receiving. Immunosuppressive therapy primarily alters cellular immunity, while humoral immunity remains relatively normal. However, the levels of antibodies formed after vaccination must be checked.[3,4]

## TIME OF VACCINATION

Live attenuated vaccines should be administered at least three weeks before the start of medication, or three months after cessation. Inactivated vaccines, polysaccharides and toxoids can be administered safely at any time in immunosuppressed patients; however, the response to the vaccine can be less pronounced than in immunocompetent patients. If administered 2 weeks before the start of immunosuppressive medication, the probability of antibody production after immunization increases.[3,4]

## INFLUENZA VACCINE

The risk of complications in cases of influenza in immunosuppressed patients with IBD is poorly documented, but is expected to be higher than in healthy individuals. International consensuses recommend vaccination for influenza annually in immunosuppressed individuals.[5] Immune response to vaccination may be diminished in patients on immunosuppression. Recent studies with H1N1 influenza vaccine suggest that patients on combined therapy with anti-TNF and immunosuppressants have reduced response with lower seroconversion rates.[6] Andrisani et al. did a survey of the immune response to influenza vaccine showing a response below the appropriate when patients were using combination therapy compared to monotherapy patients and controls.[7]

**Table 35.1** Brazilian basic childhood immunization schedule

| Age | Vaccines | Doses | Diseases prevented |
|---|---|---|---|
| At birth | BCG-ID<br>Hepatitis B vaccine (1) | 1st dose | Hepatitis B |
| 1 month | Hepatitis B vaccine | 2nd dose | Hepatitis B |
| 2 months | Tetravalent vaccine (DTP + Hib) (2) | 1st dose | Diphtheria, tetanus, pertussis, meningitis and other infections caused by *Haemophilus influenzae* type B |
| | OPV | 1st dose | Poliomyelitis (infantile paralysis) |
| | OHRV (3) | 1st dose | Rotavirus diarrhea |
| 4 months | Tetravalent vaccine (DTP + Hib) | 2nd dose | Diphtheria, tetanus, pertussis, meningitis and other infections caused by Haemophilus influenzae type B |
| | OPV | 2nd dose | Poliomyelitis (infantile paralysis) |
| | OHRV (4) | 2nd dose | Rotavirus diarrhea |
| 6 months | Tetravalent vaccine (DTP + Hib) | 3rd dose | Diphtheria, tetanus, pertussis, meningitis and other infections caused by Haemophilus influenzae type B |
| | OPV | 3rd dose | Poliomyelitis (infantile paralysis) |
| | Hepatitis B vaccine | 3rd dose | Hepatitis B |
| 9 months | Yellow fever vaccine | Initial dose | Yellow fever |
| 12 months | CRS (triple viral) | Single dose | Measles, mumps and rubella |
| 15 months | OPV | Booster dose | Poliomyelitis (infantile paralysis) |
| | DTP (triple bacterial) | 1st booster | Diphtheria, tetanus and pertussis |
| 4 to 6 years | DTP (triple bacterial) | 2st booster | Diphtheria, tetanus and pertussis |
| 10 years | Yellow fever vaccine | Booster dose | Yellow fever |

OPV: oral polio vaccine; OHRV: oral human rotavirus vaccine.

**Table 35.2** Brazilian basic immunization schedule for adolescents

| Age | Vaccines | Doses | Diseases prevented |
|---|---|---|---|
| From 11 to 19 years (on the first visit to the health service) | Hepatitis B | 1st dose | Against hepatitis B |
| | dT (double, adult--type) (2) | 1st dose | Against diphtheria and tetanus |
| | Yellow fever (3) | Booster dose | Against yellow fever |
| | CRS (triple viral) (4) | Single dose | Against measles, mumps and rubella |
| 1 month after the 1st dose against hepatitis B | Hepatitis B | 2nd dose | Against hepatitis B |
| 6 months after the 1st dose against hepatitis B | Hepatitis B | 3rd dose | Against hepatitis B |
| 2 months after the 1st dose against diphtheria and tetanus | dT (double, adult--type) | 2nd dose | Against diphtheria and tetanus |
| 4 months after the 1st dose against diphtheria and tetanus | dT (double, adult--type) | 3rd dose | Against diphtheria and tetanus |
| Every 10 years, for life | dT (double, adult--type) | Booster dose | Against diphtheria and tetanus |
| | Yellow fever | Booster dose | Against yellow fever |

There are some data on the safety of influenza vaccine in patients with IBD, particularly those taking immunosuppressants. A multicenter prospective cohort study of 575 patients evaluated local and systemic symptoms within 4 weeks after vaccination. The vaccine was well tolerated, with only 15.5% of patients reporting systemic symptoms. All symptoms resolved within 72 hours. Less than 5% of the patients had signs of disease reactivation, defined as 3 or more points on the Harvey-Bradshaw index. The authors concluded that the vaccine was well tolerated by patients, irrespective of therapy, and that the risk of disease reactivation was low.[8]

### TETANUS VACCINE

The recommendation is a booster shot every 10 years of combined tetanus and diphtheria vaccine (DT). Studies have shown that patients with IBD treated with immunosuppressants exhibit the expected levels of antibodies. Another study that stratified patients according to their immunosuppression treatment (immunosuppressant monotherapy; biological monotherapy, and combination therapy) showed that all patients who were not on immunosuppressants reached protective levels of antibodies against tetanus, compared to only 78% of patients on combined therapy (P = 0.01).[1,9]

### VARICELLA VACCINE

Immunocompromised patients should receive two doses of varicella vaccine, even when they are less than 13 years old. After exposure to the virus, and during a varicella outbreak in hospital setting, vaccination of susceptible immunocompetent individuals and children older than 1 year of age, up to 120 hours after infection, is recommended. Immunocompromised persons should receive passive immunization.[10] The immuneglobulin is prepared from the serum of donors with high antibody titers against the varicella virus. Best efficacy is observed when the administration occurs 96 hours after infection, i.e., before the first viremia. The dose is 125 IU per 10 kg, with a minimum dose of 125 IU and maximum of 625 IU, exclusively as intramuscular injections.[11,12]

## HERPES ZOSTER VACCINE

Herpes zoster, or shingles, is a reactivation of a latent infection with varicella-zoster virus. One in three patients will develop zoster at some point in life, and this risk is even greater in immunosuppressed patients.[13] Studies in patients with IBD show a higher incidence compared to controls. Treatment with corticosteroids or azathioprine/6-mercaptopurine is associated with significantly increased risk (odds ratio: 1.5; 95CI: 1.1-2.2) and RC 3.1 (95CI: 1.7-5.6), respectively.[14]

The herpes zoster vaccine (not yet available in Brazil) reduces incidence by 51%, and the occurrence of postherpetic neuralgia in 67% of patients older than 50 years. Administration in immunosuppressed patients should be carefully considered, because of its attenuated nature. Corticosteroids at doses below 20 mg/day, methotrexate < 0.4 mg/kg/week, azathioprine < 3 mg/kg/day or 6-mercaptopurine < 1.5 mg/kg/day are not a contraindication to immunization. The safety of this vaccine in patients treated with anti-TNF is unknown and therefore it should not be administered.[15,16]

## PNEUMOCOCCAL VACCINE

Pneumococcal infection is responsible for more deaths than any other bacterial disease that can be prevented by vaccination. Patients using immunosuppressants are at increased risk of more serious pneumococcal infections and complications. PCV13 pneumococcal vaccine should be administered to all patients who are or will be treated with immunosuppressants. PPSV23 should be administered to patients over 2 years of age and with plans to start immunosuppression. Patients should receive PPSV23 ≥ 8 weeks after PCV13, and a second dose of PPSV23 should be administered after 5 years.[17,18] The effectiveness of the vaccine ranges from 56 to 81% for prevention of invasive pneumococcal disease in immunocompetent patients; however, its protective capacity decreases with immunosuppression. Melmed et al. examined the immunogenicity in patients with Crohn's disease (CD) according to the treatment and found adequate response in only 45% of patients with combination therapy compared with 80 to 85% in patients who were not receiving immunosuppressants and healthy controls.[19,20] The current recommendation is that patients with IBD undergo pneumococcal vaccination with a booster injection every five years.[1,10]

### HEPATITIS B VACCINE

The risk of reactivation of hepatitis B and development of fulminant hepatitis has been described in patients with IBD receiving immunosuppression.[21,22]

Hepatitis B surface antibodies (anti-Hb), which confer immunity, should be equal to or greater than 10 mIU/mL. The response to immunization against the B virus is lower in immunosuppressed individuals than in the general population. Response rates below 50% have been reported in patients with IBD.[23-26] Modified immunization schedules with high doses have been assessed, with some studies using double doses of antigen (40 mcg) in non-responders, which increases the seroconversion rate by 30 to 50%.[27] Early vaccination is advised and, if possible, prior to beginning therapy with immunomodulators. The vaccine response should be monitored between 1 and 3 months of the last dose. Patients who do not develop adequate immune response should receive the full schedule for the second time, with double dose.[17,25]

### HEPATITIS A VACCINE

The risk of a patient with IBD acquiring hepatitis A is similar to the general population, but the complication rate is higher among adults and immunocompromised patients.[1] The prevalence of natural immunity against hepatitis A is high in Brazil, varying with age and socioeconomic conditions (between 60 and 90%).[28] It is recommended to check for the presence of antibodies against hepatitis A; if absent, vaccination is indicated.[17,29]

### HUMAN PAPILLOMAVIRUS (HPV) VACCINE

HPV is the main cause of cancer of the uterine cervix, vulva and anal canal, especially types 16 and 18.[1] Studies show a higher prevalence of abnormal cytology, dysplasia and cervical cancer in women with IBD than controls (42.5 *vs.* 7%), especially when treated with immunosuppressants.[30,31]

Marehbian et al. demonstrated, in a case-control study of 22,310 patients, increased risk of cervical dysplasia in patients on monotherapy with corticosteroids, immunosuppressants and anti-TNF medication [RC: 1.5 (95CI: 1.2-2.0)]; on the other hand, use of combination therapy increased the odds ratio to 1.8 (95CI: 1.1-3.0).[15] HPV vaccination is highly recommended for patients with IBD. Wherever possible and preferably, the anti-HPV vaccine should be given in adolescence, before the start of the sexual life, from the age of 9 years. Two vaccines are available in Brazil: one containing HPV types 6, 11, 16, 18 on a schedule of 0, 2 and 6 months, recommended for both genders from 9 to 26 years of age; the other with HPV types 16 and 18 on a schedule of 0, 1 and 6 months, recommended for girls since the age of 9 years. The Brazilian Ministry of Health includes the HPV vaccine for girls between 9 and 12 years old at the immunization schedule since March 2014.[30-33] There are no data on the immunogenicity and safety of this vaccine in patients with IBD; however, its efficacy and safety in the general population is high. Since this is a recombinant preparation, there is no reason to imagine security problems in immunosuppressed patients.[3,17,29,34]

## MENINGOCOCCAL AND *HAEMOPHILUS INFLUENZAE* TYPE B (HIB) CONJUGATE VACCINE

Meningococcal (MCC) and *Haemophilus influenzae* type B (Hib) vaccines are part of the Brazilian vaccination schedule.[11] The meningococcal conjugated vaccine is recommended in two doses in the first year of life, from 2 months of age. A booster dose should be done in the 2nd year of life, between 12 and 15 months.[10,32] The duration of protection from polysaccharide conjugate vaccines (bivalent or tetravalent) decreases with time and therefore, patients who received this type of vaccine should be revaccinated within 5 years.[12]

## YELLOW FEVER VACCINE

Composed of live attenuated viruses, some adverse events have been reported with this vaccine, such as encephalitis. Vaccination against yellow fever is contraindicated in patients receiving immunosuppressants or biological agents. Immunosuppressed patients should be discouraged from traveling to endemic areas. If travel is unavoidable, the patient should be advised to avoid mosquito bites by using all kinds of protection, including repellents.[1,3]

## MEASLES, MUMPS, AND RUBELLA (MMR) VACCINE

This is a vaccine made from live attenuated virus and which is part of the Brazilian immunization schedule for children. It is contraindicated in immunosuppressed patients. Susceptible contacts of these patients should be vaccinated. Vaccination of patients with IBD prior to the use of immunosuppressants must be considered case by case, depending on the knowledge of immunizations received during childhood.

## VACCINE RECOMMENDATION FOR INDIVIDUALS LIVING WITH PATIENTS WITH IBD

It is noteworthy that susceptible persons in constant contact with patients on immunosuppression (family contacts and health professionals) should receive vaccines for indirect protection of patients. Immunocompetent individuals living with immunocompromised patients can safely receive vaccines with inactivated virus.

Individuals living with immunocompromised persons for more than 6 months should receive influenza vaccine annually.

Immunocompetent individuals living in the same household of immunocompromised patients should receive the following live vaccines: MMR (combined measles, mumps and rubella); in children 2 to 7 months old, vaccine for rotavirus; varicella, yellow fever and oral typhoid fever. Contact between an individual who received vaccine with live attenuated virus and an immunocompromised patient should wait seven days. Oral polio vaccine should not be administered to individuals living with immunocompromised patients.

Highly immunocompromised patients should avoid changing diapers of infants who received the rotavirus vaccine for 4 weeks after vaccination.

Immunosuppressed individuals should avoid contact with people who have developed skin lesions after receiving varicella or zoster vaccine.

A situation that deserves special attention is the vaccination of babies born to mothers using immunosuppressants. Immunosuppressive drugs cross the placental barrier and can be detected in the blood of children over 6 months old. Therefore, these children should not receive vaccines with live attenuated viruses (e.g., rotavirus and oral polio vaccine).[17,35] In these cases, an inactivated poliovirus vaccine (IPV) can be used, which is indicated for patients with congenital or acquired immunodeficiency and their unvaccinated household contacts.[10,11]

## CONCLUSIONS

Detailed history on immunization and the occurrence of infectious diseases must be obtained during the first medical visit of a patient with IBD.

Serological evaluation before performing vaccination is recommended for patients with IBD, according to the history of immunization and occurrence of previous diseases such as chickenpox, mumps and hepatitis A and B.

Immunization against pneumococcal disease, influenza, hepatitis A, hepatitis B, chicken pox and HPV is advised. In immunosuppressed patients, vaccination against other encapsulated bacteria should also be considered, including group C *Neisseria meningitidis* and Hib.

Live attenuated vaccines cannot be administered when there is immunosuppression.

## REFERENCES

1. Dezfoli S, Melmed GY. Vaccination issues in patients with inflammatory bowel disease receiving immunosuppression. Gastroenterol Hepatol 2012, 8(8).504-12.

2. Lichtenstein GR, Rutgeerts P, Sandborn WJ, Sands BE, Diamond RH, Blank M et al. A pooled analysis of infections, malignancy, and mortality in infliximab- and immunomodulator-treated adult patients with inflammatory bowel disease. Am J Gastroenterol 2012; 107(7):1051-63.

3. Rahier JF, Ben-Horin S, Chowers Y, Conlon C, De Munter P, D'Haens G et al. European evidence-based Consensus on the prevention, diagnosis and management of opportunistic infections in inflammatory bowel disease. J Crohns Colitis 2009; 3:47–91.

4. CDC. National Center for Immunization and Respiratory Diseases. General Recommendations on Immunization. Recommendations of the Advisory Committee on Immunization Practices (ACIP). MMWR Recomm Rep 2011; 60(RR-2):1-64.

5. Rahier J-F, Yazdanpanah Y, Colombel JF, Travis S. The European (ECCO) Consensus on infection in IBD: what does it change for the clinician? Gut 2009; 58:1313-5.

6. Cullen G, Bader C, Korzenik JR, Sands BE. Serological response to the 2009 H1N1 influenza vaccination in patients with inflammatory bowel disease. Gut 2012; 61:385-91.

7. Andrisani G, Frasca D, Romero M, Armuzzi A, Felice C, Marzo M et al. Immune response to influenza A/H1N1 vaccine in inflammatory bowel disease

patients treated with anti TNF-α agents: effects of combined therapy with immunosuppressants. J Crohns Colitis 2013; 7(4):301-7.

8. Rahier JF, Papay P, Salleron J, Sebastian S, Marzo M, Peyrin-Biroulet L et al. H1N1 vaccines in a large observational cohort of patients with inflammatory bowel disease treated with immunomodulators and biological therapy. Gut 2011; 60:456-62.

9. Dezfoli S, Horton H, Brer D et al. Immunomodulators, but not anti-TNF monotherapy, impair pertussis and tetanus booster vaccine responses in adults with inflammatory bowel disease (IBD). Presented at Digestive Disease Week; May 19–22, 2012; San Diego, California. Abstract Su2081.

10. Bricks LF. Novas recomendações para vacinação nos Centros de Referência de Imunobiológicos Especiais (Cries). Pediatria 2006; 28(3):204-8.

11. Brasil. Ministério da Saúde. Secretaria de Vigilância em Saúde DdVeE, Programa Nacional de Imunizações. Manual dos Centros de Referência para Imunobiológicos Especiais (Cries) 2006. Available at: www.cvesaude.sp.gov.br. Available at: 5/2/2014.

12. CDC. Updated recommendations for use of meningococcal conjugate vaccines. Advisory Committee on Immunization Practices (ACIP), 2010. MMWR Morb Mortal Wkly Rep 2011; 60:72-84.

13. Harpaz R, Ortega-Sanchez IR, Seward JF. Advisory Committee on Immunization Practices (ACIP) Centers for Disease Control and Prevention (CDC). Prevention of herpes zoster: Recommendations of the Advisory Committee on Immunization Practices (ACIP). MMWR Recomm Rep 2008; 57(RR-5):1-30. MMWR Recomm Rep. 2008; 57(RR-5):1-30.

14. Gupta G, Lautenbach E, Lewis J. Incidence and risk factors for herpes zoster among patients with inflammatory bowel disease. Clin Gastroenterol Hepatol 2006; 4:1483-90.

15. Marehbian J, Arrighi HM, Hass S, Tian H, Sandborn WJ. Adverse events associated with common therapy regimens for moderate to severe Crohn's disease. Am J Gastroenterol 2009; 104:2524-33.

16. Oxman MN, Levin MJ, Johnson GR, Schmader KE, Straus SE, Gelb LD et al. A vaccine to prevent herpes zoster and postherpetic neuralgia in older adults. N Engl J Med 2005; 352:2271-84.

17. Wasan SK, Baker SE, Skoinik PR, Farraye FA. A practical guide to vaccinating the inflammatory bowel disease patient. Am J Gastroenterol 2010; 105:1231-8.

18. Fiorino G, Peyrin-Biroulet L, Naccarato P, Szabò H, Sociale OR, Vetrano S et al. Effects of immunosuppression on immune response to pneumococcal vaccine in inflammatory bowel disease: a prospective study. Inflamm Bowel Dis 2012; 18:1042-7.

19. Melmed GY, Agarwal N, Frenck RW, Ippoliti AF, Ibanez P, Papadakis KA et al. Immunosuppression impairs response to pneumococcal polysaccharide vaccination in patients with inflammatory bowel disease. Am J Gastroenterol 2010; 105:148-54.

20. Agarwal N, Ollington K, Kaneshiro M, Frenck R, Melmed GY. Are immunosuppressive medications associated with decreased responses to routine immunizations? A systematic review. Vaccine 2012; 30:1413-24.

21. Chevaux JB, Nani A, Oussalah A, Venard V, Bensenane M, Belle A et al. Prevalence of hepatitis B and C and risk factors for non-vaccination in inflammatory bowel disease patients in Northeast France. Inflamm Bowel Dis 2010; 16:916-23.

22. Pérez-Alvarez R, Díaz-Lagares C, García-Hernández F, Lopez-Roses L, Brito-Zerón P, Pérez-de-Lis M et al. Hepatitis B virus (HBV) reactivation in patients receiving tumor necrosis factor (TNF)-targeted therapy: analysis of 257 cases. Medicine (Baltimore) 2011; 90:359-71.

23. Vida Pérez L, Gómez Camacho F, García Sánchez V, Iglesias Flores EM, Castillo Molina L, Cerezo Ruiz A et al. Adequate rate of response to hepatitis B virus vaccination in patients with inflammatory bowel disease. Med Clin (Barc) 2009; 132:331-5.

24. Altunöz ME, Senateş E, Yeşil A, Calhan T, Ovünç AO. Patients with inflammatory bowel disease have a lower response rate to HBV vaccination compared to controls. Dig Dis Sci 2012; 57:1039-44.

25. Gisbert JP, Chaparro M, Esteve M. Review article: prevention and management of hepatitis B and C infection in patients with inflammatory bowel disease. Aliment Pharmacol Ther 2011; 33:619-33.

26. Gisbert JP, Villagrasa JR, Rodríguez-Nogueiras A, Chaparro M. Efficacy of hepatitis B vaccination and revaccination and factors impacting on response in patients with inflammatory bowel disease. Am J Gastroenterol 2012; 107:1460-6.

27. Gisbert JP, Menchén L, García-Sánchez V, Marín I, Villagrasa JR, Chaparro M. Comparison of the effectiveness of two protocols for vaccination (standard and double dosage) against hepatitis B virus in patients with inflammatory bowel disease. Aliment Pharmacol Ther 2012; 35:1379-85.

28. Clemens SAC, da Fonseca JC, Azevedo T, Cavalcanti A, Silveira TR, Castilho MC, Clemens R. Hepatitis A and hepatitis B seroprevalence in four centers in Brazil. Rev Soc Bras Med Trop 2000; 33(1):1-10.

29. Kotton CN. Vaccines and inflammatory bowel disease. Dig Dis 2010; 28:525-35.

30. Bhatia J, Bratcher J, Korelitz B, Vakher K, Mannor S, Shevchuk M et al. Abnormalities of uterine cervix in women with inflammatory bowel disease. World J Gastroenterol 2006; 12:6167-71.

31. Kane S, Khatibi B, Reddy D. Higher incidence of abnormal pap smears in women with inflammatory bowel disease. Am J Gastroenterol 2008; 103:631-6.

32. Sociedade Brasileira de Imunizações. Available at: www.sbim.org.br.

33. Singh H, Demers AA, Nugent Z, Mahmud SM, Kliewer EV, Bernstein CN. Risk of cervical abnormalities in women with inflammatory bowel disease: a population-based nested case-control study. Gastroenterol 2009; 136:451-8.

34. Melmed GY. Vaccination strategies for patients with inflammatory bowel disease on immunomodulators and biologics. Inflamm Bowel Dis 2009; 15:1410-6.

35. Mahadevan U, Cucchiara S, Hyams JS, Steinwurz F, Nuti F, Travis SP et al. The London Position Statement of the World Congress of Gastroenterology on Biological Therapy for IBD With the European Crohn's and Colitis Organization: Pregnancy and Pediatrics. Am J Gastroenterol 2011; 106:214-23.

# PSYCHOLOGICAL ASPECTS OF INFLAMMATORY BOWEL DISEASE

CLEIDE RODRIGUES DE CASTRO

## INTRODUCTION

Inflammatory bowel disease (IBD) is not only characterized by intestinal and extra-intestinal manifestations, but also for psychological changes, which can be reflected in relationships, social activities and work. Therefore, this chapter aims to help healthcare professionals and other interested parties to perceive the conflicts experienced by patients with IBD and the impact of the disease on their quality of life.

## INTERFERENCE OF EMOTIONAL FACTORS IN IBD

Historically, IBD was presented in the literature as a psychosomatic etiology (word derived from the Greek: *psyche* = soul + *soma* = body) characterized, in short, as a disruption of homeostasis, causing somatic manifestations. Psychosomatic approach refers to the indivisibility and interdependence of the psychological and biological aspects of humanity; a connotation that can be called holistic, as it implies the notion of the human being as a whole, one mind-body complex immersed in a social environment.[1]

According to the ICD-10 (World Health Organization – WHO), psychosomatic manifestations are classified as psychological and behavioral factors associated with disorders or diseases classified elsewhere (F54). They are characterized by psychological or behavioral influences, being a prominent factor in the etiology of physical disorders.[2] However, IBD is considered to have an unknown etiology, with affective and emotional disorders, and stressful life events appearing to be relevant in the onset and maintenance of symptoms.

Individuals living with IBD endure extremely unpleasant symptoms, including diarrhea attacks, intestinal colicky pain, bleeding and possible complications, such as strictures and fistulas, which generate a high degree of discomfort and stress. However, it can be assumed, based on clinical perception, that the very living with the disease triggers distress and anxiety.

There is evidence that symptoms of anxiety and depression are more severe during periods of active disease.[3] Among the few Brazilian studies addressing these disorders as risk factors for Crohn's disease (CD), a survey of 110 patients found that psychological disorders appear to play a role in the exacerbation of symptoms. DC activity is strongly associated with depressed mood, and depression and anxiety are conditions strongly present as risk factors for early clinical relapse in patients with inactive CD.[4]

Since most patients with IBD believe that psychosocial stress is the main reason for the worsening of their disease,[5] this factor can be defined as a threat to homeostasis of the body, whether physical or psychological, by internal or external stimuli that induce a stress response, which recruits neural and hormonal mechanisms in an attempt to restore or enhance the normal function of the body. Usually, the stress response is beneficial, enabling people to handle a range of stressful adverse situations and devise relevant solutions; however, if the response to stress is excessive or prolonged, it can be deleterious.[6]

Clinical evidence suggests a close relationship with psychological factors in relapse of patients considered in remission. A periodic evaluation survey found that being in the highest tertile of stress tripled the subsequent exacerbation rate in the medium (6-8 months) and long term (up to 5 years).[7] Preliminary results of another study, which drew an analysis of patients with IBD in remission, suggest that patients with depressive symptoms have a higher number of relapses.[8]

Recent surveys indicate that nerve connections between the brain and the gut stimulate inflammatory cells in the gut wall. In this process, the substances released worsen inflammation and provoke an increase in harmful bacteria in the intestinal mucosa. This important research shows not only that stress can worsen IBD due to delayed remission and aggression to the intestinal lining, but also indicates that relaxation and hypnosis techniques can have positive effects on these stimuli, on the patients' tolerance regarding threshold of pain or sensory awareness of their symptoms.[9]

In the context of improving the quality of life, some techniques have been developed in order to enable the treatment of stress, such as stress management training (SMT). A behavioral study showed that SMT applied to CD patients led to reduction of the levels of stress, restructuring of irrational beliefs, development of assertive behavior, decreased anxiety level, development of the ability to express anger, and clinical improvement. Patients with the same disease who did not undergo SMT showed no improvement from a psychological and clinical standpoint, after the period of 10 consecutive weeks.[10]

## IMPACT ON QUALITY OF LIFE

Often, the onset of symptoms leads the patient to an exhaustive search for a specialist who can conclude the diagnosis of IBD. In some cases, having an explanation for their symptoms and the definition of a therapeutic approach brings a sense of relief. Other patients, however, are impacted and afraid when they receive the diagnosis, because it is a chronic disease of unknown cause that creates limitations in their productive life, leading to fantasies and expectations regarding disease progression. As seen, the interpretation given to the diagnosis, prognosis and treatment is absolutely subjective.

There is not a personality type assigned to patients with IBD, although it is possible to establish, through clinical observations and perceptions, a profile of emotions and concerns common to these patients, mostly young adults, who are in the most productive period of life, seeking emotional and financial stability, independence and self-sufficiency. However, IBD can impose a dependence status due to the fecal incontinence spectrum, which, as a matter of fact, can dictate, for example, a concern or obsession with finding the nearest bathroom further damaging their self-esteem to the point of feeling different from other young people, causing repercussions in family, academic, professional, emotional and sexual relationships.[11]

In general, the phase of sexual initiation in adolescents and young adults is marked by crises of physical and psychological maturity, justified by changes in hormones and body image. It is a period of uncertainty on account of the need to be accepted, the fear of rejection and how the flares of IBD sometimes can interfere or maximize the difficulty of exposure to intimacy, because of the fear that exacerbation may occur at that time. Generally, mature couples, maintaining a stable affective sexual relationship, find it easier to deal with these issues as well as to accept them, discuss them and to overcome them.

As a result of their own experiences, considered threatening, the patient is concerned about controlling future situations, which makes the anticipatory suffering a vicious cycle, as if life revolved around their digestive tract. In fact, they commonly exhibit anticipatory anxiety disorders and eating disorders such as anorexia, motivated by food phobia related to the cycle of eating and evacuating.

A Canadian study addressed aspects of quality of life in 259 individuals affected by IBD and pointed out the top 10 concerns listed by patients: weakness, medication side effects, uncertain etiology of the disease, the possibility of surgical complications or need for bag ostomy, loss of control, loss of professional achievement, dependence on others, production of unpleasant odors and changes in body image.[12]

IBD also reflects significant impact on family ties. Those emotionally involved show powerlessness in the face of the patient's suffering and demonstrate guilt, pity, overprotection and even neglect, when in fact, they should acquire information to strengthen the coping strategy, assisting with supportive attitudes and practices. The family also needs attention from the multidisciplinary team, because it often participates in the expectation of treatment, suffers along in times of flare, gets excited when there is remission, and is distressed with the complications, feeling emotional anxiety as intense as the patient's.

Changes in daily activities because of frequent medical appointments, laboratory tests and hospitalizations can cost the patient with IBD some losses. Absence from work and possible loss of employment can occur, with social and economic repercussions, in addition to the need to maintain high treatment costs and a balanced diet.

Concerns are also related to the stigma (social label assigned to an individual) of the disease, including the fear of what co-workers are thinking while they are away several times to go to the toilet, and the risk of dismissal, because their managers could interpret it as omission or lack of commitment to work, and also distrust in the eyes of people, because of changes in body image, such as rashes and weight change. Generally, the patient imagines that others believe he or she has a contagious disease. At this time, a clinical approach makes a difference, in the sense that there are diseases more serious than IBD. When the patient accepts the diagnosis, it is easier to talk about the disease and its symptoms, and to count on the understanding and support of people in their social environment.

A US research noted significant impact on the concerns experienced by patients, especially regarding the stigma of the disease among those unable to work because of IBD. The results show that out of 211 participants, 84% reported a perception of stigma; 14% reported feeling moderately stigmatized by family; 23% by friends and 11% by a spouse or significant other. Twenty percent reported moderate stigmatization by medical service providers. Moderate stigma was also sensed among co-workers, corresponding to 28%; as well as employers, 32%.[13]

## MULTIPROFESSIONAL APPROACH

Given the complexity of the disease, a multidisciplinary approach to patients with IBD is clearly needed, including psychotherapeutic treatment and a support group, and, of course, the essential connection between doctor and patient.

The psychotherapeutic work aims at hearing the subjectivity of complaints and restoring psychic homeostasis. By assigning a symbolic meaning to the disease, in terms of conflicts, feelings and personal difficulties, the patient feels more understood and may have favorable changes both in mood as in their ability to overcome and cope life with any diseases, including IBD.

A multidisciplinary support group (with doctors, nutritionists and psychologists), in turn, aims to make patients interact with each other, encouraging the exchange of experiences through speaking and listening activities, giving the patient an opportunity to recognize their difficulties and fears as they speak and, while listening to other people, to absorb practical knowledge on all aspects of the disease, learning to cope with the feared situation.

As much as the psychotherapeutic work and the support group, the bond between doctor and patient is very important for the patient to have an active participation in the process of accepting their diagnosis and adhering to the proposed treatment, because the alliance formed through this bond is the foundation for a successful treatment.

In the approach to the diagnosis of IBD, relating the disease with psychological or emotional difficulties should be avoided, as this may reflect a negative stigma to the patient or the unfounded guilt that somehow the patient was not able to control their emotional state and caused or worsened the disease. A criterion that can be used to refer the patient to a mental health professional is the proposal to provide support in times of trouble, which may appear at any stage in life. In addition, the multidisciplinary team treating people with chronic diseases should preferably have not only scientific knowledge but, especially, availability, complicity and tolerance to deal with the ups-and-downs of flares and remissions, complex complaints and the related emotional repercussions.

## REFERENCES

1. Lipowiski ZJ. What does the word psychosomatic really mean? An historical and semantic inquiry. Psychomatic Medicine 1984; 167.

2. World Health Organization (WHO). Classificação de transtornos mentais e de comportamento da CID-10.

3. Graff LA, Walker JR, Bernstein CN. Depression and anxiety in inflammatory bowel disease: a review of comorbidity and management. Inflam Bowel Dis 2009; 15(7):1105-18.

4. Brandi MT, Ribeiro MS, Chebli LA, Franco MMC, Pinto ALT, Gaburri PD et al. Angústia pessoal psicológica em portadores de doença de Crohn no Brasil: triagem, prevalência e fatores de risco. 27. PH101. Public Med Sci Monit 2009; 2. Available at: www.medscimonit.com.

5.  Ikalcić M, Hauser G, Stimac D. Differences in the health-related quality of life, affective status, and personality between irritable bowel syndrome and inflammatory bowel disease patients. European Journal of Gastroenterology & Hepatology 2010; 22:862-7.

6.  Dunckley P, Travis S, Phil D. Is IBD associated with a stressful lifestyle? Gut 2008; 57:1386-92.

7.  Robertson DA, Ray J, Diamond I, Edwards JG. Personality profile and affective state of patients with inflammatory bowel disease. Gut 1989; 30:623-6.

8.  Mittermaier C, Beier M, Tillinger W, Gangl A, Moser G. Correlations between depressive mood and disease activity in patients with inflammatory bowel disease (IBD) – A prospective study. [Abstract]. Psychosomatic Medicine 1998; 60:96.

9.  Mawdsley JED, DS Rampton. Psichological stress in inflammatory bowel disease: new insights into pathogenic and therapeutic implications. Gut 2005; 54:1481-91.

10. Amodeo-Escribano S, Sirgo A, Amorim-Gaudencio C, Perales-Soler FJ, Lara VG, Perez-Millan JM. Análise de traços psicológicos em enfermidade inflamatória intestinal. Psicol Argum Abril 2000; 26(18):35-43.

11. Levenstein S. Psychosocial factors in peptic ulcer and inflammatory bowel disease. San Camillo-Forlanini Hospital; Copyright 2002. American Psychological Association, Inc.

12. deRooy EC, Toner BB, Maunder RG, Greenberg GR, Baron D, Steinhart AH et al. Concerns of patients with inflammatory bowel disease: results from a clinical population. Am J Gastrol 2001; 96:1816-21.

13. Tiffany H, Taft MA. Impact of stigma on patients with IBD. Inflamm Bowel Dis 2010; 5.

## BIBLIOGRAPHY

1.  Brasio KM. Eficácia do treino de controle de stress na retocolite ulcerativa inespecífica. [Tese de doutorado]. Campinas: Pontifícia Universidade Católica de Campinas, 2000; XV:234.

# PSYCHIATRIC ASPECTS OF INFLAMMATORY BOWEL DISEASE

EDUARDO DE CASTRO HUMES
RENÉRIO FRÁGUAS JUNIOR

## INTRODUCTION

The central nervous system, through the autonomic nervous system (ANS), integrates inputs coming from the digestive system. A complex network including the hypothalamus, limbic system and cerebral cortex processes such inputs and by modifying the tonus of the sympathetic (i.e. the splanchnic nerves) and parasympathetic nervous system (i.e. the vagus nerve and the sacral parasympathetic pelvic nerves), and the hypothalamic-hypophyseal-adrenal (HPA) axis, controls the digestive system. Changes in the functioning of these interactions associated with mental conditions such as increased stress and depression have been described in disorders of the digestive system, including irritable bowel syndrome (IBS) and the inflammatory bowel diseases (IBD). Although the literature is still sparse, approaches including the evaluation and treatment of mental disorders may be relevant to ensure the complete restoration of the patient's health, improve prognosis and provide a better quality of life.

## PSYCHIATRIC SYNDROMES RELATED TO INFLAMMATORY BOWEL DISEASES

### Depression

#### Epidemiology

The prevalence of depressive symptoms observed in patients suffering from IBD is higher than in the general population, ranging from 9.3 to 68%. This variation is influenced by both severity of the bowel disease and the method used to evaluate the depressive symptoms. The progression of the bowel disease is directly related to the higher prevalence of depressive symptoms. Inflammatory bowel diseases, and especially ulcerative colitis (UC), are also related to poorer scores in quality of life, and the prevalence of depressive symptoms is higher than that found in other chronic diseases.

Patients with depressive symptoms present higher attendance to medical services, as well as longer hospitalization periods. Both depression and stressful factors are associated with worse prognosis and worse rates of quality of life in IBD patients. Patients with Crohn's disease (CD) presenting depressive symptoms have higher rates of surgery. Finally, there are evidences suggesting that patients with depressive symptoms respond less to the treatment of the other medical condition.

#### Diagnosis

A topic of major impact is the frequent underdiagnosis of psychiatric syndromes by general practitioners. Besides being underdiagnosed, when diagnosed these patients are often not adequately treated, causing a worsening of quality of life and poorer overall prognosis. One of the main difficulties met by general practitioners is the recognition of depressive symptoms as part of a depressive disorder instead of a usual and normal reaction to the disease. Besides that, it is common to consider physical symptoms of depression, such as pain (or increased sensitivity to painful stimuli), changes in appetite and weight loss, as mere results of the clinical condition itself. Thus, it is recommended that general practitioners systematically perform a screening test for depression, investigating feelings of sadness (present most of the day, during most of the days) and decreased interest and pleasure in usual activities that were generally perceived as pleasant by the patient.

## Treatment

Two aspects limit the effectiveness of depression treatment by the general practitioners. The first is the use of benzodiazepines rather than antidepressants; secondly is the use of doses of antidepressants insufficient to achieve full remission of depressive symptoms (absence of all symptoms presented by a given patient).

The treatment of depression is usually based on the use of antidepressants (except for depression in bipolar affective disorder). Use of benzodiazepines should be brief and usually restricted to acute management of symptoms such as sleep disturbances, anxiety and intense anguish, in order to avoid chronic use and the risks of dependence. The use of antidepressants for the management of mood disorders in patients with IBD is associated with a reduction in the recurrence of IBD, as well as a reduction in the use of corticosteroids.

Few studies have investigated the safety and efficacy of the use of antidepressants in patients with IBD. Nevertheless, selective serotonin reuptake inhibitors (SSRIs) and serotonin and norepinephrine reuptake inhibitors (SNRIs) have been routinely used as the first-line treatment. It is worth remembering that these antidepressants may present gastrointestinal side effects, including nausea and diarrhea, which are difficult to manage in patients with IBD. These patients must be advised to take the antidepressants during their meals, to reduce the local action on serotoninergic gastrointestinal receptors. Even though bupropion use in IBD patients still requires further confirmatory studies, due to its dopaminergic and more activating action, it can also be considered as an option to the first-line antidepressants, especially for those who experience apathy, anergy or present with few anxiety symptoms.

There are also some studies on behavioral therapy and psychodynamic psychotherapy in IBD patients, and even though there is no impact on the progression of this clinical disease, it improves scores of psychiatric symptoms, quality of life and resilience. Psychotherapeutic intervention associated with the teaching of relaxation techniques has been related to lower demand for health services.

Educational strategies should take into account the disease's characteristics and specificities. Psychotherapeutic groups of IBD patients are an important tool to improve the quality of life, as well as potentially reduce psychiatric symptoms.

## Anxiety

### Epidemiology

The observed prevalence of anxiety symptoms in patients with IBD, as observed for depression, varies significantly. According to the literature, prevalence rates range between 22.5 and 33.6%. These rates are significantly greater than those reported for the general population. Anxiety is associated with increased demand for medical services, poorer quality of life, higher rates of surgical indication, as well as with the worsening of adherence to medical procedures.

### Treatment

Management of anxiety symptoms should be done as in the general population. It is based on the use of serotonergic antidepressants, SSRIs being the first line treatment. Benzodiazepines should focus on the relief of changes in sleep pattern and anxiety at the beginning of the course of treatment. The long term use of benzodiazepines is often related to the chronicity of symptoms of anxiety.

Anxious patients may benefit from psychodynamic psychotherapy and behavioral therapy. Educational strategies about the disease are also associated with a significant reduction in anxiety rates.

## PSYCHOLOGICAL REACTIONS TO IBD AND ITS TREATMENTS

Patients with IBD present more frequently difficulties to accept the diagnosis, the limitations imposed by the disease, hospitalization and medical examinations. When receiving the diagnosis, many patients have reactions similar to grief, including the five stages of Kübler-Ross (denial, anger, bargaining, depression, and acceptance).

It is possible to observe externalization of feelings of frustration and anger towards the treatment and the various diagnostic procedures to which these patients are submitted. The classic psychoanalytic/psychosomatic models linking specific conflicts or personality profiles to an inflammatory bowel disease have not been confirmed by methodologically rigorous studies. Physiological changes associated with the disease, such as the inflammatory process itself and treatment with immunomodulators, especially corticosteroids, may, however, result in changes in biological processes of the brain, changing affective responses, and facilitating behavioral changes. These findings are supported by studies reporting that patients with more aggressive symptomatic forms of the disease present more severe depressive symptoms compared to asymptomatic patients. Alexithymia (a condition characterized by difficulty in understanding and expressing emotions) is present in a significant proportion of patients with inflammatory bowel disease, and has been associated with poor quality of life.

Especially for children and adolescents, reduced independence has been described, as well as diminished perceived control, changes in self-image and in the perception of health, in addition to losses in interpersonal relationships. These difficulties are modulated by physical aspects (depending on the presence of malnutrition, for example), by disease severity (dependent on the presence of pain or need for hospitalization, for instance), and by environmental aspects such as the family's reaction.

The identification of these processes by clinicians and the appropriate referral for psychiatric and psychological evaluation are central to preserve the patient's quality of life as well as treatment adherence and, in children, to proper a healthy development. In addition to support groups, family and individual assistance is often needed.

## PSYCHIATRIC SYMPTOMS DUE TO MEDICATIONS USED IN IBD

Several drugs used in medical practice may be associated with psychiatric side effects. The mechanisms are varied and are not well established. There is a wide range of symptoms secondary to the use of such medications. In general, there is a close temporal connection between the introduction of the drug and the psychiatric symptoms. However, in some cases, the emergence of the psychiatric symptoms may occur during chronic exposure to the medication.

Psychiatric disorders that emerge soon after the use of corticosteroids, especially mood symptoms, are widely documented, particularly hippomanic, manic and depressive episodes. The emergence of manic symptoms usually occurs right after the introduction of corticosteroids, particularly when the patient is treated with high doses as well as in combination regimes (i.e. oral and rectal). Depression is most commonly related to the chronic use of these medications. Besides mood symptoms, a wide range of psychiatric symptoms has been associated with the use of corticosteroids, including cognitive changes, anxiety, psychosis and *delirium.*

Other medications used in IBD may also trigger behavioral manifestations. Metronidazol may be associated to psychosis, depressive and manic episodes. Several reports have associated cyclosporine with the onset or worsening of psychotic symptoms, as well as the emergence of depressive and anxious symptoms.

The management of these changes involves treating the symptoms using psychiatric drugs, according to the clinical presentation. However, reduction, change or even cessation of the medication must also be considered, especially in severe presentations.

## REFERENCES

1. Ananthakrishnan AN, Gainer VS, Perez RG, Cai T, Cheng SC, Savova G et al. Psychiatric co-morbidity is associated with increased risk of surgery in Crohn's disease. Aliment Pharmacol Ther 2013; 37(4):445-54.

2. Bennebroek Evertsz' F, Thijssens NA, Stokkers PC, Grootenhuis MA, Bockting CL, Nieuwkerk PT et al. Do inflammatory bowel disease patients with anxiety and depressive symptoms receive the care they need? J Crohns Colitis 2012; 6(1):68-76.

3. Bessissow T, Van Keerberghen CA, Van Oudenhove L, Ferrante M, Vermeire S, Rutgeerts P et al. Anxiety is associated with impaired tolerance of colonoscopy preparation in inflammatory bowel disease and controls. J Crohns Colitis 2013; 7(11):e580-7. DOI: 10.1016/j.crohns.2013.04.011. Epub 2013 May 9.

4. Brandi MT, Ribeiro MS, Chebli LA, Franco MB, Pinto AL, Gaburri PD et al. Psychological distress in Brazilian Crohn's disease patients: screening, prevalence, and risk factors. Med Sci Monit 2009; 15(8):PH101-108.

5. Brown ES. Effects of glucocorticoids on mood, memory, and the hippocampus. Treatment and preventive therapy. Ann N Y Acad Sci 2009; 1179:41-55.

6. Deter HC, Keller W, von Wietersheim J, Jantschek G, Duchmann R, Zeitz M; German Study Group on Psychosocial Intervention in Crohn's Disease. Psychological treatment may reduce the need for healthcare in patients with Crohn's disease. Inflamm Bowel Dis 2007; 13(6):745-52.

7. Fuller-Thomson E, Sulman J. Depression and inflammatory bowel disease: findings from two nationally representative Canadian surveys. Inflamm Bowel Dis 2006; 12(8):697-707.

8. Goodhand JR, Wahed M, Mawdsley JE, Farmer AD, Aziz Q, Rampton DS. Mood disorders in inflammatory bowel disease: relation to diagnosis, disease activity, perceived stress, and other factors. Inflamm Bowel Dis 2012; 18(12):2301-9.

9. Goodhand JR, Greig FI, Koodun Y, McDermott A, Wahed M, Langmead L et al. Do antidepressants influence the disease course in inflammatory bowel disease? A retrospective case-matched observational study. Inflamm Bowel Dis 2012; 18(7):1232-9.

10. Iglesias-Rey M, Barreiro-de Acosta M, Caamaño-Isorna F, Vázquez Rodríguez I, Lorenzo González A, Bello-Paderne X et al. Influence of alexithymia on health-related quality of life in inflammatory bowel disease: are there any related factors? Scand J Gastroenterol 2012; 47(4):445-53.

11. Kennedy AP, Nelson E, Reeves D, Richardson G, Roberts C, Robinson A et al. A randomised controlled trial to assess the effectiveness and cost of a patient orientated self-management approach to chronic inflammatory bowel disease. Gut 2004; 53(11):1639-45.

12. Koul S, Bhan-Kotwal S, Jenkins HS, Carmaciu CD. Organic psychosis induced by ofloxacin and metronidazole. Br J Hosp Med (Lond) 2009; 70(4):236-7.

13. Larsson K, Sundberg Hjelm M, Karlbom U, Nordin K, Anderberg UM, Lööf L. A group-based patient education programme for high-anxiety patients with Crohn disease or ulcerative colitis. Scand J Gastroenterol 2003; 38(7):763-9.

14. McCombie AM, Mulder RT, Gearry RB. Psychotherapy for inflammatory bowel disease: a review and update. J Crohns Colitis 2013; 7(12):935-49. doi: 10.1016/j.crohns.2013.02.004. Epub 2013 Mar 5.

15. Mikocka-Walus AA, Turnbull DA, Moulding NT, Wilson IG, Andrews JM, Holtmann GJ. Antidepressants and inflammatory bowel disease: a systematic review. Clin Pract Epidemiol Ment Health 2006; 2:24.

**16.** Mikocka-Walus AA, Turnbull DA, Moulding NT, Wilson IG, Andrews JM, Holtmann GJ. "It doesn't do any harm, but patients feel better": a qualitative exploratory study on gastroenterologists' perspectives on the role of antidepressants in inflammatory bowel disease. BMC Gastroenterol 2007; 7:38.

**17.** Nahon S, Lahmek P, Durance C, Olympie A, Lesgourgues B, Colombel JF et al. Risk factors of anxiety and depression in inflammatory bowel disease. Inflamm Bowel Dis 2012; 18(11):2086-91.

**18.** Oliveira S, Zaltman C, Elia C, Vargens R, Leal A, Barros R et al. Quality-of-life measurement in patients with inflammatory bowel disease receiving social support. Inflamm Bowel Dis 2007; 13(4):470-4.

**19.** Rubin DT, Dubinsky MC, Panaccione R, Siegel CA, Binion DG, Kane SV et al. The impact of ulcerative colitis on patients' lives compared to other chronic diseases: a patient survey. Dig Dis Sci 2010; 55(4):1044-52.

**20.** Szigethy E, Levy-Warren A, Whitton S, Bousvaros A, Gauvreau K, Leichtner AM et al. Depressive symptoms and inflammatory bowel disease in children and adolescents: a cross-sectional study. J Pediatr Gastroenterol Nutr 2004; 39(4):395-403.

**21.** Szigethy E, McLafferty L, Goyal A. Inflammatory bowel disease. Child Adolesc Psychiatr Clin N Am 2010; 19(2):301-18, ix.

**22.** Telarović S, Telarović S, Mihanović M. Cyclosporine-induced depressive psychosis in a liver transplant patient: a case report. Lijec Vjesn 2007; 129(3-4):74-6.

**23.** von Wiersheim J, Kessler H. Psychotherapy with chronic inflammatory bowel disease patients: a review. Inflamm Bowel Dis 2006; 12(12):1175-84.

# LEGAL ASPECTS OF INFLAMMATORY BOWEL DISEASE

CYNTHIA MARIA BASSOTTO CURY MELLO

Due to the importance of the subject, currently health is legally supported in Brazil, above all, by the Federal Constitution, the maximum law in the legal framework. Furthermore, it is covered in a very peculiar range of guarantees, the so-called fundamental rights of the citizen, defined as those directly related to the maximum protection of human equality, liberty and dignity - goods that cannot be sold, exchanged or bartered for.

In this context, the philosopher Kant distinguished the values to which it was possible to assign a price from those that had purely dignity:

> In the kingdom of ends everything has either a price or a dignity. When something comes at a price, you can put anything in its place as an equivalent; but when a thing is above all prices, and thus has no equivalent, it therefore has dignity.[1]

As such, it is simple to conclude that without dignity, the social life of any person is curtailed. Without fundamental rights – in which health is, therefore, included – it is certain that the human being either does not progress or achieves minimal realization. They would often not even survive, by breaching the wider principle established for centuries not merely by the law, but above all, by religion and morals: the right to life.

Therefore, the highest and mightiest law of the country dedicates a specific section to the issue, establishing health as a "right of all and a duty of the State"[2] and, consequently, places all citizens on equal footing, including those with inflammatory bowel diseases (IBD).

Before the precepts of human dignity, the right to life and health, an approach to providing access to methods of treating diseases is indispensable, through the realization of examinations and the use of appropriate medication.

Thus, to supplement everything set forth in the Federal Constitution, in 1990 the Organic Law of the Unified Health System (SUS) was created, which established the obligation of the State (understood as the Federation, the member States and municipalities) for "integrated care, including pharmaceutical".[3]

In practice, the SUS system ended up joining the trend defending evidence based medicine,[4] that is, according to which "Clinical Protocols and Therapeutic Guidelines" are adopted, which are actually a set of criteria to determine the diagnosis of each disease and the corresponding treatment with the drugs available and the correct dosages. For the patient, this directly involves lists of medicines that are dispensed to the population for free, provided the pathology is proven. However, it is easy to understand that, with the emergence of new drugs and constant technological evolution, many substances are not included in these criteria, not even as an exception, though this does not eliminate the patient's right to obtain differential treatment, provided it is approved by health and supervisory bodies and the vital need for use is proven, given the ineffectiveness of other treatments available under SUS criteria.

In such cases, the Courts have not recognized the existence of "clinical protocols" from the Departments of Health or even their regulation through internal ordinances as preventive factors for exercising full patient health and the consequent free acquisition of suitable mediation for their clinical case. This can be illustrated by the solid jurisprudence of courts in lawsuits relating to the provision of differentiated treatments.

It would be inconceivable for the rights enshrined in lower regulations such as circulars, decrees, provisional measures and ordinary laws to have immediate effect, and for constitutionally enshrined rights, inspired by the highest ethical and moral values of the nation, to be relegated to the background. With the State promising the right to health, it must enforce it, because the will of constitution, to use Konrad Hesse's expression, was for the eradication of the hardship plaguing the country.[5]

The Judiciary, exercising its high and important constitutional mission, should and can impose compliance with the constitutional provision that guarantees the right to health on the State Executive Power, otherwise it would be condoning the pain and suffering of thousands of poor and needy Brazilians who, seeking the treatment of the Unified Health Service due to lack of choice, are at the mercy of a precarious and inefficient health system, which often leads to death.[6]

Thus, although the patient with IBD is not protected by specific laws for their disease, it is quite clear that the general standards of the country are fully applicable to their questions. Indeed, this is not merely restricted to access to treatments, but is reflected in all spheres of life of the citizen.

In the criminal sphere, in analogy to crimes of discrimination related to race and color, so does the discrimination of patients with IBD can be considered a crime, with the allocation of imprisonment of up to five years and a fine as penalty.[7] In this aspect, it is important to distinguish two terms that appear similar but which, in fact, are completely different: differentiation and discrimination.

Differentiating chronic patient is not merely acceptable but sometimes necessary. This is not with the intention of harming their morale or dignity, but simply to grant them a more careful treatment due to their clinical state. Discrimination, according to the "Citizenship for All" handbook, means "conduct (acts or omissions) that violate the rights of people based on unjustified and unfair criteria such as race, gender, age, religious choice, and more".[8]

In the social security area, the right to sick pay and disability retirement can be cited. Sick pay is defined by the Social Security service itself as the "benefit granted to the insured party unable to work due to sickness or accident for more than 15 consecutive days."[9] This translates into the right for the chronic patient to take time off work, without charge, remaining paid at times when their crises worsen. Indeed, this is assured by Article 201 of the Brazilian Constitution.[10]

In relation to disability retirement, another constitutional guarantee, there is the requirement for the worker to present clinical evidence so serious as to definitively prevent them from returning to work, due to either permanently defined symptoms or very remote prospects of improvement. It is treated as

> a benefit granted to workers who, due to disease or accident, have been considered by the medical investigation of the Social Security service as incapacitated to carry out their activities or other type of work to sustain themselves.[11]

And that's not all. Although IBD does not (yet) compose the list of diseases considered disabling by law (for example, Law no. 11,052, of December 29th, 2004), it is important to note, from the perspective of the law, that the understanding is that each case must be analyzed in its uniqueness, that is, given that the determination of the individual's actual incapacity depends on the technical analysis by the investigating physician, nothing prevents other rights from being exercised by critically ill patients, such as use of the Government Severance Indemnity Fund (FGTS) or priority in judicial proceedings.

Therefore, in the eyes of the law, and based on the positive performance of the Brazilian judiciary in relation to health, it could be said that we currently count on something near to the understanding of Bobbio:

> Man himself is no longer considered as a generic being, or abstracted man, but is seen in the specificity or concreteness of his many ways of being in society, such as children, old, sick, etc.[12]

### SUGGESTION FOR WRITING MEDICAL REPORTS FOR THE PATIENT

In order to facilitate the understanding by public bodies and avoid reworking for both medical professional and the patient, the following script is recommended for the elaboration of reports, with the main points that must be addressed to as effective as possible:

- patient's name;
- age;
- name of the pathology;
- IDC;
- date of diagnosis or how long the diseases has been manifested;
- amount of time under the care of this professional;
- clinical history (more severe episodes, hospitalizations, surgeries);
- patient's current clinical condition;
- derived or correlated diseases;
- medication used (name, dose and objective);
- references to exams performed;
- patient's needs:
  - absence from work: time off from work for up to 15 consecutive days for treatment or recovery;
  - sick pay: time off from work for more than 15 consecutive days due to temporary incapacity to perform their activities, for treatment or recovery;
  - disability retirement: remaining permanently absent from work due to constant incapacity to return to their activities, with no prospect of improvement in their clinical picture;
  - for the employer's knowledge and records: priority in the use of bathrooms, for example;
  - Organic Law of Health Care (LOAS): to receive welfare benefit for maintenance of the cost of treatment or living;
  - public transportation: to receive public transport benefit to enable travel for treatment and consultations.

**Practical example**

*R. C. S., 32 years old, Crohn's disease patient (ICD-10 K.50) for 10 years, under my care since March 2006.*

*His symptoms are chronically active, having undergone two surgeries to treat perianal fistulas in May 2007 and July 2009, respectively.*

*He currently presents daily diarrheal crises, abdominal pain and malnutrition due to hypoalbuminemia as a result of the malabsorptive process induced by the disease. As a result of the social limitations imposed by the disease, he has also developed depressive symptoms, and is receiving psychiatric care.*

*He uses adalimumab (40 mg every 2 weeks) to reduce intestinal inflammatory activity; nutritional supplement for protein calorie support (1 cup, 3 times/day); and alprazolam (2 mg/day) for stabilization of his depressive symptoms.*

*As verified by the recent colonoscopy examination attached, the disease is in high activity, preventing the patient from performing their daily chores and leading to a risk of further complications. Therefore, I hereby request the withdrawal of the patient from work for a 30-day period, in order to enable the stabilization of his symptoms and return to his activities without risk.*

*I remain at your disposal.*

Date, stamp and signature of the physician.

## REFERENCES

1. Kant I, apud Hoerster N. In: Defensa del positivismo jurídico. Tradução para o espanhol de Ernesto Garzón Valdés. Barcelona: Gedisa Editorial, 2000.

2. Brasil. Constituição da República Federativa do Brasil de 1988, Título VIII, Capítulo II, Seção II, artigo 196 . Ed. Comemorativa 20 anos. São Paulo: AASP, 2008. p.132-3.

3. Brasil. Lei n. 8.088, de 31 de outubro de 1990, artigo 6°. Available at: www.planalto.gov.br/ccivil/LEIS/L8088consol.htm.

4. Drummond JP, Silva E. O que é medicina baseada em evidências? In: Medicina baseada em evidências. São Paulo: Atheneu, 1998.

5. Recurso Especial n° 577836/SC (2003/0145439-2), 1ª Turma do STJ, Rel. Min. Luiz Fux. j. 21.10.2004, unânime, DJ 28 fev. 2005.

6. Mandado de Segurança n° 1.0000.03.401817-6/000, 3° Grupo de Câmaras Cíveis do TJMG, Belo Horizonte, Rel. Maria Elza. j. 18.02.2004, unânime, DJ. 10 mar. 2004.

7. Brasil. Lei n. 7.716, de 5 de janeiro de 1989. Available at: www.planalto.gov.br/ccivil/Leis/L7716.htm.

8. Cartilha Cidadania para Todos. Conselho Estadual de Defesa dos Direitos do Homem e do Cidadão - Paraíba. Available at: www.dhnet.org.br/w3/ceddhc/bdados/cartilha14.htm.

9. Ministério da Previdência Social. Auxílio-doença. Available at: www.mpas.gov.br/conteudoDinamico.php?id=21.

10. Brasil. Constituição da República Federativa do Brasil de 1988, Título VIII, Capítulo II, Seção III, artigo 201. Edição Comemorativa 20 anos. São Paulo: AASP, 2008.

11. Brasil. Ministério da Previdência Social. Aposentadoria por invalidez. Available at: www.mpas.gov.br/conteudoDinamico.php?id=18.

12. Bobbio N. A era dos direitos. Tradução de Carlos Nelson Coutinho. Rio de Janeiro: Campus, 1992.

# MULTIDISCIPLINARY TEAM CARE FOR PATIENTS WITH INFLAMMATORY BOWEL DISEASE

CYRLA ZALTMAN

Inflammatory bowel diseases (IBD) are chronic and progressive diseases that typically affect economically active young people. They promote intestinal organ damage with clinical manifestations that affect the professional and emotional life of patients and their families (employment, studies, relationships, etc.). Long-term monitoring and the risk of complications justify the request for the numerous diagnostic procedures, and medical or surgical treatments that will take place during the life of these patients. Such complex scenario tends to be better managed by a multidisciplinary team, enabling a more realistic view of diagnosis, as well as a targeted therapeutic strategy, based on the availability of experienced professionals, equipment and organizational structure.

In addition to the fact that there are few data on the prevalence and incidence of IBD in the country,[1] these are regional figures, which do not represent the totality of the Brazilian territory. However, a gradual increase in the care of patients with IBD has been observed in different regions.[2] In populational studies performed in the European, Asian and American continents, the increased incidence and prevalence of IBD has been partly related to the improvement of diagnostic techniques.[3,4]

Although different laboratory and imaging techniques have been developed and new drugs have emerged favoring the prognosis of patients with IBD, these are not yet ideal for the diagnosis and monitoring of patients.

Crohn's disease (CD) and ulcerative colitis (UC) are considered the main components of the group of diseases known as IBD. Knowingly, they are immune disorders that are expressed through gastrointestinal and extra-intestinal manifestations. Depending on the initial clinical presentation of IBD (mild or nonspecific forms, intermittent symptoms with periods of calmness, predominance of extra-intestinal manifestations), the diagnosis can be delayed for months or even years, or even not be performed if there is no clinical suspicion.

Another aggravating factor in this context is the lack of communication among professionals from different specialties who treat patients with IBD, which hampers the optimization of clinical investigations and treatment decisions.

When making a critical analysis of various scenarios for the care of patients with IBD, it appears that patients with symptoms of moderate to high intensity tend to seek medical assistance in emergency services from public and private hospitals. In this context, after complications related to the disease (perforation, abscesses, subocclusions) or signs that mimic other conditions (ileitis mimicking appendicitis) are detected, usually these patients are hospitalized and often undergo surgery, circumstances that facilitate the diagnosis. However, in a scenario in which the patient's symptoms are less severe or even in the absence of warning signs, the trend is to start symptomatic treatment or antibiotics, if the presumptive diagnosis is gastroenteritis, delaying the final diagnosis of IBD. Research on this topic is still limited, even in countries with structured public and private health systems.[5]

A similar situation can be seen in other healthcare segments, as in primary care or consultations with general practitioners, little accustomed to this diagnosis.

In the Brazilian clinical practice, a shorter interval between the appearance of the first symptoms and the diagnosis of the disease has been observed with the introduction of new, less invasive diagnostic methods; nevertheless, there is still a delay to establish this diagnosis. The Brazilian public healthcare system (SUS, Unified Health System) for reference and counter-reference allows access to specialists, but still cannot be compared to that reported by European countries, where 70% of patients are seen by a specialist within 1 year after diagnosis, while 18% remain undiagnosed for a period of five years.[6]

In the treatment of IBD, it is not uncommon to use different therapies, combined or not, which can lead to multiple and various adverse effects, increased susceptibility to bacterial and viral infections and occurrence of neoplasia. Early detection of diseases related to the use of these drugs and the adoption of preventive measures such as vaccinations, dermatological and gynecological examinations, and investigation of latent tuberculosis, justify the involvement of different specialties and multiple healthcare professionals in each of these steps.

It is important to point out that there are differences with respect to the needs of IBD patients with established diagnosis and the objectives to be achieved by the doctors involved in their treatment. Doctors tend to have long term goals, such as: avoiding surgical procedures, inducing disease remission with acceptable adverse effects, promoting change in the course of disease, avoiding use of steroids and their side effects and inducing healing of the lesions of the mucosa. However, patients want short-term solutions and longer doctor visits to discuss different situations related to IBD, aspects that differ markedly from those sought by doctors.[7] The goals of patients include: reducing symptoms and side effects of medications, solving doubts related to cosmetic issues (ostomy, fistula, surgical scars), fertility, sexuality, and metabolic and bone disorders, and fatigue, and the need to share their anxieties during the consultations.[7,8] All these aspects should be discussed clearly and preferably approached by members of a multidisciplinary team familiar with the situations presented, increasing the patient's trust in the professionals and also adherence to treatment.

A multidisciplinary team for the care of patients with IBD is essential for greater accuracy and speed in identifying IBD, early referrals to gastroenterologists or proctologists with experience in IBD and the possibility of proper guidance to professionals from other health sectors (primary care, emergencies).

The focus of the team should be the patients and their needs, and the different professionals involved should be experienced and have well-defined roles within the group. In addition to the specific care of each professional, continuing education of patients and family members should also be a concern of the team, so that doubts about the disease and its treatment can be clarified, providing increased treatment adherence and reducing the number of unscheduled *in loco* (in units) medical visits, as well as episodes of disease relapse.[9]

The manner how each service should be organized for IBD care depends on local conditions (academic institution, referral center, public or private health unit), their degree of complexity, and the existence of interested and experienced professionals. Assembling multidisciplinary teams is more feasible in teaching hospitals or larger hospitals (tertiary or quaternary care), although this can also be achieved with professionals from different types of units (primary and secondary care) trained to perform the presumptive diagnosis of IBD and for appropriate referral to specialized hospitals when needed.

However, if the professionals working in general care units are guided on what is important in the care of patients with IBD, this process can be faster and more efficient, less fragmented, with significant decentralization so that simpler cases can continue treatment in unreferenced units.

The multidisciplinary team should be composed of experienced members with diverse, but complementary, skills and knowledge, who contribute to achieving the goals set by these professionals. The team should include the following medical specialties: gastroenterology (adults and children), coloproctology, pathology, radiology and digestive endoscopy with specific interest in IBD, in addition to social workers and nurses dedicated to patient education. A support medical team is also needed, consisting of psychologists/psychiatrists, dermatologists, ophthalmologists, rheumatologists, pulmonologists, gynecologists, and nutrologists or nutritionists involved with enteral/parenteral nutrition.

This joint effort allows more accurate diagnoses, achieving better outcomes of the disease after the introduction of appropriate therapy, reducing the number of treatments that are shorter or longer than necessary, optimizing the monitoring process (reduction in the number of tests requested, hospitalizations, emergency consultations, and surgery), and consequent cost reduction. At the same time, equally important issues should be considered in relation to the patient, including their perception about the risks and benefits of each diagnostic step, monitoring and medication, their health status, the most effective type of dialogue according to the patient's age, costs, socioeconomic, psychological (fear, shame, low self-esteem) and cultural status.[9]

Today, there are few multidisciplinary groups in the care of patients with IBD, and the time that the doctor uses to guide the patient or check whether he or she can comply with the steps required for treatment (examinations, use of prescribed medications) have proved short and insufficient.

Although usually there are no nurses specialized in IBDs, and social workers who work together in this scenario, the non-medical professionals in these two areas play a fundamental role in supporting the patient.

The role of a social worker in this team is dealing with the psychosocial effects of the disease, assisting in matters related to the acquisition of medications and pouch for ostomates, free access to care in public hospitals, documentation for sick pay, etc. The impact of their work is still not evident, but, when present, causes a visible improvement in the quality of life of patients.

As for dedicated nursing, although the function already exists and is well-organized (private office, dedicated telephone) in countries like Spain and England, these professionals practically do not exist in Brazil. Their interventions are essential to: provide technical and emotional support; counsel patients and their families; provide information; answer doubts about the disease, tests to be performed, medication and adverse effects; and assist in treatment adherence, encouraging a proactive attitude of the patient for disease control. This form of communication has a motivational aspect, unlike conventional communication, aiming at improving dialogue and collaboration between health professionals and patients.[10] Despite the importance of these professionals, there is insufficient scientific evidence to support the positive impact of their work on the care structure and the outcome of patients with IBD.[11,12]

The conclusion is that the existence of a multidisciplinary team treating patients with IBD allows a holistic, integrated, dynamic, flexible and fast approach, combining the goals of the healthcare team and those of the patients.[13,14]

## REFERENCES

1. Victoria CR, Sassak LY, Nunes HR. Incidence and prevalence rates of inflammatory bowel diseases, in Midwestern of São Paulo State, Brazil. Arq Gastroenterol 2009; 46(1):20-5.

2. Zaltman C. Inflammatory bowel disease: how relevant for Brazil? Cad Saúde Pública 2007; 23(5):992-3.

3. Berenstein C, Waida A, Swenson MS, MacKenzie M, Koehoorn M, Jackson M et al. The epidemiology of inflammatory bowel disease in Canada: a population-base study. Am J Gastroenterol 2006; 101:1559-68.

4. Vind I, Riis L, Jess T, Knudsen E, Pedersen N, Elkjaer M et al. Increasing incidences of inflammatory bowel disease and decreasing surgery rates in Copenhagen City and Country, 2003-2005: a population-based study from the Danish Crohn Colitis database. Am J Gastroenterol 2006; 104:1274-82.

5. Ananthakrishnan AN, McGinley EL, Saeian K, Binion DG. Trends in ambulatory and emergency room visits for inflammatory bowel diseases in the United States: 1994--2005. Am J Gastroenterol 2010; 105(2):363-70.

6. Wilson B, Graco M, Hommes DW, Vermeire S, Bell C, Avedano LA. European Crohns colitis patient life IMPACT survey. Abstract PO875, 19th United European Gastroenterologoly Week, Stockolm, Sweden, October 22-26, 2011.

7. IMPACT 2010-11 Crohn's and ulcerative colitis patient life impact survey. First full results: November 2011. Disponível em: www.efcca-solutions.net/impaxt/european.php. Available at: 24/01/2012.

8. Cravo M, Guerreiro CS, dos Santos PM, Brito M, Ferreira P, Fidalgo C et al. Risk factors for metabolic bone disease in Crohn's disease patients. Inflamm Bowel Dis 2010; 16:2117-24.

9. Ghosh S, D'Haens G, Feagan B, Silverberg MS, Szigethy EM. What do changes in inflammatory bowel disease management mean for our patients? J Crohns Colitis 2012; 652:S243-S249.

10. Miller WR, Rollnick S. Motivational interviewing: preparing people to change addictive behavior. New York: Guilford Press, 1991.

11. Younge L, Norton C. Contribution of specialist nurses in managing patients with IBD. Br J Nurs 2007; 16(4):208-12.

12. Hernández-Sampelayo P, Seoane M, Oltra L, Marín L, Torrejón A, Vera MI et al. Contribution of nurses to the quality of care in management of inflammatory bowel disease: a synthesis of the evidence. J Crohns Colitis 2010; 4(6):611-22.

13. Fontanet G, Casellas F, Malagelada JR. La Unidad de Atención Crohn-Colitis: 3 años de actividad. Gac Sanit 2004; 18(6):483-5.

14. Torrejón A, Masachs M, Borruel N, Castells I, Castillejo N, Malagelada JR et al. Aplicación de un modelo de asistencia continuada en la enfermedad inflamatoria intestinal: la Unidad de Atención Crohn-Colitis. Gastroenterol Hepatol 2009; 32(2):77-82.

# FREQUENTLY ASKED QUESTIONS AND ANSWERS IN CLINICAL PRACTICE

MARCO ANTÔNIO ZERÔNCIO

## 1. WHAT IS INFLAMMATORY BOWEL DISEASE?

Despite a general connotation, the term inflammatory bowel disease (IBD) refers basically to two chronic, idiopathic inflammatory diseases primarily affecting the gut: Crohn's disease (CD) and nonspecific ulcerative colitis (UC). However, other intestinal diseases of chronic nature and unknown etiology are mentioned occasionally as IBD subtypes by some authors, such as collagen colitis and lymphocytic colitis.

## 2. WHAT CAUSES INFLAMMATORY BOWEL DISEASE?

Although the pathogenesis of IBD is not fully understood, it is believed that genetic mutations related to different molecular sites that control the intestinal immune system are responsible for triggering an exaggerated inflammatory response against intraluminal antigens, including the microbiota itself. The resulting inflammation in these cases becomes chronic and relapsing, leading to tissue damage and all of its clinical consequences.[1]

### 3. WHAT ARE THE MAIN DIFFERENCES BETWEEN CD AND UC?

In UC, the inflammatory process is restricted to the mucosa and occurs continuously, always involving the rectum and a variable extent of colonic segments, sometimes reaching the cecum. In CD, the inflammatory process can affect both the mucosa and the muscularis propria/serosa layer (transmural inflammation), being more typically discontinuous and occurring in any part of the digestive tract (from mouth to anus). Because of the transmural involvement in CD, many patients can develop complications such as strictures, fistulas and abscesses. In some cases of colonic involvement, the differential diagnosis between CD and UC can be difficult, especially when there are overlapping clinical and endoscopic features between these two diseases (unclassified colitis).[2]

### 4. WHAT IS THE MOST COMMON CLINICAL PRESENTATION OF IBD?

The signs and symptoms are usually periodic, oscillating between acute and remission phases. Chronic diarrhea is the most frequent complaint. Hematochezia, tenesmus and defecation urgency can occur in both, but are most often observed in UC. Other findings include abdominal pain and weight loss. Patients with CD in the upper digestive tract can have dyspeptic signs and symptoms (gastroduodenal Crohn's disease), dysphagia and odynophagia (esophageal Crohn's disease). Fever can occur in severe acute phases, even without infection. At least half of patients have complications outside of the gastrointestinal tract (extra-intestinal manifestations), such as arthritis, pyoderma gangrenosum, erythema nodosum, uveitis, osteoporosis, primary sclerosing cholangitis and ankylosing spondylitis.[2]

### 5. ARE THERE ANY CLINICAL, LABORATORY, ENDOSCOPIC, RADIOLOGICAL OR HISTOPATHOLOGICAL SIGNS WHICH ARE PATHOGNOMONIC OF IBD?

No. The diagnosis of IBD is always based on a combination of these findings. Therefore, careful clinical judgment is required before labeling a person as CD or UC patient, especially in less typical cases of IBD.[2]

## 6. SHOULD ANTI-*SACCHAROMYCES CEREVISIAE* ANTIBODIES (ASCA) AND PERINUCLEAR ANTI-NEUTROPHIL CYTOPLASMIC ANTIBODIES (PANCA) ALWAYS BE INVESTIGATED IN CASES OF SUSPECTED OR CONFIRMED IBD?

No. ASCA and pANCA should be requested only in cases of unclassified colitis. The ASCA (+) and pANCA (-) pattern is more characteristic of CD, while the ASCA (-) and pANCA (+) pattern is more characteristic of UC. Note that some patients with unclassified colitis can have negative or positive results for both antibodies.[3]

## 7. HOW TO DIFFERENTIATE CD AND UC BASED ON COLONOSCOPY?

CD is typically discontinuous (with areas of healthy mucosa interspersing areas of inflammation), manifesting as deeper, linear and confluent aphthous ulcers or ulcerative lesions, longitudinal or transverse, with the rectum being spared in many cases. The finding of ulcers in the terminal ileum greatly assists the diagnosis of CD, and should never be mistaken for backwash ileitis, found in some cases of pancolitis in the context of severe UC.

In UC, the rectum is always affected and inflammation may extend proximally and continuously, in a symmetric and circumferential manner along the colon, sometimes reaching the cecum. Some patients with distal UC (proctitis/proctosigmoiditis) exhibit isolated focal inflammation in the cecum. Importantly, drug treatment in UC can result in a pseudosegmental macroscopic appearance of colonic lesions, which should not be mistaken for CD. Ulcers in UC are more superficial (microulceration). Diffuse loss of vascular pattern, friability and granular appearance of the mucosa constitute other findings more suggestive of UC. In long lasting UC, there may also be changes in the mucosal relief pattern, with loss of haustra in the affected segments (tubular colon).[2]

## 8. WHY IS HISTOPATHOLOGY OFTEN INCONCLUSIVE FOR IBD OR UNABLE TO DIFFERENTIATE UC FROM CD?

In clinical practice, most of the histopathological findings simply refer to the analysis of fragments of mucosa collected during endoscopy. In this context, histopathology can only show chronic signs of mucosal inflammation that can be present in both forms of IBD, such as increased cellularity in the lamina propria, glandular architectural distortion, basal lymphoplasmacytosis, and Paneth cell metaplasia. Some findings are found more often in UC, such as atrophy and crypt microabscesses. Granuloma epithelioid (noncaseating), rarely found in endoscopic biopsy specimens, is highly suggestive of CD, but can also be seen in infectious (fungi, Chlamydia, *Yersinia*) and parasitic (Schistosomiasis) colitis, sarcoidosis, foreign body reaction, and in colitis following resection of a segment.

Note that the histopathological reports of chronic nonspecific colitis lack pathological data as described above, except for a mild inflammatory infiltrate, commonly found in healthy colonic mucosa. In these patients, it is very common to see macroscopically normal mucosa on colonoscopy or only minimal changes such as edema and hyperemia. The description of chronic nonspecific colitis, therefore, should not mislead to a diagnosis of IBD in the absence of other relevant data.

Conversely, in analyzes of surgical specimens, histopathology is much richer in details, making it possible to describe transmural involvement, if present, and also, in the case of CD, to find granulomas more easily.

Despite its lack of specificity, histopathology is essential for diagnostic investigation in order to rule out diseases that mimic IBD clinically and endoscopically, including tuberculosis, lymphogranuloma, Schistosomiasis and some cancers.[2]

## 9. WHAT ARE THE METHODS FOR EVALUATING THE SMALL INTESTINE IN THE CONTEXT OF IBD?

Additional investigation of the small intestine is considered highly important to determine the extent of IBD and for the differential diagnosis of CD and UC. Some cases of rectum and colon CD can mimic UC, and the finding of concurrent small bowel lesions can be critical for a correct diagnosis of CD. The tests most widely used, depending on the availability of each service, are intestinal transit time, capsule endoscopy, enteroscopy and enterography. The intestinal transit time test, still much in demand in several centers, has low cost and risk, is reliable, but also has low sensitivity compared to all other methods. There is a growing tendency to replace the intestinal transit time test with computed tomography enterography or magnetic resonance enterography. Enterography has high sensitivity for detecting small intestinal lesions, including those of the intestinal wall and adjacent structures. Capsule endoscopy and enteroscopy (single- or double-balloon) have high sensitivity for diagnosing mucosal lesions, but these tests are still expensive and unavailable in many centers in Brazil. Capsule endoscopy is contraindicated in cases of strictures and does not allow biopsies to be performed. Enteroscopy allows biopsies, but its greater complexity confers a higher risk for the patient compared to capsule endoscopy.[4]

## 10. SHOULD UPPER GI ENDOSCOPY (EGD) BE REQUESTED FOR ALL PATIENTS WITH CD?

No. Upper digestive tract CD is uncommon (< 5% of cases); therefore, EGD should only be performed in patients presenting dyspepsia, dysphagia or odynophagia.[4]

## 11. HOW IS THE DEGREE OF DISEASE ACTIVITY DETERMINED, AND WHY IS IT IMPORTANT?

There are many methods to determine the degree of disease activity, both in UC and in CD, and most of them are used for scientific research. For rapid assessment in clinical practice, however, mild activity generally means a patient with less than 4 bowel movements a day (bloody or not), with no signs of systemic toxicity, with normal laboratory test results for inflammatory activity (erythrocyte sedimentation rate, C-reactive protein), and no weight loss. In moderate activity, the patient passes between 4 and 6 diarrheal stools a day with minimal signs of toxicity or weight loss. In severe activity, the patient has more than 6 diarrheal bowel movements a day, evident weight loss and systemic signs of toxicity manifested as fever, tachycardia, anemia, peripheral edema, and changes in laboratory tests for inflammatory activity. A patient with fulminant disease has more than 10 bowel movements daily, continuous bleeding, toxicity, abdominal distension and defense, colonic dilation on plain radiographs, and need for blood transfusion. This distinction is fundamental for a rational choice of therapy. Patients with mild forms of IBD may not need corticosteroids. Aminosalicylates may be initially used for mild to moderate UC, but they do not usually induce remission in severe disease, which may require hospitalization and intravenous corticosteroid therapy or biological therapy. In fulminant disease, a joint assessment with a surgeon is essential.[5]

## 12. HOW TO USE AMINOSALICYLATES IN UC?

Aminosalicylates (AS) (sulfasalazine [SSZ] and mesalazine [MSZ]) remain the primary therapeutic option for patients with mild to moderate UC, both orally and rectally (topical form of suppositories and enemas).

For SSZ, remission induction doses range between 4 and 6 g/day, divided into 4 doses. For maintenance of remission, an attempt to reduce the daily dose to a minimum of 2 to 3 g/day is recommended. However, in some cases, the minimally effective dose in the remission phase is practically the same as that able to induce remission and should thus be maintained.

The recommended doses of oral MSZ both to induce and maintain remission are from 2.4 to 4.8 g/day, divided into 2 or 3 doses. A relatively common error in the management of UC is to consider that there was therapeutic failure with AS before the maximum doses have been used. MSZ suppositories may aid in the treatment of proctitis (within 10 to 15 cm of the dentate line), whereas enemas may help in the treatment of left-sided colitis (extending up to the splenic flexure). Topical MSZ is more effective in reducing the inflammation than oral MSZ, and can accelerate the induction of remission as the drug is delivered in high concentrations directly at the site where action is needed. However, the oral route is still preferred by most patients because of the convenience of administration. Combination therapy (oral and topical) is more efficient than oral therapy alone and can be attempted in patients with disease refractory to oral MSZ. Topical forms are also effective in maintaining remission in cases of proctitis and left-sided colitis. In some cases, depending on patient acceptability, enemas may be used in remission of left-sided UC up to once every 3 days, with good response.[5]

## 13. HOW EFFECTIVE ARE AMINOSALICYLATES IN CD?

Several literature reviews have shown that AS are minimally effective in the active stages of colonic CD, promoting limited clinical benefit. In addition, there is insufficient evidence to justify the use of AS for maintenance of remission in CD.[6,7]

## 14. WHAT DOSES OF ORAL CORTICOSTEROIDS SHOULD BE USED IN IBD?

The choice of oral corticosteroid is prednisone, both in UC and CD. Dosage to induce remission is 0.75 to 1 mg/kg/day. However, daily doses above 40 mg for adults usually result in small clinical benefit at the expense of a significant increase in side effects. The full dose should be continued for 7 to 14 days; then, tapering by 5 to 10 mg/week should be done until a daily dose of 20 mg. From then on, reduction is that of 2.5 to 5 mg/week, until cessation. Patients with refractory disease or dependent on oral corticosteroids should be promptly evaluated for change of therapy (immunosuppressants and/or biologic therapy), according to each case. Prolonged use of corticosteroids in the attempt to maintain remission represents a serious error in the management of IBD, with resulting adverse effects always more severe than those of other available therapies, including surgery.[5,7]

## 15. WHEN SHOULD IMMUNOSUPPRESSANTS BE USED IN UC?

Thiopurines (TP) [azathioprine (AZA)/6-mercaptopurine (6-MP)] are the preferred immunosuppressants for cases of UC refractory to treatment with AS. The dose is 2 to 3 mg/kg/day for AZA and 1 to 1.5 mg/kg/day for 6-MP. Patients should be closely monitored with clinical and laboratory tests, on account of the risk of allergic reaction, pancreatitis, nausea, leukopenia, hepatotoxicity, and more.[5]

## 16. WHEN SHOULD IMMUNOSUPPRESSANTS BE USED IN CD?

TPs are the immunosuppressants of choice to treat CD, especially for the maintenance of remission, minimizing the use of corticosteroids. There is a current trend of early use of TPs in CD, often at diagnosis, because these drugs, unlike AS, are able to prevent long-term complications such as strictures and fistulas. TPs also play an important role in the prevention of postoperative recurrence of CD. Methotrexate (MTX) is an option for patients that do not respond or are intolerant to TP, with lower rates of effectiveness. In such cases, liver function should receive special attention.[6,7]

## 17. HOW TO TREAT A PATIENT WITH SEVERE UC REFRACTORY TO ORAL CORTICOSTEROID THERAPY?

Patients with severe UC refractory to maximal doses of oral corticosteroids are generally hospitalized for intravenous corticosteroid administration. Hydrocortisone is the drug of choice (300 mg/day, in 3 doses). Failure to respond to intravenous corticosteroids after 72 hours demand change to one of the following: total colectomy, infliximab or intravenous cyclosporine. The choice will depend on the experience of the service and the medical assistant, the availability of drugs, the possibility of monitoring the serum level of the drug (cyclosporine) and the choice of both the patient and their families, after extensive explanation of the benefits and risks relating to each therapy. Infliximab may still be an option for patients with severe UC refractory to oral corticosteroids and with no urgent need for hospitalization, prior to intravenous corticosteroids.

## 18. IS THERE ANY EXCLUSION DIET IN THE ROUTINE OF PATIENTS WITH IBD?

No. To date, there is no consensus evidence in the literature that certain food causes or worsens IBD. Daily practice and some publications have identified variable food intolerance in patients with IBD, which could perhaps justify individualized dietary exclusions in specific situations. There is no indication for the prescription of standardized diets that restrict certain types of foods uniformly to all patients. Overly restrictive diets may even aggravate the malnutrition of these individuals. However, there is widespread agreement that the role of diet in IBD is much more closely related to supplementation of nutritional losses, targeted towards correcting the patient's weight, dysproteinemia, anemia, osteoporosis, and vitamin and trace element deficiencies.[8]

## 19. WHEN SHOULD BIOLOGICAL THERAPY BE USED IN CD?

The classic indication of biological therapy (infliximab, adalimumab, certolizumab) in CD is for the patient presenting moderate to severe activity, who is dependent on oral corticosteroids, and refractory to the use of immunosuppressants. Over the years, the use of biological agents has been shown to be effective in several other situations (e.g., progressive clinical instability unable to wait for the slow onset of action of immunosuppressants, in cases of perianal fistulizing disease, in the presence of several adverse prognostic factors in a single patient, and in some extra-intestinal manifestations).[6,7]

## 20. HOW TO PREPARE THE PATIENT CLINICALLY FOR USE OF BIOLOGICAL AGENTS?

Biological therapy, due to its anti-TNF effect, induces a degree of immunosuppression and can predispose the patient to the development of some infections, including tuberculosis. Some cases of fulminant hepatitis B were observed, as well as rare infections such as cryptococcosis, histoplasmosis and listeriosis. Active infections should be ruled out in patients candidate to anti-TNF therapy. Chest x-ray, PPD and serology for hepatitis B should be requested. Although there is no consensus in the international literature, in Brazil, it is common also to request serology for HIV and hepatitis C virus, although these conditions do not represent absolute contraindications to the use of biological agents. Patients with pulmonary symptoms, with changes on x-ray or PPD $\geq 5$ mm, should be referred to joint assessment with a specialist. Asymptomatic patients with normal radiographs, but with PPD $\geq 5$ mm, should treat latent tuberculosis for six months with isoniazid (wait at least one month to initiate biological therapy).[5-7]

## 21. WHAT BIOLOGICAL AGENT SHOULD BE USED INITIALLY IN CD?

The two biological agents currently available in Brazil are infliximab and adalimumab. Both have the same anti-TNF mechanism of action, and the same efficacy rates. Once the need to use biological therapy is indicated, the choice of one or the other is irrelevant.[6,7]

## 22. DOES COMBINED USE OF BIOLOGIC THERAPY AND IMMUNOSUPPRESSANTS INCREASE THE EFFECTIVENESS OF THE TREATMENT?

Yes. It has been proven that the combined use of a biologic drug and an immunosuppressant increases the rates of clinical remission and mucosal healing in patients with CD in the first 6 to 12 months of treatment. In such cases, efficacy is superior to treatment with either drug alone. Despite the association of two drugs presenting immunosuppressive effect, a favorable safety profile for adverse effects with the use of combination therapy in the first 12 months was observed. However, there are reports of increased risk for the development of lymphomas in patients who received combination therapy for a period exceeding 2 years, particularly in young male patients (hepatosplenic T-cell lymphoma). Although the relative risk is increased, the absolute risk is considered low. Adverse long-term effects, resulting from deeper immunosuppression, are not known completely. Given these observations, it is believed today that combination therapy could be more properly indicated for more severe cases of CD, or when there is an association of poor prognostic factors in one individual, preferably for less than 24 months. Note that 25 to 30% of CD patients have milder forms of the disease and probably will never need a more powerful therapy with greater risks.[9]

## 23. WHAT IS THE MOST IMPORTANT GOAL IN THE TREATMENT: CLINICAL IMPROVEMENT OR TISSUE HEALING?

For many decades, the main goal of treatment in IBD involved the clinical improvement of the patient. In face of the patient's subjective opinion of clinical stability, most doctors did not seek tissue healing. However, in recent years, several studies have shown that mucosal healing correlates with better results in long term therapy (fewer recurrences, lower rates of hospitalizations, less chances of surgery and improved quality of life). This is valid both for CD and UC, for slightly different reasons. It has been shown that mucosal healing in CD is the main factor related to maintenance of remission. When this goal (mucosal healing) is reached, there is a possibility of changing the course of the CD, preventing the development of complications such as fistulas and strictures. In UC, mucosal healing being maintained not only reduces the rate of complications of the disease, but also reduces the incidence of dysplasia and colorectal cancer. The change in attitude with an emphasis on mucosal healing can lead to more frequent diagnostic tests and the adoption of more effective therapies that can cause tissue remission; the benefits for the patient appear to compensate for the additional effort.[10]

## 24. CAN INFECTIONS, NON-STEROIDAL ANTI-INFLAMMATORY DRUGS (NSAIDS) AND STRESS INFLUENCE THE COURSE OF IBD?

While no external factor has been definitively correlated with reactivation of IBD, some factors are plausibly suspected and have been investigated in several studies. Intestinal and systemic infections reinforce the inflammatory response, which could contribute to a worsening of inflammation in IBD. NSAIDs cause a loss of mucosal integrity, increasing its permeability to intraluminal antigens, a phenomenon that could be correlated with increased hyperreactivity of the intestinal immune system, as observed in IBD. The physiological response to stress involves a series of reactions in the hypothalamic-hypophyseal-adrenal axis and autonomic nervous system, capable of modifying the inflammatory response and increasing intestinal permeability. Despite methodological flaws observed in the literature and a lot of ambiguity among studies, it can be said today that there is some scientific evidence on the role of stress and NSAIDs in the course of the disease (but not in its cause), increasing the chances of reactivation or worsening symptoms during flares. As for infections, current evidence is too weak to draw further conclusions.[11]

## REFERENCES

1. Sewell GW, Marks DJB, Segal AW. The immunopathogenesis in Crohn's disease: a three stage model. Curr Opin Immunol 2009; 21:506-13.

2. Nikolaus S, Schreiber S. Diagnostics of inflammatory bowel disease. Gastroenterology 2007; 133:1670-89.

3. Plevy S. Do serological markers and cytokines determine the indeterminate? J Clin Gastroenterol 2004; 38:S51-6.

4. American Society for Gastrointestinal Endoscopy. ASGE guideline: endoscopy in the diagnosis and treatment of inflammatory bowel disease. Gastrointestinal Endoscopy 2006; 63(4):558-65.

5. The Practice Parameters Committee of the American College of Gastroenterology. Ulcerative colitis: practice guidelines in adults. Am J Gastroenterol 2010; 105:501-23.

6. The Practice Parameters Committee of the American College of Gastroenterology. Crohn's disease in adults. AM J Gastroenterol 2009; 104(2):465-83.

**7.** European Crohn's and Colitis Organisation. The second European evidence-based Consensus on the diagnosis and management of Crohn's disease: current management. J Crohns Colitis 2010; 4:28-62.

**8.** Rajendran N, Kumarf D. The role of diet in the management of inflammatory bowel disease. World J Gastroenterol 2010; 16(12):1442-8.

**9.** Colombel JF, Sandborn WJ, Reinisch W, Mantzaris GJ, Kornbluth A, Rachmilewitz D et al. Infliximab, azathioprine or combination therapy for Crohn's disease (The SONIC trial). N Engl J Med 2010; 362(15):1383-95.

**10.** Baert F, Moortgart L, Van Assche G, Caenepeel P, Vergauwe P, De Vos M et al. Mucosal healing predicts sustained clinical remission in patients with early-stage Crohn's disease. Gastroenterology 2010; 138:463-8.

**11.** Singh S, Graff LA, Bernstein CN. Do NSAID's, antibiotics, infections, or stress trigger flares in IBD? Am J Gastroenterol 2009; 104(5):1298-313.

# TREATMENT AND DIAGNOSIS ALGORITHMS

## DIAGNOSIS ALGORITHMS IN INTERNAL MEDICINE

## MALABSORPTION SYNDROME (MAS)

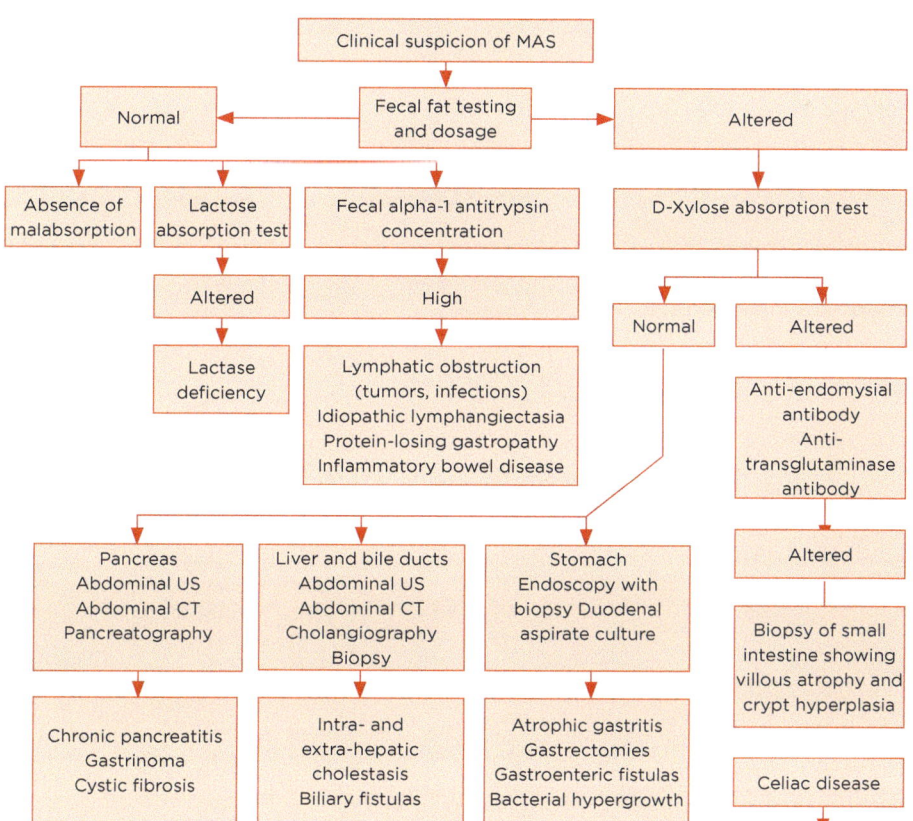

Estimation of fecal elastase levels is a good screening method for malabsorption syndrome. Therefore, low levels of fecal elastase are strongly suggestive of pancreatic failure (chronic pancreatitis, cystic fibrosis).
Source: Medicina diagnóstica. Maria Lucia G. Ferraz.

# INVESTIGATION OF CHRONIC DIARRHEAS IN ADULTS

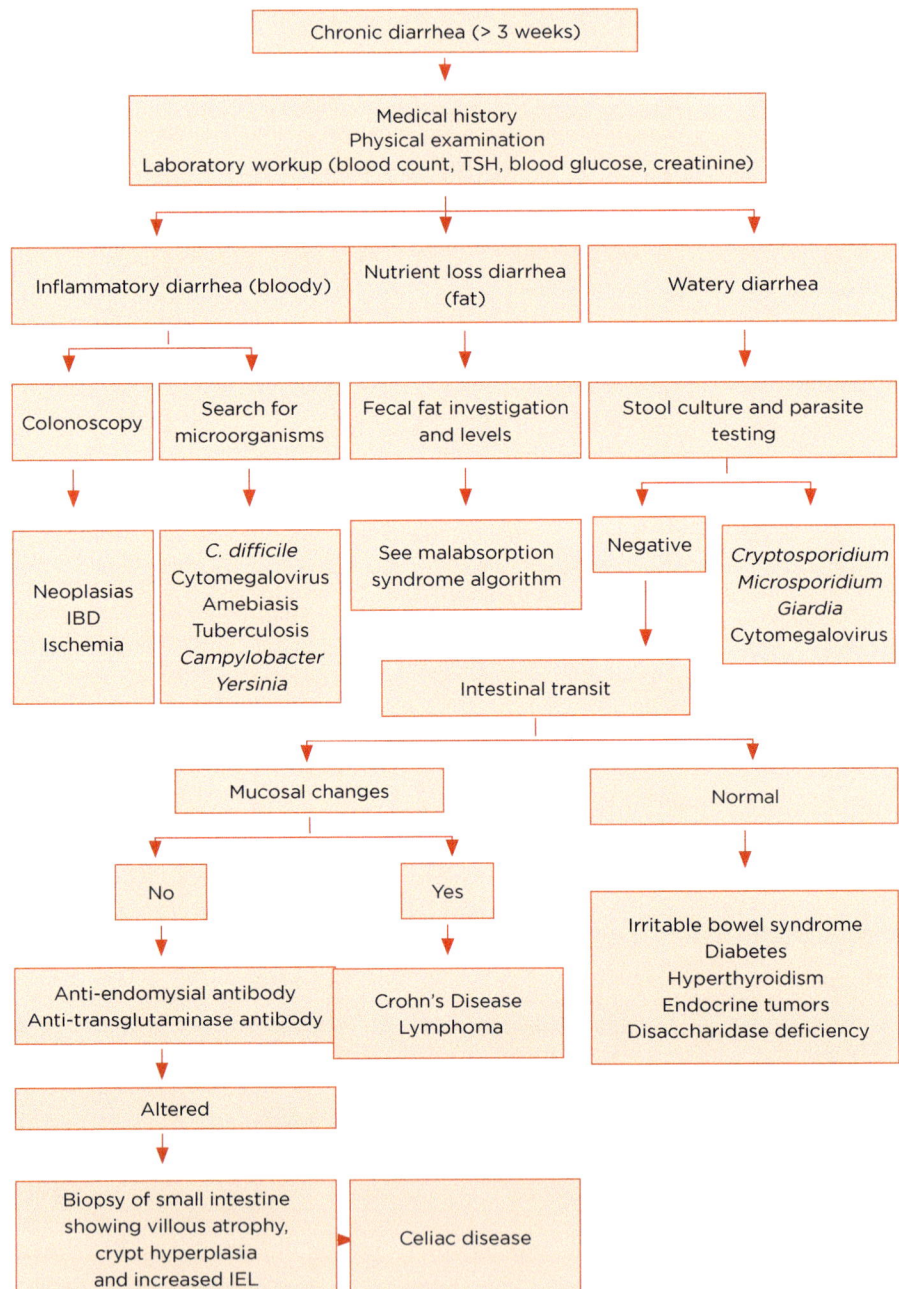

TSH: thyroid-stimulating hormone; IEL: intraepithelial lymphocytes.

# Investigation of chronic diarrheas in childhood

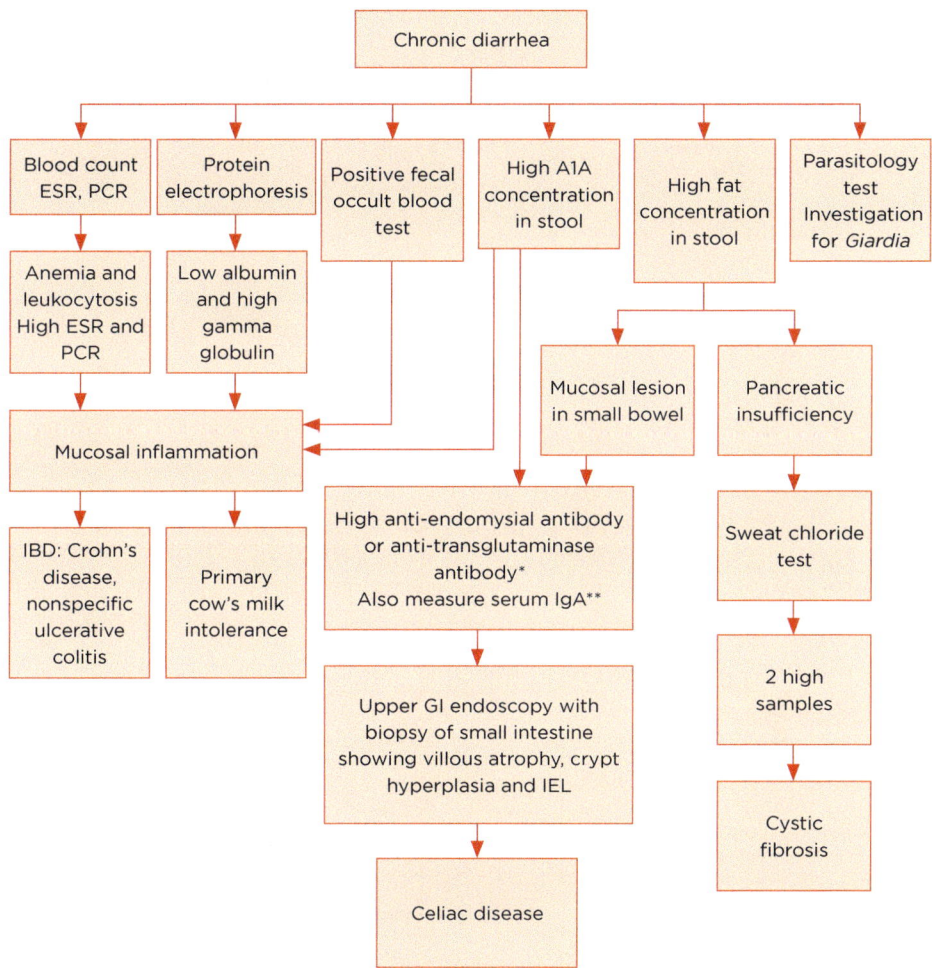

ESR: erythrocyte sedimentation rate; PCR: polymerase chain reaction; A1A: alpha-1-antitrypsin; IEL: increased intraepithelial lymphocytes.
*The anti-gliadin antibody has lower sensitivity and specificity than the others, and has not been recommended as a screening test.
**Since anti-endomysial and anti-transglutaminase are IgA antibodies, the patient must have normal serum IgA in order for the antibodies to be considered negative.
Estimation of fecal elastase levels is a good screening method for malabsorption syndrome. Therefore, low levels of fecal elastase are strongly suggestive of pancreatic failure (chronic pancreatitis, cystic fibrosis).

## Investigation of acute pain in upper abdomen

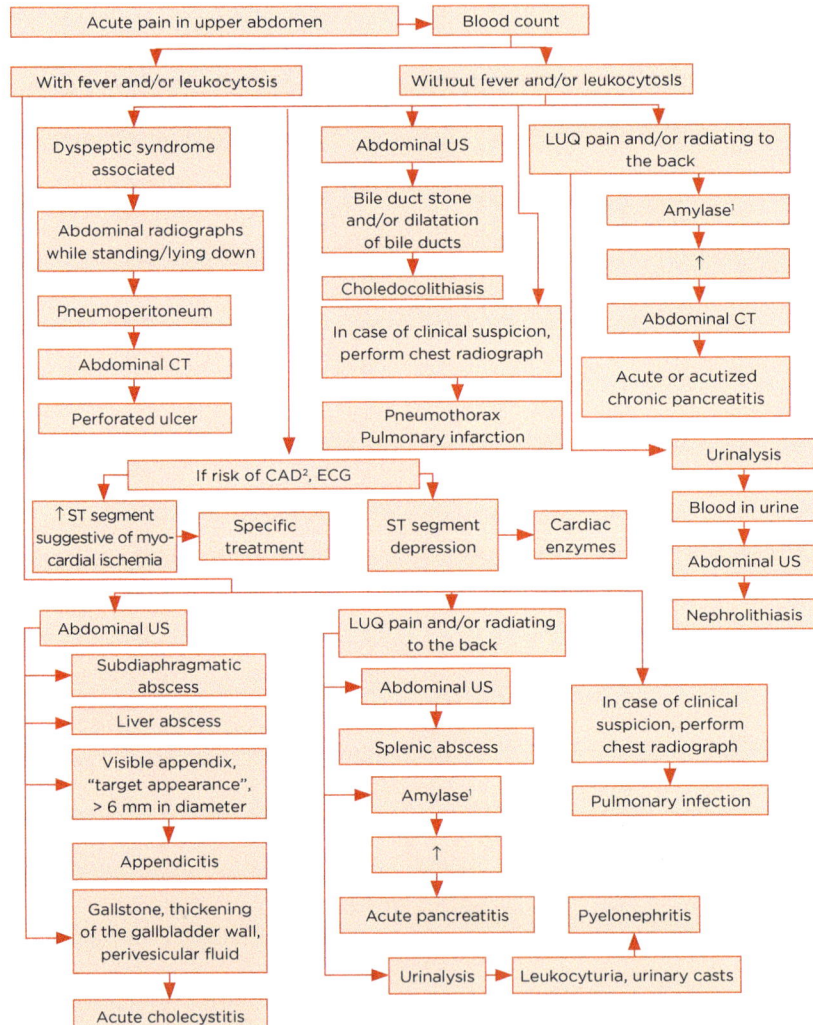

US: ultrasonography; CT: computed tomography; LUQ: left upper quadrant; ECG: electrocardiogram.
1. Causes of increased amylase: perforated peptic ulcer, intestinal obstruction, acute cholecystitis, cholangitis, acute renal failure, ruptured tubal pregnancy.
2. Risk factors for coronary artery disease (CAD): diabetes mellitus, arterial high blood pressure, dyslipidemia, smoking habit, family history, obesity.

# Investigation of acute mesogastric abdominal pain

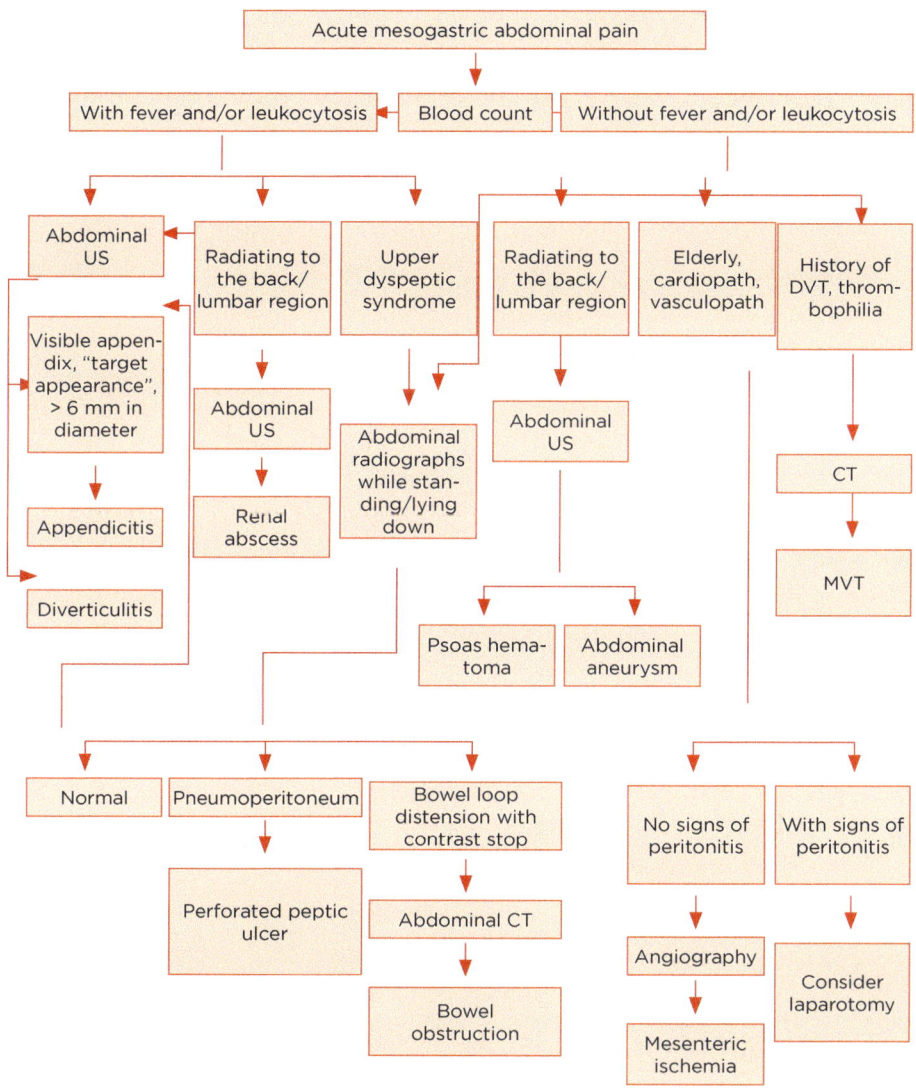

US: ultrasonography; CT: computed tomography; DVT: deep venous thrombosis; MVT: mesenteric venous thrombosis.

## Investigation of acute pain in lower abdomen

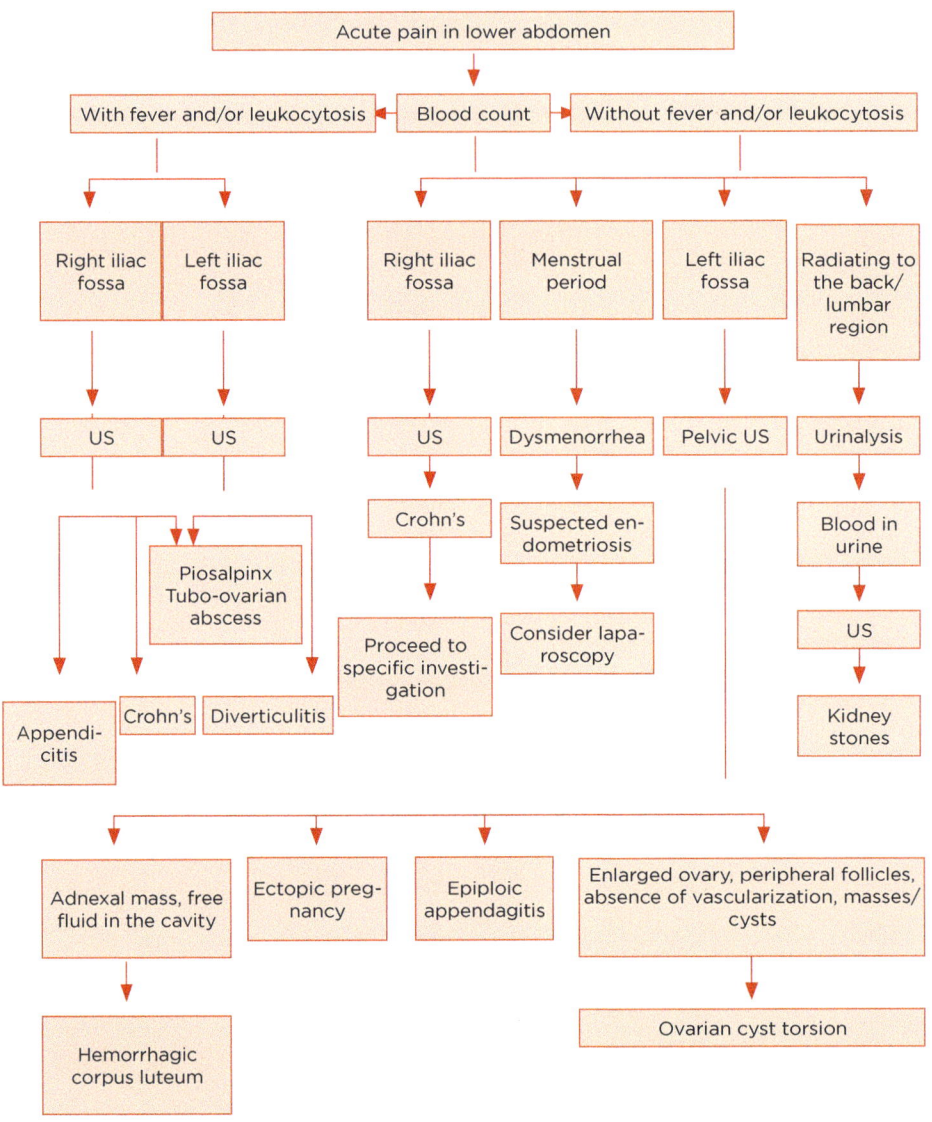

US: ultrasonography.

# Investigation of diffuse acute abdominal pain

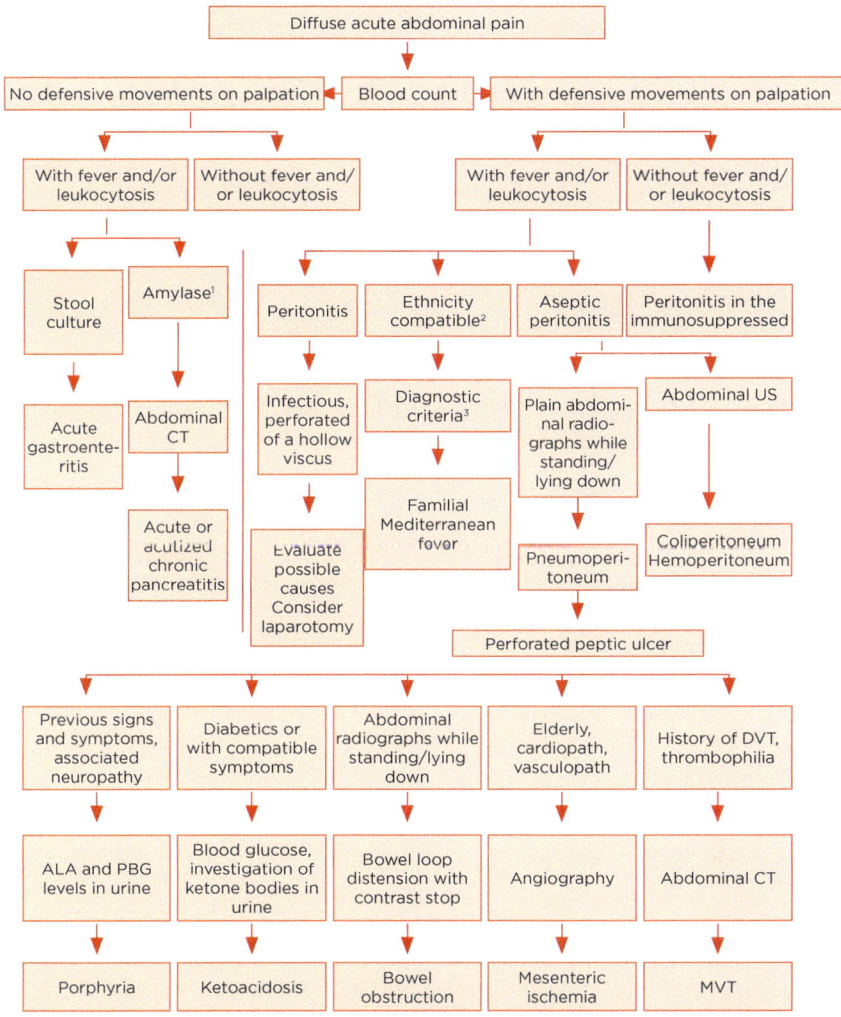

US: ultrasonography; DVT: deep venous thrombosis; MVT: mesenteric venous thrombosis; ALA: delta-aminolevulinic acid; PBG: porphobilinogen; CT: computed tomography.
1. Causes of increased amylase: perforated peptic ulcer, intestinal obstruction, acute cholecystitis, cholangitis, acute renal failure, ruptured tubal pregnancy.
2. Arab, Turkish, Sephardic Jews, Armenians.
3. Major criteria:
• typical attacks (≥3 of the same type, feverish and short – between 12 h and 3 days): generalized peritonitis, pleuritis (unilateral) or pericarditis, monoarthritis (hip, knee, ankle), fever alone;
• incomplete abdominal attacks (normal temperature, shorter or longer than typically, no peritonitis, pain is localized, affects other joints).
Minor criteria:
• incomplete attacks affecting one or more sites: thorax and joint;
• exertional leg pain;
• response to colchicine.
For a definite diagnosis of familial Mediterranean fever, 1 major criterion or 2 minor criteria are required.
Source: Livneh et al., 1997.

# Investigation of chronic abdominal pain

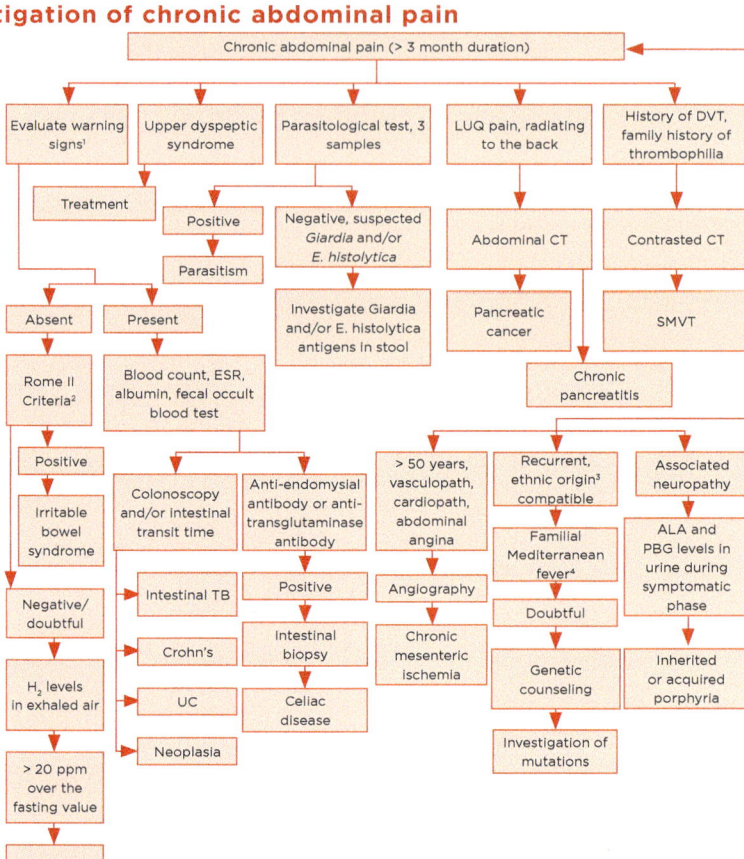

LUQ: left upper quadrant; DVT: deep venous thrombosis; CT: computed tomography; ESR: erythrocyte sedimentation rate; TB: tuberculosis; ALA: delta-aminolevulinic acid; PBG: porphobilinogen; UC: ulcerative colitis; SMVT: superior mesenteric-vein thrombosis.

1. Fever, weight loss, associated skin condition, arthritis, palpable masses.

2. At least 12 weeks (not necessarily consecutive) in the last year of discomfort or abdominal pain with two of the following symptoms:
• relief with bowel movement;
• onset of pain associated with changes in the number of daily bowel movements;
• onset associated with changes in the form and consistency of feces.

Symptoms that cumulatively aid in the diagnosis:
• abnormal bowel habits (> 3 times/day or < 3 times/week);
• abnormally shaped stool (hard or diarrhea);
• abnormal bowel movement (straining, urgency, feeling of incomplete evacuation);
• mucus in the stool;
• flatulence or abdominal distension.
Source: Olden, 2002.

3. Sephardic Jews, Arabs, Turks and Armenians.

4. Major criteria:
• typical attacks (≥ 3 of the same type, feverish and short – between 12 h and 3 days): generalized peritonitis, pleuritis (unilateral) or pericarditis, monoarthritis (hip, knee, ankle), fever alone;
• incomplete abdominal attacks (normal temperature, shorter or longer than typically, no peritonitis, pain is localized, affects other joints).

5. Minor criteria:
• incomplete attacks affecting one or more sites: thorax and joint;
• exertional leg pain;
• response to colchicine.
For a definite diagnosis of familial Mediterranean fever, 1 major criterion or 2 minor criteria are required.
Source: Livneh et al., 1997.

## Investigation of digestive bleedings

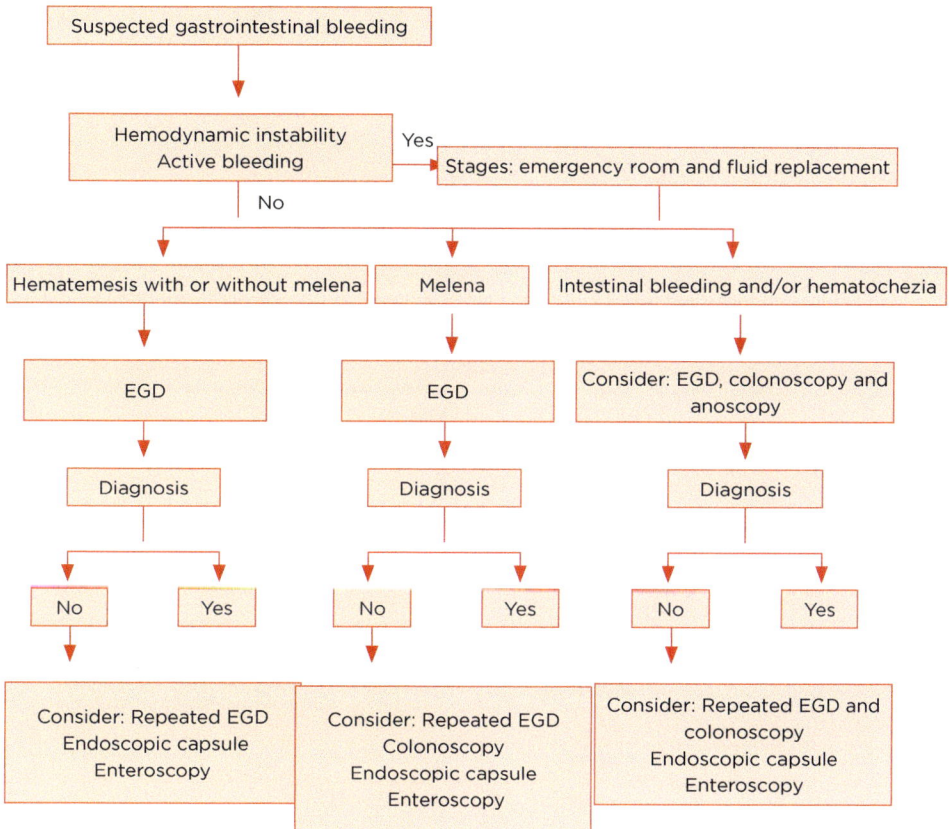

EGD: upper gastrointestinal endoscopy.

## Investigation of lower gastrointestinal bleeding (LGIB)

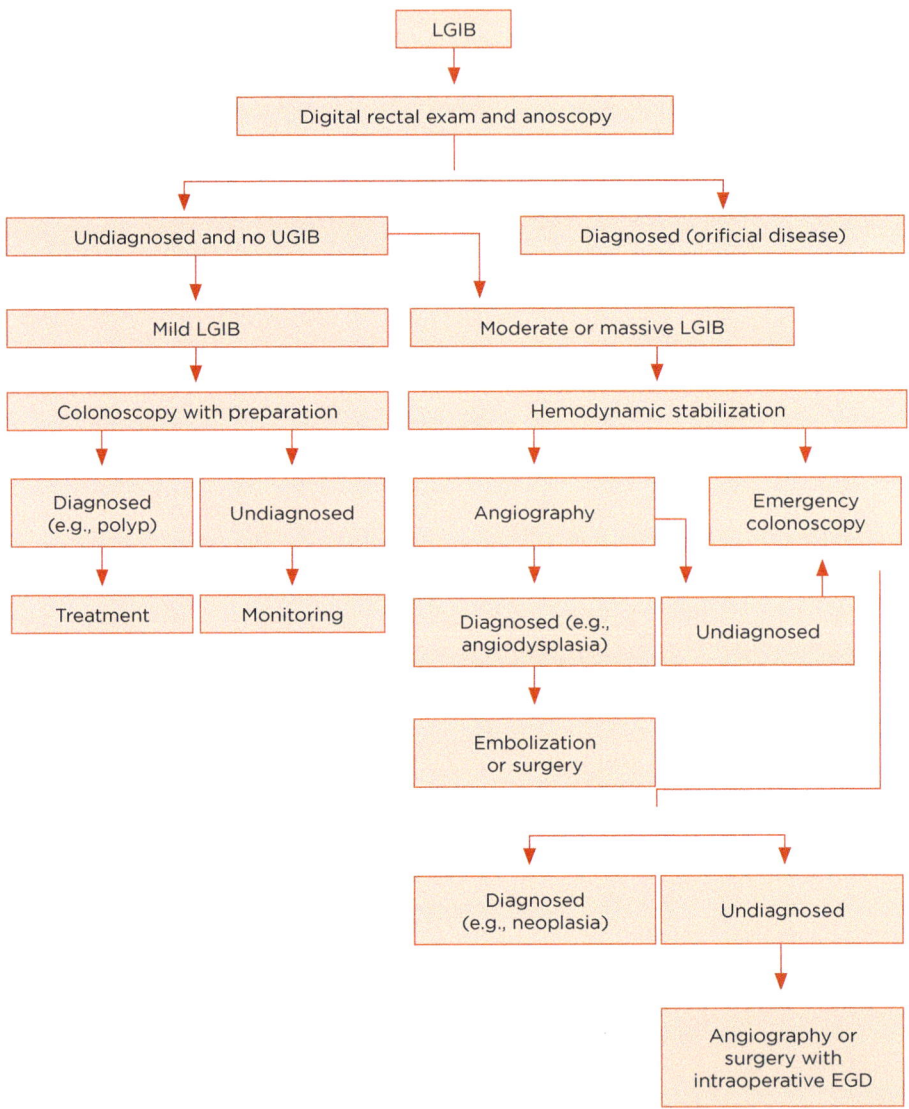

UGIB: upper gastrointestinal bleeding; EGD: upper gastrointestinal endoscopy.

# Investigation of anemia

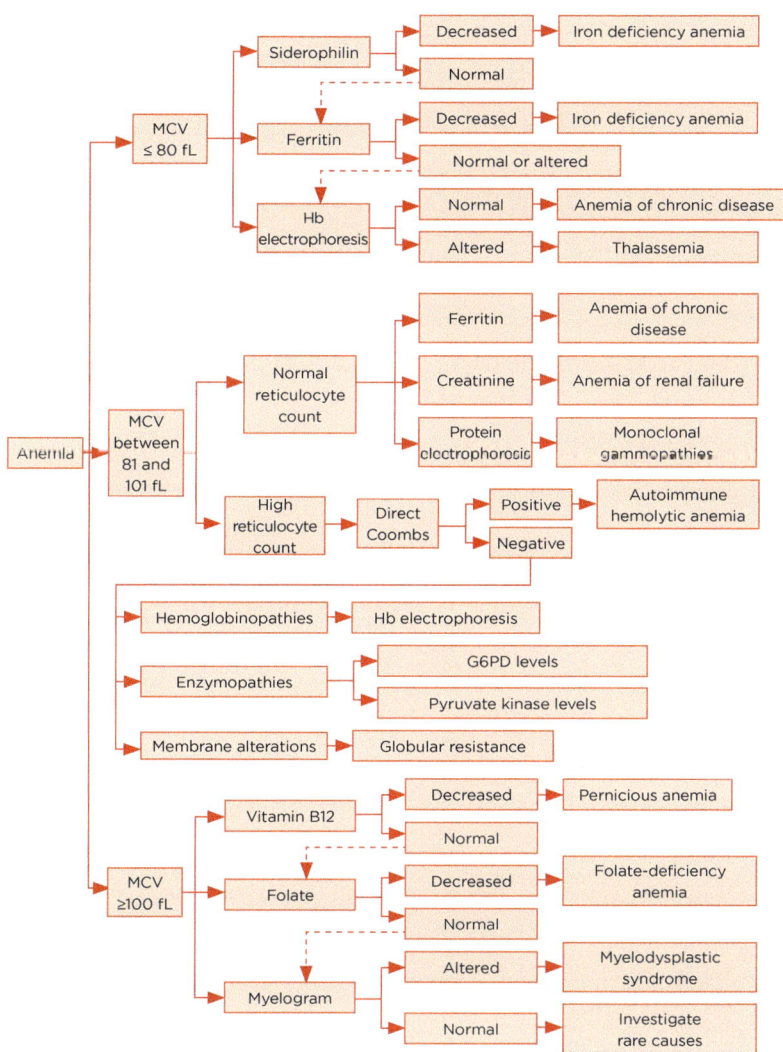

MCV: mean corpuscular volume; Hb: hemoglobin; G6PD: glucose 6-phosphate dehydrogenase.

## CROHN'S DISEASE TREATMENT ALGORITHM
### Induction of remission

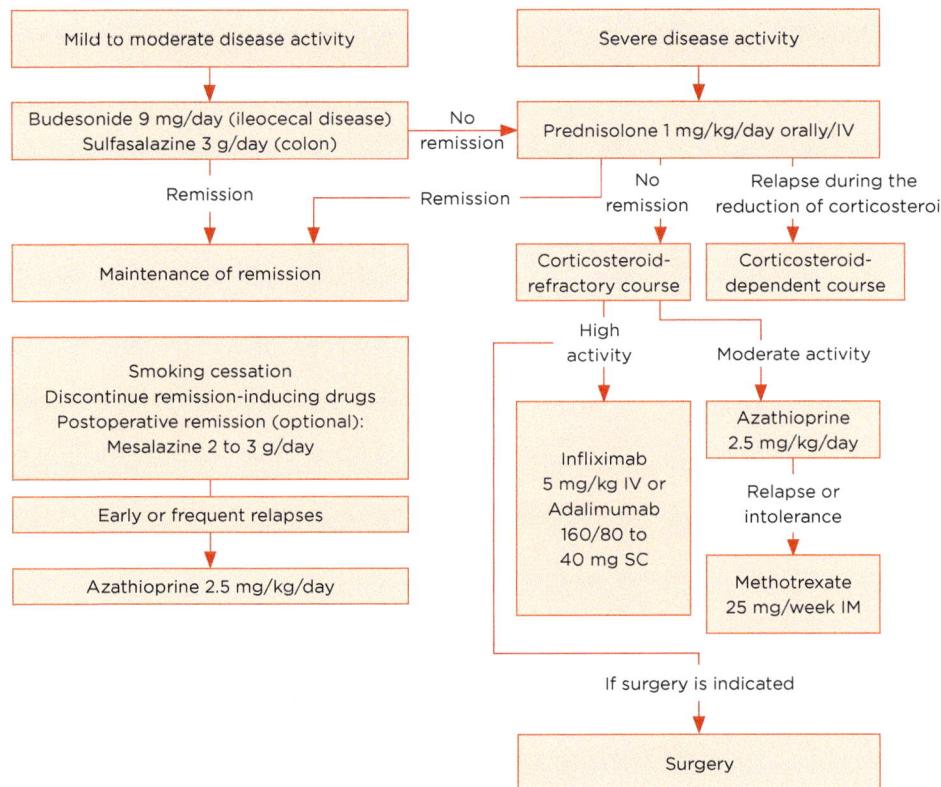

Source: Travis SPL, Stange EF, Lémann M, Oresland T, Chowers Y, Forbes A et al. For the European Crohn's and Colitis Organisations (ECCO). European evidence-based consensus on the diagnosis and management of Crohn's disease: Current management. Gut 2006; 55(Suppl.I):i16-i35.

# ULCERATIVE COLITIS TREATMENT ALGORITHM
## Induction of remission

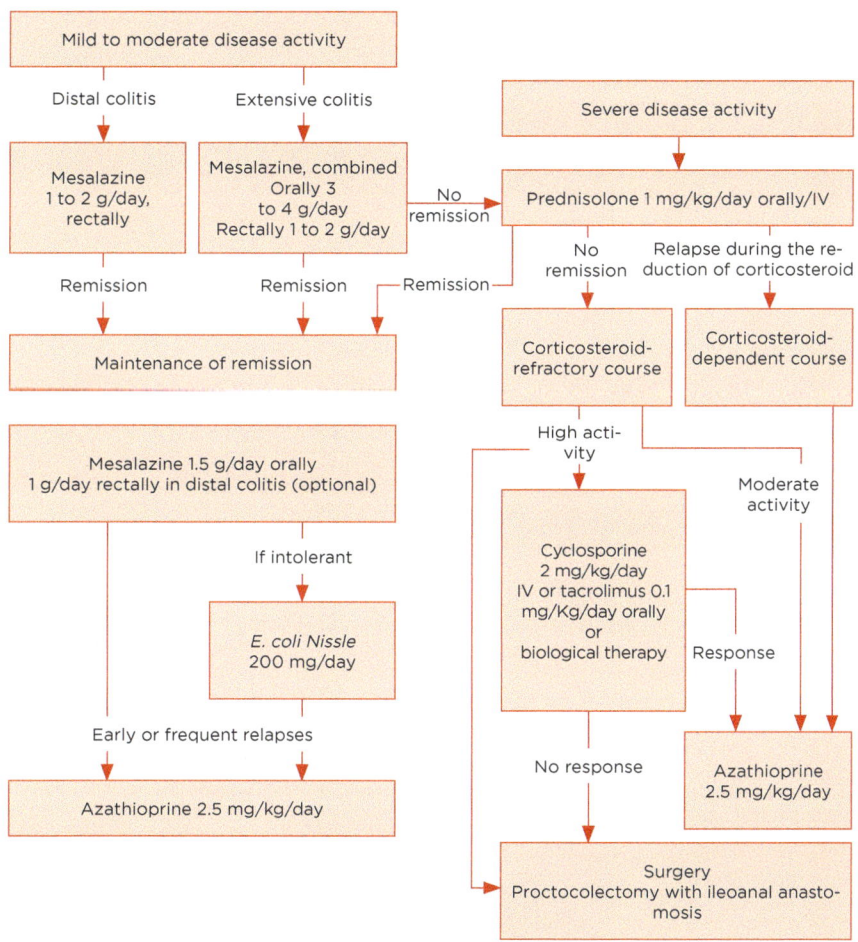

Source: Travis SPL, Stange EF, Lémann M, Oresland T, Bemelman WA, Chowers Y et al. For the European Crohn's and Colitis Organisations (ECCO). European evidence-based consensus on the diagnosis and management of ulcerative colitis: Current management. Journal of Crohn's and Colitis 2008; 02:24 AM-62.

LUCIANA DOS SANTOS
RAQUEL GUERRA DA SILVA
MAYDE SEADI TORRIANI
ELVINO BARROS

## ADALIMUMAB

**Pharmacological group:** immunosuppressant; recombinant human monoclonal antibody of immunoglobulin IgG1 humanized against TNF-alpha.

**Brand name:** Humira®.

**Presentation:** pre-prepared syringes with 40 mg of 0.8 mL.

**Uses:** moderate to severe Crohn's disease (CD) and ulcerative colitis (UC), in patients with loss of response or intolerance to conventional therapy, including corticosteroids, 6-mercaptopurine or azathioprine, or loss of response or intolerance to infliximab.

### Pharmacokinetic parameters:

- Slow absorption and distribution, with the peak concentration reached about 5 days after administration. Bioavailability of 64%.
- *Distribution*: volume of distribution (Vd) of approximately 4.7 to 6 L.
- *Metabolization*: no reports.
- *Excretion*: decreased in individuals over 40 years of age.
- *Elimination half-life*: 14 days.

**Posology:**

- *Pediatrics (DC, < 40 kg)*: 80 mg at week 0; later (at week 2), 40 mg every other week.[1] Over 40 kg, consider dosage for adolescents.
- *Adolescents and adults*: inductive dose of 160 mg (week 0, administering the 4 injections on the same day or two injections on 2 consecutive days), followed by a dose of 80 mg (week 2). Subsequently, the maintenance treatment is 40 mg (week 4).

**Dose adjustment:**

- *Liver function*: no reports.
- *Kidney function*: no reports.

**Route of administration:**

- *Subcutaneous*: preferred route of administration. Rotate the injection sites; prefer the abdomen and anterior thigh.

**Drug interactions:**

- *Natalizumab, live virus vaccines*: may have their plasma levels increased.
- *Inactivated vaccines*: their levels may be reduced.
- *Methotrexate*: may reduce the clearance of adalimumab.
- *Abatacept*: concomitant use with adalimumab can increase the risk of infection.
- *Tacrolimus, sirolimus, cyclosporine and pimozide*: there may be a reduction in the plasma levels of these medications.

**Conservation and preparation:**

- Keep refrigerated (2 to 8 ° C). The medication should not be frozen. Unused portions of the medication should be disposed of, as the product does not contain preservative. Protect from light.

**Pregnancy:** risk factor B.

**Breastfeeding:** not recommended.

**Adverse effects:** headache, rash, reactions at the injection site, upper hypercholesterolemia, nausea, abdominal pain, urinary tract infection, respiratory infections, sinusitis, flu-like symptoms, hypertension, increased alkaline phosphatase, hematuria, arrhythmia, confusion, fever, cellulitis, pancytopenia.

### Remarks:

- Defenses against malignancies and opportunistic infections (herpes) may also be impaired during therapy with adalimumab.
- Worsening of ventricular function has been reported, so it should be used with caution in patients with left ventricular dysfunction.
- Take care when administering to patients allergic to latex.

## AZATHIOPRINE

**Pharmacological group:** antimetabolite, immunosuppressant; antagonizes the metabolism of purines and can inhibit the synthesis of DNA, RNA and proteins.

**Brand names:** Imunen®, Imuran®, Imussuprex®.

**Presentation:** 50 mg tablet.

**Uses:** treatment of CD, maintenance of remission or reduction in the use of steroids, and treatment of UC.

### Pharmacokinetic parameters:

- *Bioavailability*: 47.4% (oral) and 1.3 to 5.3% (rectal).
- *Distribution (protein binding)*: 30%.
- *Metabolization*: hepatic, forming 6-mercaptopurine.
- *Excretion*: urine.
- *Half-life*: 5 hours

### Posology:

- *CD/UC*: 2 to 3 mg/kg/day.

### Dose adjustment:

- *Liver function*: no reports.
- *Kidney function (adult)*:

| ECC (mL/min) | > 50 | 10 to 50 | < 10 |
|---|---|---|---|
| Interval (h) | Standard dose | 75% of the standard dose | 50% of the standard dose |

**Route of administration:**

- *Oral*: the tablets should be taken during or soon after meals.
- *Via probe*: the tablets may be dispersed in cold water and immediately administered via probe with gastric or enteric location (monitor effects).

**Drug interactions:**

- *Enalapril, captopril, cilazapril*: can trigger severe asthma or myelosuppression.
- *Neuromuscular blockers*: may reduce the effects of these drugs.
- *Allopurinol, trastuzumab, sulfamethoxazole/trimethoprim*: may increase serum levels of azathioprine, favoring the onset of nausea, vomiting and leukopenia.
- *6-mercaptopurine, natalizumab, methotrexate and live virus vaccines*: azathioprine may increase serum levels of these drugs.
- *Cyclosporin, warfarin, phenprocoumon*: their effects may be reduced by the presence of azathioprine.

**Conservation and preparation:**

- Store at room temperature (15 to 25 °C).

**Pregnancy:** risk factor D.

**Breastfeeding:** not recommended; inconclusive data on use during breastfeeding.

**Adverse effects:** bone marrow suppression, leukopenia, thrombocytopenia, anemia, increased susceptibility to infections, liver toxicity, pancreatitis, nausea, vomiting, diarrhea, abdominal pain, fever, chills, rash, retinopathy, myalgia, alopecia, arthritis, interstitial pneumonitis, hypotension, bradyarrhythmia, pericarditis, vasculitis and skin cancer.

**Remarks:**

- Hematologic function should be monitored.
- Gastrointestinal toxicity can occur in the first weeks, but is reversible.
- Mutagenic potential.
- Available at Popular Pharmacy of Brazil.

## BUDESONIDE

**Pharmacological group:** corticosteroid.

**Brand name:** Entocort®.

**Presentations:** 3 mg extended-release capsules (for ingestion), enema for topical rectal use (dispersible tablet).

**Use:** indicated as anti-inflammatory treatment for mild to moderate CD involving the ileum or colon; not recommended for maintenance treatment of the disease and in the treatment of ulcerative colitis (topical use).

**Contraindication:** hypersensitivity to components of the formula.

**Pharmacokinetic parameters:**

- *Bioavailability*: 10 to 20% after oral administration.
- *Distribution (protein binding)*: 85 to 90%; Vd from 2 to 3 L/kg.
- *Metabolization*: hepatic.
- *Excretion*: urine (60%), feces (15.1 to 19.6%) and bile.
- *Half-life*: 2 to 3.6 hours.

**Posology:**

- *Pediatrics ≥ 6 years (CD) – Initial*: 9 mg/day, once/day, for 7 to 8 weeks. Maintenance: 6 mg/day, once/day, for 3 to 4 weeks. Maximum initial dose described: 12 mg/day.[2]
- *CD*: initial daily dose of 9 mg for 8 to 16 weeks; after the initial treatment, the dosages are reduced at the rate of 3 mg, with 6 mg used daily for 3 months.
- *UC*: 9 mg/day for 8 weeks.

**Dose adjustment:**

- *Liver function*: monitor signs and symptoms of hypercortisolism and considers reducing the dose in moderate to severe liver dysfunction.

- *Kidney function*: adjustment not required.

**Route of administration:**
- *Oral*: administration of the capsules with high-fat food affects the rate of absorption, but not the full extent. The capsules with enteric release granules should not be chewed, but rather ingested whole with water in the morning. In case of swallowing problems, the capsules may be opened and their contents mixed (without crushing or grinding beads) in apple sauce (immediate use).
- *Via probe*: not recommended.
- *Rectal*: after preparation of the dispersible tablet in the diluent solution, use is immediate.

**Drug interactions:**
- *Amiodarone*: may increase the risk of developing Cushing's syndrome.
- *Boceprevir*: increases the concentration of budesonide.
- *Amphotericin B, diuretics*: their effects may be potentiated.
- *Fluconazole, dasatinib, ritonavir, ketoconazole, itraconazole, cimetidine, clarithromycin, erythromycin, ritonavir, indinavir, saquinavir*: may increase serum levels of budesonide.
- *Aluminum hydroxide, magnesium hydroxide*: may reduce the effects of budesonide.

**Conservation and preparation:**
- Store at room temperature (15 to 30°C).

**Pregnancy:** risk factor C.
**Breastfeeding:** use with caution.
**Adverse effects:** insomnia, nightmares, nervousness, anxiety, euphoria, delirium, hallucinations, psychosis, headache, dizziness, increased appetite, hirsutism, hyper or hypopigmentation, osteoporosis, petechiae, ecchymosis, arthralgia, cataracts, glaucoma, epistaxis, amenorrhea, Cushing's syndrome, adrenal insufficiency, hyperglycemia, diabetes mellitus, growth suppression, water and sodium retention, edema, increased blood pressure, seizures,

muscle wasting, weakness, fatigue, myopathy, redistribution of body fat, increased free fatty acids, hypokalemia, alkalosis, polycythemia, leukocytosis, lymphopenia, increased susceptibility to infections, reactivation of latent tuberculosis, osteonecrosis (avascular necrosis or septic), osteoporosis, stroke, itching, weight gain, gastroenteritis and vomiting.

### Remark:
- The patient should avoid smoking and exposure to known allergens.

## CERTOLIZUMAB PEGOL

**Pharmacological group:** immunosuppressant, TNF-alpha antagonist antibody.

**Brand name:** Cimzia®.

**Presentations:** filled syringe, 200 mg in 1 mL.

**Use:** treatment of moderate to severe CD, with inadequate response to conventional therapy, either as a treatment for induction or maintenance, irrespective of previous therapies.

### Pharmacokinetic parameters:
- *Bioavailability*: 76 to 88%.
- *Distribution*: 6.4 L.
- *Metabolization*: no reports.
- *Excretion*: renal (17 mL/h).
- *Half-life*: 14 days.

### Posology:
- *Adolescents and adults*: 400 mg subcutaneously at weeks zero, 2 and 4. Maintenance of treatment with a dose of 400 mg subcutaneously every 4 weeks.

### Dose adjustment:
- *Liver function*: no reports.
- *Kidney function*: no reports.

**Route of administration:**
- *Subcutaneous*: the site of each new injection should be rotated, preferring the thigh and anterior thigh.

**Drug interactions:**
- It is not recommended to use certolizumab with abatacept.

**Conservation:**
- Keep refrigerated (2 to 8 °C). Do not freeze.

**Pregnancy:** Risk factor C.

**Breastfeeding:** there are insufficient data on use during breastfeeding.

**Adverse effects:** hypersensitivity, nasopharyngitis, urinary tract infection, arthralgia, dyspnea, rash, headache, nausea, hypertension, fever, fatigue and erythema at the injection site.

## Remarks:
- Patients receiving certolizumab have increased risk of serious infections, especially if receiving other immunosuppressive agents such as methotrexate or corticosteroids.
- There are reports of reactivation of tuberculosis with the use of certolizumab pegol.
- Consider the risk-benefit of use in patients with chronic infections.
- Use with caution in patients with heart failure or blood disorders.

### CYCLOSPORINE

**Pharmacological group:** immunosuppressant, macrolide; inhibits the production and release of interleukin-2 and inhibits interleukin-2 activation induced in T lymphocytes.

**Brand names:** Restasis®, Sandimmun®, Sandimmun neoral®, Sigmasporin®, Sigmasporin microral®.

**Presentations:** vial with 50 mg/mL in 1 or 5 mL, capsules of 25, 50 and 100 mg, oral solution with 100 mg/mL in 50 mL.

**Use:** treatment of severe or fulminant UC and severe to fulminant and fistulizing CD.

**Pharmacokinetic parameters:**

- *Absorption*: erratic and variable, occurring in the duodenum and jejunum.
- *Distribution (protein binding)*: 90%; Vd between 3.9 and 4.5 L/kg; widely distributed in tissues and cells.
- *Metabolization*: hepatic (25 metabolites).
- *Excretion*: urine and feces (6%).
- *Half-life*: 19 hours for the unmodified forms (gelatin capsules) and 8.4 hours for the modified form (microemulsion capsules).

**Posology:**

- *Pediatrics (UC)*: 4 to 10 mg/kg/day.3
- *UC*: dose of 2 to 4 mg/kg/day as a continuous infusion should be changed to oral transmission as soon as possible, at rate of 2.3 to 3 mg/kg every 12 hours.
- *CD*: the recommended dosage is 2 to 4 mg/kg/day by continuous infusion for 1 to 2 weeks and, as soon as possible, change to oral use, 6 to 8 mg/kg/day (maximum of 10 mg/kg/day) for 4 to 6 months.

**Dose adjustment:**

- *Liver function*: no reports.
- *Kidney function*: adjustment not required.

**Route of administration:**

- *Oral*: always administer in the same manner, i.e., always take this medication with food in order to prevent serum variations, if applicable. The capsules should be administered with water.
- *Via probe*: adsorption can occur with the probe material (nonspecific); administer the oral solution via probe. Pause the diet for administration of the oral solution via probe (gastric or enteric).
- *Intravenous*: administer the solution slowly (2 to 6 hours) and do not exceed a final concentration of 2.5 mg/mL in saline solution/diluent.

**Drug interactions:**

- *Hypericum*: should be avoided because it may alter the serum levels of the cyclosporin.

- *Mycophenolate*: the plasma concentration of mycophenolate can be changed, decreasing its effectiveness.
- *Simvastatin*: may increase the risk of myopathy.
- *Antacids, carbamazepine, griseofulvin, phenytoin, pyrazinamide and rifampin are medications that can decrease the plasma concentrations of cyclosporin.*
- *Digoxin, caspofungin, etoposide, fentanyl, methotrexate, minoxidil, natalizumab, salmeterol, simvastatin, atorvastatin, sirolimus, and topotecan*: cyclosporine may increase serum levels of these medications, which can lead to toxic effects.
- *Captopril, enalapril, gentamicin, amikacin, amiodarone, amphotericin B, fluconazole, bromocriptine, carvedilol, colchicine, dasatinib, melphalan, methotrexate, metoclopramide, metronidazole, nonsteroidal anti-inflammatory drugs (NSAIDs), sirolimus and trastuzumab*: may increase the plasma levels of cyclosporin; monitor serum levels.

**Conservation and preparation:**
- *Keep capsules refrigerated (2 to 8 °C), and the oral solution and vials at room temperature (15 to 25 °C) and protected from the light.*
- *Preparation of the oral solution*: the oral solution comes ready to use and is stable for 60 days at room temperature after opening the vial. Do not refrigerate.
- *Preparation of the injectable*: dilute the dose in 20 to 100 mL of 0.9% saline. The leftovers of the vials should be discarded. The solution should be prepared in polyethylene or glass containers for added stability (24 hours at room temperature). Solution prepared in PVC (polyvinyl chloride) bags (flexible) is less stable (6 hours at room temperature).

**Pregnancy:** risk factor C.
**Breastfeeding:** not recommended.
**Adverse effects:** hypertension, edema, headache, hirsutism, hypertrichosis, increased triglycerides, female reproductive tract disorders, nausea, diarrhea, dyspepsia, abdominal pain, tremor, numbness, muscle contractions, renal dysfunction, increased serum creatinine, and increased risk of infections. Less common occurrences: chest pain, arrhythmia, congestive heart failure,

peripheral ischemia, dizziness, convulsions, insomnia, depression, impaired concentration, emotional lability, encephalopathy, gynecomastia, hypo/hyperglycemia, hyperkalemia, hyperchloremic acidosis, hypomagnesemia, hyperuricemia, changes in libido, acne, gingival hyperplasia, hepatotoxicity, leukopenia, thrombocytopenia, purpura, increased risk of myelodysplasia, leukemia, lymphoma and other malignancies.

**Remarks:**
- Use with caution when the patient is using other nephrotoxic drugs.
- Some injectable may contain propylene glycol, corn oil or Cremophor® in their formulation, and these are associated with anaphylactic reactions (rare); thus, the first 30 minutes of infusion of cyclosporine should be monitored.
- Serum levels:
  - therapeutic: not fully defined, varying with the disease being treated. In general, therapeutic levels are between 100 and 400 ng/mL. Dosing is recommended prior to ingestion of the drug in the morning, "premedication" (C0).
  - toxic: not defined. Nephrotoxicity may occur at any level.
- It is important to be aware that it is a drug "critically dependent on the dose", that is, a slight reduction or increase in dose or plasma concentration results in clinically significant changes in its efficacy or toxicity.

## CIPROFLOXACIN
**Pharmacological group:** quinolone, antimicrobial.
**Brand names:** Besflox®, Cifloxatin®, Cifloxtron®, Cipro®, Ciproflonax®, Ciproflan®, Ciproflox®, Ciprofloxil®, Floxocip®, Maxiflox®, Proflox®, Quinoflox®.
**Presentations:** coated tablets of 250 and 500 mg; tablets or capsules of 250 and 500 mg; injectable solution 2 mg/mL, 100 or 200 mL.
**Use:** treatment of active CD.

**Pharmacokinetic parameters:**
- *Absorption*: fast (50 to 85%), occurring in 1 to 2 hours.
- *Distribution (protein binding)*: 20 to 40%.
- *Metabolization*: hepatic.

- *Excretion*: urine (30 to 50%) and feces (15 to 45%).
- *Half-life*: 3 to 5 hours.

**Posology:**
- Dose of 500 mg every 12 hours for 6 to 12 weeks.

**Dose adjustment:**
- *Liver function*: there is no recommendation for dose adjustment in liver dysfunction.
- *Kidney function (adult)*:

| ECC (mL/min) | > 50 | 10 to 50 | < 10 |
|---|---|---|---|
| Interval (h) | Standard dose | 50 to 75% of the standard dose | 50 % of the standard dose |

**Route of administration:**
- *Oral*: food does not affect the overall extent of absorption and can be administered whether it is taken or not. However, dairy products and beverages fortified with calcium end up chelating the antibiotic and therefore it is recommended for the medication to be administered with an interval of 1 to 2 hours with dairy products.
- *Via probe*: the oral solution can be prepared from the tablets for ease of administration via probe. Pause enteral feeding 1-2 hours prior to administration of the antibiotic. Enteral diets reduce the uptake of ciprofloxacin by 30%.
- *Intravenous*: do not exceed the maximum concentration of 2 mg/mL in the dilution. Injectable solutions come ready for use. Administration should be slow.

**Drug interactions:**
- *Hypericum*: may cause a photosensitivity reaction.
- *Cisapride, methadone*: QT interval prolongation may occur.

- *Methotrexate, ropivacaine, theophylline, tizanidine*: concomitant use of quinolones may increase the serum levels of the said drugs and trigger toxic effects.
- *NSAIDs*: increase the levels of ciprofloxacin.
- *Warfarin and oral hypoglycemic*: variations may occur in the concentrations of these drugs, whose effects should be monitored.
- *Mycophenolate, phenytoin*: there is a reduction in plasma concentrations of these medications.
- *Antacids, didanosine, calcium, iron, zinc, magnesium*: reduce serum levels of ciprofloxacin. Administer the antibiotic 2 hours before or 6 hours after these medications.
- *Caffeine*: avoid excessive use, as this can trigger cardiac effects and stimulate the CNS.
- *Statins*: increase the risk of myopathy or rhabdomyolysis.

**Conservation:**
- Tablets and injectables should remain at room temperature (15 to 30 °C), protected from the light.

**Pregnancy:** risk factor C.

**Breastfeeding:** not recommended.

**Adverse effects:** dyspepsia, nausea, vomiting, increased transaminases, abdominal pain, diarrhea, and headache. Hypersensitivity reactions such as rash, itching, fever, photosensitivity, hives and anaphylaxis are rare. Neurotoxicity may occur with changes in mental state and hallucinations, especially in the elderly and patients using maximum doses. Reversible arthralgia and arthritis may occur. Eosinophilia and leukopenia have been reported, disappearing upon suspension of the drug.

**Remarks:**
- Ingestion of 2 L of liquid to avoid crystal deposits in the urine.
- Advise the patient to use sunscreen and other accessories; avoid exposure to direct light.

### GOLIMUMAB

**Pharmacological group:** monoclonal antibody (promotes the blocking of TNF-alpha).

**Brand name:** Simponi®.

**Presentation:** injectable solution – pen applicator with 50 mg/0.5 mL.

**Uses:** Moderate to severe UC.

### Pharmacokinetic parameters:

- *Bioavailability*: 53%.
- *Distribution*: Vd between 58 and 216 L/kg.
- *Half-life*: approximately 2 weeks.
- *Maximum concentration time*: 2 to 6 days.

### Posology:

- *UC – Initial*: 200 mg subcutaneous; maintenance: 100 mg subcutaneously at week 2 and every 4 weeks.[4,5]

**Route of administration:** subcutaneously; applied to the thigh, abdomen (below the navel), and arm (upper outside).

**Drug interactions:** avoid concomitant use with abatacept due to risk of infection.

### Conservation:

- Keep refrigerated (2 to 8 °C). Prior to application, leave for 30 minutes at room temperature.

**Pregnancy:** risk factor B.

**Breastfeeding:** unknown – not recommended.

**Adverse effects:** nasopharyngitis (16%), laryngitis (16%), pharyngitis (16%), rhinitis (16%), infection (28%), reaction at injection site (6%), hypertension (3%).

### Remarks:

- Product registered at the Anvisa.
- No safety and efficacy studies in the pediatric population.

- Rotate injection sites.
- The needle cover in the applicator pen is manufactured from dry natural rubber (a derivative of latex) which may cause allergic reactions in individuals sensitive to latex.
- Discontinue use in the event of serious infection or active or latent TB.

### Hydrocortisone

**Pharmacological group:** systemic corticosteroids.

**Brand names:** Androcortil®, Ariscorten®, Benzenil®, Cortisonal®, Cortiston®, Cortizon®, Cortison®, Hidrocortex®, Hidrosone®, Solucortef®.

**Presentations:** vial with 100 and 500 mg + diluent; vial with 50 mg/mL + 2 mL of diluent; vial with 100 mg/mL + 3 mL of diluent; vial with 125 mg/mL + 4 mL diluent.

**Use:** mild to moderate active CD and UC.

### Pharmacokinetic parameters:

- *Absorption*: quick.
- *Distribution (protein binding)*: 90%.
- *Metabolization*: hepatic.
- *Excretion*: renal.
- *Half-life*: 8 to 12 hours.

### Posology:

- Rectal administration of 10 to 100 mg 1 to 2 times/day for 2 to 3 weeks; or parenteral administration of 300 mg/day, in divided doses every 8 hours or continuous infusion with response between 7 and 10 days.

### Dose adjustment:

- *Liver function*: no adjustment necessary.
- *Kidney function*: no adjustment necessary.

### Route of administration:

- *Intravenous – bolus*: the dose may be diluted in saline solution at a maximum concentration of 50 mg/mL or administered directly in 3 to 5 minutes.

- *IV/intermittent*: dilute the dose in 100 to 250 mL of 5% glucose solution or 0.9% saline and administer in 30 to 60 minutes.
- *Rectal*: administer in the form of enema, retain the solution for 1 hour (preferentially at night).

### Drug interactions:

- *Interleukin*: may decrease the effectiveness of interleukin.
- *Amphotericin B, chlorthalidone, furosemide, hydrochlorothiazide*: increased risk of hypokalemia.
- *Salicylic acid*: increased the effects of ulcers or gastrointestinal irritation.
- *Atracurium, pancuronium, rocuronium*: may reduce the effects of neuromuscular blockers, prolonging muscle weakness and myopathy.
- *Quetiapine, neostigmine, pyridostigmine, and tretinoin*: may decrease serum concentrations of these medications.
- *Carbamazepine, cholestyramine, phenobarbital, phenytoin, primidone, and rifampicin*: may reduce the effects of hydrocortisone.
- *Ciprofloxacin, levofloxacin, norfloxacin*: may increase the risk of tendon rupture.
- *Oral contraceptives, itraconazole*: may prolong the effects of hydrocortisone.
- *Vaccines*: may result in inadequate immunobiological response of the vaccine.
- *Phenprocoumon, warfarin*: may increase the risks of bleeding.
- *Indomethacin*: increases the risk of gastrointestinal perforation.

### Conservation and preparation:

- *Conservation*: the vials should be kept at room temperature (15 to 30 °C).
- *Preparation of the injectable*:
  - *Reconstitution*: reconstitute the lyophilized powder of 100 and 500 mg with the diluent supplied with the product; the resulting solution is stable refrigerated for 3 days or 24 hours at room temperature.
  - *Dilution*: dilute the dose of the medication at a concentration of 1 mg/mL in 0.9% saline or 5% glucose solution. This solution is stable for 24 hours at room temperature or refrigerated. In patients under fluid restriction, the dose of 50 mL may be diluted in 0.9% saline or 5% glucose solution and, as the solution is more concentrated, it should be used within 4 hours.

**Pregnancy:** risk factor C.

**Breastfeeding:** use with caution.

**Adverse effects:** insomnia, nightmares, nervousness, anxiety, euphoria, delirium, hallucinations, psychosis, headache, dizziness, increased appetite, hirsutism, hyper or hypopigmentation, osteoporosis, petechiae, ecchymosis, arthralgia, cataracts, glaucoma, epistaxis, amenorrhea, Cushing's syndrome, adrenal insufficiency, hyperglycemia, diabetes mellitus, growth suppression, water and sodium retention, edema, increased blood pressure, seizures, muscle wasting, weakness, fatigue, myopathy, redistribution of body fat (accumulation in the face, shoulder region [hump] and abdomen), increases in free fatty acids, hypokalemia, alkalosis, polycythemia, leukocytosis, lymphopenia, increased susceptibility infections, reactivation of latent tuberculosis, osteonecrosis (avascular or septic necrosis), osteoporosis and acne.

**Remarks:**

- In cases where treatment with high doses of hydrocortisone is continued for another 48 to 72 hours, hypernatremia may occur. In this case, it is recommended to replace hydrocortisone with methylprednisolone, which produces little or no sodium retention.
- The use of this medication should not be stopped abruptly. The doses may be reduced slowly and progressively.

## INFLIXIMAB

**Pharmacological group:** chimeric monoclonal antibody inhibiting TNF-alpha that interferes with its endogenous activity.

**Brand name:** Remicade®.

**Presentation:** vial with 10 mg/mL in 10 mL.

**Uses:** treatment of moderate to severe CD, including fistulizing disease, and moderate to severe UC in adults, with no response to treatment with corticosteroids and/or immunosuppressants, or with contraindications/intolerance to these treatments. These same criteria apply to the treatment of children between 6 and 17 years old with CD, however, in combination with classical immunosuppressants.

**Pharmacokinetic parameters:**

- *Distribution*: Vd 52.7 mL/kg.
- *Metabolization*: no reports.
- *Excretion*: no reports.
- *Half-life*: 7 to 12 days.

**Posology:**

- *Pediatrics (DC) – Initial*: 5 mg/kg; repeat 5 mg/kg/dose at week 2 and 6 after the first infusion. Maintenance: 5 mg/kg/dose (maximum of 10 mg/kg) every 8 weeks.[6]
- *Adults*: 5 mg/kg in a single dose by slow intravenous infusion (at least, 2 hours) at weeks 0, 2 and 6. Then, maintenance dose of 5 mg/kg every 8 weeks. The dose may be increased up to 10 mg/kg in patients who responded to treatment and lost the response. If there is no response by week 14, consider treatment cessation.

**Dose adjustment:**

- *Liver function*: no reports.
- *Kidney function*: no reports.

**Route of administration:**

- *Intravenous – bolus*: no.
- *IV/intermittent*: dilute the dose in 250 mL of 0.9% saline or maximum concentration of 4 mg/mL. Use 1.2 micron filters, with low protein binding.

**Drug interactions:**

- *Trastuzumab, abciximab*: the plasma concentrations of infliximab may increase in the presence of these drugs.
- *Cisapride, cyclosporin, ergotamine, fentanyl, paclitaxel, phenytoin, quinidine, sirolimus, tacrolimus, theophylline, thioridazine, and warfarin*: concomitant use lessens the effect of these drugs.
- *Vaccines*: immunobiological response to vaccines may be decreased by the use of infliximab.

### Conservation and preparation:

- *Conservation*: the vials should be stored refrigerated (2 to 8°C). Do not freeze.
- *Preparation of the injectable*:
  - *Reconstitution*: reconstitute the powder from the vial with 10 mL of water for injectables, do not shake vigorously. The reconstituted solution contains no preservative, therefore use is immediate.
  - *Dilution*: 250 mL of 0.9% saline (maximum concentration 0.4 to 4 mg/mL); this solution is stable for 24 hours refrigerated.

**Pregnancy:** risk factor B.

**Breastfeeding:** not recommended.

**Adverse effects:** headache, rash, nausea, diarrhea, abdominal pain, urinary tract infection, increased ALT, arthralgia, back pain, upper respiratory tract infection, cough, sinusitis, pharyngitis, hypertension, fatigue, fever, chills, dizziness, pruritus, dyspepsia, asthma. The use of acetaminophen and diphenhydramine 90 minutes prior to infusion may be considered in patients who have had reactions to the infusion, and corticosteroids are recommended in cases of severe reactions (prednisone, 50 mg orally in 3 doses within 24 hours before the infusion, or a single intravenous dose of hydrocortisone, 100mg, or methylprednisolone 20 to 240 mg, administered 20 minutes before the infusion).

### Remark:

- *Use with caution in patients with chronic or repeated infections, or who are predisposed to infections. There are reports of reactivation of tuberculosis (disseminated or extrapulmonary) with infliximab. Patients should be assessed for latent tuberculosis with tuberculin skin test before the start of treatment and, if present, it should be treated.*

### MERCAPTOPURINE

**Pharmacological group:** cytostatic agent, antimetabolite.

**Brand name:** Purinethol®.

**Presentations:** 50 mg tablet.

**Use:** treatment of CD and UC.

**Pharmacokinetic parameters:**

- *Absorption*: 50% absorbed variably via the oral route; bioavailability between 5 and 37%.
- *Distribution (protein binding)*: 19%; Vd of 0.9 L/kg.
- *Metabolization*: intestinal and hepatic.
- *Excretion*: renal (46%).
- *Half-life*: 21 to 90 minutes.

**Posology:**

- *Pediatrics*:
  - CD: 1.5 mg/kg/day (maximum of 75 mg/day).
  - UC: initial dose of 50 mg/day according to the response.
- *CD*: maintenance of remission or decrease in use of corticosteroids is 1 to 1.5 mg/kg/day orally with dosage adjusted as per monitoring tests.
- *UC*: the initial dose is 50 mg/day, which can be adjusted higher or lower according to clinical response and tolerance; for maintenance of the remission phase, the recommended dose is 1.5 mg/kg/day.

**Dose adjustment:**

- *Liver function*: start with low doses in patients with liver dysfunction and monitor.
- *Kidney function*: start with low doses in patients with kidney dysfunction and monitor.

**Route of administration:**

- *Oral*: should be administered on an empty stomach, 1 hour before or 2 hours after food/milk as there is a reduction of 30 to 50% in drug absorption in the presence of food and dairy products.

**Drug interactions:**

- *Allopurinol*: decreases the metabolism of mercaptopurine.
- *Warfarin*: inhibits the anticoagulant effect.
- *Azathioprine*: may cause myelosuppression, hepatotoxicity and decreased renal function.

- *Doxorubicin*: increases the risk of hepatotoxicity.
- *Methotrexate*: toxicity by mercaptopurine.
- *Mesalazine, sulfasalazine, olsalazine*: may trigger bone marrow suppression.

**Conservation:**
- Store at room temperature (15 to 25°C).

**Pregnancy:** risk factor D.

**Breastfeeding:** avoid use during breastfeeding.

**Adverse effects:** leukopenia, thrombocytopenia, hyperuricemia and hepatotoxicity.

**Remarks:**
- Hepatotoxicity and hyperuricemia effects are to be monitored with the use of the drug; hepatotoxicity may occur at any dose, but more often when exceeding a daily dose of 2.5 mg/kg.
- Formulation may contain lactose in the composition. Caution with lactose intolerant patients.

## MESALAZINE

**Pharmacological group:** derivative of 5-aminosalicylic acid (5ASA).

**Brand names:** Asalit®, Chron-ASA 5®, Mesacol®, Mesacol MMX®, Mesaneo®, Pentasa®.

**Presentations:** 500 and 1200 mg extended-release tablets; coated tablet 400 and 800 mg; suppository with 250, 500 or 1,000 mg; suspension; enema envelope 3 + diluting flask with 100 mL diluent; rectal enema 10 mg/mL in 100 mL.

**Uses:** treatment of mild to moderate symptoms of UC and maintenance of remission. Mesalazine suppositories are also used to treat distal colitis, such as proctitis, and proctosigmoiditis and CD (colitis).

**Pharmacokinetic parameters:**
- *Absorption*: rapidly absorbed orally.
- *Distribution (protein binding)*: 43%.

- *Metabolization*: hepatic.
- *Excretion*: renal (13 to 30%) and feces (72%).
- *Half-life*: 1 hour (mesalazine) and 10 hours (metabolites).

## Posology:

- *Pediatrics*:
  - CD: 50 to 100 mg/kg/day, every 6 to 12 hours, orally (maximum of 1 g/dose).[7]
  - UC: 30 to 60 mg/kg/day, every 6 to 12 hours, orally (maximum of 4 to 4.8 g/day). Enema: dose of 4 g/day. Suppository: dose of 500 mg, once or twice/day.[8]
- *UC*: 2400 to 4800 mg/day in divided doses. Maintenance of remission: 1200 to 2400 mg/day. Suppositories: 250 mg, 2 to 4 times/day; 500 mg up to 3 times/day; 1000 mg, once/day. Enema: 1 to 4 g at bedtime until the next day.

## Dose adjustment:

- *Liver function*: use with caution in patients with abnormal liver function.
- *Kidney function*: use with caution in patients with abnormal kidney function.

## Route of administration:

- *Oral*: the tablets should be swallowed whole with the aid of liquid. The contents of the granules should be emptied onto the tongue and swallowed with the aid of liquids. Do not dissolve the granules in liquid before administration.
- *Via probe*: not recommended.
- *Rectal*: maintain the retention of the suppository from 1 to 3 hours and the enema for 8 hours.

## Drug interactions:

- *Aluminum hydroxide, aluminum phosphate, magnesium hydroxide*: may alter the bioavailability of mesalazine.
- *Azathioprine, mercaptopurine, thioguanine*: may result in myelosuppression.
- *Enoxaparin, nadroparin, warfarin*: risk of bleeding.
- *Glibenclamide*: risk of excessive hypoglycemia.

- *Chickenpox vaccine*: may lead to Reye's syndrome.

**Conservation:**
- Tablets and suppositories should remain at room temperature (15 to 25°C). Do not refrigerate.

**Pregnancy:** risk factor B.

**Breastfeeding:** use with caution.

**Adverse effects:** the most common include headache, abdominal pain, belching, sore throat, chest pain, peripheral edema, chills, fever, insomnia, malaise disorders, anxiety, weakness, rash, itching, acne, constipation, diarrhea, dyspepsia, flatulence, nausea, vomiting, arthralgia, hypertension, myalgia, arthralgia, conjunctivitis, flu-like symptoms and diaphoresis. Less common effects are pericarditis, pericardial effusion, chest pain, myocarditis, ECG changes, interstitial pneumonia, asthma, sinusitis, pleurisy, fibrosing alveolitis, pancreatitis, hepatitis, jaundice, interstitial nephritis, agranulocytosis, aplastic anemia, and thrombocytopenia.

**Remark:**
- Do no administer the tablets together with antacids.

**Methylprednisolone**

**Pharmacological group:** Systemic corticosteroids.

**Brand names:** Alergolon®, Depomedrol®, Predmetil®, Solumedrol®, Solupred®, Solupren®, Unimedrol®.

**Presentations:** vial with 40, 125, 500 and 1,000 mg + diluent; vial with 40 mg/mL in 1, 2 or 5 mL.

**Uses:** treatment of moderate to severe CD and UC.

**Pharmacokinetic parameters:**
- *Absorption*: rapid effect after IM and IV administration.
- *Distribution (protein binding)*: 77%; Vd of 1.5 L/kg.
- *Metabolization*: hepatic.
- *Excretion*: urine.

- *Half-life*: 2 to 3 hours.

**Posology:**
- *Pediatrics (CD/UC)*: 0.11 to 1.6 mg/kg/day (3.2 to 48 mg/m$^2$), intramuscularly.
- *CD*: dose ranges from 10 to 40 mg per dose, intravenously or intramuscularly, based on the clinical response. If higher doses are required for severe cases, 30 mg/kg intravenously infused is recommended starting from 30 minutes every 4 or 6 hours, for 48 to 72 hours.
- *UC*: it can be administered as retention enema in a dose of 40 to 120 mg, for 2 weeks or more.

**Dose adjustment:**
- *Liver function*: no adjustment necessary.
- *Kidney function*: no adjustment necessary.

**Route of administration:**
- *Intravenous (succinate form) – bolus*: lower doses (<250 mg) may be administered within 5 minutes, without the need to dilute the dose in volume of saline solution; IV/intermittent: the IV administration of high doses (> 250 mg) should be done slowly, in 30 to 120 minutes; the dose should be diluted in a maximum concentration of 20 mg/mL (50 to 200 mL) in 0.9% saline or 5% glucose solution for infusion.
- *Rectal*: in the form of retention enema.
- *Intramuscular (acetate or succinate)*: may be administered intramuscularly; preferably avoiding the deltoid muscle.

**Drug interactions:**
- *Liposomal amphotericin, hydrochlorothiazide*: may trigger hypokalemia.
- *Aprepitant, diltiazem, itraconazole*: may increase the effects of methylprednisolone; monitor the effects of toxicity.
- *Carbamazepine, phenobarbital, phenytoin, primidone, and rifampicin*: may reduce the effects of methylprednisolone.
- *Ciprofloxacin, levofloxacin, norfloxacin*: increased risk of tendon rupture.
- *Acetylsalicylic acid*: increased risk of bleeding and stomach irritation.

- *Atracurium, pancuronium, rocuronium*: may decrease the effect of these medications, prolonging muscle weakness and myopathy.
- *Quetiapine*: may cause a decrease in the effects of quetiapine.
- *Tacrolimus*: may cause an increase in serum levels of tacrolimus, which must be monitored.
- *Dicoumarol, phenprocoumon, warfarin*: risk of bleeding.
- *Vaccines*: may cause variations in the immunobiological response.

**Conservation and preparation:**
- *Conservation*: tablets and vials should be stored at room temperature (20 to 25°C).
- *Preparation of the injectable*:
  - *Reconstitution*: acetate form (IM): 2 mL of the diluent; succinate form (IV): with 1 mL (40 mg), 2 mL (125 mg), 8 mL (500 mg) and 16 mL (1000 mg) of the diluent.
  - *Stability*: the reconstituted solutions for IV and IM use remain stable for 48 hours refrigerated while the solutions already diluted in saline solution are stable for 24 hours at room temperature.

**Pregnancy:** risk factor C.

**Breastfeeding:** use with caution.

**Adverse effects:** insomnia, nightmares, nervousness, anxiety, euphoria, delirium, hallucinations, psychosis, headache, dizziness, increased appetite, hirsutism, hyper or hypopigmentation, osteoporosis, petechiae, ecchymoses, arthralgia, cataract, glaucoma, epistaxis, amenorrhea, Cushing's syndrome, adrenal insufficiency, hyperglycemia, diabetes mellitus, suppressing growth, retention of water and sodium, edema, increased blood pressure, seizures, muscle wasting, weakness, fatigue, myopathy, redistribution of body fat (accumulation in the face, shoulder region [hump] and abdomen), increases in free fatty acids, hypokalemia, alkalosis, polycythemia, leukocytosis, lymphopenia, increased susceptibility to infections, reactivation of latent tuberculosis, osteonecrosis (avascular or septic necrosis) and osteoporosis.

**Remarks:**

- Rapid IV administration of high doses are associated with cardiovascular syncope.
- During prolonged therapy, a diet rich in protein, calcium and potassium and low sodium and with carbohydrate restriction is recommended.
- Evaluate changes in level of consciousness and headache, adrenal insufficiency (hypotension, weight loss, weakness, nausea, anorexia and lethargy).
- During therapy, monitor the patient's weight, blood pressure and pulse. Assess urinary flow rate and signs of peripheral edema.
- The diluent accompanying the product contain benzyl alcohol, which can trigger allergic reactions.
- Methylprednisolone succinate can be administered intramuscularly while the acetate form cannot be administered intravenously.

### METHOTREXATE

**Pharmacological group:** cytostatic agent; folate antimetabolite that inhibits DNA synthesis.

**Brand names:** Biometrox®, Fauldmetro®, Hytas®, Litrexate®, Miantrex CS®, Metrexato®, Lexato®, Tecnomet®, Tevametho®.

**Presentations:** 2.5 mg tablet; vial with 25 mg/mL in 1, 2, 10 or 20 mL; vial with 500 mg in 20 mL; vial with 100 mg/mL in 5, 10 or 50 mL; vial with 2.5 mg/mL in 2 mL.

**Uses:** treatment of mild to moderate CD and treatment of fistulas.

**Pharmacokinetic parameters:**

- *Bioavailability*: 76 to 100%. (IM).
- *Distribution (protein binding)*: 50%.
- *Metabolization*: hepatic.
- *Excretion*: urine (48 to 100%), feces and bile.
- *Half-life*: 3 to 12 hours.

**Posology:**

- *Pediatrics (CD)*: 17 mg/m$^2$ (11.9 to 22.5 mg/m$^2$), intramuscularly; associated with prednisolone, 1.12 mg/kg/day orally.[9]

- *CD*: the recommended dosage for the treatment of remission of induction or reduction in steroid use is 25 mg, once/week, intramuscularly or subcutaneously; for maintenance of remission, the recommended dose is 15 mg, once/week intramuscularly.

### Dose adjustment:

- *Liver function*: if bilirubin is between 3.1 and 5 mg/dL or SGOT/SGPT > 3 times the limit, administer 75% of the dose. If bilirubin > 5 mg/dL, avoid use.
- *Kidney function (adult)*:

| ECC (mL/min) | > 50 | 10 to 50 | < 10 |
|---|---|---|---|
| Interval (h) | Standard dose | 50% of the standard dose | Avoid use |

### Route of administration:

- *Oral*: may be administered with or without the presence of food.
- *Intravenous – bolus*: for low doses, slow; IV/intermittent: for intermediate doses, which are diluted in 50 to 250 mL and administered in 30 minutes or more; IV/continuous: higher doses of medication are diluted in up to 500 mL of saline solution.
- *Intramuscular*: for CD, the subcutaneous route is described as an alternative route for the administration of methotrexate.

### Drug interactions:

- *Acitretin, adapalene, azathioprine, isotretinoin*: there may be an increase in the risk of hepatotoxicity.
- *Amiodarone, amoxicillin, acetylsalicylic acid, ciprofloxacin, cyclosporin, dantrolene, diclofenac, dipyrone, doxycycline, ibuprofen, indomethacin, naproxen, nimesulide, omeprazole, penicillin G, penicillin V, phenytoin, and sulfamethoxazole/trimethoprim*: may increase the adverse effects of methotrexate (leucopenia, thrombocytopenia, anemia, nephrotoxicity, and mucositis).
- *Vaccines*: increased risk of infection.

- *Asparaginase, chloramphenicol, cholestyramine, and tetracycline*: may reduce the effects of methotrexate.
- *Hydrochlorothiazide, pyrimethamine*: increased risk of myelosuppression.
- *Warfarin*: increased risk of bleeding.

**Conservation and preparation:**
- *Conservation*: tablets and vials should be stored at room temperature (15 to 30 °C) protected from light.
- *Preparation of the injectable*:
  - *Dilution (IV use)*: the dose of medication may be diluted in 0.9% saline solution, 5% glucose solution or Ringer's Lactate solution (variable volume).
  - *Stability*: it is stable for 24 hours refrigerated or at room temperature (protected from light).

**Pregnancy:** risk factor X.

**Breastfeeding:** contraindicated.

**Adverse effects:** the most common include headache, neck stiffness, vomiting, fever, leucopenia, thrombocytopenia (peak on the 10th day), demyelinating encephalopathy, seizures, drowsiness, megaloblastic anemia, chills, hyperuricemia, defects in spermatogenesis and oogenesis. Stomatitis, mucositis, glossitis, gingivitis, diarrhea (1 to 3%), anorexia, intestinal perforation, nephropathy, renal dysfunction, pharyngitis, vasculitis, alopecia (0.5 to 3%), rash, photosensitivity, skin pigmentation changes, blurred vision, diabetes, cystitis, cirrhosis, arthralgia, pneumonitis, liver toxicity, severe infections, lymphoma, diarrhea, loss of appetite, and hair loss.

**Remarks:**
- Carcinogenic and teratogenic medication. Pregnancy should be avoided for at least 3 months after treatment of men and 1 ovulatory cycle in the case of women.
- Hydration and alkalinization of urine can prevent the precipitation of methotrexate or its metabolites in the renal tubules.

- Monitoring: laboratory control with serum levels of the drug, blood count, platelets, transaminases, alkaline phosphatase, bilirubin, lactate dehydrogenase, electrolytes, urea and creatinine. Monitor liver function and bone marrow when used in large doses.
- During treatment, the patient should not receive any type of immunization.
- Before the infusion of the drug, the patient should receive antiemetics and antacids.
- Photosensitivity reactions are rare, but it is recommended to use sunscreen (SPF 15) and avoid direct exposure to the sun without protection when using the medication.

## METRONIDAZOLE

**Pharmacological group:** nitroimidazole.

**Brand names:** Ambrosil®, Flagymax®, Flagyl®, Helmizol®, Hidazol®, Metrizol®, Metronil®.

**Presentations:** coated tablet of 250 or 400 mg; oral suspension with 40 mg/mL in 80 or 100 mL; injectable solution of 500 mg with 100 mL.

**Uses:** treatment of CD, particularly in patients with perianal and colonic disease or fistula non-responsive to previous treatments (sulfasalazine, corticosteroids).

### Pharmacokinetic parameters:

- *Bioavailability (oral)*: 100%.
- *Distribution (protein binding)*: less than 20%.
- *Metabolization*: hepatic.
- *Excretion*: urine (60 to 80%) and feces (6 to 15%).
- *Half-life*: 7 hours.

### Posology:

- Dose of 10 to 20 mg/kg/day or 250 to 500 mg/dose, 2 or 3 times a day orally.

### Dose adjustment:

- *Liver function*: reduce dose by 50 to 67% in patients with liver disease.
- *Kidney function*: does not require dose adjustments.

### Route of administration:

- *Oral*: administered with or without the presence of food. If gastrointestinal side effects occur, administer with food. The presence of food slows the absorption of the drug, but does not affect the plasma concentration. This is the preferred route for CD.
- *Intravenous*: administer in 30 to 60 minutes. The bag comes ready for use. Unused portions should be discarded.

### Drug interactions:

- *Busulfan, carbamazepine, cyclosporin, dihydroergotamine, fluorouracil, lithium carbonate, phenytoin, tacrolimus*: there may be an increase in the plasma levels of these medications, which can lead to toxic effects.
- *Cholestyramine, phenobarbital*: the effects of metronidazole may be decreased.
- *Phenprocoumon, dicoumarol, warfarin*: increased risk of bleeding.
- *Mycophenolate mofetil*: the effects of mycophenolate may be reduced in the presence of metronidazole.

### Conservation:

- Keep the tablets, oral suspension and bags at room temperature (15 to 30 °C) protected from light. Do not refrigerate.

**Pregnancy:** risk factor B.

**Breastfeeding:** not recommended.

**Adverse effects:** diarrhea, epigastric pain, nausea, dizziness, headache, reversible neutropenia, metallic taste in mouth, dark colored urine, hives, exanthema, urethral and vaginal burning, gynecomastia and, rarely, peripheral neuropathy, pseudomembranous colitis, pancreatitis, seizures, encephalopathy, cerebellar dysfunction and ataxia.

### Remarks:

- During therapy, register evacuations and evaluate edema (sodium retention).
- Advise the patient to avoid alcoholic drinks until 48 hours after the end of treatment.
- May cause color changes in urine (dark).

## NATALIZUMAB

**Pharmacological group:** monoclonal antibody (humanized recombinant antibody of anti-alpha-4-integrin).

**Brand name:** Tysabri®.

**Presentation:** injectable solution – 300 mg vial (20 mg/mL).

**Use:** moderate to severe CD.

### Pharmacokinetic parameters:

- *Distribution*: Vd of 5.2 L.
- *Excretion*: 22 mL/h.
- *Half-life*: 10 days.

### Posology:

- Dose of 300 mg, infused from 1 hour onwards; repeating every 4 weeks.

### Dose adjustment:

- *Liver function*: no reports.
- *Kidney function*: no reports.

### Route of administration:

- *Intravenous*: bolus or push – do not administer.
- *Infusion*: administer infusion from 1 hour onwards.

### Drug interactions:

- No reports.

### Conservation and preparation:

- Dilute the dose in a volume of 100 mL of 0.9% saline solution. The diluted solution remains stable for 8 hours refrigerated (2 to 8 °C).

**Pregnancy:** risk factor C.

**Breastfeeding:** unknown.

**Adverse effects:** skin rash (6%), pruritus, abdominal discomfort (11%), diarrhea (10%), nausea (17%), arthralgia (8%), pain in the limbs (16%), headache (32 to 38%), fatigue (10%) and depression (19%).

**Remarks:**

- Product registered at Anvisa.
- No safety and efficacy studies in the pediatric population.
- In CD, aminosalicylates may be continued during treatment with natalizumab; however, do not use it in combination with immunosuppressants and inhibitors of the tumor necrosis factor (TNF-alpha), at risk of developing progressive multifocal leukoencephalopathy.
- Patients should be monitored for adverse reactions during infusion and for 1 hour after it ends.
- When diluted in 100 mL of 0.9% saline solution, the product contains 17.7 mmol (or 406 mg) of sodium.

## PREDNISONE

**Pharmacological group:** systemic corticosteroids.

**Brand names:** Artinizona®, Alergcorten®, Flamacorten®, Corticorten®, Meticorten®, Precortil®, Predcort®, Prednison®, Prednax®, Predson®, Predval®.

**Presentations:** 5 and 20 mg tablets.

**Uses:** treatment of moderate to severe CD and UC.

**Pharmacokinetic parameters:**

- *Bioavailability*: 92%.
- *Distribution (protein binding)*: 70%; Vd of 0.4 to 1 L/kg.
- *Metabolization*: hepatic.
- *Excretion*: urine.
- *Half-life*: 2 to 3 hours.

**Posology:**

- Dose of 1 to 2 mg/kg/day or 40 to 60 mg/day, once or twice/day orally.

**Dose adjustment:**

- *Liver function*: does not require dose adjustments.
- *Kidney function*: does not require dose adjustments.

**Route of administration:**

- *Oral*: administered with or without the presence of food.
- *Via probe*: for administration via probe, the tablet can be crushed and the contents dissolved in a suitable volume of water (immediate use).

Drug interactions:

- *Amphotericin B, hydrochlorothiazide*: may result in a risk of hypokalemia.
- *Acetylsalicylic acid*: may result in increased risk of gastric irritation.
- *Atracurium, pancuronium, rocuronium*: may decrease the effectiveness of neuromuscular blockers.
- *Vaccines*: may result in variation of immunobiological responses.
- *Carbamazepine, phenobarbital, phenytoin, primidone, rifampin, somatropin*: may reduce the effects of prednisone.
- *Fluconazole, isoniazid, itraconazole, ritonavir*: may increase the effects of prednisone.
- *Ciprofloxacin, levofloxacin, norfloxacin*: may potentiate the effect of tendon rupture.
- *Cyclosporin, dicoumarol, phenprocoumon, warfarin*: the effects of such medication may increase.
- *Quetiapine, neostigmine, pyridostigmine, and tretinoin*: plasma levels of these drugs may decrease, as well as the desired effect.
- *Montelukast*: may result in severe peripheral edema.

**Interactions with food:**

- The presence of food does not interfere with the bioavailability of the drug.

**Conservation:**

- Keep the tablets at room temperature (20 to 25 °C).

**Pregnancy:** risk factor C in the IBD.
**Breastfeeding:** compatible.
**Adverseeffects:**insomnia,nightmares,nervousness,anxiety,euphoria,delirium, hallucinations, psychosis, headache, dizziness, increased appetite, hirsutism,

hyper or hypopigmentation, osteoporosis, petechiae, ecchymosis, arthralgia, cataracts, glaucoma, epistaxis, amenorrhea, Cushing's syndrome, adrenal insufficiency, hyperglycemia, diabetes mellitus, growth suppression, water and sodium retention, edema, increased blood pressure, seizures, muscle wasting, weakness, fatigue, myopathy, redistribution of body fat (accumulation in the face, shoulder region [hump] and abdomen), increased free fatty acids, hypokalemia, alkalosis, polycythemia, leukocytosis, lymphopenia, increased susceptibility to infections, reactivation of latent tuberculosis, osteonecrosis (avascular or septic necrosis) and osteoporosis.

**Remarks:**

- In hyperthyroidism, the dose of prednisone may need to be increased to achieve the appropriate therapeutic effects.
- The use of this medication should not be stopped abruptly. The doses may be reduced slowly and progressively.
- No kind of immunization is recommended during therapy, except in special cases.
- During prolonged treatment, a diet rich in protein, calcium and potassium is recommended, while avoiding or reducing the consumption of carbohydrates and sodium.

### SULFASALAZINE

**Pharmacological group:** anti-inflammatory; derivative of 5-aminosalicylic acid
(5-ASA); its derivative is mesalazine.
**Brand names:** Azulfin®, Salazoprin®.
**Presentation:** 500 mg gastroresistant coated tablet.
**Uses:** treatment of mild to moderate UC and CD.

**Pharmacokinetic parameters:**

- *Absorption*: 20 to 30% of orally administered drug is absorbed in the small intestine.
- *Distribution (protein binding)*: 99.3%.
- *Metabolization*: hepatic is minimal, extensive in the intestine.

- *Excretion*: urine.
- *Half-life*: 5.7 to 10 hours.

## Posology:

- *Pediatrics (UC)*:
  - *Mild exacerbation*: administer 40 to 50 mg/kg/day, every 6 hours.
  - *Moderate to severe exacerbation*: 50 to 60 mg/kg/day, every 4 to 6 hours (do not exceed 4 g/day). Maintenance dose: 30 to 50 mg/kg/day, every 4 to 8 hours (do not exceed 2 g/day).
- *Adolescents and adults*: the usual dose is 3 to 6 g/day in 4 divided doses, with food.

## Dose adjustment:

- *Liver function*: not recommended for use in patients with abnormal liver function.
- *Kidney function*: not recommended for use in patients with abnormal kidney function.

## Route of administration:

- *Oral*: administer during meals with a glass of water to reduce the effects of gastrointestinal intolerance.
- *Via probe*: not recommended.

## Drug interactions:

- *Cyclosporine*: may reduce the efficacy of cyclosporin.
- *Digoxin*: may decrease the plasma levels of digoxin, thereby reducing its effectiveness.
- *Glimepiride, glibenclamide*: may result in excessive hypoglycemia.
- *Mercaptopurine, thioguanine*: risk of potentiating the effects of myelosuppression.
- *Methotrexate may cause hepatotoxicity.*

**Conservation:**

- Keep the tablets at room temperature (20 to 25 °C) protected from light.

**Pregnancy:** risk factor B.

**Breastfeeding:** use with caution.

**Adverse effects:** the most common are headache, photosensitivity, anorexia, nausea, vomiting, indigestion, diarrhea, abdominal distension, reversible oligospermia, alopecia, anaphylaxis, aplastic anemia, ataxia, crystalluria, depression, squamous cell necrosis, hallucinations, hemolytic anemia, hepatitis, nephritis interstitial, jaundice, rash, hives, fever and pruritus.

**Remarks:**

- Sulfasalazine may impair the absorption of folate: consider folate supplementation of 1 mg/day.
- Use with caution in patients with G6PD deficiency.
- Do not administer the medication with antacids.
- Monitor stool frequency. Maintain adequate hydration (2 to 3 L/day).
- Avoid exposure to direct sunlight while using the medicine, wear clothes, sunglasses and adequate sunscreen.

## TACROLIMUS

**Pharmacological group:** immunosuppressant; inhibits T-lymphocyte activation possibly by binding to the intracellular protein FKBP-12.

**Brand names:** Prograf®, Tacrofort®.

**Presentations:** capsules of 1 and 5 mg; vial of 5 mg/mL in 1 mL.

**Use:** fistulizing CD, in association with other immunosuppressants.

**Pharmacokinetic parameters:**

- *Absorption*: incomplete and variable from the gastrointestinal tract.
- *Distribution (protein binding)*: 99%.
- *Metabolization*: hepatic (8 metabolites).
- *Excretion*: feces (92.6%) and less than 1% through urine and bile.
- *Half-life*: 17 to 31 hours.

**Posology:**

- Dose of 0.2 to 0.27 mg/kg/day, every 12 hours.

**Dose adjustment:**

- *Liver and kidney function*: patients with liver or kidney impairment should receive the lowest recommended dose and have their serum concentrations monitored.

**Route of administration:**

- *Oral*: administer the medicine on an empty stomach after fasting. Always administer at the same time every day. The presence of food delays and reduces the absorption of the drug by up to 27%. This is the preferred route for CD.
- *Intravenous*: dilute in 0.9% saline or 5% glucose solution, infusion time of 1 to 12 hours.

**Drug interactions:**

- *Aluminum hydroxide, magnesium hydroxide, magnesium carbonate, amiodarone, atazanavir, basiliximab, bromocriptine, chloramphenicol, cimetidine, clarithromycin, danazol, darunavir, dasatinib, diltiazem, erythromycin, contraceptives, fluconazole, fosamprenavir, itraconazole, lansoprazole, methylprednisolone, metoclopramide, metronidazole, nifedipine, omeprazole, posaconazole, voriconazole*: may increase serum tacrolimus levels, reaching levels of toxicity.
- *Amikacin, amphotericin B, cyclosporine, diclofenac, dipyrone, ganciclovir, gentamicin, ibuprofen, indomethacin, naproxen, tenoxicam*: may cause worsening of kidney function.
- *Amiloride, epironolactone*: may result in hyperkalemia.
- *Carbamazepine, caspofungin, efavirenz, nevirapine, phenobarbital, phenytoin, and rifampin*: may reduce the effects of tacrolimus.
- *Colchicine*: may increase plasma levels of colchicine.
- *Ziprasidone*: may result in effects of cardiotoxicity (QT interval prolongation, torsade-de-pointes, arrhythmias).

### Conservation and preparation:

- *Conservation*: the capsules and the cream should be stored at room temperature (15 to 30 °C) protected from light.
- *Preparation of the injectable*: dilute in the maximum concentration of 0.02 mg/mL in 0.9% saline or 5% glucose solution, infusion time of 1 to 12 hours. Solution stable for 24 hours. Avoid using PVC bags, as adsorption may occur.

**Pregnancy:** risk factor C.

**Breastfeeding:** use with caution.

**Adverse effects:** most commonly, they may experience chest pain, hypertension, alopecia, dizziness, headache, insomnia, tremor, itching, rash, diabetes mellitus, hyperglycemia, hyper- or hypokalemia, hypercholesterolemia, hypomagnesemia, hypophosphatemia, abdominal pain, constipation, diarrhea, dyspepsia, nausea, vomiting, urinary tract infections, anemia, leukocytosis, thrombocytopenia, ascites, arthralgia, back pain, weakness, paresthesias, nephrotoxicity, oliguria, increased creatinine, dyspnea, pleural effusion, confusion, agitation, encephalopathy, hallucinations, convulsions, depression, angina, congestive heart failure, arrhythmias, palpitations, thrombosis, coagulation disorders, jaundice, cholangitis.

### Remarks:

- In patients receiving antacids, 2 hour intervals are recommended between oral administration of tacrolimus and the antacid.

### TOCILIZUMAB

**Pharmacological group:** monoclonal antibody (a humanized monoclonal antibody anti-receptor of human IL-6 of the IgG1 subclass of immunoglobulins).

**Brand name:** Actemra®.

**Presentation:** vial of injectable solution containing 20 mg/mL (4 and 10 mL).

**Use:** CD.

**Pharmacokinetic parameters:**

- *Distribution*: in 2 phases (biphasic); with Vd of 3.5 L (central) and 2.6 L (peripheral).
- *Half-life*: 11 to 13 days (dose-dependent); in children: 28 days.

**Posology:**

- Dose of 8 mg/kg by intravenous infusion every 15 days.[10]

**Route of administration:**

- *Intravenous – bolus or push*: do not administer.
- *Infusion*: administer infusion from 1 hour onwards.

**Drug interactions:**

- *Infliximab*: concomitant use may increase the risk of infection.

**Conservation and preparation:**

- *Conservation*: intact vials should be kept refrigerated (2 to 8 ºC).
- *Preparation of the injectable – Adults*: dilute in 100 mL of 0.9% saline solution. Children: dilute in 50 mL of 0.9% saline solution Stability: tocilizumab solution is physically and chemically stable in 0.9% sodium chloride solution at a temperature between 2 and 30 °C for 24 hours, protected from light.

**Pregnancy:** risk factor C.

**Breastfeeding:** unknown – not recommended.

**Adverse effects:** hypertension (4 to 6%), rash (2 to 4%), abdominal pain (3%), changes in transaminases (0.7 48%), headache (7%), drowsiness (3%), and nasopharyngitis (7%).

**Remarks:**

- *Product registered at Anvisa.*
- *Maximum daily dose reported*: 800 mg/day.
- Therapy with tocilizumab should not be started in patients with active infections, including localized infections.
- *Tuberculosis (pulmonary or extrapulmonary)*: patients receiving tocilizumab may suffer reactivation of latent infection or new infection.

- Vaccines should not be administered in patients receiving tocilizumab.
- *Drug compatibilities*: not tested with other medications; do not administer with other medication in the same equipment line.

## TOFACITINIB

**Pharmacological group:** response modifier (inhibitor of the enzyme janus kinase).
**Brand name:** Xeljanz®.
**Presentation:** 5 mg tablet.
**Uses:** Moderate to severe UC.

### Pharmacokinetic parameters:

- *Absorption*: 74%.
- *Distribution*: Vd of 87 L; about 40% bound to plasma proteins (mainly albumin).
- *Metabolism*: hepatic (70%).
- *Excretion*: urine (30%).
- *Half-life*: 3 hours.

### Posology:

- Doses of 0.5, 3, 10 and 15 mg, twice/day for 8 weeks; positive outcomes were observed in 32, 48, 61 and 78% with the respective doses.[11]

### Route of administration:

- Administered orally without considering the presence of food.

### Drug interactions:

- Not to be used with immunosuppressive drugs as there is an increased risk of infection.
- *Fluconazole*: Concomitant administration with fluconazole may increase the plasma levels of tofacitinib.

**Interactions with food:**

- The presence of foods rich in fats decreased the Cmax by 32%, but there was no change in the AUC.

**Conservation:**

- Store at room temperature (20 to 25°C).

**Pregnancy:** risk factor C.

**Breastfeeding:** unknown – not recommended.

**Adverse effects:** infection (20%), hypertension (2%), headache (4%), diarrhea (4%) and nasopharyngitis (4%).

**Remarks:**

- Product not available in Brazil.
- During treatment with tofacitinib do not use immunization. Administer before the start of treatment.
- Monitor liver transaminases periodically, as well as cholesterol fractions, hemoglobin, platelets, neutrophils and leukocytes.

## VEDOLIZUMAB

**Pharmacological group:** humanized anti-integrin monoclonal antibody (anti-alpha(4)beta(7)-integrin).

**Brand name:** Entyvio.

**Presentation:** vial (lyophilized powder) 300 mg.

**Uses:** moderate to severe CD and moderate to severe UC.

**Pharmacokinetic parameters:**

- *Half-life*: 15 to 22 days.[12]
- *Distribution*: Vd of 5 L.

**Posology:**

- *CD– Initial (weeks 0, 2 and 6)*: 300 mg intravenously, followed by maintenance of 300 mg intravenously every 4 or 8 weeks.[13]

- *UC – Initial (weeks 0, 2 and 6)*: 300 mg intravenously, followed by maintenance of 300 mg intravenously every 4 or 8 weeks.[14]

**Route of administration:**
- Intravenous. Administer by slow infusion of 30 minutes; diluting it in 250 mL of 0.9% saline solution. Not to be administered via push or bolus.

**Drug interactions:**
- No reports.

**Pregnancy:** risk factor C.

**Breastfeeding:** no reports.

**Adverse effects:** nasopharyngitis, arthralgia, headache, nausea, abdominal pain, cough, respiratory infection.[14,15]

**Remarks:**
- Approved by the FDA in 2014.
- There are no reports of triggering progressive multifocal leukoencephalopathy with the use of vedolizumab.

**USTEKINUMAB**

**Pharmacological group:** monoclonal antibody (monoclonal antibody IgG1-kappa anti-interleukin 12/23).

**Brand name:** Stelara®.

**Presentation:** vial of injectable solution containing 45 mg/0.5 mL.

**Use:** Moderate to severe CD patients resistant to TNF-alpha antagonists.

**Pharmacokinetic parameters:**
- *Maximum concentration time*: 7 to 13.5 days.
- *Distribution*: Vd between 161 and 179 L/kg (SC).
- *Half-life*: 14.9 to 45.6 days.

**Posology:**
- *Initial (week 0)*: dose of 6 mg/kg, intravenously. Maintenance (weeks 8 and 16): 90 mg subcutaneously.[16]

**Route of administration:**

- *Subcutaneous*: administered in the thigh, abdomen, arm.
- *Intravenous*: off-label for CD.

**Drug interactions:**

- Live virus vaccines should not be administered.

**Conservation:**

- Maintain refrigerated (2 to 8 °C) and secondary packaging for protection from light.

**Pregnancy:** risk factor B.

**Breastfeeding:** unknown – not recommended.

**Adverse effects:** infection (27 to 72%), headache (5%), nasopharyngitis (8%), fatigue (3%) and pruritus (2%).

**Remarks:**

- Product registered at Anvisa.
- *Maximum dose reported:* 90 mg
- Assess risk for active and latent tuberculosis infection.
- Do not use vaccines concomitantly with ustekinumab.
- Rotate the application sites when administering subcutaneously.
- The vial stopper contains latex which may cause allergic reactions in individuals sensitive to latex.
- The product contains sucrose, so use with diabetics should be monitored.

## REFERENCES

1. Colombel JF, Sandborn WJ, Rutgeerts P, Enns R, Hanauer SB, Panaccione R et al. Adalimumab for maintenance of clinical response and remission in patients with Crohn's disease: the CHARM trial. Gastroenterology 2007; 132(1):52-65.

2. Levine A, Kori M, Dinari G, Broide E, Shaoul R, Yerushalmi B et al. Comparison of two dosing methods for induction of response and remission with oral budesonide in active pediatric Crohn's disease: a randomized placebo-controlled trial. Inflamm Bowel Dis 2009; 15:1055-61.

3. Ramakrishna J, Langhans N, Calenda K, Grand RJ, Verhave M. Combined use of cyclosporine and azathioprine or 6-mercaptopurine in pediatric inflammatory bowel disease. J Pediatr Gastroenterol Nutr 1996; 22(3):296-302.

4. Sandborn WJ, Feagan BG, Marano C, Zhang H, Strauss R, Johanns J et al. Subcutaneous golimumab induces clinical response and remission in patients with moderate-to-severe ulcerative colitis. Gastroenterology 2014; 146(1):85-95.

5. Sandborn WJ, Feagan BG, Marano C, Zhang H, Strauss R, Johanns J et al. Subcutaneous golimumab maintains clinical response in patients with moderate-to-severe ulcerative colitis. Gastroenterology 2014; 146(1):96-109.e1.

6. Stephens MC, Shepanski MA, Mamula P, Markowitz JE, Brown KA, Baldassano RN. Safety and steroid-sparing experience using infliximab for Crohn's disease at a pediatric inflammatory bowel disease center. Am J Gastroenterol 2003; 98:104-11.

7. Fish D, Kugathasan S. Inflammatory bowel disease. Adolesc Med Clin 2004; 15(1):67--90, ix.

8. Tomomasa T, Kobayashi A, Ushijima K et al. Guidelines for treatment of ulcerative colitis in children. Pediatr Int 2004; 46(4):494-6.

9. Uhlen S, Belbouab R, Narebski K, Goulet O, Schmitz J, Cézard JP et al. Efficacy of methotrexate in pediatric Crohn's disease: a French multicenter study. Inflamm Bowel Dis 2006; 12(11):1053-7.

10. Ito H, Takazoe M, Fukuda Y, Hibi T, Kusugami K, Andoh A et al. A pilot randomized trial of a human anti-interleukin-6 receptor monoclonal antibody in active Crohn's disease. Gastroenterology 2004; 126(4):989-96.

11. Sandborn WJ, Ghosh S, Panes J, Vranic I, Su C, Rousell S et al.; Study A3921063 Investigators. Tofacitinib, an oral Janus kinase inhibitor, in active ulcerative colitis. N Engl J Med 2012; 367(7):616-24.

12. Parikh A, Leach T, Wyant T, Scholz C, Sankoh S, Mould DR et al. Vedolizumab for the treatment of active ulcerative colitis: a randomized controlled phase 2 dose-ranging study. Inflamm Bowel Dis 2012; 18(8):1470-9.

13. Sandborn WJ, Feagan BG, Rutgeerts P, Hanauer S, Colombel JF, Sands BE et al.; GEMINI 2 Study Group. Vedolizumab as induction and maintenance therapy for Crohn's disease. N Engl J Med 2013; 369(8):711-21.

14. Feagan BG, Rutgeerts P, Sands BE, Hanauer S, Colombel JF, Sandborn WJ et al.; GEMINI 1 Study Group. Vedolizumab as induction and maintenance therapy for ulcerative colitis. N Engl J Med 2013; 369(8):699-710.

**15.** McLean LP, Shea-Donohue T, Cross RK. Vedolizumab for the treatment of ulcerative colitis and Crohn's disease. Immunotherapy 2012; 4(9):883-98.

**16.** Sandborn WJ, Gasink C, Gao LL, Blank MA, Johanns J, Guzzo C et al.; CERTIFI Study Group. Ustekinumab induction and maintenance therapy in refractory Crohn's disease. N Engl J Med 2012; 367(16):1519-28.

# CLASSIFICATION AND ACTIVITY INDICES

## ULCERATIVE COLITIS

### MONTREAL CLASSIFICATION

| Classification | Extent |
| --- | --- |
| E1 – ulcerative proctitis | Involvement limited to the rectum |
| E2 – left-sided UC (distal colitis) | Involvement extends up to the splenic flexure |
| E3 – extensive UC (pancolitis) | Involvement proximal to the splenic flexure |
| **Classification** | **Disease severity** |
| S0 | Clinical remission |
| S1 | Mild |
| S2 | Moderate |
| S3 | Severe |

Montreal, 2005.

## INDEX OF INFLAMMATORY ACTIVITY

## SEVERITY OF ACUTE DISEASE ACCORDING TO TRUELOVE AND WITTS

| | Mild | Moderate | Severe |
|---|---|---|---|
| 1. Number of bowel movements/day | ≤ 4 | 4 to 6 | > 6 |
| 2. Blood in the stool | ± | + | ++ |
| 3. Body temperature | Normal | Inter-mediate values | Average body temperature at night >37.5°C (99.5°F) or >37.8°C (100.04°F) in 2 out of 4 days |
| 4. Pulse | Normal | Interme-diate | > 90 bpm |
| 5. Hemoglobin (g/dL) | > 10.5 | Interme-diate | < 10.5 |
| 6. ESR (mm/first hour) | < 30 | Interme-diate | > 30 mm, first hour |

ESR: erythrocyte sedimentation rate.

Truelove SC, Witts LJ. Br Med J 1995; 2:1041-8.

# DISEASE ACTIVITY INDEX
# MAYO CLINIC SCORE

| Score | N. of bowel movements | Rectal bleeding | Endoscopic findings | Global assessment |
|---|---|---|---|---|
| 0 | Normal amount | No blood seen | Normal or inactive disease (scarring) | Normal |
| 1 | Normal amount + 1-2 BM/day | Streaks of blood with stool – less than half of BM | Mild disease (erythema, of vascular pattern, mild friability) | Mild disease |
| 2 | Normal amount + 3-4 BM/day | Obvious blood with stools | Moderate disease (evident erythema, loss of vascular pattern, erosion) | Moderate disease |
| 3 | Normal amount + > 5 BM/day | Blood alone without stools | Severe disease (spontaneous bleeding, ulceration) | Severe disease |

Schroeder KW, Tremaine WJ, Ilstrup DM. N Engl J Med 1987; 317(26):1625-9.

| Score (points) | Disease severity |
|---|---|
| 0 to 2 | Normal – remission |
| 3 to 5 | Mild activity |
| 6 to 10 | Moderate activity |
| 11 to 12 | Severe activity |

## CROHN'S DISEASE

### MONTREAL CLASSIFICATION

| Age at diagnosis (A) | | | |
|---|---|---|---|
| A1 | 16 years or younger | | |
| A2 | 17 to 40 years | | |
| A3 | Over 40 years | | |
| **Location (L)** | | **Upper GI tract (L4)** | |
| L1 | Terminal ileum | **L1 + L4** | Terminal ileum + upper GI tract |
| L2 | Colon | **L2 + L4** | Colon + upper GI tract |
| L3 | Ileocolon | **L3 + L4** | Ileocolon + upper GI tract |
| L4 | Upper GI tract | | |
| **Behavior (B)** | | **Perianal disease (p)** | |
| B1 | Non-stricturing Non-penetrating | B1p | Non-stricturing + perianal |
| B2 | Stricturing | B2p | Stricturing + perianal |
| B3 | Penetrating | B3p | Penetrating + perianal |

Montreal, 2005.

### HARVEY-BRADSHAW INDEX OF INFLAMMATORY ACTIVITY

| | Score |
|---|---|
| General well-being (very well = 0; slightly below par = 1; poor = 2; very poor = 3; terrible = 4) | 0 to 4 |
| Abdominal pain (none = 0; mild = 1; moderate = 2; severe = 3). | 0 to 3 |
| Number of liquid stools per day | no./day |
| Abdominal mass (none = 0; dubious = 1; definite = 2; definite and tender = 3) | 0 to 3 |
| Complications: arthralgia/arthritis, uveitis/iritis, erythema nodosum, aphthous stomatitis, pyoderma gangrenosum, anal fissure, fistula, abscess, etc. | score 1 per item |

< 8 = inactive/mild; 8 to 10 = mild/moderate; > 10 = moderate/severe.

Harvey RF, Bradshaw, Lancet 1980; 1:514.

## INDEX OF INFLAMMATORY ACTIVITY

## CROHN'S DISEASE ACTIVITY INDEX (CDAI)

| | Multiplied by |
|---|---|
| Number of liquid stools in the previous week | 2 |
| Abdominal pain (none = 0; mild = 1; moderate = 2; severe = 3). Consider the total sum of individual data in the previous week | 5 |
| General condition (excellent = 0; well = 1; poor = 2; very poor = 3; terrible = 4). Consider the total sum of individual data in the previous week | 7 |
| N. of associated symptoms/signs – list by categories: a) Arthritis/arthralgia; b) Iritis/uveitis; c) Erythema nodosum/pyoderma gangrenosum/aphthous stomatitis; d) Anal fissure, fistula or perirectal abscess; e) Other fistula; f) Fever | 20 (maximum score = 120) |
| Use of antidiarrheal drugs (no = 0; yes = 1) | 30 |
| Abdominal mass (none = 0; dubious = 2; definite = 5) | 10 |
| Hematocrit deficit. For Male – 47 Ht, For Female – 42 Ht (subtract instead of adding if the patient's Ht is > the standard) | 6 |
| (Weight/Usual weight) × 100 Weight*: percentage below the expected (subtract instead of adding if the patient's weight is higher than expected) | 1 |
| Total (Crohn's Disease Activity Index) = < 150 = Remission  150 to 250 = Mild  250 to 350 = Moderate  > 350 = Severe | |

Best WR et al., 1976.

## CID 10

## INTESTINAL INFECTIOUS DISEASES

| | |
|---|---|
| **A0** | **Infectious diseases** |
| A02.0 | Salmonella enteritis |
| **A03** | **Shigellosis** |
| A03.9 | Shigellosis, unspecified |
| **A04** | **Other bacterial intestinal infections** |
| A04.9 | Bacterial intestinal infection, unspecified |
| **A05** | **Other bacterial foodborne intoxications, not elsewhere classified** |
| A05.9 | Bacterial foodborne intoxication, unspecified |
| **A06** | **Amebiasis** |
| A06.9 | Amebiasis, unspecified |
| **A07** | **Other protozoal intestinal diseases** |
| A07.1 | Giardiasis [lambliasis] |
| A07.9 | Protozoal intestinal disease, unspecified |
| **A08** | **Viral and other specified intestinal infections** |
| A08.4 | Viral intestinal infection, unspecified |
| **A18.3** | **Tuberculosis of intestines** |

## HELMINTHIASES

| B65 | **Schistosomiasis [bilharziasis]** |
|---|---|
| B65.1 | Schistosomiasis due to *Schistosoma mansoni* |
| **B76** | **Hookworm diseases** |
| B76.0 | Ancylostomiasis |
| **B77** | **Ascariasis** |
| B77.9 | Ascariasis, unspecified |
| **B78** | **Strongyloidiasis** |
| B78.0 | Intestinal strongyloidiasis |
| B78.9 | Strongyloidiasis, unspecified |
| **B80** | **Oxyuriasis** |
| **B82** | **Unspecified intestinal parasitism** |
| B82.9 | Intestinal parasitism, unspecified |

## MALIGNANT NEOPLASMS

| C17 | **Malignant neoplasm of small intestine** |
|---|---|
| C17.0 | Duodenum |
| C17.1 | Jejunum |
| C17.2 | Ileum |
| C17.3 | Meckel diverticulum |
| C17.8 | Overlapping lesion of small intestine |
| C17.9 | Small intestine, unspecified |
| **C18** | **Malignant neoplasm of colon** |
| C18.0 | Cecum |
| C18.1 | Appendix |
| C18.2 | Ascending colon |
| C18.3 | Hepatiflexure |
| C18.4 | Transverse colon |
| C18.5 | Spleniflexure |
| C18.6 | Descending colon |
| C18.7 | Sigmoid colon |
| C18.8 | Overlapping lesion of colon |

| C18.9 | Colon, unspecified |
|-------|--------------------|
| **C19** | **Malignant neoplasm of rectosigmoid junction** |
| **C20** | **Malignant neoplasm of rectum** |
| **C21** | **Malignant neoplasm of anus and anal canal** |
| C21.0 | Anus, unspecified |
| C21.1 | Anal canal |
| C21.2 | Cloacogenizone |
| C21.8 | Overlapping lesion of rectum, anus and anal canal |
| **C26** | **Malignant neoplasm of other and ill-defined digestive organs** |
| C26.0 | Intestinal tract, part unspecified |
| C26.8 | Overlapping lesion of digestive system |
| C26.9 | Ill-defined sites within the digestive system |

## BENIGN NEOPLASMS

| **D12** | **Benign neoplasm of colon, rectum, anus and anal canal** |
|---------|-----------------------------------------------------------|
| D12.0 | Cecum |
| D12.1 | Appendix |
| D12.2 | Ascending colon |
| D12.3 | Transverse colon |
| D12.4 | Descending colon |
| D12.5 | Sigmoid colon |
| D12.6 | Colon, unspecified |
| D12.7 | Rectosigmoid junction |
| D12.8 | Rectum |
| D12.9 | Anus and anal canal |

## DISEASES OF THE DIGESTIVE SYSTEM

| **K12** | **Stomatitis and related lesions** |
|---------|------------------------------------|
| K12.0 | Recurrent oral aphthae |
| **K14** | **Diseases of tongue** |
| K14.0 | Glossitis |

## DISEASES OF ESOPHAGUS, STOMACH AND DUODENUM

| K20 | **Esophagitis** |
|-----|-----------------|
| K21 | **Gastro-esophageal reflux disease** |
| K21.0 | Gastro-esophageal reflux disease with esophagitis |
| K21.9 | Gastro-esophageal reflux disease without esophagitis |
| K29 | **Gastritis and duodenitis** |
| K29.7 | Gastritis, unspecified |
| K30 | **Dyspepsia** |

## DISEASES OF APPENDIX

| K35 | **Acute appendicitis** |
|-----|------------------------|
| K37 | **Unspecified appendicitis** |
| K38.9 | **Disease of appendix, unspecified** |

## NONINFECTIVE ENTERITIS AND COLITIS

| K50 | **Crohn disease [regional enteritis]** |
|-----|----------------------------------------|
| K50.0 | Crohn disease of small intestine |
| K50.1 | Crohn disease of large intestine |
| K50.8 | Other Crohn disease |
| K50.9 | Crohn disease, unspecified |
| K51 | **Ulcerative colitis** |
| K51.0 | Ulcerative (chronic) pancolitis |
| K51.1 | Ulcerative (chronic) ileocolitis |
| K51.2 | Ulcerative (chronic) proctitis |
| K51.3 | Ulcerative (chronic) rectosigmoiditis |
| K51.4 | Inflammatory polyps |
| K51.5 | Left-sided colitis |
| K51.8 | Other ulcerative colitis |
| K51.9 | Ulcerative colitis, unspecified |

| K52 | **Other noninfective gastroenteritis and colitis** |
|---|---|
| K52.0 | Gastroenteritis and colitis due to radiation |
| K52.1 | Toxigastroenteritis and colitis |
| K52.2 | Allergiand dietetigastroenteritis and colitis |
| K52.8 | Other specified noninfective gastroenteritis and colitis |
| K52.9 | Noninfective gastroenteritis and colitis, unspecified |

## OTHER DISEASES OF INTESTINES

| K55 | **Vascular disorders of intestine** |
|---|---|
| K55.0 | Acute vascular disorders of intestine |
| K55.1 | Chronivascular disorders of intestine |
| K55.2 | Angiodysplasia of colon |
| K55.9 | Vascular disorder of intestine, unspecified |
| K56.0 | **Paralytiileus** |
| K56.I | **Intussusception** |
| K56.2 | **Volvulus** |
| K56.5 | **Intestinal adhesions [bands] with obstruction** |
| K56.6 | **Other and unspecified intestinal obstruction** |
| K57 | **Diverticular disease of intestine** |
| K57.2 | Diverticular disease of large intestine with perforation and abscess |
| K57.3 | Diverticular disease of large intestine without perforation or abscess |
| K57.9 | Diverticular disease of intestine, part unspecified, without perforation or abscess |
| K58 | **Irritable bowel syndrome** |
| K59.0 | **Constipation** |
| K59.1 | **Functional diarrhea** |
| K59.3 | **Megacolon** |
| K59.4 | **Anal spasm** |
| K59.8 | **Other specified functional intestinal disorders** |
| K60 | **Fissure and fistula of anal and rectal regions** |
| K60.0 | Acute anal fissure |
| K60.1 | Chronianal fissure |
| K60.2 | Anal fissure, unspecified |

| K60.3 | Anal fistula |
|-------|--------------|
| K60.4 | Rectal fistula |
| K60.5 | Anorectal fistula |
| **K61** | **Abscess of anal and rectal regions** |
| K61.0 | Anal abscess |
| K61.1 | Rectal abscess |
| K61.2 | Anorectal abscess |
| **K62** | **Other diseases of anus and rectum** |
| K62.0 | Anal polyp |
| K62.1 | Rectal polyp |
| K62.2 | Anal prolapse |
| K62.3 | Rectal prolapse |
| K62.4 | Stenosis of anus and rectum |
| K62.5 | Hemorrhage of anus and rectum |
| K62.7 | Radiation proctitis |
| K62.8 | Other specified diseases of anus and rectum |
| K62.9 | Disease of anus and rectum, unspecified |
| **K63** | **Other diseases of intestine** |
| K63.0 | Abscess of intestine |
| K63.1 | Perforation of intestine (nontraumatic) |
| K63.2 | Fistula of intestine |
| K63.5 | Polyp of colon |
| K63.9 | Disease of intestine, unspecified |
| **K81** | **Cholecystitis** |
| K81.0 | Acute cholecystitis |
| K81.1 | Chronicholecystitis |
| **K85** | **Acute pancreatitis** |

## OTHER DISEASES OF THE DIGESTIVE SYSTEM

| | |
|---|---|
| **K90** | **Intestinal malabsorption** |
| K90.0 | Celiadisease |
| K90.4 | Malabsorption due to intolerance, not elsewhere classified |
| K90.9 | Intestinal malabsorption, unspecified |
| **K91** | **Postprocedural disorders of digestive system, not elsewhere classified** |
| K91.4 | Colostomy and enterostomy malfunction |
| K91.9 | Postprocedural disorder of digestive system, unspecified |
| **K92** | **Other diseases of digestive system** |
| K92.0 | Hematemesis |
| K92.1 | Melena |
| K92.2 | Gastrointestinal hemorrhage, unspecified |
| **K93** | **Disorders of other digestive organs in diseases classified elsewhere** |
| K93.0 | Tuberculous disorders of intestines, peritoneum and mesenteriglands (A.18.3) |
| K93.1 | Megacolon in Chagas disease (B.57.3) |
| **I84** | **Hemorrhoids** |
| I84.0 | Internal thrombosed hemorrhoids |
| I84.1 | Internal hemorrhoids with other complications |
| I84.2 | Internal hemorrhoids without complication |
| I84.3 | External thrombosed hemorrhoids |
| I84.4 | External hemorrhoids with other complications |
| I84.5 | External hemorrhoids without complication |
| I84.6 | Residual hemorrhoidal skin tags |
| I84.7 | Unspecified thrombosed hemorrhoids |
| I84.8 | Unspecified hemorrhoids with other complications |
| I84.9 | Unspecified hemorrhoids without complication |

## SYMPTOMS, SIGNS AND ABNORMAL CLINICAL AND LABORATORY FINDINGS, NOT ELSEWHERE CLASSIFIED

### Symptoms and signs involving the digestive system and abdomen

| | |
|---|---|
| **R10** | **Abdominal and pelvipain** |
| R10.0 | Acute abdomen |
| R10.1 | Pain localized to upper abdomen |
| R10.2 | Pelviand perineal pain |
| R10.3 | Pain localized to other parts of lower abdomen |
| R10.4 | Other and unspecified abdominal pain |
| **R11** | **Nausea and vomiting** |
| **R14** | **Flatulence and related conditions** |
| **R15** | **Fecal incontinence** |
| **R19** | **Other symptoms and signs involving the digestive system and abdomen** |
| R19.0 | Intra-abdominal and pelviswelling, mass and lump |
| R19.4 | Change in bowel habit |
| R19.5 | Other fecal abnormalities |
| R19.6 | Halitosis |
| R19.8 | Other specified symptoms and signs involving the digestive system and abdomen |
| **Z93** | **Artificial opening status** |
| Z93.1 | Gastrostomy status |
| Z93.2 | Ileostomy status |
| Z93.3 | Colostomy status |
| Z93.4 | Other artificial openings of gastrointestinal tract status |

Source: ICD-10. International Statistical Classification of Diseases and Related Health Problems th Revision (ICD-10) http://apps.who.int/classifications/icd/browse//en#/I